CARDIOLOGY RESEARCH AND CLINICAL DEVELOPMENTS

CIRCADIAN CARDIOLOGY WITH FOCUS ON BOTH PREVENTION AND INTERVENTION

Cardiology Research and Clinical Developments

Additional books in this series can be found on Nova's website under the Series tab.

Additional e-books in this series can be found on Nova's website under the e-book tab.

CARDIOLOGY RESEARCH AND CLINICAL DEVELOPMENTS

CIRCADIAN CARDIOLOGY WITH FOCUS ON BOTH PREVENTION AND INTERVENTION

KRASIMIRA HRISTOVA
ABDULLA SHEHAB
GERMAINE CORNELISSEN
AND
R. B. SINGH
EDITORS

NOVA BIOMEDICAL
New York

Library of Congress Cataloging-in-Publication Data

ISBN: 978-1-63463-956-9
Library of Congress Control Number: 2014960313

Published by Nova Science Publishers, Inc. † New York

Contents

Preface

Cardiovascular diseases have become a major cause of deaths in both middle and higher income countries despite significant increases in evidence-based drug therapy and interventions as well as cardiovascular surgery. It seems logical to educate the world about actual causes of cardiovascular diseases, which are important in the pathogenesis of these problems. The science of chronocardiology gives due consideration to our cardiovascular functions according to time structures which appear to be related to light and dark cycles of 24 hours, due to day and night. All the organisms on the Earth have evolved to predict these cycles by developing an endogenous circadian clock, which is synchronized to external time cues. The organisms, including human beings, ensure that physiological processes are carried out at the right time of the circadian cycle. The circadian clock resets physiological and metabolic processes, including sleep-wake cycles, cardiovascular activity, endocrine system, body temperature, renal activity, gastrointestinal tract motility, and metabolism. 10–20% of all cellular transcripts are known to be cyclically expressed, most of which are tissue-specific. The central circadian clock is located in the suprachiasmatic nuclei (SCN) of the brain anterior hypothalamus which is composed of multiple, single-cell oscillators synchronized to generate circadian rhythms. "Circadian" is a term coined by Dr. Franz Halberg after he demonstrated the partly built-in nature of about 24-hour cycles that persist in the absence of external time cues, assuming a period slightly but statistically significantly different from precisely 24 hours, a phenomenon known as free-running. This indirect evidence for an endogenous circadian system preceded by a few decades the discovery of clock genes that placed chronobiology on a modern molecular basis. We take this opportunity to pay tribute to Franz Halberg, the Father of Chronobiology by emphasizing how this science could be used for prevention of cardiovascular diseases by chronotherapy of drugs, foods and exercise as well as with greater efficacy of chrono-coronary intervention. The first four articles by his close associates, including his daughters, Francine Halberg and Julia Halberg, represent but a small sample of the many colleagues around the world who were inspired by Franz and benefited from his science and his scholarship. As an illustration of the great respect in which we all hold him and as recollections from memories shared over the past 64 years of Franz's career at the University of Minnesota, reminiscences have been collected to provide a glimpse into Franz's life, to help the reader better understand the force of character and intellectual clarity of this great man.

The fifth article by Franz Halberg, Kuniaki Otsuka, Yoshihiko Watanabe, George Katinas, Zhengrong Wang, and Germaine Cornélissen reports on the cosmos with aeolian cycles, tipping the scale between death and survival. In the sixth article, Douglas W. Wilson, Keith Griffiths, Fabien de Meester, Agnieszka Wilcyznska, Toru Takahashi and Ram B. Singh pay tribute to Franz Halberg, emphasizing chronobiology and chronomics research during four decades from Cardiff, Durham and Tsim Tsoum. Shabnam Omidvar, Sumit Karn, Shahla Shafiee, R.B. Singh, Miki Tokunaga, H.S. Buttar, and D.W. Wilson mention proatherogenic risk factors and under-nutrition among adolescents in South East Asia with emphasis on "When to Eat and What to Eat?", reported by Halberg in 1976. In the next article, Miroslav Mikulecký reports on "Tides at Solid Earth: Biomedical Effects (An 'Anatomy' of Tidal Wave)". In the ninth article, D.G. Gubin, G. Cornélissen, D. Weinert, A.S. Vetoshkin, L.I. Gapon, N.P. Shurkevich, F.A. Poshinov and N.V. Belozerova report on circadian disruption and vascular variability disorders (VVD) emphasizing mechanisms linking aging, disease state and arctic shift-work as well as chronotherapy. The next piece is an original article on personalized circadian timing of exercise which a study on himself by R.B. Singh under guidance of Halberg, in which R.B. Singh, Franz Halberg, Germaine Cornelissen, Jarmila Siegelova, Krasimira Hristova, Eri Toda, Toru Takahashi, Jan Fedacko, and Kuniaki Otsuka report for the first time that exercise done in the morning causes less adverse effects on heart rate and MESOR of systolic and diastolic blood pressures. In the last article related to chronocardiology, D.W. Wilson, F. De Meester, R.B. Singh and Elena Gerasimova focus on the origin of life, the evolution of the circadian system and the role of circadian rhythms in cardiovascular diseases. In articles 12 to 23, various aspects of prevention are highlighted; these are followed by a review on arrhythmias by NS Neki. The last section contains two editorials, one highlighting coronary intervention by Fedacko et al. and heart devices by Singh et al. This book is excellent for health professionals interested in prevention and for those who are interested in how to prevent the need for intervention.

Dr. Krasimira Hristova,
Dr. Abdulla Shehab,
Dr. Germaine Cornelissen
and Dr R. B. Singh*

* Correspondence: R. B. Singh, MD, FICC, Halberg Hospital and Research Institute, Moradabad, India; Email: rbs@tsimtsoum.ne

Editor Contact Information

Krasimira Hristova, MD, PhD, FICC
National Heart Hospital, Sofia, Bulgaria
Email: khristovabg@yahoo.com

Abdulla Shehab, MB ChB, PhD, FRCP
UAE University, Fujairah, UAE
Email: a.shehab@uaeu.ac.ae

Germaine Cornelissen, PhD, FICC
Halberg Chronobiology Center,
University of Minnesota Medical School, Minneapolis, US
Email: corne001@umn.edu

R. B. Singh, MD, FICC
Halberg Hospital and Research Institute, Moradabad, India
Email: rbs@tsimtsoum.net

In: Circadian Cardiology with Focus on Both Prevention ... ISBN: 978-1-63463-956-9
Editors: K. Hristova, A. Shehab, G. Cornelissen et al. © 2015 Nova Science Publishers, Inc.

Chapter 1

In Memoriam, Franz Halberg, MD (5 July 1919 – 9 June 2013)

Germaine Cornelissen*[*1], Francine Halberg[1], Julia Halberg[1], Othild Schwartzkopff[1] and R. B. Singh[2]
[1]Halberg Chronobiology Center, University of Minnesota, Minneapolis, MN, US
[2]Halberg Hospital and Research Institute, Civil Lines, Moradabad, India

Time skipped a beat. On Sunday June 9, 2013, a legend, great scientist, and exceptional human being left us. His close associates also lost a mentor, an inspiration, and a truly staunch friend. Franz Halberg's passing shy of his 94th birthday leaves a void that cannot be filled.

Figure 1. Franz Halberg (1919-2013).

* Correspondence: Germaine Cornelissen, PhD. Halberg Chronobiology Center, University of Minnesota, MMC8609. 420 Delaware Street SE. Minneapolis, MN 55455, USA. Tel. 612-624-6976, Fax. 612-624-9989. Email. corne001@umn.edu Website. http://www.msi.umn.edu/~halberg/.

Franz Halberg (Figure 1) will be remembered for founding the fields of chronobiology [1, 2], chronomics [3-5] and chronobioethics [6, 7]. These new transdisciplinary scientific disciplines could not have flourished without Franz Halberg's unveiling of lawful variations as a function of TIME within the physiological range and his vision that they had far-reaching implications. An essential element was the key principle Franz followed throughout his life's work, namely to "measure in time what is measurable and render measurable what as yet is not". Toward this goal, he not only gathered a critical mass of data, himself, but with a steadily increasing network of colleagues worldwide, he also developed inferential statistical methods for their analysis and interpretation [8-18].

By adding TIME to the existing body of knowledge in all of biology and medicine, and by recognizing the crucial role this new element plays in all matters of life, Franz Halberg developed the new science of chronobiology. By insisting on an inferential statistical foundation, details of a rich time structure were revealed akin to the finer spatial resolution obtained with a microscope. His methodical scrutiny of periodicities shared between biological systems and their broad environment, seen (photic) and unseen (non-photic) influences from the Sun and the cosmos led to chronomics in a way reminiscent of discoveries enabled by the advent of the telescope.

Born on July 5, 1919 in Romania, Franz studied the adrenal as a physician scientist in post-World War II Innsbruck, Austria. He continued this work at Harvard Medical School, where he held a World Health Organization fellowship in clinical endocrinology in 1948. In 1949, he moved to the University of Minnesota, which saw his breakthrough experiments that led to the important discovery that circadian rhythms are partly endogenous and can be manipulated by environmental synchronizers, notably the lighting and feeding schedules [19, 20]. Franz coined the term circadian, after documenting that biologic rhythms tip the scale between health and disease and even between life and death [21, cf. 22, 23]. His results were widely published, including a 1969 citation classic [1]. By 1958, Franz had recognized the important role played by the cell's RNA and DNA cycles, which he was first to demonstrate as complementing the hypothalamic-pituitary-adrenal system as mediator of photic inputs [24]. He subsequently added pineal feed sidewards and the understanding that there are endogenous physiologic networks that respond to the cosmos [25, 26].

Beyond circadians, Franz demonstrated that many other built-in cycles resonate in part with their counterparts in our broad environment. His recent work focused on building a growing edifice of shared periodicities with bridges across disciplines [27-29], addressing wide-ranging applications from the optimization of individualized health care to concerns for the health of societies. He strived to understand how to enhance positive thoughts and emotions as a scaffold for tolerance and love by seeking optimal configurations of the time structured realm of the mind, what he called the chronousphere [30]. He was a scholar in the true sense of the word, combining science, philosophy, poetry, and spirituality, laying the foundation of chronobioethics.

With applications in all fields of medicine and biology, Franz's legacy is far-reaching. He will be remembered for showing that timing cancer treatment according to marker rhythms improves outcomes both in terms of heightened efficacy and lesser undesired side effects [31-34], for showing that a calorie is different whether it is consumed at breakfast or dinner [20, 35-37], and for his work in preventive cardiology by screening for abnormal patterns of blood pressure variability [38-41]. Corresponding circadian timed treatment more than halved the risk of stroke and other adverse cardiovascular events. "Primum nil nocere" (above all, do no harm) prompted Franz to advocate the individualization of treatment, guided by marker rhythms [32, 42].

Franz's lifetime accomplishments are summarized in his over 3,500 scientific publications, in cooperation with colleagues from around the world. Many worldwide indeed call him their mentor and turned to him for advice, from study design and data analysis to the interpretation of results in the time dimension. Minnesota Medicine called him Father Time [43], and colleagues in Russia and Azerbaijan honored him as Lord of Time [44].

Franz's endeavors earned him numerous awards. Apart from holding professorships in Laboratory Medicine and Pathology, Physiology, Biology, Bioengineering and Oral Medicine at the University of Minnesota, he was a honorary member of the Romanian Academies of Science and Medical Sciences and a Corresponding Member of the French National Academy of Medicine. Franz received honorary doctorates from the University of Montpellier (France), Ferrara (Italy), Tyumen (Siberia), Brno (Czech Republic), L'Aquila (Italy), and People's Friendship University of Russia (Moscow, Russia). He was also an elected member of the prestigious Leibniz Society and of the International Academy of Science. His achievements in the new field of chronomics earned him the O.Yu. Schmidt Medal and diploma for outstanding merits in development of geophysics, the first such award given to a non-physicist.

Until his last breath, Franz strived to introduce timing for diagnosis, prognosis, treatment, and first and foremost prevention into clinical practice. At 93 years of age and still active 7 days a week in the Halberg Chronobiology Center at the University of Minnesota, which continues his work, he was the last recipient of a lifetime career award from the National Institutes of Health.

Figure 2 is Franz's last contribution, made from his hospital bed after he suffered a stroke shortly after midnight on 17 May 2013. It summarizes his current endeavors and illustrates the two major lines of research at the Halberg Chronobiology Center. First, Vascular Variability Anomalies (VVAs) identified by the Chronobiologically-interpreted Ambulatory Blood Pressure Monitoring (C-ABPM) are associated with a large increase in cardiovascular disease risk, as shown in several outcome studies [45]. In this transverse approach (Figure 2, left), an elevated blood pressure (MESOR-Hypertension) is associated with less than 10% adverse events within 6 years, in the absence of other VVAs. In this same study, the presence of 1, 2 or 3 additional VVAs (for a total of 2 to 4 VVAs) raises the incidence of adverse events to 29%, 53% and 100%, respectively [38]. Using his own record, Franz asked the question whether this transverse approach could be applied longitudinally. To answer his question, daily sphygmochrons in April and May 2013 were prepared.

As shown in Figure 2 (top right), while there were several days without any VVA, on other days, 1, 2 or 3 VVAs were present. There were 3 VVAs on 3 separate occasions, and during the 2-week span preceding the stroke, there were 2 days with 4 VVAs. A self-starting cumulative sum (CUSUM) control chart [46, 47] on the number of daily VVAs detects on May 3 an increase in risk that is estimated to have been triggered around April 19 (Figure 2, right bottom). It thus appears that the longitudinal record mimics the results from the cross-sectional study on the left. Figure 2 thus illustrates an approach that could be routinely applied, thus providing a basis to determine both the sensitivity and specificity of the proposed procedure from longitudinal records of patients with adverse cardiovascular outcomes. Franz's 25-year record can already provide some tentative answers related to any false positive warnings, a task we are working on. We are also investigating the relevance of the emergence of systolic ecphasia, often present starting in late February 2013 and lasting until a few days before the stroke, a VVA that did not occur in the previous 25 years of Franz's record. Hopefully we will be able to apply the results of this work to stroke prevention.

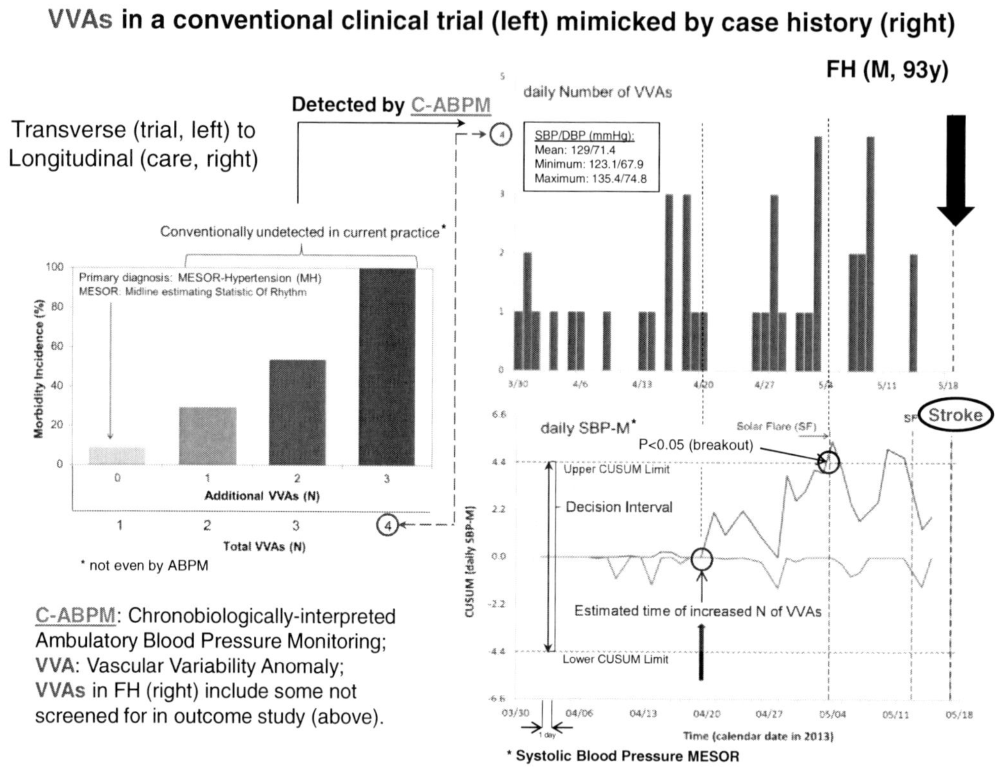

Figure 2. Increase in cardiovascular disease risk related to the presence of multiple Vascular Variability Anomalies (VVAs) observed in a cross-sectional outcome study (left; data from Prof. K Otsuka) mimicked in the longitudinal record of Franz Halberg preceding a stroke (right, top), interpreted by a self-starting cumulative sum (CUSUM) control chart (right, bottom). These results, Franz's last contribution, illustrate the two main avenues of research ongoing in his Halberg Chronobiology Center at the University of Minnesota, namely Chronobiologically-interpreted Ambulatory Blood Pressure Monitoring (C-ABPM) for personal health and for the study of influences from our broad environment.

Second, Figure 2 illustrates the interest Franz had in understanding how space and terrestrial weather may affect human physiology and pathology. At the beginning of May 2013, Franz suffered a bad headache and more severe neck and shoulder pain than usual, resulting in a day with 4 VVAs on May 3, 2013. Upon inquiring about the status of space weather at the time, Franz found out that the National Oceanic and Atmospheric Administration (NOAA) had issued a warning that read "STRONG FLARE: For the second time in three days, an active region just over the sun's east limb has exploded, producing a strong solar flare and CME [*Coronal Mass Ejection*]. The blast on May 3rd registered M5." Another strong solar flare occurred a few days later, on May 13, 2013. Again, it was associated with Franz suffering severe neck pain. Was Franz magneto-sensitive? Did the flares precipitate an event? While we cannot answer this question, again Franz had proposed a way to conduct a rigorous inquiry once data become available from a large enough number of subjects. His vision was to use C-ABPM not only for personal health, but also as a means to learn more about environmental influences and use it for avoiding natural and societal cataclysms [48].

It was Franz Halberg's recognition that variability and not homeostasis is the essence of life that put chronobiology on the map [1, 19, 21]. The discovery of important physiological problems and their mechanisms is certainly one of Franz's major achievements. But the science of chronobiology would not have developed without his changing the very fundamentals of medicine and biology still prevailing today. The impact he had in science stemmed not only from his original findings but primarily from his vision of their implications that led beyond a scientific breakthrough to a new way of thinking. Franz's incredible persistence and intellectual clarity in the face of entrenched thinking in fields that transcend disciplinary boundaries sets him apart as a truly great scientist. As we say farewell with a great sense of loss, we remember Franz Halberg as a trail blazer who leaves a remarkable legacy that will live forever.

References

[1] Halberg F. Chronobiology. *Annu Rev Physiol* 1969; 31: 675-725.

[2] Cornelissen G, Halberg F. Introduction to Chronobiology. Medtronic Chronobiology Seminar #7, April 1994, 52 pp. (Library of Congress Catalog Card #94-060580; http://www.msi.umn.edu/~halberg/)

[3] Halberg F, Cornelissen G, Katinas G, Hillman D, Schwartzkopff O. Season's Appreciations 2000: Chronomics complement, among many other fields, genomics and proteomics. *Neuroendocrinol Lett* 2001; 22: 53-73.

[4] Halberg F, Cornelissen G, Otsuka K, Katinas G, Schwartzkopff O. Essays on chronomics spawned by transdisciplinary chronobiology: Witness in time: Earl Elmer Bakken. *Neuroendocrinol Lett* 2001; 22: 359-384.

[5] Halberg F, Cornelissen G, Otsuka K, Schwartzkopff O, Halberg J, Bakken EE. Chronomics. *Biomed & Pharmacother* 2001; 55 (Suppl 1): 153s-190s.

[6] Cornelissen G, Bingham C, Siegelova J, Fiser B, Dusek J, Prikryl P, Sonkowsky RP, Halberg F. Cardiovascular disease risk monitoring in the light of chronobioethics. *Chronobiologia* 1994; 21: 321-325.

[7] Halberg F, Bingham C, Siegelova J, Fiser B, Dusek J, Prikryl P, Sonkowsky RP, Cornelissen G. "Cancer marker chronomes" assessed in the light of chronobioethics. *Chronobiologia* 1994; 21: 327-330.

[8] Halberg F, Tong YL, Johnson EA. Circadian system phase -- an aspect of temporal morphology; procedures and illustrative examples. Proc. International Congress of Anatomists. In: Mayersbach H v. (Ed.) *The Cellular Aspects of Biorhythms, Symposium on Biorhythms*. New York: Springer-Verlag; 1967. p. 20-48.

[9] Halberg F, Johnson EA, Nelson W, Runge W, Sothern R. Autorhythmometry -- procedures for physiologic self-measurements and their analysis. *Physiol Tchr* 1972; 1: 1-11.

[10] Halberg F. Chronobiology: methodological problems. *Acta med rom* 1980; 18: 399-440.

[11] Bingham C, Arbogast B, Cornelissen Guillaume G, Lee JK, Halberg F. Inferential statistical methods for estimating and comparing cosinor parameters. *Chronobiologia* 1982; 9: 397-439.

[12] Nelson W, Cornelissen G, Hinkley D, Bingham C, Halberg F. Construction of rhythm-specified reference intervals and regions, with emphasis on "hybrid" data, illustrated for plasma cortisol. *Chronobiologia* 1983; 10: 179-193.

[13] Halberg F, Bingham C. The scope and promise of chronobiology and biostatistics: interpenetrating, inseparable disciplines. *Proc. Biopharmaceutical Section, Am Statistical Assn*, Chicago, Illinois, August 15-18, 1986. Alexandria, VA: American Statistical Association; 1987. p. 11-32.

[14] Halberg F, Shankaraiah K, Giese AC, Halberg Francine. The chronobiology of marine invertebrates: methods of analysis. In: Giese AC, Pearse JS, Pearse VB. (Eds.) *Reproduction of Marine Invertebrates*, Vol. IX: General Aspects: Seeking Unity in Diversity. Palo Alto, CA: Blackwell Scientific Publications/Pacific Grove, CA: Boxwood Press, 1987. p. 331-384.

[15] Halberg F, Bingham C, Cornelissen G. Clinical trials: the larger the better ? *Chronobiologia* 1993; 20: 193-212.

[16] Halberg F, Cornelissen G, Bingham C, Witte H, Ribary U, Hesse W, Petsche H, Engebretson M, Geissler H G, Weiss S, Klimesch W, Rappelsberger P, Katinas G, Schwartzkopff O. Chronomics: Imaging in time by phase synchronization reveals wide spectral-biospheric resonances beyond short rhythms. ("Wenn man über kurze Rhythmen hinausgeht") In memoriam – lost future: Dr.-Ing. habil. Dr. rer. nat. Barbara Schack: 1952-2003. *Neuroendocrinol Lett* 2003; 24: 355-380.

[17] Cornelissen G, Halberg F. Chronomedicine. In: Armitage P, Colton T. (Eds.) *Encyclopedia of Biostatistics,* 2nd ed. Chichester, UK: John Wiley & Sons Ltd; 2005. p. 796-812.

[18] Refinetti R, Cornelissen G, Halberg F. Procedures for numerical analysis of circadian rhythms. Biological Rhythm Research 2007; 38 (4): 275-325. http://dx.doi.org /10.1080/09291010600903692. NIHMSID: 249029.

[19] Halberg F. Some physiological and clinical aspects of 24-hour periodicity. *Journal-Lancet* (Minneapolis) 1953; 73: 20-32.

[20] Halberg F, Visscher MB, Bittner JJ. Eosinophil rhythm in mice: Range of occurrence; effects of illumination, feeding and adrenalectomy. *Amer J Physiol* 1953; 174: 109-122.

[21] Halberg F. Physiologic 24-hour periodicity; general and procedural considerations with reference to the adrenal cycle. Z Vitamin-, Hormon-u Fermentforsch 1959; 10: 225-296.

[22] Ertel RJ, Halberg F, Ungar F. Circadian system phase-dependent toxicity and other effects of methopyrapone (SU-4885) in mice. *J Pharmacol exp Ther* 1964; 146, 395-399.

[23] Halberg F. Organisms as circadian systems; temporal analysis of their physiologic and pathologic reponses, including injury and death. In: *Symposium on Medical Aspects of Stress in the Military Climate*, Walter Reed Army Institute of Research, Washington DC, 22-24 April 1964. Washington DC: U.S. Government Printing Office 1965-778-714; 1965. p. 1-36.

[24] Barnum CP, Jardetzky CD, Halberg F. Time relations among metabolic and morphologic 24-hour changes in mouse liver. *Amer J Physiol* 1958; 195, 301-310.

[25] Halberg F, Guillaume F, Sanchez de la Pena S, Cavallini M, Cornelissen G. Cephalo-adrenal interactions in the broader context of pragmatic and theoretical rhythm models. *Chronobiologia* 1986; 13: 137-154.

[26] Halberg F, Cornelissen G, Katinas GS, Hillman D, Otsuka K, Watanabe Y, Wu J, Halberg Francine, Halberg J, Sampson M, Schwartzkopff O, Halberg E. Many rhythms are control information for whatever we do: an autobiography. *Folia anthropologica* 2012; 12: 5-134. http://ttk.nyme.hu/blgi/Knyvek%20kiadvnyok/FOLIA%20 ANTHROPOLOGICA/folia12.pdf

[27] Halberg F, Cornelissen G, Sothern RB, Katinas GS, Schwartzkopff O, Otsuka K. Cycles tipping the scale between death and survival (= "life"). *Progress of Theoretical Physics* 2008; Suppl. 173: 153-181.

[28] Halberg F, Cornelissen G, Schwartzkopff O. Early glimpses of lunisolar and reciprocal biospheric cycles with overlapping uncertainties. *World Heart J* 2011; 3 (3): 187-231.

[29] Halberg F, Sothern RB, Watanabe Y, Hilllman D, Best WR, Schwartzkopff O, Cornelissen G. Signatures of space weather in the aging human blood circulation? *World Heart J* 2011; 3: 43-62.

[30] Halberg Franz, Cornelissen G, Beaty L, Schwartzkopff O, Halberg Francine, Revilla M, Siegelova J, Chibisov S, Breus T. Aligned physiological-epidemiological-physical environmental monitoring of variability in Vernadsky's noösphere: a chronousphere. *Proceedings, Space Weather Effects on Humans in Space and on Earth*, Moscow, 4-8 June 2012, p. 28-30.

[31] Halberg F, Haus E, Cardoso SS, Scheving LE, Kühl JFW, Shiotsuka R, Rosene G, Pauly JE, Runge W, Spalding JF, Lee JK, Good RA. Toward a chronotherapy of neoplasia: Tolerance of treatment depends upon host rhythms. *Experientia* (Basel) 1973; 29: 909-934.

[32] Halberg F. Biological as well as physical parameters relate to radiology. Guest Lecture, *Proc. 30th Ann. Cong. Rad.*, January 1977, Post-Graduate Institute of Medical Education and Research, Chandigarh, India, 8 pp.

[33] Halberg F, Gupta BD, Haus E, Halberg E, Deka AC, Nelson W, Sothern RB, Cornelissen G, Lee JK, Lakatua DJ, Scheving LE, Burns ER. Steps toward a cancer chronopolytherapy. In: *Proc. XIV International Congress of Therapeutics.* Montpellier, France: L'Expansion Scientifique Française; 1977. p. 151-196.

[34] Halberg F, Nelson W, Cornelissen G, Haus E, Scheving LE, Good RA. On methods for testing and achieving cancer chronotherapy. *Cancer Treatment Rep* 1979; 63: 1428-1430.

[35] Halberg F, Visscher MB. A difference between the effects of dietary calorie restriction on the estrous cycle and on the 24-hour adrenal cortical cycle in rodents. *Endocrinology* 1952; 51: 329-335.

[36] Halberg F. Chronobiology and nutrition. *Contemporary Nutrition* 1983; 8: #9, 2 pp.

[37] Halberg F, Haus E, Cornelissen G. From biologic rhythms to chronomes relevant for nutrition. In: Marriott BM. (Ed.) *Not Eating Enough: Overcoming Underconsumption of Military Operational Rations.* Washington DC: National Academy Press; 1995. p. 361-372.

[38] Halberg F, Cornelissen G, Otsuka K, Siegelova J, Fiser B, Dusek J, Homolka P, Sanchez de la Pena S, Singh RB, BIOCOS project. Extended consensus on means and

need to detect vascular variability disorders (VVDs) and vascular variability syndromes (VVSs). *World Heart J* 2010; 2 (4): 279-305.

[39] Halberg F, Cornelissen F, Halberg Francine, Kessler T, Otsuka K. Measuring mental strain by duration of blood pressure overswing (CHAT): case report. *World Heart J* 2010; 2 (2): 141-167.

[40] Halberg F, Cornelissen G, McCraty R, Czaplicki J, al-Abdulgader AA. Time structures (chronomes) of the blood circulation, populations' health, human affairs and space weather. *World Heart J* 2011; 3: 73-114.

[41] Halberg F, Cornelissen G, Hillman D, Beaty L, Hong S, Schwartzkopff O, Watanabe Y, Otsuka K, Siegelova J. Chronobiologically interpreted ambulatory blood pressure monitoring in health and disease. *Global Advances in Health and Medicine* 2012; 1 (2): 64-88.

[42] Watanabe Y, Halberg F, Otsuka K, Cornelissen G. Toward a personalized chronotherapy of high blood pressure and a circadian overswing. *Clin Exp Hypertens* 2013; 35 (4): 257-266.

[43] Kiser K. *Father Time. Minnesota Medicine* 2005 [November]: 26-29, 42-43.

[44] Lord of Time. London: Science without Borders/International Publishing House SWB; 2011. 45 pp.

[45] Halberg F, Powell D, Otsuka K, Watanabe Y, Beaty LA, Rosch P, Czaplicki J, Hillman D, Schwartzkopff O, Cornelissen G. Diagnosing vascular variability anomalies, not only MESOR-hypertension. *Am J Physiol Heart Circ Physiol* 305, in press.

[46] Hawkins D.M. Self-starting CUSUM charts for location and scale. *The Statistician* 36: 299-315, 1987.

[47] Cornelissen G, Halberg F, Hawkins D, Otsuka K, Henke W. Individual assessment of antihypertensive response by self-starting cumulative sums. *J Medical Engineering & Technology* 1997; 21: 111-120.

[48] Halberg F, Cornelissen G, Schwartzkopff O, Khalilov E, Khalilov T, Damirov F, Wang Z, Watanabe Y, Otsuka K, Siegelova J, al-Abdulgader AA. Preventive cardiology concerns avoidance of personal and societal health-related and natural cataclysms. In: Halberg F, Kenner T, Fiser B, Siegelova J. (Eds.) *Noninvasive Methods in Cardiology,* October 17, 2011, Brno, Czech Republic. Brno: Faculty of Medicine, Masaryk University. p. 33-44.

In: Circadian Cardiology with Focus on Both Prevention ... ISBN: 978-1-63463-956-9
Editors: K. Hristova, A. Shehab, G. Cornelissen et al.

Chapter 2

Editorial: The 'West' Has Made the Worst Model of Health and Is Pushing the World to the Slow Moving Disaster on Total Health

Toru Takahashi, PhD*[*1], R. B. Singh, MD, PhD[2], Kuniaki Otsuka, MD, PhD[3], R. R. Watson, PhD[4], Prof. D. W. Wilson, MD, PhD[5], and Fabien De Meester, PhD[6]

[1]Graduate School of Human Environment Science, Fukuoka Women's University, Japan
[2]Internal Medicine, Halberg Hospital and Research Institute, India
[3]Department of Medicine, Tokyo Women's Medical University, Medical Center East, Tokyo, Japan
[4]Department of Health Promotion Sciences, Health Science Center, School of Medicine, Tucson, AZ, US
[5]School of Medicine, Pharmacy and Health, Durham University, UK
[6]Tsim Tsoum Institute, Krakow, Poland

Introduction

Most experts believe that health is not limited to wellbeing of physical health alone but it is also the well-being of social, mental and spiritual health which may depend on environment. Therefore, we need to find out the cause of the disease and treat the patient as a whole, rather than disease alone. These basic principles of health care are ignored by most of the countries, which are considered to be leaders in developing the present model of health care. The United Nations high level Meeting dominated by experts from the western world proposed that poverty is the major cause of deaths due to non-communicable diseases (NCDs)

* Corresponding author: Dr Toru Takahashi, 1-1-1, Kasumigaoka, Higashi-ku, Fukuoka, 813-8529, Japan. E-mail: takahashitoru71@nifty.com, Tel: +81-092-661-2411, Fax: +81–92–661–2415.

(UN-HLM) [1]. Surprisingly, WHO is a party to conclusions made at UN-HLM with little consideration that unhealthy behaviors, among individuals and populations with underlying lack of general and health education are major causes of deaths rather than family income as such [3]. Millions of deaths occur every year due to lack of health education and poor health policies, because people and populations living in poverty and therefore scant resources, have no opportunity for health education due to poor public health services [1-5]. Various social markers of health and attributes of social class can influence the levels of physical inactivity, dietary patterns, salt intake, alcohol consumption, tobacco use and psychosocial stress, hence social health appears to be important in the pathogenesis of NCDs [3-6].

Mortality and burden of disease estimates for WHO Member States in 2008 revealed that 36·1 million deaths per year occur as a result of NCDs [1]. These estimates represent almost two out of three deaths per year worldwide. Some experts misinterpret the WHO estimates and propose that, of total deaths, 22·4 million arise in the poorest countries, and 13.7 million in high-income and upper-middle-income countries and therefore poverty may be the major cause of NCDs [1]. The causes of death among *Homo sapiens, Homo erectus* and Hunter-gatherer societies were mainly injury and accidents, whereas among western *Homo economicus* communities, 80% of deaths occur due to NCDs [1-3]. Many affluent countries in the various regions have increasingly seen inequities in people's life conditions and declining social mobility and social cohesion resulting in to unhealthy diet and lifestyle leading to poor total health [3-7]. Progress can be made in reducing health inequities based on global evidence and recommended policies and the health divide across all countries, including those with low incomes [1-4]. The objective should be to find out the attributes of social class and social determinants of health, across the life course, and in wider social and economic spheres, to achieve greater health equity by alteration in policy [2-4].

The Health and the Race to Follow the Western World

There is a race in the whole world to follow Western countries, particularly, the United States in developing war heads as well as industry, fashion and films, even in lower middle income countries like China and India. The budget for health expenditure in India and China is hardly one third of that in United States but it is being ignored by politicians in both the countries. After its victory in the World war II, United States, made tremendous progress, leading to eradication of communicable diseases and good physical health at the cost of poor social, mental, and spiritual health. This false progress in the United states created over confidence among policy makers in many countries to follow the Western world model of health, as if money can provide everything including, much better health than a good public health policy. In this connection, the report of the US National Research Council and Institute of Medicine, *U.S. Health in International Perspective: Shorter Lives, Poorer Health 2013.* (http://www.nap. edu/catalog.php?record_id=13497;For the US prevention strategy see http://www. healthcare. gov/prevention/nphpphc/strategy/report.pdf.) appears to be quite surprising.

It was interesting to note that despite spending more on health care per person than other high-income countries, many people in United States, die sooner, are least likely to reach the age of 50 years, and have higher rates of disease or injury. Unfortunately, Americans are less healthy from birth to 75 years of age than people in 16 other economically wealthy countries, if judged by health alone. This disadvantage in physical health has been getting worse for 30 years, more among women compared to men. This report compared comprehensive mortality and morbidity data from United States with affluent democratic countries; Australia, Canada, France, Italy, most of the Nordic countries, Spain, and the UK. Life expectancy is shorter at birth for American men, than for men in any of the other 16 countries, and American women fare little better. Surprisingly, Denmark is the only country that has a lower life expectancy for women at birth. Infant mortality and low birth weight; injuries and homicides; teenage pregnancies and sexually transmitted infections; HIV/AIDS prevalence; drug-related deaths; obesity, and diabetes; heart disease; chronic lung disease; and disability are more common in United States than in other Western countries. The Heart Disease and Stroke Statistics--2013 Update by American Heart Association indicates that the burden of cardiovascular diseases and type 2 diabetes mellitus is increasing despite decrease in coronary disease [3].

It is surprising again that a college education, higher incomes, and knowledge on healthy behaviors in United States were not found protective against this health disadvantage and this applies uniformly; to those with health insurance, as well as to those without. Similar observations have been made in India and elsewhere reporting that economic status is not the cause of risk of deaths due to non-communicable diseases (NCDs), although it may buy few years more due to drug therapy and surgery [2]. Despite a high budget on health care, US experts think that Affordable Care Act (ACA 2012) aimed at paying additional fee for health education may be a practical approach for prevention of NCDs. This capitalist approach via insurers, is open to bias because many US physicians have now increased risk of poor mental health. It may be good to provide funds for technological therapies as well as for health education but does not treat the cause of the lack of health behavior and provide affordable prudent foods and physical activity to achieve total health.

Despite better health in France, Italy and Nordic countries, inequities, both between and within countries, across 53 Member States of the WHO European region persist. The European Union has commissioned to support the development of the new health policy framework for Europe: Health 2020, because these experts believe that the Western world need to do modify existing health policy, related to social causes of health. The focus of European Union is on the global evidence regarding socioeconomic policy related determinants of health across the life course and recommends policies, to ensure that progress can be made in reducing health inequities and health divide across all countries of the world. The EU is collaborating with Health Canada, India, China, and many other countries, to develop global health agenda for providing better health.

We must emphasize about the utility and necessity of total health once again because physical wellbeing of the people and populations is not independent of social, mental, and spiritual wellbeing. If people and populations have satisfactory mental and spiritual health, they are more likely to have positive effect on self-motivation for better health behavior. A good total health is more important for policy makers then for people because the fate of the population health is in the hands of these experts.

We at the Tsim Tsoum Institute are of the firm opinion that health care of the population should not be considered a commercial commodity (www.tsimtsoum.net). Therefore buying of health behavior by a policy such as Affordable Care Act would be only partly useful.

Weapons of mass destruction can protect and prevent wars and injury, as in case of India and Pakistan but are not an absolute solution for prevention of wars and deaths due to injury. Because North Korea like Pakistan have poor mental and spiritual health and can strike at any time increasing the world toll of deaths due to injury. We neither support communism nor communalism but strongly believe that Budhism in Japan and Japanese diet are basic to better mental and spiritual health as well as physical health in Japan. In Islamic countries, where communalism is not so much strict, social health appears to be excellent. The world has to learn from Japan how to improve total health, although social health is deteriorating and diets and policy for health behavior alteration are exploited due to westernization [2]. Japan can lead the world, if and only if they can provide health behavior at affordable cost to all citizens Many affluent countries such as Australia, Canada, Nordic countries, Spain etc do not have giant war heads, but are least inclined to consider health as a commodity, and hence have better public health policies. It is unfortunate that now UK and most of the middle and lower middle income countries are inclined to have policies common in US. Corporate hospitals, health insurance companies, drug industry, health technologies; all generating employment to people, prosperity to doctors and researchers and funds to politicians. This approach is quite successful in the extension of life and better quality of life but has also increased the burden of sick populations in several of these nations [1-3]. The prevalence of a healthy lifestyle among individuals with cardiovascular disease in high-, middle- and low-income countries revealed that only little is known about adoption of healthy lifestyle behaviors among individuals with a cardiovascular diseases in communities across a range of countries worldwide [8]. This large, prospective cohort study involving 153 996 adults, aged 35 to 70 years, from 628 urban and rural communities in 3 high-income countries (HIC), 7 upper-middle-income countries (UMIC), 3 lower-middle-income countries (LMIC), and 4 low-income countries (LIC) is excellent." Among 7519 individuals with self-reported coronary disease (past event: median, 5.0 [interquartile range {IQR}, 2.0-10.0] years ago) or stroke (past event: median, 4.0 [IQR, 2.0-8.0] years ago), 18.5% (95% CI, 17.6%-19.4%) continued to smoke; only 35.1% (95% CI, 29.6%-41.0%) undertook high levels of work- or leisure-related physical activity, and 39.0% (95% CI, 30.0%-48.7%) had healthy diets; 14.3% (95% CI, 11.7%-17.3%) did not undertake any of the 3 healthy lifestyle behaviors and 4.3% (95% CI, 3.1%-5.8%) had all 3. Overall, 52.5% (95% CI, 50.7%-54.3%) quit smoking (by income country classification: 74.9% [95% CI, 71.1%-78.6%] in HIC; 56.5% [95% CI, 53.4%-58.6%] in UMIC; 42.6% [95% CI, 39.6%-45.6%] in LMIC; and 38.1% [95% CI, 33.1%-43.2%] in LIC).

Levels of physical activity increased with increasing country income but this trend was not statistically significant. The lowest prevalence of eating healthy diets was in LIC (25.8%; 95% CI, 13.0%-44.8%) compared with LMIC (43.2%; 95% CI, 30.0%-57.4%), UMIC (45.1%, 95% CI, 30.9%-60.1%), and HIC (43.4%, 95% CI, 21.0%-68.7%). "Among a sample of patients with a CVD event from countries with varying income levels, the prevalence of healthy lifestyle behaviors was low, with even lower levels in poorer countries indicating that health education via changing a policy could be important. However these low income populations continue to have occupational, cultural and traditional health behavior which has been protective against CVDs. However, the threat of policies from Western world is rapidly emerging because of lack of government policies in changing the health behavior in these countries.

The present capitalist model of health has failed and pushing the world to the slow moving disaster against total health. It seems that the policy makers in the Western World have no other way than to damage the making of health policy for their people and population because of the bias due to poor mental and spiritual health. We need to educate the policy makers about, how to provide prudent foods at affordable cost by the food industry and farmers and low cost high alcohol wines, to decrease the consumption of highly concentrated alcoholic beverages [9]. Free community centers for exercise, yoga and meditation as well as tax relief to hotels, motels and resorts providing these services free of cost in a no tobacco world would be a great step forward in providing better total health as well as for prevention of NCDs. Further steps may be to develop housing inclined to physical activity without the need for automobiles for routine household shopping. The challenges of how to develop "Functional Food Security" and inexpensive spare time and occupational physical activity in a no tobacco environment, across whole world for sustainable human development and prevention of CVDs and other chronic diseases may be accomplished by these steps [3-7,9,10].

Figure 1. Effect of body mind index on body composition.

Conflict of interest has not been declared by the authors.

Acknowledgments

This article are due to the International College of Nutrition and International College of Cardiology and The Tsim Tsoum Institute, for providing logistic support to write this article.

References

[1] Beaglehole R, Bonita R, Alleyne G, et al, for the Lancet NCD Group. UN High-Level Meeting on Non-Communicable Diseases: addressing four questions. *Lancet* 2011; 378: 449-55.

[2] Singh RB, Takahashi T, Nakaoka T, Otsuka K,Toda E, Shin HH, Kyu Lee M, Beeharry V, Hristova K, Fedacko J, Pella D, De Meester F, Wilson DW, Juneja LR. Nutrition in transition from Homo sapiens to *Homo economicus. The Open Nutra J* 2013;6:6-17.

[3] Horton R. Non-communicable diseases: 2015 to 2025. *Lancet* 2013;381:509-510.

[4] The "Heart Disease and Stroke Statistics - 2014 Update" *Circulation*. 2014; 129: e28-e292 Published online before print December 18, 2013, doi: 10.1161/01.cir.0000441139.02102.80

[5] Clark H. NCDs: a challenge to sustainable human development. *Lancet* 2013;381:510-511.

[6] Editorial. Wealth but not health in USA. *Lancet* 2013;381:177.

[7] Moodie R, Stuckler D, Monteiro C, Sheron N, Thamarangsi, Lincoln P, Casswel S on Behalf of the Lancet NCD Group. Profits and pandemics: prevention of harmful effects of tobacco, alcohol, and ultra-processed food and drink industries. *The Lancet*, Early Online Publication, 12 February 2013,doi:10.1016/S0140-6736(12)62089.

[8] The PURE Investigators. Prevalence of a healthy lifestyle among individuals with cardiovascular disease in high-, middle- and low-income countries: The Prospective Urban Rural Epidemiology (PURE) Study. *JAMA*. 2013 Apr 17;309(15):1613-1621

[9] Singh RB, De Meester F, Wilczynska A, Takahashi T, Juneja L, Watson RR. Can a changed food industry prevent cardiovascular and other chronic diseases. *World Heart J* 2013;5:1-8.

[10] Antman EM, Jessup M. Clinical Practice Guidelines for Chronic Cardiovascular Disorders: A Roadmap for the Future. *JAMA*. 2014;311(12):1195-1196. doi:10.1001/jama.2014.1742

In: Circadian Cardiology with Focus on Both Prevention ... ISBN: 978-1-63463-956-9
Editors: K. Hristova, A. Shehab, G. Cornelissen et al.

Chapter 3

Reminiscences: In Memoriam of Franz Halberg (5 July 1919 – 9 June 2013)

Germaine Cornelissen**, Mary Sampson, Cathy Lee Gierke, Larry A. Beaty and Othild Schwartzkopff
Halberg Chronobiology Center,
University of Minnesota, Minneapolis, MN, US

At the time of his death on 9 June 2013, just a month shy of his 94th birthday, Franz Halberg was one of the longest-serving faculty members at the University of Minnesota. He joined the faculty here in the mid-20th century and stayed into the second decade of the third millennium. Here, Franz pursued the innovative study, first of innate cycles in living things, and later how they are influenced by changes in their environment and react to stimuli from the cosmos. He would become known as the "Father of Chronobiology" [1, 2].

The term "circadian" was coined by Franz Halberg more than half a century ago. It led to the fields of chronobiology and chronomics, and the vast scientific legacy Franz left us, a path that started back during his tenure at the Medical University of Innsbruck, Austria. We are very fortunate that back in 1948 Franz accepted a WHO fellowship to work at Harvard University, and in Dr. George W. Thorn's laboratory at the Peter Bent Brigham Hospital, instead of joining Sir Alexander Fleming, who had extended an invitation to Franz to work with him in London. Indeed, it was the fact that Franz found flaws in the "Thorn Test" he had been assigned to perfect that led him to his pioneering work in chronobiology. The hot news in science at the time was the discovery that the adrenocortical hormone cortisone was an effective treatment for severe rheumatoid arthritis. Cortisone was scarce, and Franz was asked to test different substances for cortisone-like activity. Conventional wisdom suggested that the best test would gauge the effect of epinephrine on eosinophil cell counts, supposedly requiring the presence of adrenal glands. A marked drop of eosinophils 4 to 6 hours after

* Correspondence: Germaine Cornelissen, Halberg Chronobiology Center, University of Minnesota, MMC 8609. 420 Delaware Street SE, Minneapolis, MN55455, USA. Tel. 612-624-6976, Fax. 612-624-9989, Email. corne001@umn.edu.

injecting epinephrine to stimulate the adrenal cortex, not seen in patients with adrenal insufficiency, was hoped to provide an inexpensive test to reach a rapid diagnosis at a time when hormone analysis was costly and time-consuming. Franz could not confirm the result, and reported his findings to Thorn. Thorn admired Halberg's sticking to his guns, but argued that all the other researchers whose results upheld the epinephrine test could not have been wrong while Halberg was right [1-3]. Yet, right he was, and the epinephrine test was reconsidered and eliminated within a year. Halberg's results in the mouse were soon amply extended to humans by William R. Best [4]. Franz's demonstration that results varied as a function of the time of day at which the test was performed cost him the renewal of his WHO fellowship. Today, this circadian stage-dependence is readily understood as adrenal cortical hormones are highest around the time of awakening and lowest when going to sleep, the circadian pattern being synchronized by the lighting and feeding schedules.

An invitation to join the faculty at the University of Minnesota came shortly thereafter, through the auspices of Maurice B. Visscher, whom Halberg had met when Visscher was part of a Unitarian medical mission to post-war Europe. Visscher's department included a division of cancer biology headed by John J. Bittner. In examining whether an adrenocortical activation, then assessed by eosinophil depression, can be an answer for treating breast cancer and for prolonging life, Franz solved another puzzle, namely that both the lighting and feeding schedules were important in determining the phase of the circadian rhythm in circulating eosinophils [1, 5].

By then, Franz had realized the importance of circadian rhythms as an indispensable control. To learn more about them, he turned to rectal temperature as a marker rhythm that could be more readily recorded around the clock. This work led him to find that the core body temperature of mice continued to show cyclic rises and falls even after they had been surgically blinded. Visscher at first dismissed as "paranoia" Franz's insistence that this was important. Franz lost his laboratory space and for a while had to work out of a makeshift laboratory in a paint room at the Cambridge State School and Hospital 30 miles north of the Twin Cities. He continued to buck conventional wisdom, however. He asked his straitlaced department head, Cyrus P. Barnum, professor of biochemistry and an expert on nucleic acids, for a technician to test his belief that RNA and DNA showed circadian periodic changes. So ingrained was the position that RNA and DNA were constant that Barnum was reluctant to provide a technician with government grant money, but so intrigued was Barnum by Halberg's hypothesis that he offered to act as the technician himself. Franz's demonstration of circadian rhythms in RNA and DNA formation and in mitoses laid the foundation for cancer chronotherapy and provided a hint for the existence of an RNA world preceding our DNA world. During his long tenure, a great many students and research associates from around the world passed through our laboratory. They came first to Millard Hall and to Franz's makeshift laboratory in Cambridge, Minnesota. Later they found him in Lyon Laboratories, a space purpose-built for this research with funding from NASA's Biosatellite program. Finally, at his 80th birthday and the turn of the 21st century, moving to the Mayo Memorial Building, with the generous support of Earl Bakken, founder and chairman of the board of Medtronic, Franz continued his research with a few dedicated colleagues who stuck with him through thick and thin. As we continued to learn from Franz, we invariably enjoyed his enthusiasm, his passion and his vision of a world he wanted to leave in a slightly better shape than he found it. We shall miss his modesty, his humor, his huge scholarship, and his unwavering support and friendship. From the many of us who had the privilege to know

Franz, condolences and reminiscences listed below provide a glimpse into the full life of this great scientist and most generous man.

*** Reminiscences ***

... My deep condolences to you and to all of Franz's friends. It is great loss for us all. I call myself part of Franz's group. He is spirit father. May God bless Franz and shower his spirit with mercy and love. Please be informed that the life message of Franz is on our shoulders now. We shall continue. This is the best love that we send to his spirit.

Abdullah al-Abdulgader
Director, Prince Sultan Cardiac Center
Al-Ahsa, Saudi Arabia

It is really very sad to hear about it.... Please accept condolences from myself personally (who knew him and was co-author, even 1 paper) as well as from Shamakhy Astrophysical Observatory (Azerbaijan) and Science Development Foundation (Azerbaijan). We will always remember him, his energy, his wishes, and his contribution in creating a new scientific direction, his willing to bring all of us together through joint researches. I wish him a Paradise.

Elchin Babayev
Vice-Director, N. Tusi Shamakhy Astrophysical Observatory
Baku, Azerbaijan

Please, convey our heartfelt condolences to Othild and let her know, please, that we were very sad to hear today that Franz left us on June 9. This is a big loss for chronobiology which was essentially coined by Franz' life-time's work.
Hella and myself can see in our ongoing studies on melatonin and the solar cycle how right his vision was and there is certainly so much to discover still which is beyond the narrow limits set by most colleagues. Franz will always remain in our heart since we also know and feel how much he loved and supported our work on the pineal gland.

PD Dr. Christian Bartsch and Hella Bartsch
Center for Research in Medical and Natural Sciences
University of Tübingen, Tübingen, Germany

I was so sorry to hear of your irretrievable loss. Accept our most sincere condolences. I and my husband Sergei Chernouss personally knew Franz Halberg and very cordially treated him. Infinitely sorry that such a great scientist passed away. We feel deeply for your sorrow. You can always count on our support.

Professor Natalia Belisheva
Polar Geophysical Institute, Kola Science Centre
Russian Academy of Science, Apatity, Russia

… we are mourning the great loss together with you. Franz was a great scientist, a kind teacher and a reliable friend for us. He will live forever in our memories and hearts.

Professor Mikhail Blank
Central Research Institute of Roentgenology and Radiology
St. Petersburg, Russia
with Leonid and Olga Blank and Daria Denisova

… we are in great shock about the very sad news! Last night we spoke so intensely about Franz and yourself and discussed what we can do to help for his dream to become reality. With love,

Vera Brandes
Research Program Music Medicine, Paracelsus Medical University
Salzburg, Austria
and
Dieter Broers
International Council for Scientific Development
Erlangen, Germany

I send you my condolences and great sorrow and also promise that we never forget what great figure Franz was in science and like Human-being. He was our teacher, our friend and our leader. Very many aspects of our science we started to understand due to his and his co-workers activity and discoveries and explanation. I promise that we in Russia -- all his followers -- will continue the science, which he loved so much, and I hope that you too will help us as he helped.

Tamara K. Breus
Space Research Institute, Academy of Sciences
Moscow, Russia

I am very sorry to hear of your recent loss, and so surprised to hear that Prof. Franz Halberg passed away. I would like to offer my sympathy and condolences to you and your laboratory and Franz's family. My thoughts and prayers are with you and your family at this difficult time.

Naoto Burioka
Professor, 3rd Department of Internal Medicine
Tottori University, Yonago, Japan

I join you with the great sorrow: for me also Franz was a second father, as you know. The life circumstances caused a certain estrangement between me and the Minnesota friends, but I did not forget all the wonderful moments passed with Franz, Erna and you. Unfortunately life is so short, as Franz used to say.

Franca Carandente
Professor of Internal Medicine,
University of Milan, Milan, Italy

We have met in 1985 in Prof. Halberg's lab and I hope this will find you in good health. It was very sad for all of us down here, in Brazil, knowing Prof Halberg passed away.

José Cipolla-Neto, MD, PhD
Professor of Physiology, Institute of Biomedical Sciences
University of São Paulo, São Paulo, Brazil

The news I read is very sad in that Franz was regarded by me as a father in life and science. His presence in my scientific and human life has been determinant for my scientific development. I remember the first day I arrived to Minnesota. It was August 1, 1980. I never could think that there was a scientific Lab so organized and in the forefront in the field of Chronobiology. The experience has been amazing as it is amazing the long period of time in which Franz, you and Bob remained in touch with me. I cannot forget that this time culminated with the eponym of Cugini's syndrome. I cannot forget the "Womb to Tomb Project" in which you introduced me as a sign of confidence in my scientific and human person. ... A few persons in the world can have the opportunity and privilege to be the pupil of a great teacher, scientist and human being. Franz was used to say: "Tempora mutantur and non mutamus in illis".

The death is the last mutation of this finite cycle of immanent life and the beginning for the eternal cycle of the transcendent life.

Pietro Cugini
Professor of Medicine, University of Rome
Rome, Italy

Thank you very much for your message. Its contents are very sad and, frankly, hard to believe, particularly given Franz's enormous vitality. Under the circumstances, there is little chance that words of consolation can have any effect, especially upon those who were close to him. I would only like to point out that usually the passing is more painful for those who stay than for those who go.

Please do not forget that he was lucky enough to be passionate about what he was doing, and that he could do it longer than most of us. His life span has exceeded the life expectancy for people born in 1919 by 67% (ref. "Life Expectancy in the United States", Congressional Report RL32792, 2006)!

I am sure that for Franz, his life was but one cycle of his existence, and he is already preparing his comeback.

Jerzy Czaplicki
Professor, Institute of Pharmacology and Structural Biology, CNRS
Paul Sabatier University, Toulouse, France

Thank you for your sad message! I miss his friendship and help!

Franz Daxecker
Professor, University of Innsbruck
Innsbruck, Austria

I feel the vacuum here, so do I imagine it on your side. Be sure I am with you with my heart.

Fabien De Meester
President and CEO, Tsim-Tsoum Institute, Krakow, Poland

... je reçois l'avis de décès de Franz Halberg. Dans ces moments les mots sont peu de chose. Nos pensées sont de tout coeur avec vous deux pour surmonter la douleur de la perte de ce grand homme.
[...I receive the news of the death of Franz Halberg. At such times words are inadequate. Our thoughts are with both wholeheartedly to overcome the pain of the loss of this great man.]

Dr. Alain Delcourt
Kraainem, Belgium

I am very sorry to know the loss of Franz Halberg, a pioneer in many new directions of cancer research. A great friend and influential supporter for all of us. This volume of IN VIVO will be dedicated to his memory.

John Delinassios
Editor, International Institute of Anticancer Research,
Attiki, Greece

My deep sorrow and condolences to you, Othild and all of the folks at The Halberg Center!!! Please send also my condolences to his relatives.
I will miss him very much. He was my best mentor, advisor to my DM/PhD in Bulgaria, keen supporter of my career and professional development as a doctor, chronobiologist, chronoepidemiologist and medical statistician.
We will all miss Franz very, very much.

Borislav Dimitrov
Senior Lecturer in Medical Statistics, University of Southampton
Southampton, UK

Thank you very much for this very sad information. I always admired his activities and enthusiasm for the Chronobiology. I want to say "thank you, Franz, for all you have done; and also for your friendship".

Peter Engel
Professor, Institute for Occupational Physiology and Rehabilitation Research,
University of Marburg, Marburg-Kassel, Germany

Even though, after reading Hans Wendt's recent message about Franz' stroke, I was somehow prepared that he was in much danger, I received your message of his demise with sorrow and grief. I remember with gratitude that Franz often included me and my work in his own endeavors. I remember that he often said that he wanted to learn something from me even though I could only learn much from him. I am sure that his admirable contribution to our knowledge about effects on humankind of the cycles of the universe will be acknowledged in due course by more mainstream researchers.

Suitbert Ertel
Professor, University of Göttingen
Göttingen, Germany

... only now I read your email with the mournful news of Franz' death. In such a case it is very difficult to translate in simple words what is felt in one's own heart ... Franz was a true friend for us and when a friend, as he was for us, passes away, a portion of us also dies. This is now the frame of my spirit. I am now only a child unable to express an unacceptable grief... Hold Franz always together. This, perhaps, is the way to defeat his death.

Piero Faraone
Vice-President, International Committee for Research and Study of Environmental Factors, Milan, Italy

It was with great sadness that we received the devastating news about Franz.
We loved him very much and we were very often thinking and talking about him, about you and about Othild, but we are not very diligent writers. We consider all three of you as our close friends and in fact part of family. We will miss Franz very deeply; he will always stay in our hearts.

Peter Gaudenz Fedor-Freybergh & Lili Maas
Editor & Art Director, Neuroendocrinology Letters
Stockholm, Sweden

We are mourning with you and your working group for a great scientist and a loyal friend. It is a comfort that Franz will continue to live on in the Olympus of great spirits.

Eva-Maria & Jürgen Freytag
Institut für Bibliothekswissenschaft und wissenschaftliche Information
Berlin, Germany

It is very sad to hear that Franz passed away a short time ago. I was aware of his enormous contribution to science, in particular concerning the development of new concepts, such as chronobiology, thus opening new fields of research, but recently I had also the privilege and honor to interact with him and his group in a collaborative research.

I became immediately aware of his great interest and attention towards areas of research that could benefit by interactions between different areas and between researchers with quite different background, His "intellectual curiosity", still highly present at his age, has certainly contributed to maintain an open and active mind throughout the whole life, and to apply the emerging concepts to clinical problems.

We all shall miss him, but at the same time we'll have a clear example of a great mind and person to be remembered.

Claudio Galli
Professor, University of Milan
Milan, Italy

It is a great sorrow that our Franz passed away. He was GREAT indeed and we will forever remember him as the ideal of human and scientist. His remarkable legacy will be investigated many years still. It was so great that we met all of you in Istanbul. I didn't think that it was our last meeting... We shall miss him too! ... It's necessary to continue the work together and it will be good memory of our Franz.

Marina Gigolashvili
Professor, E.K. Kharadze Abastumani Astrophysical Observatory
Tbilisi, Georgia

I'm so sorry. Franz Halberg was a great scientist, all the time asking more and more from himself and all the time quite open to learn a little bit more from people coming with new ideas, joining people from different scientific fields.
An extraordinary man that will be alive in our mind forever!
Franz was a very good friend, a unique scientist in international scientific community.

Paulo Gomes
Professor, Comissao de Coordenaçao e
Desenvolvimento Regional do Norte Portugal
Porto, Portugal

Je viens d'apprendre aujourd'hui la disparition de du Pr Franz Halberg. Je voulais humblement me joindre à vous et à la famille du Dr Franz Halberg pour cette douloureuse perte.
[I have just learned today about the death of Dr. Franz Halberg. I join you and his family in sorrow for this painful loss.]

Laurent Gouthière
Research Associate, Expert Soft Tech
Caen, France

Zum Heimgang des von mir sehr verehrten Franz spreche ich Ihnen, all seinen Mitarbeiterinnen und Mitarbeitern und besonders Othild meine herzliche Anteilnahme aus. Die vielen Diskussionen, die ich mit Franz geführt habe, waren für mich stets von großer Bedeutung, und die Zusammenbeit mit ihm hat mir viele Freude bereitet und auch sein Besuch hier bei uns in Niemegk. Dafür bedanke ich mich bei ihm ganz besonders. Er wird bei mir immer in sehr angenehmer Erinnerung bleiben. Bitte verzeihen Sie mir, daß ich nicht in englisch schreibe, aber es gelingt mir nicht, meine Anteilnahme in einer anderen Sprache angemessen auszudrücken.
[On the occasion of Franz's passing, let me say that I greatly admired you and all of his associates. To Othild in particular I extend my heartfelt condolences. The many discussions I had with Franz for me were always of great importance, and my collaboration with him and his visit here in Niemegk gave me a lot of joy. For that I thank him very much. He will always be with me in a very pleasant memory.
Please forgive me for not writing in English, but I am unable to express my sympathy in another appropriate language.]

Armin Grafe
Physicist, Geomagnetisches Observatorium Niemegk
Potsdam, Germany

I certainly had the privilege and luck to meet Franz (although unfortunately only on very few occasions). In addition, we had several contacts by e-mail. But, his charisma, his youth, his positive feeling and enthusiasm in "reading the Book written by Mother Nature" (just to use a Leonardo's expression), were immediately perceived. He immediately fascinated everybody who can understand these rare qualities.

When a person like Franz passes away, it just like we miss a part of ourselves. It is a part of the best "components" of our living and beliefs that is passing away. We feel that time is elapsing. We miss a friend, we miss a light which has been one of the several unconscious lights, the guide for our living, ... It is a part of our own enthusiasm for life that arrived to its end .... We get abruptly older, and our thoughts, our same living, must unavoidably change. We need to think, to reconsider the meaning of what we do .....Franz will always live in the heritage of thought that he left, to you, to all friends, and also to all people who never met him, and that never will even read his papers .... The thought of humankind, the true "civilization", is made by leaders (by their thought, rather than by their papers) who in general are appreciated and considered in their real value only after some time. Indeed, people need time before understanding the innovative ideas of true leaders. I always considered Franz as an important precursor of future fundamental developments of knowledge. The conventional separation of knowledge into disciplines is an unfortunately unavoidable need required by the limit of human mind. But the largest number of people (also of scientists) feel happy about getting some level of knowledge they reached, they become self-referencing, they close themselves inside their "shell" and feel to be "important" and "authoritative". Only very few people spend the whole time of their life to improve their knowledge, to attack new frontiers, to challenge new landscapes of natural science, ... But, how can I express by a few words the wealth of thought of a person like Franz? Yes, I will always be honored to be (very modestly) eventually useful (just a little bit ....) to your effort to continue the Franz challenge! I wish you all good luck, and all my sincere appreciation. I heartily hope that you afford to find several new "proofs" in order to clarify the mystery of the relations between biology and environment!

Giovanni Gregori
Associate Scientist, Istituto di Acustica O.M. Corbino,
National Research Council, Rome, Italy

It is the saddest news I learned today from your message. I, my mother and people who had privilege of knowing Franz deeply regrets our common loss.

To me and my mom, Svetlana, Franz always have been my American Father and his passing brought us the saddest time since we lost Gennady almost 5 years back, and whom Franz and you so cordially honored just few weeks ago. There are but few persons whose lifetime impact is so immense that they truly deserve to live several lives instead of one, not matter how long it was.

Franz is among few remarkable and great human beings who left us an empty hollow that cannot be ever refilled. And our world changed and will never be the same since. His bright and unique personality will always be with us and he will live in our hearts and minds. The best thing we all can do now to honor Franz's memory is to continue to do our best in defending and further developing his countless ideas and visionaries.

With deepest sorrow, and personal condolences to you and Othild,

Denis Gubin and family
Professor, Tyumenskii Gosudarstvennyi Meditsinskii Institut
Tyumen, Russia

Je suis navré par ce qui vous arrive, je comprend la tristesse que tu exprimes dans ton annonce. Bien que je n'aie que très peu connu Halberg -- je ne l'ai vu qu'une fois lors d'une de ses 1ères visites à Jean De Prins -- je réalise à présent la grande importance qu'il a eue dans vos vies à toi et à Francis. C'est un moment où la mémoire ranime de nombreux souvenirs, et montre ce qui a vraiment été important. Avec toute ma sympathie.
[I am saddened by what just happened and understand the sorrow you expressed in your announcement. Although I knew Halberg only slightly -- I once met him during one of his first visits to Jean De Prins -- I now realize the great importance he had in your and Francis' lives. This is a time when the memory revives many remembrances, and shows what was really important. With all my sympathy.]

Jean Luc Guisset
Senior Research Associate, Universite Libre de Bruxelles
Brussels, Belgium

I feel with you, Professor Franz is missing also for me.

Anna Gvozdjakova
Professor, Comenius University Medical School
Bratislava, Slovakia

This is sad news and it brings up a lot of memories from the many decades I had been in contact with him. And for certain I shall not forget this great scientist and good friend.

Rüdiger Hardeland
Professor, Zoologisches Institut der Universität Göttingen
Göttingen, Germany

bei unserer Rückkehr nach einer längeren Russlandreise erreichte uns Eure traurig stimmende Nachricht vom Ableben unseres lieben Franz. Wir fühlen mit Euch den Schmerz durch den Verlust eines guten, weisen Menschen, der mit uns seit Jahren in herzlicher Freundschaft verbunden war, auf die wir bis zu unseren Lebensende stolz sein werden. Einen solchen Menschen wie Franz, mit seiner einzigartigen Genialität und Kreativität und seiner unermüdlichen Aktivität, wird niemand der ihn kannte vergessen. Der von Franz entwickelten Wissenschaft von den Rhythmen des Menschen und der Natur, welche für die gesamte Menschheit von großer Bedeutung ist, gehört die Zukunft. Der Geist unsers lieben Franz ist unsterblich. Wir wünschen Euch in der schweren Zeit der schmerzerfüllten Trennung Kraft und Gesundheit und verbleiben, wie immer in freundschaftlicher Verbundenheit

[The sad news of the demise of our beloved Franz reached us on our return after a long trip to Russia. We feel the pain of the loss of a good, wise man who was connected with us for years in warm friendship, for which we will be proud for the rest of our life with you. No one who knew him will forget such a man as Franz, with his unique genius and creativity, and his tireless activity. The future belongs to the science developed by Franz of the rhythms of man and nature, which is for all humanity of great importance. The spirit of our dear Franz is immortal. We wish you in the difficult time of painful separation strength and health and remain, as always in friendly communion.]

Karl & Lena Hecht
Professor, Humboldt-Universität (Charité)
Berlin, Germany

Merci pour l'information que m'avait, en effet, déjà transmise Jacqueline. Le décès du Professeur Halberg est un événement dans le monde de la chronobiologie. Jacqueline m'a également informé que vous aviez maintenant en charge la direction du laboratoire du Professeur Halberg. Je vous en félicite et vous souhaite beaucoup de réussite dans cette mission qui va, certainement, vous occupez beaucoup.
[Thank you for the information which Jacqueline had already provided. The death of Professor Halberg is an event in the world of chronobiology. Jacqueline also informed me that you have succeeded Prof. Halberg as director of his laboratory. I congratulate you and wish you a lot of success in this mission which surely occupies much of your time.]

Bernard Hecquet
Chronobiologist, Lille, France

Franz was more than a brother to me; he was always approachable concerning new chronobiological questions. Unfortunately the distance was too long as to cooperate more closely on a scientific basis. ... You wrote that at 93 years of age he was still active 7 days a week in the Halberg Chronobiology Center. That is why we suppose that he did not have to suffer from a longer disease.
Franz will be in my heart as long as I live.

Theodor Hellbrügge
Professor, Founder of Theodor-Hellbrügge-Stiftung
Munich, Germany

Let me express my deepest sympathies to you. I was so very sad to hear about Franz's passing away. It was the honor and the pleasure to know Franz and I shall miss him so much! I believe we keep his legacy alive and his work will continue in our cooperation in future.

Pavel Homolka, MD
Functional Diagnostics and Rehabilitation Clinic, St. Anna Hospital
Masaryk University, Brno, Czech Republic

It is a great loss that Franz passed away. He can look back onto a very fulfilled life though! I thank him and all of you for the kindness you spread when I was in Minnesota!

Harald Hörmann, MD
Medizinische Universität Innsbruck
Innsbruck, Austria

Mit großer Betroffenheit haben wir vom Ableben unseres Mitglieds Franz Halberg Kenntnis nehmen müssen. Ein unersetzlicher Verlust hat uns getroffen. Wir trauern um den Vater der Chronobiologie, um einen aktiven Streiter für neue Erkenntnisse, um einen anregenden Diskussionspartner und um einen nimmermüden Wahrheitssucher. Seine Arbeiten und vor allem seine Publikationen in den Organen der Sozietät werden immer wieder herangezogen werden, wenn Forschungen auf dem Gebiet der Chronobiologie zu initiieren sind und bereits Erreichtes nicht vergessen werden soll. Ich bin überzeugt, dass seine engen Mitarbeiterinnen und Mitarbeiter im Sinne von Franz schöpferisch weiter arbeiten werden, um auch Vorhaben zu verwirklichen, an denen er gern noch selbst mitgearbeitet hätte. An Ideen fehlte es nie bei ihm. Er drängte stets auf die weitere Entwicklung der Wissenschaft. Stillstand hasste er. Meine Frau Helga als Ethikerin und ich als Wissenschaftsphilosoph verlieren nicht nur einen stets für Philosophie und Ethik aufgeschlossenen kreativen Wissenschaftler, sondern auch einen Freund, mit dem man über viele Probleme sich austauschen konnte. Er legte stets den Finger in die Wunde, wenn er Lücken in der Argumentation bemerkte. In unserem umfangreichen elektronischen Informationsaustausch ging es um die Philosophie der Zeit, um die philosophische Entwicklungstheorie, um die Zyklizität des Geschehens. Stets bedachte er ethisch-moralische Auswirkungen der Forschung auf das menschliche Individuum. Er war ein Ganzheitsdenker, der Patienten nicht auf eine bestimmte Anomalie reduzierte, sondern sich um Rhythmen im Lebensalltag ebenso kümmerte, wie um langfristige Prognosen und Therapien. Spirituelles Verhalten analysierte er mit seinem Team chronobiologisch ebenso akribisch, wie den Zusammenhang von Geo- und Biozyklen, von mentalen Prozessen und Krankheitsbildern. Es lohnte sich für uns, mit ihm über Interdisziplinarität und Transdisziplinarität im Zusammenhang der Wissenschaften zu debattieren, da ihn nicht die allgemeine theoretisch-philosophische Reflexion allein interessierte, sondern vor allem die praktische Lösung von Problemen. Als ich meinte, wir seien doch Zyklizitäts-Theoretiker, ergänzte er berechtigt, wir seien doch auch Zyklizitäts-Praktiker. Das war er auf jeden Fall in der Chronomedizin, auch wenn er sich über die Auswirkungen der Chronomik auf die Organisation des Gesundheitswesens Gedanken machte.

Wir trauern um einen hervorragenden Wissenschaftler und um einen Humanisten, der uns stets aufs Neue überraschte, wenn er Probleme aufwarf und nach Lösungen suchte. Dazu brauchte er den Gedankenaustausch mit anderen. Gegenüber neuen Ideen war er stets aufgeschlossen. Er wird uns fehlen.

[We take note with great sadness of the death of our Member Franz Halberg. He is irreplaceable. We mourn the father of chronobiology, an active fighter for new insights who could lead a stimulating discussion, and a seeker of truth. His work and especially his publications in the Society's journal will always be used when initiating research in the field of chronobiology, and his achievements will not be forgotten. I am convinced that his close associates will continue to work within the meaning of Franz's creativity in order to pursue the work Franz started. Franz never lacked new ideas. My wife Helga as an ethicist and I have lost not only an open-minded creative scientist concerned with Philosophy and Ethics, but also a friend that cannot be replaced. He always put the finger in the wound, if he noticed gaps in the argument. Our extensive electronic exchange of information related to the philosophy of time, to the philosophical theory of evolution, and to the cyclicity of events. He always pondered the ethical-moral impact of research on the human individual. He was a "global thinker": patients were not reduced to a specific anomaly, but well cared for with consideration for rhythms in daily life in terms of prognosis and therapy. He and his team also analyzed chronobiologic spiritual behavior meticulously, such as the connection of nature and biological cycles, of mental processes and pathologies. It was well worth it for us to debate with him on the interdisciplinarity and transdisciplinarity of scientific disciplines because the general theoretical and philosophical reflection was not considered without concern for a practical solution of the problems. As I said, we were but cyclicity theorists, but he added that we are also cyclicity practitioners. He was definitely in chronomedicine, even when dealing with healthcare, chronomics was on his mind. We mourn a great scientist and a humanist who always surprised us when he posed problems, searching for solutions. For this he needed to exchange ideas with others. Franz was always open to new ideas. He will be missed.]

Herbert Hörz
President, Leibniz-Sozietät e.V.
Berlin, Germany

It is really sad and shocking to hear of Dr. Halberg's passing. I have no words when I think how hard and painful time you have. The only thing I could do is to pray for him to rest in peace. Over more than fifteen years, we are communicating with him, I regret that I could not meet him even once.

Tomoko Iizuka
A&D Ltd.
Tokyo, Japan

I am very sorry that Professor Franz Halberg died. I realise how much he had done for the entire field of chronobiology. We'll miss him deeply.

Helena Illnerova
Past President, Academy of Sciences of the Czech Republic, Prague, Czech Republic

I am deeply sad and very very shocked. I knew his age, I am aware that we do not know what the next minute brings, but still, it is a very bad feeling to be faced with the cruel reality – I shall never meet him again.

He meant a lot for me, he was a unique, fantastic person, a real humanistic great person. He had a special personality, with wisdom, broad knowledge -- his ideas, his empathy, his care, his "father"-like behaviour, his suggestive style – and so much more, I will never forget.

Now I also feel the same as you, great sorrow, great loss. Very very sad moment. Sitting here and crying. I wish I could be there to place a flower with "You always live in our heart -- Your Hungarians".

Rita Jozsa
Professor, University of Pecs
Pecs, Hungary

Let me send my deep condolences! Franz was my chronobiological father, my teacher and friend. Everybody who was acquainted with Franz will keep Franz in his mind up to his own last day.
In Russia people say: let Earth be like down to him.

George S. Katinas
Professor, Yaroslav Mudry Novgorod State University
Russia

Information about the death of Professor Franz Halberg has caused me and my family great grief for a lost friend and colleague. I ask you to accept from me personally, from my wife Tamila and sons Anar and Farid deepest condolences. It is an irreparable loss for all of us and for modern science. His many friends and students will always keep in heart the good memory of this wonderful man and a great scientist.

Elchin S. Khalilov
Chairman of the International Committee GEOCHANGE,
Chairman of the Board WOSCO, Germany, Munich

With great sorrow we were informed by Prof. Cornelissen that Prof. Franz Halberg deceased on Sunday 9th June 2013. Franz was one of our most outstanding academy members and (besides the Nobel Laureates) one of only 5 full members. We mourn the loss of a great scientist and a personal friend!

Walter Kofler
President, International Academy of Science
University of Innsbruck, Innsbruck, Austria

I was deeply shocked to learn about the sad demise of our most revered teacher and guide, Prof Franz Halberg. I express our heartfelt condolences to the bereaved family and to all who received his fatherly love and scholarly guidance. I will never forget his visit to Amritsar-India. I pray to God almighty to grant peace to the departed soul. I think we should all collectively do something to immortalise his personal and scientific achievements.

Adarsh Kumar
FACC, Government Medical College Hospital
Amritsar, India

Je suis évidemment attristé par le décès de Franz, qui m'a appris la chronobiologie et fait partager sa vision de notre discipline. J'ai l'intention de faire paraître l'obituary que tu m'as envoyé dans Rythmes, la revue de la SFC, dont le prochain numéro attend mon éditorial.

[I am obviously saddened by the passing of Franz who taught me chronobiology and shared with me his vision of our discipline. I plan on publishing the obituary you sent me in "Rythmes", the journal of the French Society for Chronobiology (SFC); its next issue awaits my editorial.]

Francis & Monique Lévi
INSERM, U Rythmes Biologiques et Cancers,
Hôpital Paul Brousse
Villejuif, France

Oh no!! il fallait s'y attendre vu son age, mais je suis sous choc comme à la mort de papa, je ne veux pas y croire, je ne peux pas continuer sans lui, même si je ne le voyais pas souvent, savoir que le Prof. Halberg était là me rassurait; je t'embrasse très fort, j'essaierai de t'appeler, bises

[Oh no!! It was to be expected given his age, but I am in shock as I was on the death of my father, I do not want to believe it, I cannot go on without him. Even if I did not see him often, I was reassured that Prof. Halberg was there. I embrace you most heartily and would try to call you. Kisses.]

Cristina Maggioni, MD
University of Milan
Milan, Italy

I am saddened to hear about Franz. I should imagine how difficult it has been for you to come to terms with the demise of your closest collaborator and friend. Although I only had the opportunity to interact with him over a short period, I was deeply impressed by his kind personality and his scientific vision. Please accept my sincere condolences, and also for his family. I feel honoured to have had the opportunity to interact with Franz. It is only a few weeks ago that we exchanged emails about how to proceed with the complexity analysis of long-term physiological data I got from you. I am ever so sorry that we did not have chance to finish the study, and would like to continue as my humble homage to Franz. The sad news found me during a short business trip, this weekend my thoughts have been about Franz's exceptional personality and contributions.

Danilo Mandic
Professor, Communication and Signal Processing Research Group
Department of Electrical and Electronic Engineering, Imperial College, London

Dearest all friends and workers in the science of chronobiology in Minneapolis, dearest all chronocosmobiologists around the world! We, with my Junior, are shocked by the quite unexpected bad news. We just prepared to write to his great jubilee on July 5. It is impossible to describe sufficiently the huge lost for science, for cosmopolitan humanism, and for all Franz so magnanimously donated his excellent personality to us all and to the world.

He was an absolutely good man in the best sense of the word. His noble-minded spirit did always emanate the enormously great love to humans, including all of ourselves. In fact, I never recognized such personality, supergiant in all dimensions, as such like our dearest Professor Franz. It is the great honour for ourselves to may be joined to the deep mourning worldwide – from the United States to Russia, from Europe to Japan, to India, etc. We are, as so many scientists on this planet, much indebted to Professor Halberg for all his leading ideas, for his support in concrete situations, for his unlimited love.

Dear Franz, we thank You wholeheartedly for all that, and for much more. In these sorrowful moments, You are entering the eternal cathedral of the world top science. Good bye, dear Franciscus. We have unfortunately to say it although we are still not realizing fully this terribly bad news.

But we believe we will meet You, dearest Franz, again in one of the future Universes which can -- or must? -- once come.... At least, it is not excluded - according to the laws of probability, taking into account the never ending number of future Cosmoses which will come .... Why could one of them be not exactly the same as this one we presently live? This is our secret hope, dearest Germaine and dearest mates in Minnesota and worldwide. Sincerely Franz, and also Yours.

Miroslav Steven Mikulecky Sr. and Miroslav Mikulecky Jr.
Professor Emeritus, Comenius University Medical School
Bratislava, Slovakia

I am and will always be with you. May his spirit stay and thrive within us forever.

Gen Mitsutake
Division of Neurocardiology and Chronoecology
Tokyo Women's Medical University, Japan

I am indeed sorry to hear of Franz's passing but grateful that we had the benefit of his wise counsel for so long.

Karin Olson
Professor, University of Alberta
Edmonton, Alberta, Canada

I am deeply sad that Franz has passed away. I am surprised at receiving your E-mail, because Franz has a lot of energy and I believe he is invulnerable. I miss him so deeply. … I will go ahead with him (with his heart) our project on writing a book. Could you please let us join forces to support him for the last work for him.

Kuniaki Otsuka, MD
Head, Daini Hospital, Tokyo Women's Medical University, Tokyo, Japan

All my respect to Dr Halberg's family, team and friends. I hope time can cure the pain.

Mauricio Pacheco de Andrade
Professor, University of Sao Paulo
Sao Paulo, Brazil

I join with my colleagues at Raipur [India], prominently Dr. Arti Parganiha, who knew Franz in offering our deepest sympathy on his passing. I knew Franz since 1979, when he visited Varanasi [Banaras Hindu University] and delivered the keynote address during the inauguration of the 3rd National Conference on Chronobiology organized by the Indian Society for Chronobiology. I considered myself fortunate having him as one of the adjudicators of my PhD thesis. I owe a lot to him and learnt many things from him. He also visited our lab at Raipur along with Othild and was our most cherished guest. We have many unfinished work and I trust we can complete those work and that will be a real tribute to Franz.

Atanu Kumar Pati
Professor, School of Life Sciences, Ravishankar Shukla University
Raipur, India

… I always knew this would happen but it was a thought that was continuously removed. Franz for me was like a rock immortal, immovable and firm that bind me in times of trouble. I have known Franz Halberg thanks to Professor Tarquini in 1991 and since then I considered him as one of the most important people I've ever met in my academic career. I will never forget his scientific curiosity nor the incredible intelligence and his ability to work (seven day a week as you remarked). But most of all I was impressed by his incredible and unsurpassed knowledge in all fields of human knowledge. For me it was an inexhaustible source of knowledge, but above all an example to follow as a man and scientist. To Franz and you are bound several exciting moments of my career as a researcher. Only a few days ago my wife Cecilia and I watched some pictures of our Chronobiology meeting in Gubbio. Even though our collaborations in recent times have been less close (only to our fault), I assure you that what I have learned from him is always an important part of my scientific background as demonstrated by several recently publications in chronobiologic fields that I will never have conceived without the fundamental contribution of Franz. Surely I will never be up to the genius of Franz but be sure that I will work to keep alive his valuable scientific legacy.

Federico Perfetto
Professor, Department of Internal Medicine, University of Florence
Florence, Italy

He was an inspiration to me and to an entire generation. My condolences.

Michael Persinger
Professor, Laurentian University, Sudbury
Ontario, Canada

Je suis très triste d'apprendre le décès de Franz. C'est une perte terrible pour la science mais son oeuvre reste un grand classique pour la médecine.
Comme les étoiles éteintes, la vie et l'oeuvre de Franz continueront à nous éclairer longtemps encore.

Mes condoléances, chère Germaine, à toi qui l'a tant aidé dans son oeuvre et dans sa vie, à Francis. A Othild et aux enfants de Franz. A tous ses collaborateurs et ses amis, dont j'avais la chance de faire partie.

Franz restera pour moi un modèle d'inventivité, d'amitié, de générosité.

C'est un grand savant, un grand homme qui nous quitte.

[I am very sad to learn of the death of Franz. This is a terrible loss for science, but his work remains a classic for medicine.

As extinguished stars, the life and work of Franz will continue to enlighten us for a long time to come.

My condolences, dear Germaine, to you who has helped him so much in his work and in his life, to Francis. Also to Othild and to Franz's children. To all colleagues and friends with whom I had the chance to take part.

Franz remains for me a model of creativity, friendship, and generosity.

A great scientist and a great man has left us.

Olga Quadens
Professor Emeritus, Aerospace Institute
Castelre, The Netherlands

The terrible news just reached me. I want to tell you as well as all yours how much I am sad. I remember the dinner, in May 2012, when we met again in Minneapolis. It was a real pleasure to see that our friendship was still as warm as it always was since 48 years.

Alain Reinberg
Professor of Medicine, CNRS,
Unité de Chronobiologie
Fondation Adolphe de Rothschild, Paris, France

Sorry for delaying my answer, but I do not know really what to say. It is true that Franz had a very long life, but it is difficult for me to imagine he is not here anymore. I suppose it is almost impossible for you.

Miguel Revilla
Professor of Mathematics, University of Valladolid
Spain

I was sad to hear of the passing of Franz. I have seldom encountered such an enthusiastic and motivated individual during my career as a space scientist.
In particular, he had a rare ability to convince people to help him in solving the problems that he focused on with his broad overview of the physical and biomedical aspects of nature. I am sure that Franz realized that he had lived a productive and satisfying life - certainly I saw his life in that way.

Gordon Rostoker
Professor, University of Alberta, Edmonton
Alberta, Canada

My sincere condolences. For the short time I had to deal with Mr. Halberg I found him a very wise and interesting person. I hope his wide work helps to advance in our knowledge and connection to nature and thus to Universe.

Alberto Saco Alvarez
University of Vigo
Vigo, Spain

I was feeling really deep sorrow when I got your information about Franz' passing and immediately remembered the wonderful days we passed together in Azerbaijan in 2007. We had lots of talks and he impressed me so much with his enthusiasm, his friendly and modest character and his enormous knowledge and culture. He was one of the last professors as I understand them - there is hardly anybody left who is willing to fight the windmills and not selling himself or ducking away. I loved his little literature quotes in his e-mails and admired his endless fortitude. When I most needed it, Franz was the first and only one to write me an unsolicited letter of recommendation. I have had contact with him only for a short time but I am feeling to have lost a friend too. Please hug Othild strongly for me and salute me Franz when you next visit his grave. In deepest sympathy, mourning with you and Othild

David Schnaiter
Innsbruck Medical University
Innsbruck, Austria

Collaborating with you and Franz was a very stimulating experience, I will miss it. Whenever you find interest and time to continue our work, I would be happy!

Christian Schubert
Professor, Psychoneuroimmunology Laboratory, Innsbruck Medical University
Innsbruck, Austria

Franz made great contributions into various scientific directions and we could think that he would always be there. It is very difficult to imagine chronomics and some other sciences founded by Franz Halberg without him. Being a man of indomitable energy and unmortified spirit, Franz was also a generator of new ideas and inexhaustible toiler. …

Franz Halberg's life will be continued in the results of his scientific activity.

Alexander Sidorin
Professor, Schmidt United Institute of Earth Physics
Moscow, Russia

We are very sad about your Franz information. … Franz was our teacher; we must deal in his sense. Please be active to prolong Franz's scientific work.

MUDr. Jarmila Siegelova, DrSc.
Professor, Department of Physiotherapy and Rehabilitation
St. Anna Teaching Hospital, Masaryk University
Brno, Czech Republic

Désolés d'apprendre la triste nouvelle, nous imaginons combien ce doit être difficile pour vous de perdre un homme exceptionnel comme lui. J'imagine qu'il devait être un peu comme un 2ème père.
[We are sorry to learn the sad news, we imagine how difficult it is for you to lose an exceptional man like him. He must have been like a second father to you.]

Danielle & Yves Simonsen-Lasudry
Kraainem, Belgium

Biology will never be the same after the death of Franz. His creative genius was to dissect and unravel numerical data into statistically relevant rhythmic components whose understanding then provided a new basis for the assumption of the normal value on which medical practice depends.

Hugh Simpson
Professor, University Department of Surgery
Glasgow, Scotland

With tears and extreme sorrow, I am writing you as I have been talking about Dr Halberg almost daily, several times a day about his teachings and blessings.

R. B. Singh
Director, Halberg Hospital and Research Institute
Centre of Nutrition and Heart Research
Moradabad, India

Shocked to know the sad demise of our Mentor, Teacher, Father Figure and Guardian. No words. Confused. My family joins me in sending our deep condolences and we all pray for eternal peace to the departed soul.

R. K. Singh
Professor, Department of Biochemistry
King George's Medical University, Lucknow, India

Franz's passing is like the end of an era of great vision and opening of new frontiers.

Alexander Stoynev
Professor, Medical University
Sofia, Bulgaria

With deep regret we received the news of the death of Franz. We lost a friend, teacher and colleague. We hope that the work begun by Franz will be continued.

Elena Syutkina, with Maria Mitish,
Artak Abramyan, Aleksander Grigoriev
Professor, Institute of Pediatrics,
Scientific Center for Children's Health
Academy of Medical Sciences
and
Anatoly V. Masalov
Professor of Physics,
Lebedev Physical Institute
Moscow, Russia

The news of the death of Franz is a big pain for me, was one of the most important meetings and lucky in my life, (with you), which he, as well as Brunetto I learned to love the study of the rhythms biological and science in general. In addition to a great scientist was a great man and probably the smartest I've known, always making me feel old compared to his innovative ideas and his incredible enthusiasm. Much of what I've done (perhaps not much) I owe it to him and you. To ask me condolences to Othild, a big hug,

Roberto Tarquini
Professor, Department of Internal Medicine
University of Florence, Florence, Italy

It was our pleasure to work with outstanding scientist and a founder of modern chronobiology at his first visit in Moscow and later at many congresses on chronobiology. In our memory Franz was an exceptional friend, colleague and outstanding scientist who will be deeply missed by all who knew him.

Dr. Tchoubarov and Dr. Davidova
Department of Pharmacology, Pharmaceutical Faculty
I.M. Sechenov 1st Moscow State University, Moscow, Russia

It is really sad to hear the news. I owe him a lot. I will pray to God for his calm days in Heaven. Remembering Prof. Halberg,

Keiko Uezono
Professor of Medicine Emerita, Institute of Health Science
Kyushu University, Fukuoka, Japan

With deep sorrow I have obtained your information that Franz passed away. I hoped to meet him this year in September or October, because I got a collaboration contract on proton physics in San Diego which will start at September. Then taking a flight via Minneapolis would always be possible. Now only a common continuation paper of our publication is only possible 'In Memoriam Franz Halberg'. I am deeply sorry since I hoped that he will reach the Hundred. We all shall miss him.

Waldemar Ulmer
Gelnhausen, Germany

... uno dei più grandi scienziati del Novecento, Franz Halberg è salito in cielo a 93 anni. Di straordinario c'è a parte il suo sterminato contributo in aree strategiche della ricerca biologica e medica che ci hanno consentito di decifrare i bioritmi che ci governano in tutte le implicazioni psiconeuroendocrine, circolatorie, ecc. la sua grande capacità di collaborazione interdisciplinare e internazionale che hanno fatto di Franz un modello di scienziato per le giovani generazioni e non solo.

La sua passione per la scienza è stata totale e di recente ha licenziato uno degli ultimi contributi alla cronobiologia da presentare a COSMOS AND BIOSPHERE.

Il grande Maestro Franz imparava tutti i giorni da tutti i giovani che ovunque nel mondo incontrava e coltivava come figli della scienza. Una grande fortuna averlo conosciuto e fatto qualche piccola collaborazione con lui in aree poco esplorate come l'influsso del ciclo lunare sulle maree e sulla vita in tutte le sue manifestazioni. Il suo sorriso, la sua disponibilità, la sua apertura alla conoscenza dei fenomeni della Natura sono la grande eredità che ci lascia con i suoi quattromila articoli che sono un patrimonio da cui ripartire per continuare la grande avventura della conoscenza sui segreti della vita a cui Franz ha dato molto e che dobbiamo continuare a esplorare forti del suo luminoso esempio che ci guiderà nel cammino. Addio Franz!!!

[... one of the greatest scientists of the 20th century, Franz Halberg, has passed away at 93 years of age. Apart from his huge contributions in strategic areas of biological and medical research which have enabled us to decipher the biorhythms that govern us in all psychoneuroendocrine, circulatory and other issues, his great capacity for interdisciplinary and international collaboration made Franz a scientist for the younger generation and not only his contemporaries.

His passion for science was total and he had recently sent us his last contribution in Chronobiology to be presented at our meeting on the COSMOS AND BIOSPHERE.

The great master Franz learned every day from all young people everywhere in the world who met him and whom he cultivated as sons of science. It is a great fortune if you have known and cooperated with him in the little explored areas such as the influence of the lunar cycle on the tides and life in all of its manifestations. His smile, his willingness, his initiation to the knowledge of the phenomena of nature are his great legacy that leaves us with its nearly four thousand articles from which to continue the great adventure of knowledge about the secrets of life to which Franz gave a lot and that we must continue to explore his bright strong example that will guide us in the journey. Good bye Franz!!!]

Vincenzo Valenzi & Gualtiero Valeri
Professors, University of Rome "La Sapienza"
Rome, Italy

Really sorry to know the demise of Dr. Franz Halberg. We have a great loss of an eminent scientist who was the founder of cosinor models. May his soul rest in peace. He remains in our thought.

Dr. P. Venkatesan
Professor, Loyola College (Autonomous)
Chennai, India

I am so sorry to hear the sad news that Franz passed away. Dr Halberg is not only the father of chronobiology, but also our tutor, dedicating his whole life to chronobiology. We all miss him. Dr Halberg created and cultivated chronobiology, and meanwhile he worked so hard to rise it up. Chronobiology is his interest and life. Now chronobiology has grown up, and is being applying in different fields, but Franz left us. However, his spirit and contributions, especially his chronobiology, will be with us forever. Please let us carry forward Franz's spirit and contribute to Chronobiology to keep his legacy alive.

Zhengrong Wang
Professor and Director, School of Basic Medicine and Forensic Medicine
Sichuan University, Chengdu, China

It was terrible for me to have received this bad news whereas only one month back Franz and I have still been in contact ... Therefore I regret infinitely that our contact has been interrupted in such a brutal manner as so more Franz had not given any precision about his own health but was waiting to have soon his 94th birthday ...

Raymond Wegmann
Institute of Medical Histochemistry
Universities of Paris

Even if I knew it had to happen, it fills me great sorrow and emptiness. I feel with you, but I also know that you are the one that has the knowledge and guts to keep his spirit alive and continue the work and research in chronobiology.

Andi Weydahl
Professor, Finnmark College
Alta, Norway

I share your sorrow deeply; Franz was someone I had absolute trust in as a personal and fatherly friend; as a brilliant scientist at the cutting edge of all he undertook; as someone who made an enormous impact on me and my work and my place in Society. Words are insufficient at this time in your (and my own) hour of need. You served him and humanity well and he always appreciated "his indispensable right hand 'man" and for me to have him and your good self to honour me very recently in a publication means so much more to me now.

I promised to bring a bottle of red wine when I came to Minneapolis next and I will live up to my promise. To me he lives as an 'Immortal Great' just as those like Galileo Galilei, Nicolaus Copernicus, Watson and Crick, and other Great Scientists and to both of us he is an immeasurable loss. To think of the joy he gave us is perhaps his passing wish, and that we, being his true disciples, will continue his work ourselves and teaching others his incredible insight though we can but lie in his shadow. When you feel composed, please pass my condolences to the family. Meanwhile, if there is anything we can do please let us know. My broken heart is now held together by his strength, his science, his friendship, and his humanity. God be with him and yourself and his family.

Will be in touch.

Best regards

Douglas W. Wilson
President Emeritus,
International Society of Clinical Biostatistics
Durham University, Stockton-on-Tees, England

It was a great shock to learn of the death of Franz Halberg. He was a brilliant and world-renowned scientist and a man of integrity, dedication and wisdom. I had the privilege of knowing him and I am sure he will be greatly missed by all who knew him.

I would like to express my sincere condolences to his family and friends.

I deeply mourn the loss of Franz Halberg.

Herbert Witte
Professor, Friedrich-Schiller University
Jena, Germany

It is the best arrangement that Halberg Chronobiology Center continues to develop the new science under your directorship although it was not the time for him to leave from the center. Recently, the great progresses of modern science and technology have innovated and are continuously improving conditions and environments to develop chronobiology in health and medicine.

Jinyi Wu
West Yak Industry Group
Chengdu, China

It is with great sorrow that we have heard the sad news of the untimely passing of Franz, a great professor. I am certain that he passing must be a great loss for the world, as he accomplished many big achievements and had made so much contribution to the world. I tender deepest sympathy and condolences to Dr. Othild Schwartzkopff and you and your colleagues.

Nobuhiko Yasui
A&D Ltd.
Tokyo, Japan

With great grief I find out about death of Franz! I loved Franz as my teacher as my friend and as a kind man.

Rina Zaslavskaya
Professor of Cardiology, Hospital #60
Moscow, Russia

Please, accept my heartfelt sympathies for the loss of Franz. I cannot still imagine that he passed away with all his energy but the time is the most powerful agent as we know and learned from him.

I am convinced that he will support our research in chronobiology even from the chronobiological heaven.

Michal Zeman
Professor, Comenius University
Bratislava, Slovakia

I am very very sad to hear the bad news that Franz passed away. He is a great man and his leaving is a big loss for all chronobiologists over the world.
We will all remember him and his great achievememts in chronobiology forever. Would you please convey my deep-hearted condolences to his family.
We will miss him forever,

Ziyan Zhao
Professor, Anti-Senility Research Centre,
Shandong Academy of Medical Sciences
Jinan, Shandong, China

For many years, even after I retired officially from NOAA, I corresponded with Franz and tried to follow his many papers. When he asked, I gave him all the geomagnetic information I had at my disposal. I think he appreciated my help. He was surely a kind and thoughtful person and must have been exciting to work with.

Joe H. Allen
Former Director,
World Data Center for Solar-Terrestrial Physics,
NOAA, Boulder, Colorado

I came to meet Prof Franz Halberg, father of CHRONOBIOLOGY, sometime in 1976-1979 in Germany and elsewhere in Europe, when he was introducing chronobiology to the European medical profession in an attractive popular way, which proved that Chronobiology is a question concerning all the medical fields, not only the military bunker. I am very sorry to learn about Franz's death. The medical profession loses by his death a great modernizer.

Dr. Moneim Attia
Carle Foundation Hospital
Urbana, Illinois

My condolences to you and to his family. Dr. Halberg was an inspiring figure. I appreciated my contact with him and with you.

David E. Axelrod, Ph.D.
Professor of Genetics, Nelson Biolabs
Rutgers University, Piscataway, New Jersey

I am saddened to learn that Franz has passed away. I believe I only met him once at Hines VA, introduced by the late Dr. Gene Kanabrocki, several years before I began cooperating with Franz, you, and the rest of the chronobiology lab. However, we have since had many communications concerning my long-term blood pressure records and other matters, mainly over the internet, that I feel I knew him, his interests, and his innovations in chronobiology very well. He acknowledged the last update of my BP records shortly after he was hospitalized for a stroke, so I realized he was sick and had hoped he would recover. I know this has been difficult for you, and express my condolences to you, Franz's family, and his other colleagues. He will be missed by all.

Professor William R. Best
Hines VA Medical Center
Hines, Illinois

I am very sad to hear of Franz' passing. He was a remarkable scientist, and will be missed professionally and personally.

Professor Katarina Borer
Division of Kinesiology, University of Michigan
Ann Arbor, Michigan

I was sad to hear of Franz's passing. He contributed so much to the world and to me personally.

Brian Brockway
President and CEO, VivaQuant™
Saint Paul, Minnesota

I cannot honestly say that I understand what this news means nor how you feel. During that same week you found yourself unable to write, I was free to imagine conversations in my head and emails I would write to update you both on the metabolomics. For me he was alive, and that was a week that I had thanks to you.

You both have always been there for me. You knew him most deeply. Over these years I have found him to be the most empirical person I have ever met, and willing to consider something even if it did not agree with his paradigm. I found him to be kind and generous and curious. It was my privilege to know him.

Timothy Bromage
Professor, Department of Biomaterials and Biomimetics,
Director, Hard Tissue Research Unit, New York University, New York, NY

Sorry to hear this sad news. Dr. Halberg will be deeply missed here at DiagnosTechs as our business model is based and thrived on his discovery. May he rest in peace!

Ning Cegielski
DiagnosTechs
Kent, Washington

Thank you for sending this sad news. Please accept my sympathy, and please pass it on to Othild and others in your group.

Mark Engebtretson
Professor of Physics, Augsburg College
Minneapolis, Minnesota

I was greatly saddened to learn of Franz's passing. He is very dear to me and my family for championing our search for solutions to our health problems. We will always have gratitude for him and the center for the investment that has brought us to a better quality of life!

The StarTribune's obituary employed the perfect term to describe Franz, for he was, indeed, a maverick -- not only in the sense of his extraordinary discernment and pursuit to understand the cosmos, but also in his extraordinary humility and commitment to service! Many have and will continue to benefit from his independence. Best wishes to each of you as your carry on his legacy in your uniquely individual ways.

Judy Finley
Prayer Resources Inc. Memphis, Tennessee

I want to express my deepest sympathy to the family and co-workers of Franz. Knowing that he was aging and experiencing health issues has not diminished the shock of his passing. Franz was a wonderful servant who took out of his demanding schedule the time to minister to my family. Words cannot adequately convey the confidence I felt knowing that he was just a few clicks away on email and that he genuinely had our family's best interest at heart. All of the accolades that he accumulated over his lifetime were correctly given and yet these do not express adequately the personal touch that he exhibited.
In a day and age when doctors herd people through their offices so that they, at the end of the day, have met their financial income quotas, Franz was above all that. He was personal and available. He took time to research and explain what was happening to my wife and family. He educated each of us on a very rare condition. May I express a heart-felt thank you.

Rob Finley
Prayer Resources Inc.
Memphis, Tennessee

Eternal rest with God for a great and good Man and Scientist. ... I hope you are well and now will carry on his awesome legacy and wish we may partner in the future ...
As he once said --- 'and the world turns'......

Joe Franchetti
CEO, 3Fs & TBD Co.
Philadelphia, Pennsylvania

Thank you for sending notice of Franz's passing. He was an amazing scientist who will be missed by us all.

Martin Frank
Executive Director, American Physiological Society
Bethesda, Maryland

Thank you so much for sending this sad news on to me. In the short times I got to know Franz, a little, I was always taken by his kindness and gentle nature. Just recently I have been thinking of one of his important goals and how it might be attained; we will carry on and I will keep moving towards the goals Franz set. I shall miss his discussion and I shall miss him.

David Gerdt
President, Empirical Technologies Corp.
Charlottesville, Virginia

Franz deserved five Nobel Prizes. One of the great tragedies of all time because -- as I was told -- he did things his way.

David Goodman
Newport Neuroscience Center
San Marcos, California

Our heartfelt condolences to the family and friends.

Alok and Sunita Gupta
Associate Professor,
Pennington Biomedical Research Center
Louisiana State University at Baton Rouge, Louisiana

Thank you for sending me the sad news. I have known Franz for almost six decades, always as a mentor, later as a colleague and friend. My sincere condolences. You must feel the loss more than anyone.

Richard Harner
President, BrainVue Systems
Philadelphia, Pennsylvania

Although I never met him or you in person, I feel very privileged to have been able to discuss our work with both of you and to have the benefit of your advice on our circadian rhythm studies. My condolences to you and his family and friends.

Helen L. Hellmich, PhD
Professor, University of Texas Medical Branch
Galveston, Texas

The end of an era...inevitable, yet daunting to consider. It is great that we could all get together in Minneapolis last year to talk about the history of chronobiology.

William Hrushesky
Medical Director, Oncology Analytics
Plantation, Florida

I am saddened to hear of Franz' passing. Although I had just begun to know him, our collaboration has been a true honor for me. His credentials told the tale of his many accomplishments. But in meeting him, I became profoundly aware of his deep humanity. He was a great scientist, but also a great soul.

Thomas Kazlausky
President, Ambulatory Monitoring Inc.
Ardsley, New York

The passing of Dr. Halberg saddens me deeply. I am very sorry for your loss. It was always very clear to me that Dr. Halberg was loved, liked, and respected by so many. May his memory live on not only through his work, but through the hearts of people like you, as well.

Louis Maurici
Nova Science Publishers
Hauppauge, New York USA

Count us in as partners in supporting you in keeping his legacy alive, and expanding. … We will miss him very much.

Rollin McCraty
Executive Vive President and Director of Research,
Institute of HeartMath, Boulder Creek, California, USA

Please accept my most sincere condolences on the passing of Franz Halberg. May he rest in peace with God's mercy.

Kilmer McCully
Chief of Pathology and Laboratory Medicine Services
United States Department of Veterans Affairs
West Roxbury, Massachusetts

Abba and I are so saddened by the unfortunate news. Franz had been saying that he did not have too much time left, and wanted us to be efficient to carry out mission that he started decades ago.

I am sorry that we made him disappointed. He has been such an inspiration, a great friend, a generous father of the field. He will continue to be our role model. We miss him very much. Please let us know what we can do.

Please feel free to call me any time, and I will check with you once we get over the acute sorrow.

Weihong Pan
Professor, Pennington Biomedical Research Center
Louisiana State University, Baton Rouge, Louisiana

I am very sad to know that Franz left all of us. I talked to him a few months ago. He was full of great spirits. One should realize that he lived more than one life working too many long hours everyday. I am fortunate to have worked with him though only for a short time. I share the sorrow with you.

Edapallath Radha
Food and Drug Administration
Rockville, Maryland

After meeting Halberg in 2002, I had many intellectual exchanges with him via e-mail, telephone, and in person. When in Washington D.C. for a conference in 2007, he took me out for dinner at the Cosmos Club, a private social club for men and women distinguished in science, literature and the arts. After a wonderful dinner accompanied by excellent wine and illuminating conversation, I knew that I was in the presence not only of a legend in chronobiology but also of a man of taste and manners. With Franz Halberg's death, the world lost not only the foremost advocate of chronobiology but also one of the last few true gentlemen.

I am so sorry to hear the news! I was really planning on showing up in Minnesota for Franz's 100th birthday! He will definitely be missed! Yours in sadness,

Roberto Refinetti
Professor, Circadian Rhythm Laboratory
University of South Carolina

I am so sorry to hear about Franz. He was a lovely man, always, very generous and kind to me. I am proud to have cited our twins reared apart collaboration in my book.

Nancy Segal
Director, Twin Studies Center
California State University-Fullerton

I am much saddened by the news about Franz. I appreciated his enthusiasm, and his continuing productivity as he got older was a great model for those of us not too far behind him in age.

Sam Silverman
Professor, Department of Physics, Boston College
Chestnut Hill, Massachusetts

This is a great loss to all of us who have been touched by Franz's genius and endless curiosity. I hope you'll continue in his footsteps to push the international scientific community to do more in the field of chronobiology. I believe that this is Franz's legacy.

Professors Assia Shisheva & Ognian Ikonomov
Wayne State University, Detroit, Michigan

Thank you for taking the time to notify me of Franz's passing away. We three children of Henry Nash Smith were very fond of Franz, and I am glad he was able to continue working into the ripe old age of 94. Herr Halberg came to visit me when I was having a difficult time healing after having my tonsils out in Minnesota. He took my little wrist in his big hand and checked my pulse with great solemnity, as he watched his pocket watch.

My mom and dad spoke of him often and with great fondness. He seemed very pleased that my father, an English professor and author in the field of American studies, had managed to furnish Hans with a term that perfectly described the rhythmicity aspect of his biological studies and experiments. Clams on the east coast transported to the west coast, as I recall, continued to open and close in accordance with the tides of the Atlantic at first, but then gradually changed their behavior to match that of their Pacific west coast neighbors.

This reminds me of learning how to adapt to the many schools we attended as we moved about the country while we were growing up.

Someone in your offices also helped me with some German terminology for strange plants in a children's book I was writing, "Professor Pepperhouse and the Magic Mustard Sauce". The book has had to be modified to "Pepperhouse" after everyone complained about not being able to read the long-worded names for the magic plants. The plant names remain in English now. The book is being line-edited, and I hope will come out this year.

Thanks again for your notification, and I join you in mourning the loss of our dear Herr Professor Franz Halberg.

Janet Smith
Bella Roma Music
Berkeley, California

This leaves an enormous gap that never can be filled.

Novera Herbert Spector
Professor, American Insitute for Neuroimmunomodulation Research
Carlsbad, California

Thanks for telling me about Franz. He was a great scientist and a great human being.

Richard Watson
Professor, Washington University of St. Louis

This is very sad news...Sincere condolences!

Bi-Botti "Celestin" Youan, PharmD, MBA, PhD
Professor, University of Missouri, Kansas City

Very sorry to hear of the passing of the great scientist. I will pass this on to the American Physiological Society.

Irving H. Zucker, Ph.D., F.A.H.A.
Chairman, Department of Cellular and Integrative Physiology,
University of Nebraska Medical Center

I was saddened to hear the news, and I would like to express my commitment to helping in any way I could to continue the great work that Dr. Halberg did on campus.

Mustafa al'Absi, Ph.D.Professor of Behavioral Medicine
and Director of Research,
University of Minnesota Medical School, Duluth, Minnesota

What great sadness to receive your message. Franz had such a powerful will and determination that it is hard to believe that he has passed on. My own memories go back over 50 years. He has left an enduring, world-wide legacy. Thank you for letting us know of this inevitable event.

Frank Barnwell, Professor of Ecology,
Evolution and Behavior, University of Minnesota

I once told Franz that I was sure he had arranged it so that he would never die. He would not believe me; it was sad beyond words to hear that the most amazing man I've ever known insisted on being correct this one last time at age 93.9. Unselfish and concerned for others until the end, his last instructions to me were to not go visit him in the hospital, but instead to continue to work to bring his life's love and advancements of chronobiology to bear on problems of the world in my own tiny, tiny way. Though it is hardly a sufficient way to honor Dr. Franz Halberg, his life, and his accomplishments, this I shall do, as will everyone that doesn't want to have missed the significance of what he taught all of us in his marvelous world-wide, lifetime-wide network of colleagues who ever worked with or were inspired by him.

Larry Beaty
Engineer
Twin Cities IEEE Phoenix Project

I was very sorry when I heard of Franz's death. He contributed so much in his life that his influence will last for centuries.
I hope chronobiology can survive at the U of M.

Kit Bingham
Professor Emeritus, School of Statistics
University of Minnesota

I am sure you will continue his good works at the center. But I do understand how you will miss his vision. Best wishes,

Henry Blackburn
Professor of Epidemiology
University of Minnesota

Very sad to hear the news of Dr. Halberg's death. Just got the news in an email that was forwarded to me by Jurgen Fohlmeister in Physiology.
My sincere condolences to you for the loss of a colleague and a friend. He was an amazing person.

George R. Bloom
Department of Integrative Biology and Physiology
University of Minnesota

Condolences on the passing of Franz. A very unique man who forged his own pathway through science.

Thomas J. Bouchard Jr.
Director, Minnesota Center for Twin and Adoption Research
University of Minnesota

An American in Paris.
This is how I came to meet Franz.
That day changed my life.
It was the prelude to a lasting association in his laboratory, the acceptance in a professional family close to home and around the world.
It was the honor and privilege of working with a truly great scientist and a wonderful human being.
But I should not say "working".
For Franz and for all of us who were privileged to participate in his studies, it was not "work", but fun.
And fun it has been for me for the past 37 years.
It is difficult to say goodbye.
But Franz's legacy will live on and be for us a beacon guiding us to follow in his footsteps.
For all this, I shall ever be so grateful.

Germaine Cornelissen
Professor of Integrative Biology and Physiology
University of Minnesota

My sincere condolences regarding the sad loss of Prof Halberg. He was the pioneer of chronobiology. I am totally convinced with all your expertise and knowledge that you will continue the Franz Halberg center for chronobiology.

Daniel Duprez
Professor of Medicine, Donald and Patricia Garofalo Chair in Preventive Cardiology,
University of Minnesota

I am terribly sorry to learn of Franz's untimely loss. Please accept my deepest sympathy. He will be remembered forever.

Apostolos Georgopoulos
Regents Professor of Neuroscience
University of Minnesota

A special part of my life: I remember well meeting Dr. Halberg in late 1963, before the prefix "chrono" was applied to "biology", when time in time-series data was customarily dealt with cumbersomely. Together we decided that even the data we manually collected back then should be recorded as calendar-minute. So I soon made programs (now common-place) that convert between Gregorian-calendar and linear (required by analysis computations) time, making the data handling easier and more reliable. I have worked with Dr. Halberg intermittently over the almost fifty years since then on data from many interesting projects. Although he knew so much more than I do, the teaching each other was very special to me. He always guided me in what I should do, and his absence leaves an empty void in my life.

Dewayne Hillman
Halberg Chronobiology Center
University of Minnesota

I'm so sorry to hear about Franz's passing. I will miss him as well but I know you will miss him greatly as for you he was an extraordinarily special colleague. I wish the best as you ably carry on his work.

Dana Johnson
Professor of Pediatrics
Director, International Adoption Clinic
University of Minnesota

We were privileged that Professor Halberg made the University of Minnesota his academic home. ... You have my heartfelt condolences ...

Eric W. Kaler
President, University of Minnesota

My sincere condolences to you, the family and all the colleagues who had worked with Franz over the years. I was very fortunate to have known him myself, and appreciate the loss to the U and the world of science and medicine.
My very best wishes to you and what you are doing to build on the Halberg legacy.

Mostafa Kaveh
Associate Dean, Science & Engineering Administration
University of Minnesota

My deepest condolences to you, Othild and other members of the family! He was a fatherly figure to me and I will miss his guidance!

Prateek Khasigawala
Halberg Chronobiology Center
University of Minnesota

I am still in shock and shaken so badly with Dr. Halberg's news. I do not know how you are going through nowadays without him. ... I know he is up there to look over us but it is just woeful to think about him not there anymore. It would have been very difficult for you to inform us the news - even receiving the news was too much in pain. ... I hope to remember Dr. Halberg's caring and hard work for Chronobiology that we have to be strong to carry on his legacy.

Jong Ye Lee
Halberg Chronobiology Center
University of Minnesota

I dedicate my master's thesis to Dr. Franz Halberg (7/5/1919 – 6/9/2013), in memory of his amazing journey into biological rhythms, and his foresight in understanding how fundamental their complex inter-weavings are to our health, and to every system in the body. I was very honored to meet him and discuss chronobiology with him, and to have him take an interest in my study, even offering to review that thesis from his hospital bed! He continued doing the work he saw as so important until the very end, sharing his zeal for chronotherapy, and chronomics in general, with the world. What an impact he has made on the world!

Non est vivere sed valere vita est (Martial).

Cathy Lee Gierke
Halberg Chronobiology Center
University of Minnesota

We first apologize for not being in touch with you or Franz for quite some time. We just read your July 3 e-mail about Franz' passing and are of course saddened by the news. If it would serve any purpose at this time, Walt could write recollections of working with Franz during the many years since he first met him in the early 1950's. We shall be making a memorial contribution to the Halberg Center in his honor.

Walt and JoAnn Nelson
Halberg Chronobiology Center
University of Minnesota

I'm sorry to hear about the passing of Dr. Halberg, as I know he meant so much to our community and that you worked very closely with him for years.

John Popovich
Medical student, University of Minnesota

It is really a great loss to all his admirers. We too share with you the grief. Our prayers are with you and your associates.

Gundu H.R. Rao
Emeritus Professor, University of Minnesota

This is very hard, isn't it? I imagine that his passing is even harder for you than for me, even though he has been a close friend for over 40 years. There is the personal loss, and also the knowledge that Franz was a unique and irreplaceable human being and scientist.

Philip Regal
Professor of Ecology, Evolution and Behavior
University of Minnesota

MSI has put a notice on our web page, www.msi.umn.edu, under the "Recent News" section. Again my deepest condolences to you and Dr. Halberg's family.

Nancy Rowe
Scientific Computing Consultant, Minnesota Supercomputing Institute (MSI)
University of Minnesota

I started working in the chronobiology labs on my twentieth birthday in 1978. At first I planned to work here only temporarily, but 35 years later I'm still on board. Working for Dr. Halberg has been a fascinating experience. Over his long life his path crossed those of many of the biggest names in medicine; he often reminisced how his studies took place in the pre-sulfa-drug, pre-antibiotic era of the 1930s and he was acutely aware that the crisis or lysis (i.e., the turn toward death or recovery) of a case of pneumonia could be expected on the 7th day after the onset of symptoms.

Younger clinicians who no doubt take for granted the existence of wonder drugs and modern medical techniques have plenty to learn from an earlier generation's experience. He remained busy right up to the very last day of his life. May his experience and contributions continue to inspire generations of clinicians to come.

Mary Sampson
Halberg Chronobiology Center
University of Minnesota

"With you at my side I was able to pursue my goal during the rest of my life", I recall Franz saying to me not long ago. Franz continued his work in chronobiology, culminating in proving the connection of our physiology to chrono-astrobiology: the exploration of our connection with the stars and the universe, an area of study he entered into recently.
My initial connection with the Halberg family dates back over 40 years having met shortly after I settled in St. Paul, Minnesota, from Sault Ste. Marie, Ontario, Canada in 1968 -- widowed with three children -- to continue my practice in pediatrics. We met shortly after I had settled in St. Paul, where we played tennis together. Following my retirement, I had the good fortune to renew my previous friendship with the Halberg family I had lost contact with over the years. The circumstances under which we reconnected were rather unusual: during a visit to my sister in Hannover, Germany, I met Gwen Künkel, a classmate and friend of mine during my school years in Germany. She told me on our last visit together, just a minute before my train left that her late husband Helmut Künkel and Franz Halberg had worked together and were friends, and Franz's daughter Francine had been their house guest. This chance mention spurred me to reconnect with the Halbergs in 1993. Unfortunately, Franz's wife Erna passed away from cancer in 1994, as did my second husband in December 1994. Since then, I have changed my field of interest to chronobiology and chrono-astrobiology.

Over the remaining years, I made many journeys with Franz to far-away countries, carrying his message and his work and widening my perception of a human common denominator: our intensive connection and dependence on our environment, known as chrono-astrobiology. Franz, I miss you!

Othild Schwartzkopff, MD, FAAP
Professor of Pediatrics emerita
Clinical Director, Halberg Chronobiology Center
University of Minnesota

I was saddened to hear about Franz's death. He was a great scientist, dedicated to his work, and also a real gentleman.

Douglas Wangensteen
Professor of Integrative Biology & Physiology
University of Minnesota

Franz was truly a great man. I will miss him, but will think of him often.

Bob Wilson
Professor, Department of Cardiology
Academic Health Center, University of Minnesota

References

[1] Halberg F, Cornelissen G, Katinas G, Syutkina EV, Sothern RB, Zaslavskaya R, Halberg Francine, Watanabe Y, Schwartzkopff O, Otsuka K, Tarquini R, Perfetto P, Siegelova J. Transdisciplinary unifying implications of circadian findings in the 1950s. J *Circadian Rhythms* 2003; 1: 2. 61 pp. www.J CircadianRhythms.com/content /pdf/1740-3391-1-2.pdf

[2] Halberg F, Cornelissen G, Katinas GS, Hillman D, Otsuka K, Watanabe Y, Wu J, Halberg Francine, Halberg J, Sampson M, Schwartzkopff O, Halberg E. Many rhythms are control information for whatever we do: an autobiography. *Folia anthropologica* 2012; 12: 5-134. http://ttk.nyme.hu/blgi/Knyvek%20kiadvnyok/ FOLIA%20 ANTHROPOLOGICA/folia12.pdf

[3] Rosch PJ. In Memoriam: Franz Halberg 1919-2013. *AIS Health and Stress,* August 2013. www.stress.org

[4] Best WR, Muehrcke RC, Kark RM. Clinical and statistical evaluation of four-hour eosinophil response tests. *J Laboratory and Clinical Medicine* 1951; 38: 790-791.

[5] Cornelissen G. When you eat matters: 60 years of Franz Halberg's nutrition chronomics. *The Open Nutraceuticals* J 2012; 5 (Suppl 1-M1): 16-44.

In: Circadian Cardiology with Focus on Both Prevention ... ISBN: 978-1-63463-956-9
Editors: K. Hristova, A. Shehab, G. Cornelissen et al.

Chapter 4

Riding the Chronobiology Roller Coaster with Franz Halberg for Nearly Half a Century: A Memoir

Robert B. Sothern[*]
The Rhythmometry Lab, College of Biological Sciences,
University of Minnesota, St. Paul, MN, US

A random encounter impacted and shaped my life nearly every day from the late 1960's to today. In order to pay for my college education as a 19-year-old freshman at the University of Minnesota in the fall of 1965, I first applied for a part-time job in a library, but it had just been filled, so I next fortuitously applied for a job posted by Dr. Walter Nelson in the Periodicity Analysis Laboratory. He needed someone to tabulate data from printed dots on hundreds of scrolls of paper that indicated the body temperatures of mice that had been implanted with sensor transmitters and their body rhythms were recorded every 10-minutes via telemetry in various around-the-clock 24-hour experiments. Walt Nelson was the chief chemist in this lab headed by Dr. Franz Halberg in the Dept of Laboratory Medicine and Pathology, whose lab was being completely remodeled and various employees were scattered in temporary rooms and facilities. I have a vague memory of a tall impressive figure breezing through our rooms very briefly in Dec of that year to converse with Dr. Nelson, but I didn't know who it was until the Chronobiology Laboratories were ready for occupancy sometime in 1966, at which time I finally met and started interacting more closely with Franz Halberg. This was an exciting time for a young student, with the opportunity to learn how to use a computer that filled a room and learn more about biological rhythms in both mice and humans, including myself. The hours were long, often around-the-clock, as well as over most weekends, as the excitement of the rhythm research took precedence over everything else over the next two decades in Halberg's lab and the ensuing decades beyond as our collaboration continued.

[*] Correspondence: Robert B. Sothern, Ph.D., Research Assoc II/Research Asst Prof. The Rhythmometry Laboratory - Room 470 BioSci. Depts of Plant Biology and Lab Medicine & Pathology. College of Biological Sciences and Medical School. 250 Biological Sciences Center, University of Minnesota. 1445 Gortner Ave, St. Paul, MN 55108-1095 USA.. Email: sothe001@umn.edu.

In the Summer-Fall of 1967, I volunteered to participate in a 3-week isolation experiment in "Aschoff's Bunker" in Erling-Andechs, Germany, and I began performing self-measurements of about a dozen physiological and psychological variables about 6 times each day during waking on May 11, 1967 for the isolation study that was carried out in Sept of that year [1]. After four months in Germany, I was so enthused by learning first-hand more about biological rhythms due to this isolation experiment and the personal experience in monitoring my own body rhythms by phase-shifting them after flights to and from Central Europe and documenting that they continued under conditions of constant light and total isolation from all time cues (my free-running body temperature rhythm in isolation was 25.0 hours in length), that I never stopped self-measuring and my time series now extends beyond 46.5 years[1]. Among many publications over the years, Halberg's team and I described an annual cycle [2], a 1.3 year cycle [3], a decadal cycle [4], and recently an about-20 year solar-associated cycle in my long time series of blood pressure [5], as well as found many of these longer periodicities in blood pressure data from several others, including my parents, Margaret and Bruce, who self-measured their blood pressure and heart rate twice a day for 19 years and 28 years, respectively [6]. One of Halberg's final requests was to analyze my self-measurement series for any about 50-year solar-associated rhythms (personal communication).

Back in 1967 I didn't realize I was in the presence of two of the spirited titans in the field of biological rhythms: Franz Halberg in Minnesota [7] and Jürgen Aschoff in Germany [8]. When Halberg gave me a tie to present to Aschoff after I arrived at the latter's institute for the isolation study, Aschoff asked me how things were back in the 'Halbergian Circus', which of course as a youngster I had no idea what he was implying. Now I know it was the passion, zeal and exigency of Halberg's ever present pursuit, and many would say obsession, of new ideas in the field of rhythms as they could (and should) be applied to nearly every aspect of life on Earth. Of course Halberg's broad-minded fixation on the importance of biologic rhythms on every aspect of life on earth had varying effects on others, who could be enthusiastic, equivocal or dismissive.

Halberg pioneered development of new statistical techniques to analyze and display characteristics of biologic rhythms (especially the 'cosinor' method of analyzing data for rhythms and their 3 main characteristics (MESOR [Middle Estimated Statistic Of Rhythm], amplitude, acrophase) [9] and was constantly coining new terminology, such as introducing and defining the word "circadian" in 1959 [10] and renaming the field of biologic rhythms "chronobiology" in 1969 [11]. The reason for emphasizing the word 'circa' rather than using "diurnal" or "nocturnal" is the fact that although the daily period of a rhythm can be precisely 24 hours under normal synchronizing environmental conditions (e.g., alterations of light and/or temperature between day and night), it is only approximately (i.e., circa) 24 hours when organisms are isolated from these time cues. Even under synchronized conditions, however, the prefix 'circa' can be applied to biological rhythms since there is invariably some margin of error due to internal and external phenomena that do not always recur at identical sequences at identical intervals [12].

[1] All original data in Excel spreadsheets is accessible on a University of Minnesota website for scientific purposes by contacting the author (RBS). Information available 5-6 times each day for over >4.5 decades (between the ages of 20 and 67 years) includes oral temperature, pulse, blood pressure, respiration, peak flow, 1 minute time estimation, mood and vigor ratings, morning body weight and sleep-wake times, among many other variables and calculations (e.g., mean arterial pressure, pulse-pressure, pulse-pressure product, etc.).

To this day, there is sometimes hesitation to use the word 'circadian' in describing 24 hour patterns under synchronized conditions if monitoring has only been done during 'constraining' synchronized conditions and not continuous ones, such as constant light and/or darkness [13].

It is herein perhaps useful for posterity and as a benefit for current chronobiologists and others investigating characteristics of circadian rhythms to reproduce Halberg's original definition of 'circadian' for those not able to easily access his original text written in 1959 [10]:

> "... the term "circadian" was derived from "circa" (about) and "dies" (day); it may serve to imply that certain physiologic period close to 24 hours, if not of exactly that length. Herein, "circadian" might be applied to all "24-hour" rhythms, whether or not their periods, individually or on the average, are different from 24 hours, longer or shorter, by a few minutes or hours. "Circadian" thus would apply to the period of rhythms under several conditions. It would describe: 1. rhythms that are frequency synchronized with "acceptable" environmental schedules (24-hour periodic or other) as well as 2. rhythms that are "free-running" from the local time scale, with periods slightly yet consistently different from 24 hours (e.g., in relatively constant environments)."

The following definition of "circadian" was adopted in 1977 by the International Committee on Nomenclature of the International Society for Chronobiology [14]:

> "Circadian: relating to biologic variations or rhythms with a frequency of 1 cycle in 24±4 h; circa (about, approximately) and dies (day or 24 h). Note: term describes rhythms with an about 24-h cycle length, whether they are frequency-synchronized with (acceptable) or are desynchronized or free-running from the local environmental time scale, with periods slightly yet consistently different from 24-h, ...)."

This latter definition is used as a guideline in the journal Chronobiology International when using the word 'circadian' (M. H. Smolensky, editor, personal communication).

Halberg also gave labels to rhythms of various lengths (i.e., for rhythm domains such as ultradian, infradian, circaseptan, circavigintan, circatrigintan, circamensual, circasemiannual, circannual, transannual, circadecennian and beyond, etc.) and introduced terms such as chronotherapy, chronotolerance, chronotoxicity, chronopharmacology and many more [14], including 'chronomics' (mapping time structures) [15]. More recently he coined terms that described rhythmic characteristics of blood pressure and hypertension for use in diagnosis, including MESOR-Normotension (MN), MESOR-hypertension (MH), Excessive Pulse Pressure (EPP) and Circadian Hyper-Amplitude-Tension (CHAT) [16].

Of note, Halberg's passion, or some would say obsession or fixation, with looking for and finding rhythms in every variable in every living thing was once dismissed as "Halberg's Paranoia" [17], a term first applied while he was a Research Fellow at the Harvard Medical School in 1948-49, where he discovered a circadian rhythm in eosinophils in mice [18] and later a built-in adrenocortical cycle in humans, as well as mice [19]. Upon further study he noted circadian rhythms in other circulating blood cells types and rightly predicted that circadian rhythms characterized every other body function, including all organic compounds, hormones, trace elements, physiologic and psychologic variables, and even RNA and DNA cycles [17]. Thus, at the cellular level, Halberg anticipated the endogenous mechanisms of molecular clocks more than 50 years ago with the demonstration of built-in nucleic acid rhythms in the cell that could be manipulated by the lighting regimen [20, 21] and that persisted under conditions of light and continuous darkness (i.e., without an external synchronizing light-dark signal) [22]. In fact, molecular clocks based in mRNA levels that underlie biologic rhythms, now identified as “clock genes”, have since been shown to be present in all organs and individual cells throughout the body and are central for the temporal coordination of internal biological processes among themselves and with external environmental cycles [23, 24]. Relationships between circadian clocks and normal vs. abnormal organ physiology can have significant impact relevant not only to cardiovascular health [25], but also to the general treatment and prognosis of human disorders and disease [26].

On a personal anecdote, the term “Halberg’s Paranoia” may well have saved my life and career after I graduated from the University of Minnesota in June 1969 with an undergraduate degree. For the previous two years I had been self-measuring my own physiologic and psychologic body functions, and recording body weight and sleep-wake times (among many other variables) and in July 1967 began collecting a 30ml urine sample around-the-clock from each voiding time in order to also follow rhythms in this body fluid. But during this time the Vietnam War was still underway and I had a student deferment from being drafted into the military, but a draft number was assigned to every eligible male in the USA and my number was very high. Upon graduation I received notification to report for a physical exam to see if I was fit to serve and upon learning about this, Halberg called the draft board and explained my value to his lab and the long medical time series that I was collecting. He subsequently learned that one way to avoid military service was to have a chronic medical problem, whereupon he said I was suffering from a condition known as “Halberg’s Paranoia” that he also suffered from and this “condition” had interfered with “normal” activities during his previous employment at Harvard Medical School and could not be cured! The only way I could be cured was to obtain a long time-series of health data and look for patterns that might help explain and cure the problem, so they would have to guarantee that I could continue collecting data on myself to this end. I subsequently received a “work deferment” reclassification that remained in effect until the end of that war in 1972. We were never able to find a cure for my affliction (i.e., “Halberg’s Paranoia”) and I collected >31,000 urine samples over the ensuing 17 years and >82,000 self-measurements over 46.5 years! To the very end, Halberg continued to investigate all oscillations (from ultra-low to ultra-high) found on earth to the outer fringes of the solar system as a universal panacea [27]. He once gave an invited lecture to the FDA entitled “Neglect of circadian rhythms: ignorance, indolence, or criminal negligence?” [17], yet there still seems to be more and more evidence that this is still the case for many medical conditions, especially when survival of leukemic mice and tumor-bearing rats was doubled in the laboratory [28, 29] and survival and the number of cures was greatly increased in human cancers and other diseases when chronotherapy was implemented [30-33].

I would be remiss in not mentioning the person who was at Franz Halberg’s side until her passing on Sept 8, 1993 and that was his wife Erna, who was also a dear friend, colleague, and in many aspects like a second mother, to me. Since entering the world of chronobiology at the University of Minnesota in 1965, I have never had the pleasure of working so closely with anyone else as charming, warm-hearted and sacrificing as this wonderful scientist. Erna was one of the first persons to perform self-measurements over several years, and indeed this is how I first met her when she regularly delivered her data sheets and those of her 2 daughters to me for data entry and analysis back in the 1960's.

Through the many ups and downs that go with research in biological rhythms, from sometimes grueling 24h studies on animals or ourselves, the many deadlines, the long hours that included weekends, the late night meals, the need for an extra pair of hands in almost every area of running a world-renowned lab, Erna was always there to help in a self-sacrificing way with nary a complaint. I still remember the little white mouse that made a mad dash to freedom after being weighed during one of our studies by running up the sleeve of her lab coat and trying to hide near the heart of a true animal lover. That was really good for a laugh!

Not only was Erna a dedicated and supportive wife to Franz and doting mother to her daughters Francine and Julia, but also she was an unwavering friend to everyone in the lab, always taking time to inquire about mutual interests, remember mutual friends and families, and offer words of often much-needed encouragement and advice.

Scores of international students and visitors were taken under her wing as she helped them acclimate to American life and a world of intense chronobiological research in her husband's lab during their stay in Minnesota. She was a totally dedicated and respected chronobiologist in her own right, working long hours six or seven days a week as a devoted colleague of her famous husband. I remember that she told me that she and Franz were performing around-the-clock studies on mice on their wedding day, taking only a short time away from the lab between testing times to exchange their marriage vows at a city hall before they quickly returned to the mouse study!

It is often said that behind every successful man, there is a strong, wise and hardworking woman, and Prof. Jürgen Aschoff acknowledged this in his closing remarks at the "Symposium on Chronobiological Aspects of Endocrinology" held in Capri, Italy in May 1974, when he said:

> "There remains to me one last task. Already the very first whoever lectured on rhythms, Archilochus, invited ladies to his conferences... I am sure all gentlemen here enjoyed having ladies at the Conference, those who helped organizing, those who gave lectures, and those who watched their husbands being important and who so often have to compete with their husbands mistress, the sciences. The more important he is, the more she sometimes suffers. Therefore, roses to Mrs. Halberg as representative for all the others. Gentlemen: to all the ladies, cheers!"

This was followed by a prolonged, and well-deserved, standing ovation for Erna after she was presented with a huge bouquet of roses! [34][2]

Following her diagnosis of ovarian cancer in Dec 1990, Erna received chronotherapy with some of the very drugs that we had successfully given to tumor-bearing rats and mice in the lab. We had learned that if each drug was given at its best circadian stage, more cures were induced and a protocol was subsequently designed and applied to humans with equal success [28, 29, 31-33].

For more than 2.5 years, chronotherapy with both older and newer compounds was administered to Erna while she collected her own urine and saliva for determination of tumor

[2] Prof. Aschoff also used this after dinner speech to show a very funny picture of Halberg staring down a "cosinor monster." [34]

markers in order to learn more about the timing of each drug and its effect on the tumor. While she eventually succumbed to her illness, chronotherapy may have given Erna more time to enjoy those things she cared deeply about, including her family and friends, books and nature [35]. Erna would remark each morning what a beautiful day it was and how much she enjoyed each one and encouraged others to do the same in spite of any seen or unseen dilemmas or problems. Her passing was a big blow to not only Franz, but to all of us that knew her.

Figure 1. Franz Halberg in his "War Room" with his right-hand man Walt Nelson in 1981.

Figure 2. Franz Halberg discussing results with MD, PhD student Salvador Sanchez in 1981.

Chronobiologists and other rhythm research enthusiasts have now lost both members of a renowned husband and wife team in the field of Chronobiology that may be impossible to ever replace. I'm sure that I share with others my deep condolences to their daughters and husbands, Francine E. Halberg, M.D. and Terry Kessler, and Julia U. Halberg, M.D. and

Mark Chatterton, as well as Franz's second wife, Othild Schwartzkopff, M.D., and the surviving members of the Halberg Chronobiology Center, including its new Director, Germaine Cornélissen, Ph.D., and long-time office manager extraordinaire, Mary Sampson. Appended are some photographs of Franz Halberg for posterity that show him with family members and many of his colleagues who in one way or another will never forget him!

Figure 3. Franz Halberg with a group of employees, colleagues and visitors in the Chronobiology Lab in 1985.

Back Row: Franz Halberg, Ed Wink, Nelson Marques, Mary Sampson, Germaine Cornelissen, Mirian Marques, Scott Mason, Robert Sothern, Frank Barnwell.

Front Row: Erna Halberg, Wolfgang März, Andi Weydahl, Ramon Hermida.

Figure 4. Franz Halberg on his 69th birthday on July 5, 1988 during a meeting held at Copper Mountain, Colorado. His dear friend Prof. Otto Schmitt is over his shoulder in the background.

Figure 5: Franz Halberg at the U of M Campus Club on his 80th birthday in July 1999 with Othild Schwarzkopff and his daughters Julia and Francine.

Figure 6. Franz Halberg with colleagues after a farewell luncheon at the U of M Campus Club in June 2007 for Dr. George Katinas, who was returning to St. Petersburg, Russia after several years in Minnesota. (from left to right: Germaine Cornelissen, Othild Schwartzkopff, Franz Halberg, Robert Sothern, George Katinas, Mary Sampson).

References

[1] Sothern RB. Chapter 12: Autorhythmometry. Essay 12.1 Self-Measurements in "Aschoff's Bunker". In: Koukkari WL, Sothern RB. (Eds.) *Introducing Biological Rhythms.* New York: Springer Science; 2006, p. 540-544.

[2] Sothern RB. Circadian and circannual characteristics of blood pressure self-measured for 25 years by a clinically-healthy man. *Chronobiologia 1*994; 21(1&2): 7-20.

[3] Halberg F, Cornelissen G, Schack B, Wendt HW, Minne H, Sothern RB, Watanabe Y, Katinas G, Otsuka K, Bakken EE. Blood pressure self-surveillance for health also

reflects 1.3-year Richardson solar wind variation: spin-off from chronomics. *Biomed Pharmacother* 2003; 57 (Suppl 1): 58s-76s.

[4] Halberg F, Cornelissen G, Sothern RB, Hillman D, Watanabe Y, Haus E, Schwartzkopff O, Best WR. Decadal cycles in the human cardiovascular system. *World Heart J* 2012; 4(4): 263-287.

[5] Sothern RB, Cornelissen G, Halberg F, Nikityuk BA, Bakken EE, Chibisov SM, Kharlitskaya EV, Agarwal RK. Individualized circadecadal (about 20-year) cycle in human systolic (S) and diastolic (D) blood pressure (abstract). In: *Proc 6th Intl Scientific and Practical Conference, Health and Formation in the 21st Century.* Moscow, Dec 8-10, 2005, p. 33-35.

[6] Halberg F, Cornelissen G, Stoynev A, Ikonomov O, Katinas G, Sampson M, Wang Z, Wan C, Singh RB, Otsuka K, Sothern RB, Sothern SB, Sothern MI†, Syutkina EV, Masalov A, Perfetto F, Tarquini R, Maggioni C, Kumagai Y, Siegelova J, Fiser B, Homolka P, Dusek J, Uezono K, Watanabe Y, Wu J, Sonkowsky R, Schwartzkopff O, Hellbrügge T, Spector NH, Baciu I, Hriscu M, Bakken E. Season's Appreciations 2002 and 2003. Imaging in time: The transyear (longer-than-the-calendar year) and the half-year. *Neuroendocrinol Lett* 2003; 24(6): 421-440.

[7] Halberg F. Temporal coordination of physiologic function. *Cold Spring Harbor Symp Quant Biol* 1960; 25: 289–310.

[8] Aschoff J. Exogenous and endogenous components in circadian rhythms. *Cold Spring Harbor Symp Quant Biol* 1960; 25: 11-28.

[9] Sothern RB. Chapter 13. Chronobiometry: analyzing for rhythms. Analyzing time-series by curve fitting. In: Koukkari WL, Sothern RB. (Eds.) *Introducing Biological Rhythms.* New York: Springer Science; 2006, p. 589-600.

[10] Halberg F. Physiologic 24-hour periodicity; general and procedural considerations with reference to the adrenal cycle. *Z für Vitamin-, Hormon u Fermentforsch* 1959; 10(3/4): 225–96.

[11] Halberg F. Chronobiology. *Ann Rev Physiol* 1969; 31: 675–725. Review.

[12] Halberg F, Cornelissen G. Chronobiology: rhythms, clocks, chaos, aging, and other trends. In: Maddox GL. (Ed.). *Encyclopedia* of Aging, 3rd ed. New York: Springer, 2001. p. 196-201.

[13] Sothern RB. Chapter 2: General Features of Rhythms: Terminology and Characteristics. Essay 2.1: Use of the term "Circadian". In: Koukkari WL, Sothern RB. (Eds.) *Introducing Biological Rhythms.* New York: Springer Science; 2006, p. 23-25.

[14] Halberg F, Carandente F, Cornélissen G, Katinas GS. Glossary of chronobiology. *Chronobiologia* 1977; 4(Suppl): 1–189.

[15] Halberg F, Cornelissen G, Wang Z, Wan C, Ulmer W, Katinas G, Singh R, Singh RK, Singh RK, Gupta BD, Singh RB, Kumar A, Kanabrocki E, Sothern RB, Rao G, Bhatt ML, Srivastava M, Rai G, Singh S, Pati AK, Nath P, Halberg F, Halberg J, Schwartzkopff O, Bakken E, Shastri VK. Chronomics: circadian and circaseptan timing of radiotherapy, drugs, calories, perhaps nutriceuticals and beyond. *J Exp Ther Oncol* 2003; 3(5): 223-260.

[16] Halberg F, Cornelissen G, Otsuka K, Siegelova J, Fišer B, Dušek J, Homolka P, Sánchez de la Peña S, Singh RB, and the BIOCOS Project. Extended consensus on need and means to detect vascular variability disorders (VVDS) and vascular variability syndromes (VVSS). *World Heart J* 2010; 2: 279-306.

[17] Halberg F, Cornelissen G, Katinas G, Syutkina EV, Sothern RB, Zaslavskaya R, Halberg Fcn, Watanabe Y, Schwartzkopff O, Otsuka K, Tarquini R, Frederico P, Siggelova J. Transdisciplinary unifying implications of circadian findings in the 1950s. *J Circadian Rhythms* 2003; 1(1): 2. (61 pp)

[18] Halberg F, Visscher MB. Regular diurnal physiological variation in eosinophil levels in five stocks of mice. *Proc Soc Exp Biol Med* 1950; 75(3): 846-847.

[19] Halberg F. Some physiological and clinical aspects of 24-hour periodicity. Lancet 1953; 73(1): 20-32.

[20] Halberg F, Barnum CP, Silber RH, Bittner JJ. 24-hour rhythms at several levels of integration in mice on different lighting regimens. *Proc Soc Exp Biol Med* 1958; 97(4): 897–900.

[21] Halberg F, Halberg E, Barnum CP, Bittner JJ. Physiologic 24-hour periodicity in human beings and mice, the lighting regimen and daily routine. In: Withrow RB, ed. *Photoperiodism and related phenomena in plants and animals.* Ed Publ No. 55. Washington DC: AAAS; 1959: 803–878.

[22] Halberg F, Barnum CP. Continuous light or darkness and circadian periodic mitosis and metabolism in C and D8 mice. *Amer J Physiol* 1961; 201: 227–230.

[23] Reppert SM, Weaver DR. Molecular analysis of mammalian circadian rhythms. *Ann Rev Physiol* 2001; 63: 647-676.

[24] Reppert SM, Weaver DR. Coordination of circadian timing in mammals. Nature 2002; 418(6901): 935-941.

[25] Sothern RB, Cornelissen G, Yamamoto T, Takumi T, Halberg F. Time microscopy of circadian expression of cardiac clock gene mRNA transcription revisited: chronodiagnostic and chronotherapeutic implications. *Clin Ter* 2009; 160(2): e25-e34.

[26] Sothern RB, Yamamoto T, Cornelissen G, Takumi T, Halberg F. Central and peripheral circadian clock genes, their statistical analysis for rhythms, and relationship to health and disease. *Scripta Medica* 2009; 82(3): 133-163. Review.

[27] Halberg F, Cornelissen G, Katinas G, Tvildiani L, Gigolashvili M, Janashia K, Toba T, Revilla M, Regal P, Sothern RB, Wendt HW, Wang Z, Zeman M, Jozsa R, Singh RB, Mitsutake G, Chibisov SM, Lee J, Holley D, Holte JE, Sonkowsky RP, Schwartzkopff O, Delmore P, Otsuka K, Bakken EE, and the Intl BIOCOS Group. Chronobiology's progress: season's appreciations 2004–2005. Time-, frequency-, phase-, variable-, individual-, age- and site specific chronomics. *J Appl Biomed* 2006; 4: 1–38.

[28] Haus E, Halberg F, Scheving LE, Pauly JE, Cardoso S, Kühl JFW, Sothern RB, Shiotsuka RN, Hwang DS. Increased tolerance of leukemic mice to arabinosyl cytosine with schedule adjusted to circadian system. *Science* 1972; 177(43): 80-82.

[29] Sothern RB, Levi F, Haus E, Halberg F, Hrushesky WJM. Control of a murine plasmacytoma with doxorubicin-cisplatin: Dependence on circadian stage of treatment. *J Natl Cancer Inst* 1989; 81: 135-145.

[30] Hrushesky WJM. (1985) Circadian timing of cancer chemotherapy. *Science* 228: 73-75.

[31] Rivard GE, Infante-Rivard C, Dresse MF, Leclerc JM, Champagne J. (1993) Circadian time-dependent response of childhood lymphoblastic leukemia to chemotherapy: a long-term follow-up study of survival. *Chronobiol Intl* 10(3): 201-204.

[32] Sothern RB. Chapter 11. Clinical Medicine. Timing Treatment: Chronotherapy. In: Koukkari WL, Sothern RB. (Eds.) *Introducing Biological Rhythms.* New York: Springer Science; 2006, p. 475-491.

[33] Lévi F, Focan C, Karaboué A, de la Valette V, Focan-Henrard D, Baron B, Kreutz F, Giacchetti S. Implications of circadian clocks for the rhythmic delivery of cancer therapeutics. *Adv Drug Deliv Rev* 2007; 59(9-10): 1015-1035.

[34] Aschoff J. Speech after dinner. In: Aschoff J, Halberg F, Ceresa F. (Eds.) Chronobiological Aspects of Endocrinology. Capri, April 8-10, 1975. *Chronobiologia* 1974; 1(Suppl. 1): 483-495.

[35] Kennedy BJ. A lady and chronobiology. *Chronobiologia* 1993; 20: 139-144.

In: Circadian Cardiology with Focus on Both Prevention ... ISBN: 978-1-63463-956-9
Editors: K. Hristova, A. Shehab, G. Cornelissen et al.

Chapter 5

The Cosmos with Aeolian Cycles, Tipping the Scale between Death and Survival: An Indispensable Control

Franz Halberg[1], Kuniaki Otsuka[2], Yoshihiko Watanabe[2], George Katinas[1], Zhengrong Wang[3] and Germaine Cornélissen[*1]

[1] Halberg Chronobiology Center, University of Minnesota, Minneapolis, MN, US
[2] Tokyo Women's Medical University, Medical Center East, Tokyo, Japan
[3] Health Ministry Key Laboratory of Chronobiology, Pre-clinic and Forensic Medical School, Sichuan University, Chengdu, Sichuan, P. R. China
Support: Halberg Chronobiology Fund,
University of Minnesota Supercomputer Institute, MN, US

Abstract

The cycles in space climate are critical in discussing global warming. Two or more cycles' periods are described as congruent when their CIs (95% confidence intervals) overlie or overlap and the amplitudes' CIs' lower limits are positive. Some congruent cycles in organisms, counterparts of the environmental day and the seasons, relate to electromagnetic radiation in the visible domain; these are the usually environmentally synchronized socio-photo-thermo-periodisms (photics). The biosphere also resonates with or is pulled or driven by nonstationary, environmental nonphotic cycles (nonphotics) -- particle emissions from the sun and the wider cosmos, cosmoheliogeomagnetics, ultraviolet flux, gravitation, acoustics and others. Nonphotics, a set of in part transdisciplinarily-novel spectral components can be intermittent; when present, they coexist, compete with and may replace signatures of photic cycles, monitored, e.g., in blood pressure (BP) and heart rate (HR) for vascular disease detection and prevention. Nonphotics of, e.g., about (~) 1 week (circaseptans), and ~16 months (transyears), can

* Correspondence: Germaine Cornelissen, Halberg Chronobiology Center. University of Minnesota – MMC 8609. 420 Delaware Street SE. Minneapolis, MN 55455, USA. Tel. 612-624-6976. Fax. 612-624-9989. Email. corne001@umn.edu.

modulate and sometimes override society's (photic) schedules, such as in dying suddenly either unintentionally or by one's own will or at the hand of others. Nonphotics persist but are damped in physiology or in terrorism when their environmental counterpart is not detected.

Systematic long-term, eventually lifelong BP and HR monitoring is now automatically possible and is advocated for self-help in preventive health care and for science. A website (http://www.sphygmochron.org/) planned by the Phoenix Project (www.phoenix.tc-ieee.org) for analyses notably of BP and HR, including archival data, may facilitate a project on The BIOsphere and the COSmos, BIOCOS (corne001@umn.edu), which provides cost-free analyses and monitors at a cost reduction of 80% in exchange for the data. Biospheric monitoring complements the records from purely physical tools. It does so for surveilling the variable sun, by validating in the biosphere the reality of intermittent, aeolian environmental spectral components that can be more consistent than their physical counterparts once they are coded in genes, thereby indicating the relevant associations of space weather with human health and ecology. We invite readers, as a dividend from self-surveillance for preventive health care, to monitor themselves for comparison with solar variability, thereby producing a transdisciplinary gauge of individuals' and societies' health and disease.

Keywords: Aeolian; Chronomics; Congruence; Transdisciplinary

Remove and Replace Revisited, with Circaseptans

About 7- and about 3.5-day rhythms -- circaseptans and circasemiseptans -- opened a new chapter in chronomics, the aligned recording of variables in and around us. As in the case of circadians [1], a remove-and-replace approach proved to be helpful, except that this time the sun, not the surgeon (in blinding) or genetics (in the case of congenitally blind ZRD mice), did the removal and the cosmos the replacement: when solar activity had an about 7-day component in its (Walsh) spectra [2], the circaseptan in human HR was amplified and when it was lacking in the sun's record, the corresponding HR component was damped, but not lost, showing that it was also partly built-in [3, 4]. This genetic coding of circaseptans is compatible with resonance; hence it is hardly surprising to find an occasional selective pull of a biological circaseptan acrophase by a geomagnetic cycle with a proximal frequency, e.g., for HR but not for BP [5].

Biological Cis-Half-Years and Circasemiannuals, and Influence of Terrestrial-Space Weather

Rieger et al. [6] documented an about 5-month (~0.42-year) cycle in hard solar flares. This was found to be a wobbly phenomenon by Ballester et al. [7]. We found congruence with it in sudden cardiac death and in several kinds of cardiac arrhythmia [8], again as an intermittent phenomenon. We called this component a “cis-half-year” since “cis” denotes “on this side” of the half-year (or of the year, in the case of cisyears).

Armin Grafe [9] not only recognized the solar origin of the physical half-year, but had looked into its biological associations (albeit not yet with congruences) to learn about a half-year pattern in human epilepsy (personal communication). Subsequently, Russell and McPherron [10] modeled this cycle, which indeed has physiological and pathological counterparts [11]. A half-yearly cycle in circulating melatonin, found at noon at high latitudes [12], was also detected in nightly values at middle geographic latitudes [13].

Additional analyses further suggested the presence of cis-half-years [14]. These results render this variable of particular interest as a mechanism underlying effects of terrestrial magnetism on mammals [15]. Both the midline-estimating statistic of rhythm (MESOR) and the circadian amplitude of the rhythm in melatonin content *in situ* are increased in the hypothalamus and dampened in the pineal by magnetic storms in the rat [16]. Human nocturnal melatonin excretion at all latitudes investigated is also decreased during magnetic storms [17, 18].

Circannual Variation

Julius Bartels, the late mathematician turned opinion-leading geophysicist, advocated the use of inferential statistics. He did not, however, refrain from stacking data to describe a yearly variation. A half-yearly pattern, but no annual variation, is seen (in Figure 4 of Bartels [19]).

Zieger and Mursula [20] describe an annual variation in near-earth solar wind speed and in the geomagnetic activity (Kp), along with an IMF sector component with a scholarly literature review. Yearly data of the aa index, greatly antedating Kp, were analyzed over the span 1868-2004 (137 years), fitting a 1-year cosine curve by least squares to each separate calendar year's data in order to obtain estimates of the MESOR and circannual amplitude (A1y) and acrophase as well as a measure of the coefficient of determination or percentage rhythm (PR).

Three new time series (N=137 each) were thus formed, consisting of yearly estimates of the MESOR, A1y and PR. Results shown in Figure 1 reveal differences in the most prominent spectral peak with a period of about 6.3 years for PR, 10.9 years for the MESOR and 15.4 years for A1y.

The following components are linearly statistically significant:

1. PR is characterized by components with periods of about 10.4 years (P=0.037), 6.3 years (P=0.002), 5.2 years (P=0.030), 3.4 years (P=0.024) and 3.1 years (P=0.022);
2. The MESOR is characterized by components with periods of 12.1 years (P=0.014) and 10.9 years (P<0.001). The latter component corresponds to the solar activity cycle, expected to also characterize the average aa values (or the yearly MESORs);
3. A1y is characterized by components with periods of 72.1 years (P=0.020) and 15.4 years (P=0.003). The latter component is the most prominent and is seen to modulate the circannual amplitude of aa. In a cosinor periodogram of daily aa data, it would be seen as a double peak at frequencies of 1.065 and 0.935 cycles per year, the latter corresponding to a near-transyear component often seen in spectra of helio- and geomagnetic indices, albeit not invariably as a symmetrical peak bracketing the year,

as would be anticipated if it corresponded only to a modulation in amplitude, phase, or frequency.

Biological circannuals and their modulations of circadian rhythms have been described by Halberg et al. [21] and Watanabe et al. [22].

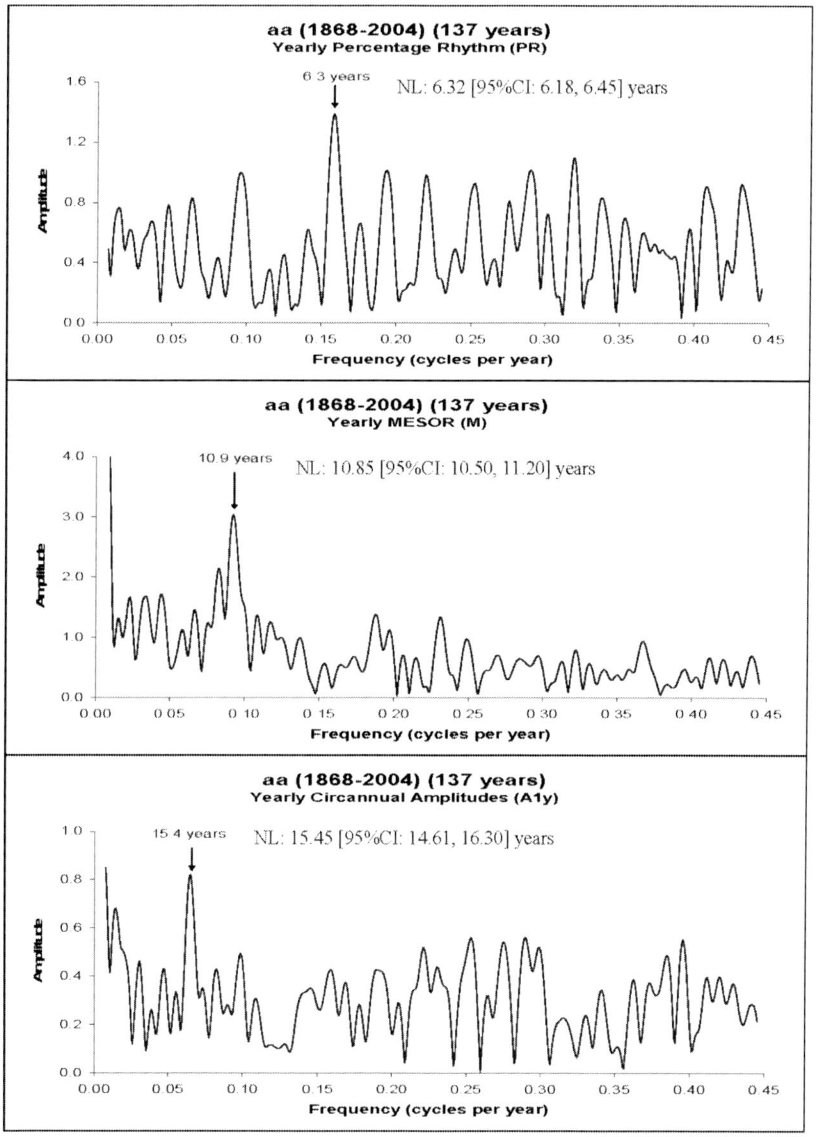

Figure 1. When the percentage rhythm (PR) computed by analyses of consecutive 1-year data sections is analyzed, the largest nonlinearly validated period differs from that in a time series consisting of MESORs (middle) or circannual amplitudes (bottom). The longest major period could be a harmonic of the Hale cycle in PR on top; the major period in the middle, perhaps a harmonic of the Schwabe cycle; and that at the bottom could correspond to Markov and Sivaraman's global solar cycle [64]. © Halberg Chronobiology Center.

Cisyears and Transyears

Cisyears and transyears are cycles with a period somewhat shorter or longer than a year, respectively. Because these nonphotic components drift in their period over a wide range, we distinguish a spectral region for near-transyears (1.00y < [τ-CI] < [τ+CI] < 1.20y, where τ stands for the period length) from one for far-transyears (1.20y < [τ-CI] < [τ+CI] < 1.90y).

Prabhakaran Nayar [23] reviewed the literature of physical far-transyears; his wavelet analyses uncovered an ~33-year variation with a biological published counterpart [24], another example of transdisciplinarily congruent environmental and biological periods, where at least the physical component is independently validated and confirmed, biologically (or vice versa, as the case happened to be for circaseptans: [5]). Far-transyears with a period of ~1.3 years were found in dozens of long biologic time series covering a decade or longer [25-27].

Transyears were later found in population statistics, related to natality in the Philippines [28], and to sudden cardiac death [29], suicide [30], and terrorism [31-33]. In the absence of a calendar-year component, the transyear in the worldwide pattern of terrorist activity persists but is damped when it is no longer detected in solar wind speed (SWS) or aa [33].

All of these findings rest on the shoulders of geophysicists who had earlier found spectral components longer than a year [34-37]. Corresponding results available from investigations of cosmic rays [38, 39] include transyears in a variety of transdisciplinary time series, notably one in helioseismics at the solar equator, discovered by Howe et al. [40]; such a component (not commented upon) among others is also apparent in a spectrum of geoseismics [41].

Far-transyears in the antipodal index aa, recorded since 1868 could not be consistently detected (with statistical significance) in the late 19th and early 20th centuries, in keeping with the more general demonstration by Clilverd et al. [42] that from solar cycle 14 on, solar activity was increased (cf. [43]). Wolf's relative sunspot numbers allow us to go further back in viewing solar variability, with its uncertainty [24, 44].

The methodology of gliding spectra with varying intervals visualizes, for instance, in the case of telescope-assessed solar magnetism, the role of the length of the interval used for analysis in the transyear region.

Relatively long intervals serve to clearly resolve an oscillation with a given average frequency, whereas shorter intervals are needed to scrutinize the details of behavior in time. For a span of four decades, Figure 2 shows gliding environmental spectral windows for alignment with psychophysiological ones [45, 46].

Multidecadal Congruence

Congruence in the transyear range is also found in the multidecadal range, as discussed elsewhere [47-52]. In one longitudinal series of self-measurements spanning more than 4 decades, the longest periods in systolic and diastolic BP and in HR have overlapping or overlying CIs with those of different periods of the various environmental variables, each with distinct congruences. Among others, competing effects of the sun and the earth influence different physiological variables. If one biological variable does not lock into the solar wind's pattern at one frequency, it may still lock into one in geomagnetics if it happens to be a nearby frequency or vice versa [53].

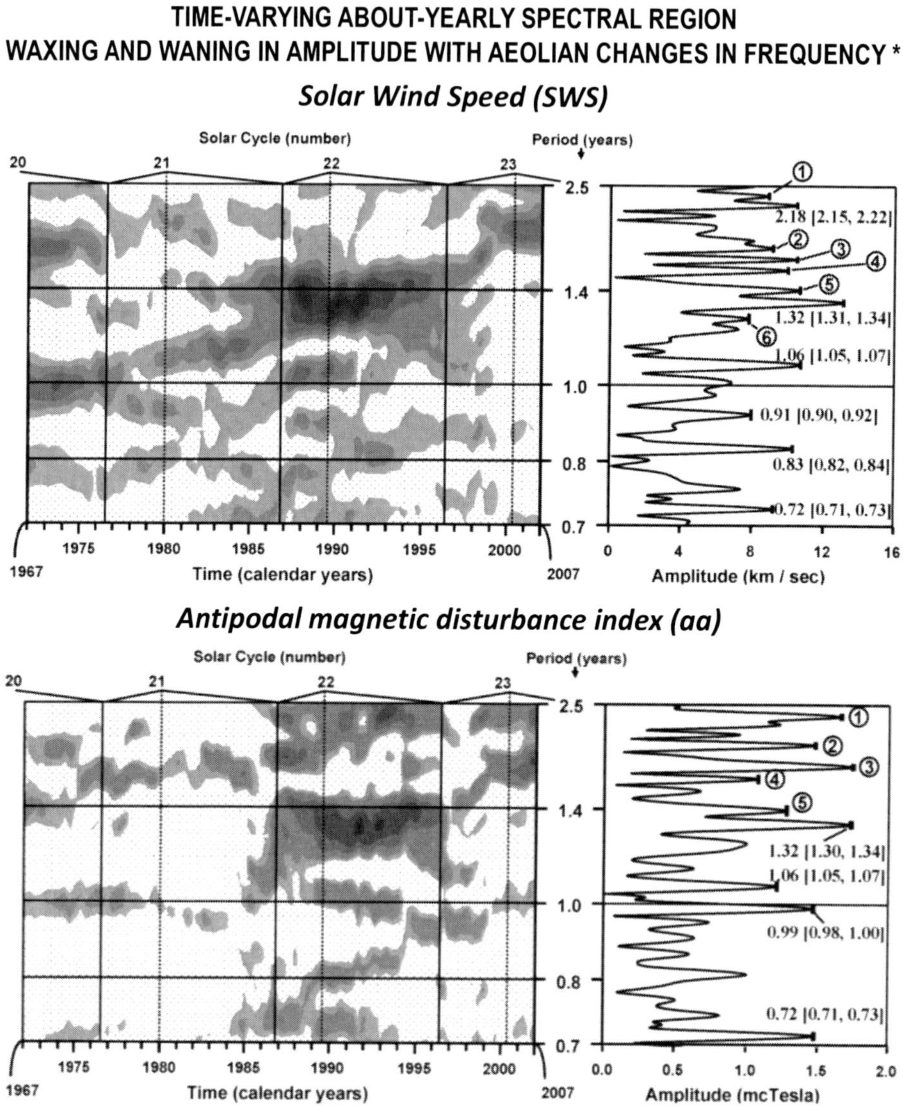

* Weekly averages of data from http://omniweb.gsfc.nasa.gov during 1967-2006. Gliding spectra: 10-year interval; 4-month increment; 0.05 harmonic increment. Darker shading corresponds to larger amplitude. Global spectra list periods of peaks with their 95% confidence intervals (CIs), shown as bars. CIs for peaks identified by circled numbers are: for SWS: 1. 2.33[2.28-2.37], 2. 1.69[1.67-1.72], 3. 1.60[1.58-1.62], 4. 1.52[1.50-1.54], 5. 1.39[1.37-1.41], 6. 1.24[1.23-1.26]; for aa: 1. 2.32[2.27-2.35], 2. 1.92[1.89-1.95], 3. 1.71[1.69-1.74], 4. 1.61[1.58-1.64], 5. 1.39[1.37-1.42].

Figure 2. Time course (left) and global summary (right) during 4 decades of wobbly behavior of transyear and cisyear in solar wind speed (top) and the geomagnetic index aa (bottom) during about 4 decades. Note prominent amplification during the years from ~1985 to ~1995. © Halberg Chronobiology Center.

Biospheric Consequence of the Subtraction of a Solar Far-Transyear?

The global mapping in the longest available series and in series of corresponding length and density, when feasible, of each new transdisciplinary spectral component does not only contribute useful information, but it should prompt us to study in the biosphere effects of amplification and damping, including disappearance of environmental spectral components, with the extreme of a remove-and-replace approach [54, 55], as soon as we find an opportunity offered by the environment, in particular the variable sun. In this context, the aeolian nature of environmental variables such as the solar wind's speed and of geomagnetics (Figure 2) offers an opportunity to learn how human physiology and/or pathology is influenced by terrestrial-space weather. For instance, a transyear was present as a broad band around 1.3 years in SWS during about the first half of a 16-year record (a span during which an elderly person, FH, happened to monitor his systolic BP) [29]. This SWS-band first bifurcated and then disappeared from the record of the solar wind. With a lag of about 1 transyear, part of a similar broad band component of ~1.3 years in systolic BP was also lost. Does the lag of a transyear in the partial loss of the 1.3-year band in BP indicate a physiological resonance of this variable for an added cycle? Does the partial loss indicate driving? Does the partial, albeit damped persistence mean partial endogenicity? More cases are certainly needed to reapply the “subtraction” (and to seek “addition”) at more environmental counterparts of biological cycles than a transyear in one case and circaseptans in five others [3].

Conclusion

For the prevention of the individual's cardiovascular ills, the Phoenix Project of the Institute of Electrical and Electronics Engineers in Minnesota (http://www.phoenix.tc-ieee.org) advocates a website (http://www.sphygmochron.org/) that can monitor solar variability, as BIOCOS (corne001@umn.edu) now does, by providing free time series analyses for all participants in self-health care who donate their data (e.g., time series of BP and HR) for, i.a., monitoring solar variability [56].

Dokutchaev deserves credit for emphasis on the need for transdisciplinarity, suggesting that finding laws in biology only, as Darwin did, or in chemistry only, as Lavoisier did, was not enough [26, 57]. In the transdisciplinary sense of Dokutchaev, as well as of Galileo and Humboldt, we can try to go beyond what is regular [58]; even when the endpoints are highly variable, we need to render them measurable in all their aspects now dealt with in artificially separated disciplines, usually without any uncertainty estimate, e.g., in the solar wind of the physicist as in the patient's blood pressure of a physician. We need to realize not only that we measure in time and only thus meaningfully, but also that whatever we single out for our assessing endeavor, there will always remain, in the background and very often in the foreground, an immense residual, dynamic complementary system that is not only intra- or inter-disciplinary, but transcends more than a few disciplinary boundaries [59-62].

We conclude theoretically that nothing in transdisciplinary science, including disciplines dealing with human affairs, can be understood except in the light of the cosmos and the biosphere and a noosphere that have evolved in it. In that information base, biospheric evolution stands tall, yet on the shoulders of the cosmos, and remains incomplete until a broad spectrum of transdisciplinary cycles, the predictable element in time structures, is mapped and thus can be taken into account. Those very cycles may help physicists backtrack to see with what time structure the solar wind blew billions of years ago, when other tools [63] may not suffice. The same nonphotics will serve to assess the effects of space weather in individuals' and populations' health care. For this purpose, and for basic aims, the systematic physiological government-supported monitoring seems overdue to match the surveillance in physics begun by Gauss, Humboldt and Sabine.

References

[1] Halberg F, Cornelissen G, Katinas G, Syutkina EV, Sothern RB, Zaslavskaya R, Halberg, Francine, Watanabe Y, Schwartzkopff O, Otsuka K, Tarquini R, Perfetto P, Siegelova J. Transdisciplinary unifying implications of circadian findings in the 1950s. *J Circadian Rhythms* 2003; 1: 2. 61 pp. www.JCircadianRhythms.com/content /pdf/1740-3391-2-3.pdf

[2] Vernova YeS, Pochtarev VI, Ptitsyna NG, Tyasto MI. Short-period variations in the rate of change of solar activity as a geosensitive parameter. *Geomagnetism and Aeronomy* 1983; 23: 425-427.

[3] Cornelissen G, Halberg F, Wendt HW, Bingham C, Sothern RB, Haus E, Kleitman E, Kleitman N, Revilla MA, Revilla M Jr, Breus TK, Pimenov K, Grigoriev AE, Mitish MD, Yatsyk GV, Syutkina EV. Resonance of about-weekly human heart rate rhythm with solar activity change. *Biologia* (Bratislava) 1996; 51: 749-756.

[4] Halberg F, Engeli M, Hamburger C, Hillman D. Spectral resolution of low-frequency, small-amplitude rhythms in excreted 17-ketosteroid; probable androgen induced circaseptan desychronization. *Acta endocrinol* (Kbh) 1965; 50 (Suppl 103): 5-54.

[5] Halberg F, Breus TK, Cornelissen G, Bingham C, Hillman DC, Rigatuso J, Delmore P, Bakken E, International Womb-to-Tomb Chronome Initiative Group. *Chronobiology in space*. University of Minnesota/Medtronic Chronobiology Seminar Series, #1, December 1991, 21 pp. of text, 70 figures.

[6] Rieger A, Share GH, Forrest DJ, Kanbach G, Reppin C, Chupp EL. A 154-day periodicity in the occurrence of hard solar flares? *Nature* 1984; 312: 623-625.

[7] Ballester JL, Oliver R, Carbonell M. The near 160 day periodicity in the photospheric magnetic flux. *Astrophys Journal* 2002; 566: 505-511.

[8] Cornelissen G, Benser M, Halberg F. Patterns of incidence of tachyarrhythmias recorded in implantable cardioverter-defibrillators during 2001-2005. PS-004, *Proceedings, 2nd World Congress of Chronobiology*, November 4-6, 2007, Tokyo, Japan. p. 63.

[9] Grafe A. Einige charakterische Besonderheiten des geomagnetischen Sonneneruptionseffektes. *Geofisica Pura e Applicata* 1958; 40: 172-179.

[10] Russell CT, McPherron RL. Semiannual variation of geomagnetic activity. *J Geophys Res* 1973; 78: 92-108.

[11] Cornelissen G, Halberg F, Pöllmann L, Pöllman B, Katinas GS, Minne H, Breus T, Sothern RB, Watanabe Y, Tarquini R, Perfetto F, Maggioni C, Wilson D, Gubin D, Otsuka K, Bakken EE. Circasemiannual chronomics: half-yearly biospheric changes in their own right and as a circannual waveform. *Biomed & Pharmacother* 2003; 57 (Suppl 1): 45s-54s.
[12] Martikainen H, Tapanainen J, Vakkuri O, Leppaluoto J, Huhtaniemi I. Circannual concentrations of melatonin, gonadotrophins, prolactin and gonadal steroids in males in a geographical area with a large annual variation in daylight. *Acta endocrinol* (Copenhagen) 1985; 109: 446-450.
[13] Tarquini B, Cornelissen G, Perfetto F, Tarquini R, Halberg F. Chronome assessment of circulating melatonin in humans. *In vivo* 1997; 11: 473-484.
[14] Cornelissen G, Tarquini R, Perfetto F, Otsuka K, Gigolashvili M, Halberg F. Investigation of solar about 5-month cycle in human circulating melatonin: signature of weather in extraterrestrial space? *Sun and Geosphere* 2009; 4: 55-59.
[15] Otsuka K, Cornelissen G, Weydahl A, Holmeslet B, Hansen TL, Shinagawa M, Kubo Y, Nishimura Y, Omori K, Yano S, Halberg, F. Geomagnetic disturbance associated with decrease in heart rate variability in a subarctic area. *Biomed & Pharmacother* 2001; 55 (Suppl 1): 51s-56s.
[16] Jozsa R, Halberg F, Cornelissen G, Zeman M, Kazsaki J, Csemus V, Katinas GS, Wendt HW, Schwartzkopff O, Stebelova K, Dulkova K, Chibisov SM, Engebretson M, Pan W, Bubenik GA, Nagy G, Herold M, Hardeland R, Hiither G, Poggeler B, Tarquini R, Perfetto F, Salti R, Olah A, Csokas N, Delmore P, Otsuka K, Bakken EE, Allen J, Amory-Mazaudier C. Chronomics, neuroendocrine feedsidewards and the recording and consulting of nowcasts -- forecasts of geomagnetics. *Biomed & Pharmacother* 2005; 59 (Suppl 1): S24-S30.
[17] Burch JB, Reif JB, Yost MG, Keefe TJ, Pitrat CA. Reduced excretion of a melatonin metabolite in workers exposed to 60 Hz magnetic fields. *Am J Epidemiol* 1999; 150: 27-36.
[18] Weydahl A, Sothem RB, Cornelissen G, Wetterberg L. Geomagnetic activity influences the melatonin secretion at latitude 70°N. *Biomed & Pharmacother* 2001; 55: 57-62.
[19] Bartels J. L'activité du magnétisme terrestre et ses relations avec les phénomènes solaires. In: de Valbreuze R. (Ed.) *Comptes rendus des travaux de la 11ème section et de la 13ème section*, Congrès international d'électricité, Paris 1932. Paris: Gauthier-Villars, 1932. p. 77-88.
[20] Zieger B, Mursula K. Annual variation in near-Earth solar wind speed: evidence for persistent north-south asymmetry related to solar magnetic polarity. *Geophys Res Lett* 1998; 25: 841-844.
[21] Halberg F, Cornelissen G, Haus E, Northrup G, Portela A, Wendt H, Otsuka K, Kumagai Y, Watanabe Y, Zaslavskaya R. Clinical relevance of about-yearly changes in blood pressure and the environment. *Int J Biometeorol* 1996; 39: 161-175.
[22] Watanabe Y, Cornelissen G, Halberg F. Thousands of blood pressure and heart rate measurements at fixed clock hours may mislead. *Neuroendocrinol Lett* 2003; 24: 339-340.
[23] Prabhakaran Nayar SR. Periodicities in solar activity and their signature in the terrestrial environment. *ILWS Workshop*, Goa, February 19-24, 2006. 9 pp.

[24] Halberg F, Cornelissen G, Katinas G, Tvildiani L, Gigolashvili M, Janashia K, Toba T, Revilla M, Regal P, Sothern RB, Wendt HW, Wang ZR, Zeman M, Jozsa R, Singh RB, Mitsutake G, Chibisov, S.M., Lee J, Holley D, Holte JE, Sonkowsky RP, Schwartzkopff O, Delmore P, Otsuka K, Bakken EE, Czaplicki J, International BIOCOS Group. Chronobiology's progress: season's appreciations 2004-2005. Time-, frequency-, phase-, variable-, individual-, age- and site-specific chronomics. *J Appl Biomed* 2006; 4: 1-38. http://www.zsf.jcu.cz/vyzkum/jab/4_1/halberg.pdf

[25] Cornelissen G, Masalov A, Halberg F, Richardson JD, Katinas GS, Sothem RB, Watanabe Y, Syutkina EV, Wendt HW, Bakken EE, Romanov Y. Multiple resonances among time structures, chronomes, around and in us. Is an about 1.3-year periodicity in solar wind built into the human cardiovascular chronome? *Human Physiology* 2004; 30: 86-92.

[26] Halberg F, Bakken EE, Katinas GS, Cornelissen, G, Zaslavskaya RM, Blank MA, Syutkina EV, Breus TK, Watanabe Y, Masalov A, Chibisov SM. Chronoastrobiology: Vernadsky's future science? Benefits from spectra of circadians and promise of a new transdisciplinary spectrum of near-matching cycles in and around us. Opening keynote, Proceedings, III International Conference, *Civilization diseases in the spirit of V.I. Vernadsky*, People's Friendship University of Russia, Moscow, Oct. 10-12, 2005, p. 4-22.

[27] Halberg F, Cornelissen G, Wendt HW, Schwartzkopff O, Berger S, Schweiger HG. Erweiterung transdisziplinärer photisch-nichtphotischer Spektren: Sich gegenseitig stützende Nah- und/oder Ferntransjahre in Sonnenmagnetik, Sonnenwind, Interplanetarmagnetik, Erdmagnetik und Biologie. Drei Magneotzyklizitäten im Autismus (Zusammenfassung). In: Kautzleben H, Hamel J. (Eds.) "*Albert Einstein in Berlin*": Kolloquium am 17. März 2005. Sitzungsberichte der Leibniz-Sozietät 2005; 78/79: 209-214.

[28] Mikulecky M, Florida PL. Daily birth numbers in Davao, Philippines, 1993-2003: Halberg's transyear stronger than year. *Abstract, 26th Seminar, Man in His Terrestrial and Cosmic Environment*, Upice, Czech Republic, May 17-19, 2005.

[29] Halberg F, Cornelissen G, Otsuka K, Fiser B, Mitsutake G, Wendt HW, Johnson P, Gigolashvili M, Breus T, Sonkowsky R, Chibisov SM, Katinas G, Siegelova J, Dusek J, Singh RB, Berri BL, Schwartzkopff O. Incidence of sudden cardiac death, myocardial infarction and far- and near-transyears. *Biomed & Pharmacother* 2005; 59 (Suppl 1): S239-S261.

[30] Halberg F, Cornelissen G, Berk M, Dodd S, Henry M, Wetterberg L, Nolley E, Beaty L, Siegelova J, Fiser B, Wolff C, BIOCOS project. Solar signatures in Australian suicide incidence: gender differences in prominence of photic vs. nonphotic spectral components. In: Halberg F, Kenner T, Fiser B, Siegelova J. (Eds.) *Proceedings, Noninvasive Methods in Cardiology,* Brno, Czech Republic, October 4-7, 2008. p. 44-62.

[31] Cornelissen G, Halberg F, Wendt HW, Sothern RB, Chibisov SM, Kulikov SI, Agarwal RK. Weak magnetoperiodism rather than socio-photo-thermoperiodism characterizes human terrorism detection of about 1.3-year aeolian transyear but not precise 1.0-year cycle. *Proc. 8th International Congress "Health and education millennium",* Moscow, Russia, November 14-17, 2007, p. 77-80.

[32] Halberg F, Cornelissen G, Sothern RB, Chibisov SM, Wendt HW. Do unseen, very weak magnetic mechanisms contribute to terrorism in wobbly spectral windows? Bulletin of the People's Friendship University of Russia, *Proceedings, 8th International Congress "Health and education millennium"*, Moscow, Russia, November 14-17, 2007, pp. 63-66.

[33] Halberg F, Cornelissen G, Sothern RB, Katinas GS, Schwartzkopff O, Otsuka K. Cycles tipping the scale between death and survival (= "life"). *Progress of Theoretical Physics* 2008; Suppl. 173: 153-181.

[34] Delouis H, Mayaud PN. Spectral analysis of the geomagnetic activity index aa over a 103- year interval. *J Geophys Res* 1975; 80: 4681-4688.

[35] Fraser-Smith AC. Spectrum of the geomagnetic activity index Ap. *J Geophys Res* 1972; 77: 4209-4220.

[36] Shapiro R. Interpretation of the subsidiary peaks at periods near 27 days in power spectra of geomagnetic disturbance indices. *J Geophys Res* 1967; 72: 4945-4949.

[37] Silverman SM, Shapiro R. Power spectral analysis of auroral occurrence frequency. *J Geophys Res* 1983; 88: 6310-6316.

[38] Cornelissen G, Halberg F, Kudela K, Chibisov SM (presenter). Transdisciplinary transyears and some of their uncertainties in cosmic rays. In: *Proceedings, International Symposium, Problems of ecological and physiological adaptation, PFUR*, Moscow, 30-31 Jan 2007. Moscow: People's Friendship University of Russia; 2007. p. 519-524.

[39] Valdes-Galicia JF, Pérez-Enriquez R, Otaola JA. The cosmic-ray 1.68-year variation: a clue to understand the nature of the solar cycle? *Solar Physics* 1996; 167: 409-417.

[40] Howe R, Christensen-Dalsgaard J, Hill F, Komm RW, Larsen RM, Schou J, Thompson MJ, Toomre J. Dynamic variations at the base of the solar convection zone. *Science* 2000; 287 (5462): 2456-2460.

[41] Duma G, Vilardo G. Seismicity cycles in the Mt. Vesuvius area and their relation to solar flux and the variations of the earth's magnetic field. *Phys Chem* Earth 1998; 23: 927-931.

[42] Clilverd MA, Clark TDG, Clarke E, Rishbeth, H. Increased magnetic storm activity from 1868 to 1995. *J Atmos Solar-Terr Phys* 1998; 60: 1047-1056.

[43] Ouattara F, Amory-Mazaudier C, Legrand J-P, Simon P. On the geomagnetic activity change from 1900 to 2000. *Geophysical Research Abstracts* 2007; Vol. 9, 04849, 2007. SRef-ID: 1607-7962/gra/EGU2007-A-04849.

[44] Cornelissen G, Halberg F, Gheonjian L, Paatashvili T, Faraone P, Watanabe Y, Otsuka K, Sothern RB, Breus T, Baevsky R, Engebretson M, Schröder W. Schwabe's ~10.5- and Hale's ~21-year cycles in human pathology and physiology. In: Schröder W. (Ed.) *Long- and Short-Term Variability in Sun's History and Global Change*. Bremen: Science Edition; 2000. p. 79-88.

[45] Halberg F, Cornelissen G, Grambsch P, McCraty R, Beaty L, Siegelova J, Homolka P, Hillman DC, Finley J, Thomas F, Kino T, Revilla M, Schwartzkopff O. Personalized chronobiologic cybercare; other chronomics' progress by transdisciplinary cycles' congruences: Season's Appreciations 2009. *J Appl Biomed* 2011; 9: 1-34.

[46] Cornelissen G, Grambsch P, Sothern RB, Katinas G, Otsuka K, Halberg F. Congruent biospheric and solar-terrestrial cycles. *J Appl Biomed* 2011; 9: 63-102.

[47] Halberg F, Cornelissen G, Bernhardt K-H, Sampson M, Schwartzkopff O, Sonntag D. Egeson's (George's) transtridecadal weather cycling and sunspots. *Hist Geo Space Sci* 2010; 1: 49-61.
[48] Halberg F, Cornelissen G, Sothern RB, Hillman D, Watanabe Y, Haus E, Schwartzkopff O, Best WR. Decadal cycles in the human cardiovascular system. *World Heart J* 2012; 4: 263-287.
[49] Haus E, Halberg F, Sackett-Lundeen L, Cornélissen G. Differing paradecadal cycles, semidecadal/decadal amplitude ratios and vascular variability anomalies in the physiology of a physician-scientist. *World Heart J* 2012; 4: 141-163.
[50] Syutkina EV, Cornelissen G, Mitish MD, Krylova OS, Narogan MV, Masalov AV, Halberg F. Decadal solar activity cycles modulate neonatal health. In: Grigoriev AI, Zeleny LM. (Eds.) *Proceedings, International Conference, Space Weather Effects in Humans: In Space and on Earth,* Space Research Institute, Moscow, Russia, June 4-8, 2012. pp. 782-788.
[51] Halberg F, Powell D, Otsuka K, Watanabe Y, Beaty LA, Rosch P, Czaplicki J, Hillman D, Schwartzkopff O, Cornelissen G: Diagnosing vascular variability anomalies, not only MESOR-hypertension. *Am J Physiol Heart Circ Physiol* 2013, 305:H279-H294.
[52] Halberg F, Cornelissen G, Schwartzkopff O, Khasgiwala P, Hillman D, Sothern RB, Katinas G, Hong S, Siegelova J. Decadal and multidecadal cycles in the cardiovascular system, relating to diagnosis and treatment? In. Kenner T, Cornelissen G, Siegelova J, Dobsak P (Eds.) *Noninvasive methods in cardiology* 2013, Brno, Czech Republic, 2013. p. 69-78.
[53] Halberg F. Historical encounters between geophysics and biomedicine leading to the Cornélissen-series and chronoastrobiology. In: Schröder W. (Ed.) *Long- and Short-Term Variability in Sun's History and Global Change. Bremen*: Science Edition; 2000. p. 271-301.
[54] Cornelissen G, Otsuka K, Halberg F. Remove and replace for a scrutiny of space weather and human affairs. In: Grigoriev AI, Zeleny LM. (Eds.) *Proceedings, International Conference, Space Weather Effects in Humans: In Space and on Earth*, Space Research Institute, Moscow, Russia, June 4-8, 2012. pp. 508-538.
[55] Cornelissen G, Watanabe Y, Otsuka K, Halberg F. Influences of space and terrestrial weather on human physiology and pathology. In: Rosch P. (Ed.) *Bioelectromagnetic and subtle energy medicine*, in press.
[56] Halberg F, Cornelissen G, Hillman D, Sothern RB, Nolley ES, Beaty LA, Schwartzkopff O, Otsuka K, Chibisov SM, Valenzi V, Pantaleoni G, Singh RB. Bakken's prehabilitation in the service of a budding chronoastrobiology. Invited lecture, VII International Crimean Conference "*Cosmos and biosphere*", Sudak, Crimea, Ukraine, October 1-6, 2007. p. 10-13.
[57] Bailes KE. Science and Russian Culture in an Age of Revolutions: V.I. Vernadsky and his Scientific School, 1863-1945. Bloomington/Indianapolis: Indiana University Press; 1990. 238 pp. Dokuchaev, p. 182-183.
[58] Halberg F, Cornelissen G, Schwartzkopff O, Bakken EE. Cycles in the biosphere in the service of solar-terrestrial physics? In: Schroeder W. (Ed.) *Case studies in physics and geophysics.* Bremen: Wilfried Schroeder/Science Edition, 2006c, p. 39-87. (Beiträge zur Geophysik und Kosmischen Physik/Joumal for the History of Geophysics and Cosmical Physics, Special issue, 2006/2. ISSN 1615-2824]

[59] Halberg F. Some physiological and clinical aspects of 24-hour periodicity. *Journal-Lancet* (Minneapolis) 1953; 73: 20-32.

[60] Halberg F. Chronobiology. *Annu Rev Physiol* 1969; 31: 675-725.

[61] Roederer JG. Are magnetic storms hazardous to your health? *Eos, Transactions, American Geophysical Union* 1995; 76: 441,444-445.

[62] Gregori GP. Climate and the atmospheric electrical circuit: the electromagnetic coupling between Sun and Earth. www.europeana.eu/portal/record/2023823/istituti_ProdottoDellaRicerca_html_cds_001_id_5423.html

[63] Usoskin IG, Mursula K, Solanki S, Schüssler M, Alanko K. Reconstruction of solar activity for the last millennium using 10Be data. *Astronomy & Astrophysics* 2004; 413: 745-751.

[64] Markov VI, Sivaraman KR. New results concerning the global solar cycle. *Solar Physics* 1989; 123: 367-380.

Updated Conference Report (Vienna, 2007)

In: Circadian Cardiology with Focus on Both Prevention ... ISBN: 978-1-63463-956-9
Editors: K. Hristova, A. Shehab, G. Cornelissen et al. © 2015 Nova Science Publishers, Inc.

Chapter 6

Tribute to Professor Franz Halberg: Chronobiology and Chronomics: Four Decades of Research from Cardiff, Durham and Tsim Tsoum

***Douglas W. Wilson*[1], *Keith Griffiths*[2], *Fabien de Meester*[3], *Agnieszka Wilcyznska*[3], *Toru Takahashi*[3,4] *and Ram B Singh*[*3,5]**

[1]School of Medicine, Pharmacy and Health, Durham University, UK
[2]Formerly Director Cancer Research, Tenovus Institute for Cancer Research, Cardiff University, Laurel Cottage, Castleton, Cardiff, UK
[3]The Tsim Tsoum Institute, Krakow, Poland
[4]Department of Nutrition, Graduate School of Human Environment Science, Fukuoka Women's University, Japan
[5]Halberg Hospital and Research Institute, Moradabad, India

Introduction

This special paper extols the scientific and human qualities of an exceptional remarkable man who developed chronobiology and took the science from everyday experiences to those of the chronomics of evolution and the impact of space weather on health.

It has tended to mainly to focus on his early work which motivated the authors of the Tenovus Institute, as they extended their analytical work on quality control of hormone assays to temporal biological assessment and interpretation but threads of advances in chronobiological science, in middle, and recent years with Tsim Tsoum, will also be highlighted.

* rbs@tsimtsoum.net.

The Early Years Of Collaborative Work with Franz Halberg –Tenovus Institute for Cancer Research (1976-1981)

While most researchers in the world of yesteryear, including DW in 1975 [1], generally assumed homeostasis in biology and medicine, Franz Halberg echoed the thoughts and more so of the eminent nineteenth century physiologist Claude Bernard [2-5] that life cannot solely be explained by an internal principle of action but exists as a conflict, not a battle, with the exterior physico-chemical conditions to which it is exposed. Logically, it seemed reasonable that the integration of neural, cellular and humoral variations of their components that constitute temporal aspects of physiology manifested in body rhythms in health are indeed a 'consensus partium in tempore' [6] and this suggested that the characterization and quantification of body rhythms were different in health and disease, which may allow the identification of rhythmic changes that mark the progression of a healthy subject to one with overt disease. Furthermore, Franz developed the idea, analogous to a biopsy, of the chronopsy such as, *i.a.*, the haemopsy, salopsy and thermopsy; temporal phenomena in blood, saliva and temperature respectively which may enable the clinician to define rhythms in health, screen for disease or even assess the efficacy of a particular chronotherapeutic regimen.

Halberg recognized that these coordinating or oscillatory systems had a wide frequency range [7], e.g., beta waves of the brain (50Hz), circadian rhythm [8] of many physiological variables [9], circaseptan (human kidney transplant rejection), circatrigintan, and circannual rhythms (e.g., prolactin) [9], which were documented in these early years. In the case of the human being, it is the influence of a spectrum of rhythms that determines health, work efficiency, moods, psychomotor skills and other activity. Franz clearly stated that in defining health and disease, and pursuing preventative health maintenance and treatment, rhythms cannot be ignored [10]. To Franz, the sylleptic ambiguity of the so-called Normal range was re-examined and replaced by time-qualified reference ranges (*vide infra*) and one of his oft cited references was the diagnosis of Cushing's or Addison's disease [11]. These issues will now be discussed.

Chronobiological Principles in Medicine

Time, which $>10^{-43}$ s is a dimension of the universe [12], is also a dimension of evolution [9, 13,14] and life (e.g., 14-16): a cursory study of physiological function in single cell organisms or the more complex mammalian forms of life, immediately reveals that biological processes are integrated in time. Furthermore, an evolutionary species having an internally adapted system is most likely to survive since priorities are related to survival. For example the search for food and avoidance of predators is a consequence of time-related attributes contained within the organism, possibly imprinted by emanated solar energy frequencies received on Earth [13, 17]. Halberg in his many publications showed that growth, maturation and aging are all chronobiological consequences of life (chronome [18]), particularly from conception with subsequent modulation by the external environment. In health it is desirable to know the most beneficial times to work, eat, travel, and receive treatment. At a fundamental level, Halberg showed how the brain, heart and endocrine system, not to mention an almost infinite number of biological sub-systems, govern mood, psychomotor skills, health and disease. An illustrative example of internal timing is plasma cortisol production which is superimposed by neuroendocrine rhythms of adrenocorticotrophic hormone, governed by neural circadian rhythms of corticotrophic releasing hormone. Clinical diagnosis, the time at which a physiological measurement is taken is often crucial. Franz recognized that depending on the time of measurement, e.g., plasma cortisol, a patient could be hypo-, hyper-, or euchronic. Indeed, clinical intervention by a low salt diet and a change in the phase could bring about a fall in blood pressure and reduce the risk of stroke [11]. Franz Halberg and his team were responsible for guidance on the interpretation of breast skin temperature [19-38], data gathered by the Chronobra [19, 36] and Simpson made major contributions, apart from that cited, to the pathology of the breast at risk and the use of the instrument [39-42]. Simpson hypothesized that the hyperplastic foci associated with the pericancerous tissue of the breast might be the origin of preneoplasia that leads progressively to breast cancer. If this preneoplastic stage is reversible on admission of anti-estrogens then the Chronobra data may be first used to detect preneoplasia and secondly, possibly to monitor the efficacy of treatment by quantifying the degree of modulation of breast skin temperatures (intervention - chronoprotopathology). Logically, the outcome of chemotherapy depends on the availability of the drug and its toxicity [43] but Halberg working with Scheving [44, 45] and Haus [45] sought to use chronobiological principles to extend the therapeutic efficacy of drugs and associated chronotolerance. Oncological agents depend on the biological and biochemical processes of cell division and these processes have been found to be circadian-stage dependent. In this pioneering age, Scheving and aforementioned colleagues demonstrated improved 'cure' rates for L1210 leukemic mice using cytosine arabinoside [46] and Deka et al. [47] exploited circadian-stage dependent differences between host and tumor in the radiotherapy of circumoral cancer. Of fundamental importance, underpinning a chronobiological experiment design, are chronobiometry methods for analyzing data which Halberg and colleagues pioneered [45].

Nature of the Data

In chronobiology data in the time domain are almost invariably positive e.g., leucocyte counts, plasma prolactin or salivary steroid concentrations, diastolic and systolic blood pressure, body temperature, kidney transplant rejection, survival time, *etc.* Certain data may be obtained in analogue but now usually in digital form. Franz Halberg pioneered chronobiological engineering [48] and data analysis with important advances made in the 1970s by De Prins and Cornelissen [49, 50], and others [51]. Thus computation of fast Fourier transforms and subsequent analysis of the data by linear and nonlinear least squares gave investigators new insights into the interpretation of findings.

Statistical Analysis of Chronobiological Data

In 1977, Halberg et al. published a glossary of chronobiological terms [10] indispensable to early and contemporary researchers engaged on endocrine and related projects at the Tenovus Institute. Fundamental was the definition of a biorhythm in simplistic terms, i.e., a statistically validated physiological change with reproducible waveform which requires time series analysis for its interpretation. It was evident from the work of Halberg that classical methods of analyzing chronobiological data that measured location, dispersion, correlation and statistical inference may be inadequate and misleading with respect to biological inferences drawn. An example of time series plots was demonstrated by Phillips et al. [52] for breast skin temperatures sampled for 4 days at 30-minute intervals. The difference between the two series may be attributable to the, albeit, large tumor. Power spectrum analysis allowed the identification of periodicities in the data which were compared for particular groups under study e.g., breast skin temperatures [53]. Once periodicities are revealed regression models based on the cosine function allowed computation of parameters of rhythms present in the data.

Calculation of Rhythm Parameters

Simple Cosinor

Nelson, Tong, Lee and Halberg [54] published the statistical analysis of a simple sinusoidal rhythm which comprised the MESOR (Midline Estimating Statistic Of Rhythm a rhythm-adjusted mean or mean of the fitted rhythmic model), amplitude (numerically equal to half the maximum excursion of the fitted rhythm), and the acrophase (the lag from some reference time, such as midnight, to the crest of the fitted cosine function of period τ). This method, together with summary statistics associated with the fit of the data, formed the basis of early Institute analyses, such as breast skin temperature rhythms. This was accompanied by a polar plot of the data in the form of a clock, called a cosinor diagram. In this diagram the length of the pointer represents an estimate of the amplitude, the time indicated by the pointer represents the acrophase and the associated ellipse is the joint confidence region of the uncertainty of these two parameters. Certain statistical tests such as the significance of a rhythm based on the zero-amplitude test and rhythm comparison tests for different time series of the same physiological variable were also devised. The amplitude and acrophase cannot be assessed separately since they are not independent. It is noteworthy that the above analysis applied to individual and population cosinors. Thus Halberg and his team provided the tools [10, 54] to analyze data which the Institute welcomed. Also developed was a graphical technique based on the cosinor model called serial section analysis which was used to evaluate breast skin temperature circadian rhythms through the menstrual cycle rather like a histologist examines tissue. At autopsy (to see oneself is an impossibility) the pathologist examines the macroscopic aspects of the body at post-mortem and this is analogous to the graphical record of time series data. Detailed examination of serial sections of say liver tissue reveals features that are microscopic in nature; from a data perspective the cosinor model is made to fit sections of the time series, each section of the data being a specified width and moved at fixed increments along each data series; this is analogous to the simple moving average [55].

In brief, the serial section diagram frames shows the times series (chronogram), the statistical significance of the fit, the MESOR plus amplitude and their standard deviation, and the acrophase and its statistical limits, and the final frame represents the sampling frequency. Professor Halberg's contribution to the study of ovulation showed breast skin temperature rhythm parameters changing during the menstrual cycle, particularly during ovulation. Power spectra may indicate more than one periodicity in the data and in the case of uncorrelated error in the fitted error sequence, data can easily be analyzed by including the additional components in the cosine/sine form of the cosinor-type model.

Extended Cosinor with Correlation

The method of least squares for the simple cosinor or extended model with additional periodicities would probably give consistent parameter estimates but would be seriously biased if the error terms were correlated. A study of the autocorrelation and partial autocorrelation functions suggested, for the breast skin temperature data, a low-order autoregressive model [53]. Knowledge of the correlation between residual errors allowed the calculation of unbiased estimates of rhythm parameters using Kalman filters or by using the Gram-Schmidt orthogonalization procedure; and the fit of the extended model with correlation was excellent. Quite simply, Halberg chose as one of several complementary approaches to use a decimated cosinor analysis.

Decimated Cosinor

Probably mindful of his classical education in which the Roman army punished its errant cohort, which committed serious offences such as mutiny, by dividing it into groups of ten and then by drawing lots would execute 1 in 10 soldiers, this approach consists of analyzing by regression only data in a series which were uncorrelated. In other words, if correlograms, which are graphs of a particular set of autocorrelation coefficients versus time, are plotted for autoregressive data, then a lag must be reached when no correlation exists [56]. However, the limitations of this method depend on the sampling frequency and period.

Other Variant Methods of Analysis

There were other variants of the cosinor model such as a distorted model where the time difference between maximum and minimum differs from half the period and/or when the maximum and minimum values of the function are at different levels from the MESOR. A generalized least squares with constrained optimization is another method the aim of which is to fit a function to a set of individual data points using the method of least squares and a number of harmonics and the Fourier coefficient vector but this often leads to divergence and oscillations. The use of the calculated quadratic content of the signal and technique are described by De Prins and Lechien [57].

Power spectrum was an important analytical tool. Although much work was carried out on circadian rhythms, Professor Halberg was well aware of other interacting rhythms such as ultradians, circaseptan, circatrigintan, circannual periodicities, and a Fourier transform of the function gives rise to a power spectrum which is a plot of the Fourier transform squared versus frequency from which rhythm parameters from selected periodicities can be calculated using least squares analysis. Details are beyond the scope of this report. Thus, for breast skin temperature rhythms, 24-hour and 12-hour components were present in nearly all series, ultradians in some, and there were some series which showed little periodicity but were suggestive of an autoregressive component. Thus, with Halberg as a co-author, breast surface skin temperature rhythms were comprehensively analyzed. One more hurdle required attention, *viz.* the comparison of two series, e.g., for left and right breast: the differential cosine model. Halberg had clearly demonstrated the difference between two series could be expressed as a single cosinor provided that correlation between the series was ignored. In the model concerned with fitting the paired cosine functions which allowed for correlation between residual errors, rhythm parameters were calculated separately. However, when comparing rhythm parameters for two series, cross-covariance must be taken into consideration, since if the covariance in rhythm parameters is positive, this variance associated with the difference in rhythm parameters is reduced and vice-versa. Details of such estimates of rhythm parameters have been published [53].

Graphical Representation of Time Series Data

The serial section composite graphical display from an incremental moving window of selected width e.g., 24-hour and 120-hour respectively has been described but folded representation of data into one period width as bands ('desms') is a useful feature.

Plexogram

The plexogram is simply a plot of mean and standard deviation of values of chronobiological data obtained for different cycle stages studied; data being derived from each cycle stage in successive though equal classes. In this way, for example, the circadian variation of data collected over several days can be folded into just one day where pertinent features of rhythmic change can be identified.

The chronodesm is a time-qualified reference range tool used by Halberg and team [58-60] to *i.a.* assess health and disease in the individual or group of individuals with a specified set of characteristics that qualifies them as a homogeneous set. Values of the measured variable for each time span from individuals within the group, or from successive days in the case of a single individual, allowed the chronodesm to be constructed from a knowledge of the mean and standard deviation, and a multiplier that allows a tolerance band about the mean to contain a certain percentage of the population with a specified probability.

Halberg devised the chronodesm which is a time-qualified reference range associated with a single cosinor model fitted to the data. In this case the tolerance bands are calculated from the proportion of the residuals and the confidence coefficient typically at 5%.

Application of Chronobiological Principles, Devised by Professor Halberg and Team, to Institute Studies

Steroid Rhythms in Saliva

It was certainly apparent that estrogens (and possibly phytoestrogens) may play a major role in the development and growth of human breast cancer. Equally, it was reasonable to presume that a detailed study of patients with carcinoma of the breast, together with good analytical techniques for the measurement of hormones, would ultimately be of value in identifying some degree of hormone imbalance that was associated with the disease, and thereby provide the means by which the 'high risk' group of women could be selected. There have been many studies in which the concentrations of various hormones have been measured as single samples of plasma removed from women with breast cancer, or those considered at risk, and the concentrations compared with results from a so-called 'Normal' population. The results from such studies have never been particularly promising, especially those concerned with the analysis of estrone, estradiol, and progesterone.

Analysis of single plasma samples is obviously not ideal for the study of hormones that undergo circadian and circatrigintan rhythmic change. Even circannual rhythms must be considered for certain hormones, e.g., prolactin [9]. Furthermore, it is also well recognized that for some steroid hormones in plasma, the free non-protein bound fraction is the biologically active moiety that should be measured such as that found in saliva. Undoubtedly Professor Halberg stimulated interest in biological rhythms, of which salivary steroids are mentioned herein; Professor Halberg's contribution is stated above in relation to hormones and breast skin temperature rhythms.

Breast Cancer

In the 1970s there was evidence that adrenal androgens were implicated in the etiology of breast cancer but the role of cortisol remained obscure. Though it was generally accepted that women with metastatic disease had elevated plasma concentrations of cortisol, there was little evidence to support the view that there was hyperactivity of the adrenal cortex in primary disease. Saliva was obtained from postmenopausal, pre-mastectomy, before cancer was confirmed.

Samples were taken every two hours during their wake spans (07:00 -23:00) throughout 2 days with a further sample taken at 07:00 on the third day. Data for salivary cortisol comprised 19 samples each for 50 breast cancer and 50 controls, for cortisol and dehyroepiandrosterone sulphate.

Circadian rhythms for both hormones were demonstrated and chronodesms for each hormone folded over from 07:00-23:00 indicated support for the original Bulbrook concept [61-62] that low androgen secretion, relative to cortisol production, was possibly a factor in the development of breast cancer [63].

Prostate Cancer

At the time, the basis of most endocrinological studies concerning the prostate gland relied on the fact that growth, maintenance and functional activity of the gland depended largely on androgenic hormones secreted by the testes. The clinical behavior of prostatic cancer also reflects androgen stimulation, but their role may be permissive, not exciting, or else estrogen, now at a higher plasma ratio of estrogens to androgens in the older man.

A chronobiological approach was needed because when plasma androgen and estrogen concentrations were measured in prostatic cancer, benign prostatic hyperplasia, and in age-matched controls, no significant differences were observed [64]. Though salivary estradiol assays were not developed yet for technical methodological reasons, salivary androgens were investigated at this time.

Studies of testosterone rhythms were undertaken in healthy young men; in a group of healthy young men who were first resident in the UK and then went on an expedition to Spitsbergen (Svalbard archipelago); and patients with prostatic cancer.

Testosterone in Young Healthy Men

This may be the first occasion when the circadian rhythms of testosterone were measured in saliva using the cosinor methodology developed by Halberg and his team [65-67]. Saliva samples (n=92) were taken from six healthy young men at two hourly intervals during the wake span (06:30 – 23:30) for two days which demonstrated that salivary testosterone exhibited a circadian rhythm (P=0.001) with the following characteristics: MESOR (184 pmol/L, SEM 7.0), Amplitude (76, 95% confidence interval: 53 - 99), Acrophase (-103, 95% confidence interval: -85 to -125). It is pertinent to note that one subject had an irregular profile that was modelled for fundamental, second and third harmonics using constrained optimization methods developed by De Prins (*vide supra*). MESOR, amplitude and (acro)phases for the 3 respective components were 72.5; amplitudes 29.6, 28.4, 2.89; phases -0.34, -0.90, and -0.35; data in pmol/L; this is possibly unreported data.

The Spitsbergen (Svalbard archipelago, 15 degrees East by 78 degrees North, Norway) study was designed in 1979 to investigate biological "clocks" governing adrenal and testis using salivary testosterone and cortisol measurements and the rhythmometric method of analysis developed by Halberg and his team. It was generally accepted that plasma testosterone concentrations were highest in the morning and lowest during the evening. The mechanism(s) responsible for the 'nocturnal' rise were not understood, although the variation during the daytime may be a reflection of changes in the rate of secretion of testosterone by the testis.

Various factors have been implicated in this process such as episodic secretion of luteinizing hormone, the nocturnal increase in plasma prolactin or changes in blood flow through the gonads related to the sympathetic nervous system and the adrenal medulla at different times of the day.

Perturbation of normal sleep or alteration of the daily sleep pattern had been shown to affect the characteristics of the circadian variation of plasma cortisol and testosterone. In Spitsbergen, the effect of continuous daylight and a phase shift in the sleep-wake schedule on circadian secretions of cortisol and testosterone, as adjudged from saliva samples were studied, *i.a.* the gain further insight into the mechanism(s) concerned with 'nocturnal' change in their secretions. Rhythms of salivary cortisol and testosterone concentrations were quantified for healthy young men: firstly while on a usual sleep-wake schedule in the UK, secondly during a similar schedule in continuous daylight, thirdly after an 8-hour phase shift under the immediately latter conditions, and finally after 1 month following a return to the UK (Phase E). Saliva samples were taken during normal working hours for 2 days (Phase A), 1 week before departure, similarly 1 (Phase B) and 2 (Phase C) weeks after arrival in Spitsbergen (wake span, 08:00 – 24:00); after 3 weeks their schedule was changed by 8 hours (Phase D), with a wake span from 00:00 to 16:00. Further methodological details are not the subject of this paper.

Cosinor analysis of cortisol rhythms (3 subjects showed a significant circadian rhythm) did not reveal any statistically significant differences in MESOR values for any of the study phases, except that values appeared higher during the first week in Spitsbergen (N.B. possibly a time of more stressful activity). Amplitudes, too, were not statistically different to each other.

The acrophase for cortisol in the phase-shift was 03:00. Those exhibiting significant circadian rhythms in cortisol also demonstrated a significant circadian rhythm in testosterone in the majority of study phases investigated (11/15). The acrophases of salivary testosterone rhythms characterized in Britain and for phases B and C were estimated; those for phase D had a 7-hour phase shift. MESOR values of salivary testosterone were significantly lower than those in the UK. No trends in amplitudes were apparent. In conclusion, results suggested that the early morning rise in salivary testosterone was related to activity whereas the change in acrophase for salivary cortisol was more associated with re-synchronization of the endogenous rhythm of cortisol secretion. Some other studies on salivary testosterone are documented elsewhere (68-69).

Plasma Testosterone in a Patient before and after Treatment with Cyproterone Acetate

Studies of testosterone rhythms in the saliva of prostatic cancer patients [70], probably previously not undertaken, before and 6 days after treatment with 100 mg cyproterone acetate, twice daily, were undertaken.

In both situations there was a statistically significant circadian rhythm (P=0.001) with comparable acrophase (-147° degrees, 95%CI: -120° to -178°); following treatment there was a decrease in MESOR: 4.3 (SEM 0.19) vs. 2.5 (SEM 0.10) nmol/L, and a less pronounced decrease in amplitude: 1.3 (95% CI: 0.76 - 1.88) vs. 1.0 (95% CI: 0.69 - 1.22).

It is fitting that this section should close with a quotation from Professor Franz Halberg in 1977, viz. "the study of biologic time structure should provide more satisfactory data for more effective use of therapeutic agents, for defining states of health and disease, and – what is more important – for preventive measures. Chronobiology implies not only the study of rhythms; from a viewpoint of public health, its aim is timely intervention.

Furthermore, chronobiology may become a powerful tool for a medically guided self-maintenance of personal health, enabling the individual to act in advance of catastrophic events that lead death and disability; an ulcer, a stroke, a coronary attack, or even the overt manifestations of a cancer. We may thus substitute a chronobiologic ounce of prevention for many pounds of homeostatic cure."

This quotation does not implicate or absolve the clinician's responsibility for deaths arising from untimely administration of therapeutic agents, when perhaps sound chronobiological, chronotoxicity and chronobiological investigations may have led to decreased mortality.

The Middle Years [Tenovus Institute (1982-1991); and the Faithful Collaboration Bestowed on DWW by Franz Halberg and Germaine Cornelissen (1992-2007)]

Prolactin and Breast Skin Temperatures

Following a seminal publication by Halberg, Cornelissen and co-workers on international geographic studies of oncological interest on chronobiological variables, particularly prolactin in 1981 [9], a hormone of mutual interest [23, 33], prolactin rhythms were measured in 7 postmenopausal diurnally-active and nocturnally-resting women with primary breast cancer (and 1 with benign breast disease). Plasma samples for cortisol and prolactin assay were taken over a 1- to 2-day timespan; breast skin temperatures were taken at 30-minute intervals for 96 hours. Given the relatively sparse sampling, statistically significant circadian rhythms ($P=0.05$) were found in only two cancer and 1 benign breast disease subjects, however, in all subjects, plasma prolactin peaked during the early morning hours. Cortisol and oral temperature served as internal physiological markers of anticipated rhythmicity in adrenocortical function and metabolism.

In all patients studied, the tumor site had an increased MESOR and reduced amplitude of its circadian rhythm compared to a similar site on the contralateral breast. Thermal asymmetry was also observed between left and right breast. Thus a cancer in the left breast may re-enforce lower amplitude and higher MESOR but in the right breast may nullify differences in left and right breast. It was concluded that subjects needed to be monitored on at least two occasions before tumor abnormalities due to tumor or preneoplasia became discernible.

Breast Skin Temperatures in Relation to Ovulation

Breast skin temperatures were monitored at 30-minute intervals throughout wake span for whole or part of the menstrual cycle. These data were modelled using techniques previously described and the main finding was a peri-ovulating surge in breast temperature, possibly detected within 24 hours [32, 52].

Breast Skin Temperatures: Overt and Occult Benign Disease and Occult Primary Malignant Breast Disease

Extensive time series analysis of a small number of subjects failed to reveal any consistent and highly statistically significant differences in rhythm parameters for the two-component model (12 and 24 hours) with the exception that the percentage variance explained and the initial variance appeared reduced in overt benign breast disease and occult primary breast cancer.

Increased MESOR and decreased amplitude of the fundamental harmonic for two patients with occult primary breast cancer were in keeping with the data reported for postmenopausal women with primary breast cancer, *vide supra* [53].

Chronobiotechnology and Chronobiological Engineering

Professors Halberg, Scheving and Dr Ehret, all of whom are unfortunately not with us any more, held at innovative NATO conference on this topic at Cardiff, UK on April 21-25, 1985. The foresight reflected in presentations brought technology to the fore in order to suit the needs of the chronobiologist [48].

The Halberg Influence

As with many great scientists, the seeds are sown that brings one to the forefront of one's mind, and researchers pursue their own course of investigation. This was the case with the Institute's saliva programme but the Institute was always mindful of the value of adopting a chronobiological approach to studies of human development. To the true scientist, 'No man can reveal to you aught but that which already lies half asleep in the dawning of your knowledge [71]' and such was the case with DWW's early childhood and his love of country and marine life around him; it was first Professor Griffiths and then Professor Halberg who taught me good science. 'If he is indeed wise he does not bid you enter the house of his wisdom, but rather leads you to the threshold of your own mind [71].'

Hormone Rhythms and Breast Cancer Epidemiology

Studies were conducted using saliva in numerous studies on breast cancer and breast disease [72-77]. Fundamental to the epidemiology of breast cancer was human development in pre- and post-menarchal girls and 'normal' premenopausal women. Circadian and circatrigintan time series of salivary progesterone in premenarchal and adolescent girls [78-83] and premenopausal women [84] were investigated as possible determinants of breast cancer risk; though somewhat restrictive, as estradiol assays were not. One subject, CVE Wilson whose data have been published elsewhere [74], collected saliva daily for nearly 3 years spanning pre- and post-menarche and in each menarchal state, progesterone was also verified using mass spectroscopy. Circadian variation in salivary progesterone appeared to be more random than systematic and estimates of total daily output are better represented by pooled samples collected at 2-hour intervals. Different patterns of circatrigintan rhythms were recognized and similar to those found in fertile and infertile women. Although difficult, it was still considered that salivary progesterone could be used as a tool for identifying individuals at risk of breast cancer.

Greenland Study of Salivary Cortisol for Wake-Sleep Phase Shifts

In the Greenland study [85] there were 5 phases in which saliva samples were collected 3-hourly. Phase 1 in the UK – sampled for 3 days; in Summer, Arctic travel is best done at night when snow surface is firmer and sampling carried out after 8 days in Greenland, bed rest was 17:00 – 00:00; else 3-hour sampling for either 1 or 2 days; Phase 3 similar to Phase

2, 10 days after Phase 2 – heavy loads and difficult travelling; Phase 4 – 15 days after Phase 3, samples collected under difficult conditions on ice cap between 02:00 and 14:00. Phase 5 was the end of the journey and sampling was similar to Phase 4. Significant rhythms of cortisol were identified. Data for the rhythm parameters have been reported elsewhere but in Phase 3 the acrophases occurred from about 04:00 to 05:00. The important message was that salivary cortisol could be collected under difficult conditions, a long way from the laboratory, and sample collection achieved when blood sampling is nigh impossible or difficult.

Circadian Rhythms of TSH, Prolactin and Cortisol in Surgically Treated Microprolactinoma Patients

Pharmacological levels of glucocorticoids inhibit TSH release, and circadian rhythms of TSH and cortisol show a reciprocal relationship and the latter was thought possibly a primary determinant of TSH change. To explore this, 7 subjects studied during their post-operative assessment following transethmoidal adenomectomy for microprolactinomas, were evaluated before and during metyrapone blockade [86]. Blood was taken hourly for 24 hours before and on the third day of metyrapone treatment and the Group cosinor methodology [10, 54] used to evaluate cortisol, 11-deoxycortisol, TSH and prolactin circadian change on 3-hour interval samples.

Group mean (SD) for TSH (Mu/L) for control and metyrapone treatment respectively were: MESOR 1.5 (0.26), 1.29 (0.48); amplitude 0.46 (0.26), 0.23 (0.13); and acrophase -49° (9.8°) and -62° (2.7°); the rhythm was significant, and there was a small but significant difference in amplitude and acrophase ($P<0.05$) but no change in the pattern of circadian change. All subjects showed a normal circadian rhythm which was strikingly blunted in morning samples after awakening during metyrapone treatment. Circadian rhythm for prolactin was apparently unaffected by treatment. It appeared that the variation in endogenous cortisol was not the sole factor involved in determining the TSH rhythm.

Clinical Applications of Neural Steroids in Saliva

In a position paper the role of chronobiology in laboratory measurements of salivary neutral steroids was introduced. This paper which very substantially referenced Professor Halberg's contributions, which demonstrated his influence, discussed perceptions of time and space; concepts of time-related changes in biology; statistical procedures used in chronobiological aspects of medicine, *vide supra*, *etc.* These concepts were related to adrenal activity (cortisol: circasemidian in the newborn infant [66, 87], circadian in the young boy or girl, healthy adult, and circadian rhythms of 17α-OH Progesterone [88, 89]; testicular activity (testosterone: circadian rhythms in pubertal development in boys [69, 90], and healthy young men and data were analyzed, *i.a.* using cosinor to extract values for rhythm parameters. The conclusion was that salivary steroid assays were valuable tools for use in clinical studies in endocrinology and many other compounds such as melatonin, drugs, and phytoestrogens were potentially on the horizon.

Value of Steroid Hormones in Chrono-Epidemiology

The above latter study some hormones for which reflecting those pioneering epidemiological studies by Halberg [9], addressed issues of accuracy and precision using mass spectroscopy, stability of steroids in saliva on short- and long-term storage, biologically active moiety, ease of multiple sampling, as they related to the adrenal, testis, and ovary, This chrono-epidemiology concerned both current and future work at the time.

Cortisol: Mental Health

Numerous studies had implicated glucocorticoids with mental health well-being. It was suggested that glucocorticoid metabolites bind to GABA receptors and simulate the inhibitory effects of GABA, and restrain behavioral response giving rise to depression [91]. Lower GABA function is associated with over-arousal and gives rise to anxiety and epilepsy.

Irrespective of the GABA hypothesis, glucocorticoid therapy restores some commonplace symptoms e.g., loss of memory, fatigue, and increases sensitivity implicating anxiety and epilepsy receptors and other neurotransmitter systems and there is a similarity between both diseases. Even now, some 25 years later a review suggests that GABA receptors could be implicated in the development of both disorders. Monitoring of cortisol in Addison's disease and Cushing's syndrome would be useful especially when correlated with clinical symptoms. ACTH cannot be measured in saliva.

Cortisol: Stress

At that time, the Tenovus researchers used the term 'stress' in much the same way as most, namely that 'any stimulus or succession of stimuli of such magnitude as to tend to disrupt the homeostasis of the organism'; the later seminal paper by Halberg complimenting Cannon, Selye, and Claude Barnard for their indelible contributions to medicine and the relative constancy of homeostasis; but he pointed out the importance of time, at the same time their weaknesses principally of not being time-qualified, nor the recognition that stress could be analogously described by a troop of soldiers marching in lockstep across a rickety bridge can be the "stress", while the "strain" is the structural weakening of the bridge, possibly leading to its collapse; however the effect of timing is important and discussed no further vis-à-vis stress/strain [92]. Recovery from surgery, intensive care, military training, occupations such as air traffic control, shift-work, public transport drivers, urban environments, student examinations, are among examples of stressors. Stress may also be psychosocial such as divorce, financial difficulty, and family bereavement/disharmony. A general neglected area is emotion and stress and their adverse effects on the autoimmune disease such as lupus erythematosis, myasthenia gravis and rheumatoid arthritis; these were among examples that were considered for future work.

Cortisol: Exercise and Sports Medicine

In studies of non-elite marathon runners [93] circadian rhythms of cortisol were equivocal for 3 days prior cf. 2 days post-race. On race day at 21m into the race the cortisol concentrations were nearly 500% of the value on the representative non-race day, values high enough to inhibit testicular activity as testosterone production. Thus saliva monitored before, during and after exercise could be a useful tool in determining training regimens [94] and assessing fitness [65, 85, 93] together with ambulatory blood pressure and heart rate monitoring, and classic treadmill parameters [95].

Cortisol: Human Development

Described elsewhere are reference ranges [96]. A range of rhythms can be explored for cortisol, e.g., ultradian, circasemidian, circadian, circatrigintan and circannual rhythms could now be monitored; and reference ranges for chronological age, sex, ethnicity, geographic domicile, etc., could be established for clinical and epidemiological studies of human development [97].

Cortisol: Breast Cancer

It was noted in the study by Read et al. [63] that MESOR values for a group of postmenopausal women were 8.82 and 5.82 nmol/L for cancer and controls respectively which tended to give some support – at that time – that the adrenal gland was implicated in the development of breast cancer, supporting the work of Bulbrook [98]. Other steroid hormones have been discussed earlier i.e., progesterone and breast cancer risk; testosterone and human development and prostate cancer.

Cardiology: Challenges in Social and Clinical Medicine

A memorable meeting took place in c1982 at The Department of Cardiology, Northwick Park Hospital, Harrow, England, between Professor Halberg (who with his team, autorhythmometrically determined blood pressure ranges and rhythms – published in 1973 [99]), DWW, and Dr Edward Raftery. Dr Raftery's major contribution to clinical science was that of ambulatory blood pressure monitoring, initially for arrhythmias. Raftery and Frank Stott developed a device for intra-arterial ambulatory monitoring of blood pressure, and used it to establish the circadian variation of blood pressure, both in clinically healthy subjects and patients with hypertension in the late 1970s. With Tultul Lahiri he played a major role in the development of non-invasive cardiological techniques [100]. Paradoxically, he died of a self-diagnosed myocardial infarction in 1995 [101]. He had an interest in clinical pharmacology and specialized in the evaluation of anti-hypertensive agents, for which ambulatory monitoring was key. This meeting again brought stimulus in chronobiology to the Tenovus Institute.

In 1986, in another seminal paper as first author, Halberg [102] indicated that methods and concepts had entered medical semeiotics and clinical pathology. He indicated quite clearly the cost disadvantages when otherwise preventable cardiovascular disease was allowed to develop. In the context of social medicine he highlighted the cost-effective need for self-help, for education, for family involvement in BP and HR monitoring, even at a time when automated equipment was reasonably expensive. He stated 'It is more important to detect and assess any blood pressure excess and to reduce or eliminate it according to chronobiologic criteria. ..... There is a need to explore the role of timing of excess pressure within a day. Weighting by the timing of the largest excess should eventually enter into the computation of excess, gauged by the hyperbaric index. Moreover, consideration of when an excess occurs, e.g., for treatment timing, is also indispensable.' Even then the essential ingredients of prevention/intervention programme were being laid down as the global shadow of increasing cardiovascular, and diabetes risk, was to reach epidemic proportions. Later papers on reference standards for blood pressure involved Franz Halberg [103].

More advanced chronobiological developments were added and enshrined in The RESOLUTION arising from a meeting in Brussels in 1995 on blood pressure, proposed for adoption by DW [18], echoed the need to store, analyze and use blood pressure and related variables/parameters for future research into the long-term cardiovascular and related outcomes of individuals. The Frontispiece depicts Baron Münchhausen's horse, severed in two by the town gate, which when drinking at the well, the water (data) spilled onto the ground and was lost, as is often at the present time when single time-unqualified blood pressure measurement are taken.

Already, blue sky, yet fundamental research, was being conducted in relation to solar imprinting of rhythms in life and more importantly on the human chronome. In 1993, at Puschino Science Center, Breus, Halberg and Cornelissen [104], and other papers presented at the International Symposium on Relations of Biological and Physicochemical Processes with Solar Activity and Other Environmental Factors, demonstrated a statistical association between magnetic field vectors arising from solar winds and mortality from cardiovascular disease. Now the prospect loomed of having weather maps for cardiovascular events, particularly in the Northern Hemisphere in winter; thus the environmental impact of internal, terrestrial, and solar risk of cardiovascular events had a more holistic global meaning that required attention.

Recent Years – Professor Halberg and His Team Working with Tsim Tsoum Institute: RB Singh (President), F De Meester (Founder), A Wilcyznska, T Takahashi and Durham University (DW: 2007- present)

The true impact of Professor Halberg's influence at Durham took place when DW was employed in the Department of Medicine and Health and when DW reported back to Professor Hungin, the proceedings of a meeting in 2009 on 'The study of time structures in and around us with emphasis on vascular variability disorders.', Hilton Garden Inn, Shoreview, Minnesota, USA. In this over100-page report Professor Cornelissen (USA) discussed the importance of screening for Vascular Variability Disorders; followed by Professor Singh's (India) contribution relating dietary interventions for pre-habilitation regarding vascular variability disorders.

The Czech team comprising Professor Siegelova and Drs Dusek and Homolka discussed various programs of blood pressure monitoring either in the long term, as part of a Health Watch programme, and the need for 7-day/24-hour ABPM interpreted chronobiologically. Larry Beaty (MN, USA) described his work directed toward an inexpensive chronobiologic home ABPM. The contributions by Drs Gupta and Pan (USA) from the Pennington Biomedical Research Center, Baton Rouge, discussed Cardiovascular Disease Risk: glycemia, inflammation and circadian blood pressure variability, and telemetry for metabolic studies. Dr Kino, from NIH, discussed his work searching for the genes for sensing magnetic field in humans. DW (UK) gave a talk on chronobiologic perspectives in endocrine cancer. Professor Halberg on the occasion of his 90th birthday celebration year clearly enthralled those in attendance with a closing address entitled "A unified transdisciplinary science – the next frontier". This was an exciting stimulating read and thence departmental support was given at every opportunity where there was mutual need.

Franz Halberg and Germaine Cornelissen together with DW and others produced another seminal paper [17], somewhat shortened due to space, describing the rules for chronobiology. Adumbrative inexact descriptions of congruence of biospheric and environmental (e.g., photic and non-photic electromagnetic signals) now follow: Rule 1- congruence in period can be documented by overlapping CIs of infradian periods in various regions of spectra; Rule 2- a selective assortment of biospheric-cosmic pairings; Rule 3- a consistent additive or subtractive behaviour – i.e., damping or amplification of biospherical rhythms; and Rule 4- concerned with genetic coding, quantified by cosinor and revealed at the molecular level due to the removal of an environmental component. A more complete description is later found elsewhere [105].

Halberg made an emphatic plea that enough blood pressure data should be collected that allows objective characterization of the temporal series investigated, e.g., MESOR, amplitude or acrophase. In this way the subject can compare their data to others or their self on another occasion. His final remarks in this paper concerned the combination of VVD that become VVSs, vascular variability syndromes (106), detection of which depends on education of the public at large, and inclusion of same in the business plan of health care professionals.

This timespan of Professor Halberg's influence, and that of Germaine Cornelissen, would not be complete without the influence/legacy that enabled the Tsim Tsoum partnership to develop and produce many commentary papers on global cardiovascular and related risk.

The timing of dietary intake has long been an important aspect of human development and, for example, cardiovascular disease, diabetes mellitus, and related health issues. Franz researched early works, for example those of Santorio Santorio of Padua in the early 1600s [107] and Virey of Paris [108]. In 1978, he produced, with his colleagues, and important paper on eating behavior and biorhythms [109] of a US Army study at Natick Research and Development Command, Natick, Massachusetts, which has 153 references documenting eating behavior and circadian change and related information.

In another part of the world, Fabien De Meester, notably an egg food producer (1992-present), was focusing on the Global Health Problem relating *i.a.* to cardiovascular disease as was Professor RB Singh and others whose publications testify to the effort of raising awareness of the area of growing concern [110, 111], and it was he who had the Halberg connection for FDM, and the Tsim Tsoum Institute was founded in 2009.

Since this time, Franz Halberg, who introduced DWW to RBS, and his team made major contributions to diet and cardiovascular disease, mental health and related issues which are beyond the scope of this report but papers and presentations are listed herein [112-127] to mark just some important contributions made through Franz's office.

Conclusion

This paper has described in some detail, the considerable influence that Professor Franz Halberg has played on our teams in developing chronobiology as a respectable science; and he has played a major role in some of the programmes undertaken by the authors. His genius will be missed but not forgotten [128]. We believe that he would have wished us all to promote the expanding field of chronomics which he and Professor Cornelissen developed so well: and in that way we can justly inherit his legacy in the service of mankind, though we will never stride with his seven league boots. Thank you, Franz.

Acknowledgments

The erstwhile generous support of the Tenovus Organization, and the past support by many in the Tenovus Institute who recognize the immense contribution that Professor Halberg has made in the field of chronobiology is recognized.

References

[1] Wilson DW, Gaskell SJ., Kemp KW. *Proc. 8th Tenovus Workshop: Quality Control in Clinical Endocrinology*. 1981. Cardiff: Alpha Omega Pub. ISBN: 9780900663093.

[2] Bernard C. De la diversité des animaux soumis à l'expérimentation. De la variabilité des conditions organiques dans lesquelles ils s'offrent à l'expérimentateur. J. de l'Anatomie et de la Physiologie normales et pathologiques de l'homme et des animaux 1865:2: 497-506.

[3] Bernard C. Leçons sur les phénomènes de la vie communs aux animaux et aux végétaux. 1885. J.B. Bailliere, Paris.

[4] Forster M. "Claude Bernard". 1899. Logmans, Green and Co., New York.

[5] Halberg, F. Claude Bernard and Experimental Medicine. Eds. F Grande and MB. Visschor, 1967. Schenkman Publishing Company, Inc., Cambridge Massachussetts.

[6] Halberg F. Temporal coordination of physiologic function. *Cold Spr. Harb. Symp. Quant. Biol.* 1960:25: 289-310. Long Island Biological Association, New York.

[7] Halberg F, Ahlgren A. Chronobiology: a review. In: Biological Rhythms. 1980. Documenta Geigy, Ciba-Geigy Ltd., Basel, Switzerland, pp. 1-3.

[8] Halberg F, Halberg E, Barnum CP., Bittner JJ. Physiologic 24- hour periodicity in human beings and mice, the lighting regimen and daily routine. In: *Photoperiodism and Related Phenomena in Plants and Animals*, Withrow R.B. (ed.), Ed. Publ. No. 55. 1959 Am. Assn. Adv. Sci., Washington, D.C.

[9] Halberg F, Cornelissen G, Sothern R B, Wallach L A, Halberg E, Ahlgren A, Kuzel M, Radke A, Barbosa J, Goetz F, Buckley J, Mandel J, Schuman L, Haus E, Lakatua D, Sackett L, Berg H, Wendt HW, Kawasaki T, Ueno M, Uezono K, Matsuoka M, Omae T, Tarquini B, Cagnoni M, Garcia Sainz M, Perez Vega E, Wilson D, Griffiths K, Donati L, Tatti P, Vasta M, Locatelli I, Camagna A, Lauro R, Tritsch G, Wetterberg L. International geographic studies of oncological interest on chronobiological variables. In: *Neoplasms, Comparative Pathology of Growth in Animals, Plants and Man,* H. Kaiser, ed., 1981, Williams and Wilkins, Baltimore, pp. 553- 596.

[10] Halberg F, Carandente F, Cornelissen G, Katinas GS. Glossary of chronobiology. *Chronobiologia* 1977: 4 (Suppl. 1), 1977.

[11] Bartter FC. Periodicity and medicine. In: Scheving LE, Halberg F, Pauly JE, editors. *Chronobiology*. 1974. Tokyo: Igaku Shoin Ltd.; pp. 6–13.

[12] Hawking SW, Penrose R. **The nature of space and time.** 1996. *Scientific Am. 1995: 275; 44-49.*

[13] Halberg, F, Cornelissen G., Katinas G, Appenzeller O, Otsuka K, Southern, R., Tarquini, R., Perfecto, F., Maggioni, C., Wilson, D., Schroder, W, Schwartzkopff O, Kessler, T, Wang Z, Burioka N, Watanabe Y, Bakken E. System times and time horizons for biospheric near-matches of primarily non-photic environmentl cycles. *Biomed. Pharmacother.* 2002:56, 266-272.

[14] Halberg F, Cornelissen G, Matalova A, Siegelova J, Sonkowsky R, Chibisov SM, Ulmer W, Wilson D, Revilla M, Katinas GS, Homolka, P., Dusek, J., Fiser, B. and Sanchez de la Pena, S., Schwartzkopff, O, Beaty LA. Mendel's legacy. Omnis cyclus e cosmo: Mendel's chronoastrobiological legacy for transdisciplinary science in personalized health care. In: *Symposium on non-invasive methods in cardiology 2009.* Brno, Czech Republic: Masaryk University, Faculty of Medicine, pp. 77-111.

[15] Darwin C. On the Origin of Species by Means of Natural Selection. *Or The Preservation of Favoured Races in the Struggle for Life.* 6th ed, 1872. Murray, London.

[16] De Mairan JJO. Observation Botanique. Des Sciences, p35. Houghton Library, (1729). Harvard University, Cambridge, Massachusetts, USA.

[17] Halberg F, Cornelissen G, Wilson D, Singh RB., De Meester F. Goodbye to "true" blood pressure: chronobiology reveals treatable vascular variability disorders, VVDs and much more. 2009. *The Biologist* 4: 209-214.

[18] Halberg F, Cornelissen G, International Womb-to-Tomb Chronome Initiative Group. Resolution from a meeting of the International Society for Research on Civilization Diseases and the Environment (New SIRMCE Confederation), Brussels, Belgium, March 17-18, 1995: Fairy tale or reality? Medtronic Chronobiology Seminar #8,April 1995, 12 pp. text, 18figures. http://www.msi.umn.edu/~halberg/resol.html

[19] Wilson DW, Green D, Halberg E, Simpson HW, Griffiths K. An ambulatory monitoring device for the measurement and storage of breast skin temperature. Bioengineering and the Skin 1979:2: 42-47.

[20] Taggett-Anderson MA, Cornelissen G, Halberg F, Simpson H, Wilson D, Griffiths K. Self-measurements of breast surface temperature: mammary thermopsy before a biopsy. *Chronobiologia* 1979: 6: 161-162.

[21] Wilson DW, Groom GV, Pierrepoint CG, Phillips M, Fahmy DR, Simpson H, Halberg F, Griffiths K. Breast skin temperature rhythms throughout the menstrual cycle. *J. Endocrinol.* 1979:81: 118P- 119P.

[22] Fanning R, Halberg F, Halberg E, Cornelissen G, Wilson D, Griffiths K, Simpson H. Circatrigintan modulation of circadian breast surface temperature. *Minn. Acad. Sci., 48th Annual Spring Meeting, Mankato State University*, Mankato, Minnesota, 1980: April 25-26, p. 5.

[23] Phillips MJ., Wilson DW, Riad-Fahmy D, Groom GV., Phillips MEA, Pierrepoint C G, Simpson HW, Holliday HW, Blamey RW., Halberg F, Griffiths K. Prolactin and breast skin temperature rhythms in relation to breast cancer. *Proc. endocrinol.* Soc. 1980:87: 28P-29P.

[24] Rao K, Halberg F, Osgood DW., Gupta BD, Chatterjee B, Brockway B, Wilson D, Griffiths K, Fundakowski R, Kennedy BJ., Frantz ID. Tumor temperature monitoring yields circadian markers for experimental and clinical chronotherapy. *Proc. AACR/ASCO* 1980:21: 196.

[25] Halberg E, Fanning R, Halberg F, Cornelissen G, Wilson D, Griffiths K, Simpson H. Toward a chronopsy: Part III. Automatic monitoring of rectal, axillary and breast surface temperature and of wrist activity; effects of age and of ambulatory surgery followed by nosocomial infection. *Chronobiologia* 1981:8: 253-271.

[26] Simpson HW, Griffiths K, Halberg F, Mutch F, Wilson D, Henery R, Chalmers D. Breast surface thermal ultradians: preliminary observations, from a 4-hour study of one 18-year-old female (topographical symmetry and asymmetry of parameters). *Int. J. Chronobiol.* 1981:7: 317.

[27] Simpson HW, Griffiths K, Mutch F, Wilson D, Halberg F, Gautherie MA. short history of breast thermorhythmometry. In: *Proc. Int. Symp. Biomedical Thermology,* 1981. Strasbourg, France, June 30-July 4, pp. B15-B16.

[28] Simpson H, Mutch F, Giles C, Grant JK., Griffiths K, Halberg F, MacEwan J, McNicol AM, Turbitt M, Wilson D. A fresh approach to human breast cancer. In: *Commentaries on Research in Breast Disease*, vol. 2, R.D. Bulbrook, D.J. Taylor eds., 1981. Alan Liss Publ., New York, pp. 133-164.

[29] Mutch F, Simpson HW, Wilson D, Griffiths K, Halberg F. Age distribution of epitheliosis in cancer mastectomies: possible relevance to 'false-positive' thermograms. In: *Biomedical Thermology*, M. Gautherie, E. Albert eds., Strasbourg, June 30-July 4, 1981, Alan R. Liss, New York, 1982pp. 313-320.

[30] Simpson H.W., Mutch F., Halberg F., Griffiths K., Wilson D. Bimodal age-frequency distribution of epitheliosis in cancer mastectomies: relevance to preneoplasia. *Cancer 1*982;50: 2417-2422.

[31] Simpson HW, Wilson D, Griffiths K, Mutch F, Halberg F, Gautherie M. Thermorhythmometry of the breast: a review to 1981. In: *Biomedical Thermology*, M. Gautherie, E. Albert eds., Strasbourg, June 30-July 4, 1981, Alan R. Liss, New York, 1982, pp. 133-154.

[32] Wilson DW, Griffiths K, Halberg F, Simpson HW, Griffiths R, Kemp KW, Nix ABJ, Rowlands RJ. Breast skin temperature rhythms in relation to ovulation. *Chronobiologia* 1983:10: 231-243.

[33] Wilson DW, Phillips MJ, Holiday HW, Blamey RW, Simpson HW, Pierrepoint CG, Halberg F, Griffiths K. Prolactin and breast skin temperature rhythms in post-menopausal women with primary breast cancer. *Chronobiologia* 1983:10: 171.

[34] Wilson DW, Simpson HW, Halberg F, Griffiths K. Comparison of circadian rhythms of breast surface temperature in clinically healthy women and in patients with benign breast disease or primary breast cancer. *Chronobiologia* 1983:10: 171.

[35] Wilson DW, George D, Mansel RE, Simpson HW, Halberg F, Griffiths K. Circadian breast skin temperature rhythms: overt and occult benign and occult primary malignant breast disease. *Chronobiol. Int.* 1984:1: 167-172.

[36] Wilson DW, Simpson HW, Griffiths K, Halberg F. An ambulatory monitoring device for measuring breast surface temperature: the chronobra. Abstracts from the International Workshop on Chronobiological Technologies, Como, Sept. 27-28, 1984. *Chronobiologia* 1984:11: 405-407.

[37] Simpson HW, Gruen W, Halberg E, Halberg F, Knauth P, Land DV, Moog R, Nougier JJ, Reinberg A, Smolensky M, Wilson D. Human thermometry in health and disease: the chronobiologist's perspective. In: *Chronobiotechnology and Chronobiological*

Engineering, Scheving L.E., Halberg F., Ehret C.F. eds.,1987. Martinus Nijhoff, The Netherlands, pp. 141-188.

[38] Simpson HW, Cornelissen G, Katinas G, Halberg F. Meta analysis of sequential luteal-cycle-associated changes in breast tissue. *Breast Cancer Res. Treat.* 2000;63:171–173. doi: 10.1023/A:1006434318939.

[39] Simpson H W, Candlish W, Jensen H, Mutch F, Pauson A, Griffiths, K, Wilson D. Breast thermal conductivity measured by the chronobra; a promising non-invasive technique for the measurement of cancer risk in the pre-neoplastic breast, 1985. *Biennial Breast Cancer* Conference. London.

[40] Simpson HW, Griffiths K, McArdle C, Pauson AW, Hume P, Turkes A. The luteal heat cycle of the breast in health. *Breast Cancer Res. Treat.* 1993;27:239–245. doi: 10.1007/BF00665693.

[41] Simpson HW, McArdle C, Pauson AW, Hume P, Turkes A, Griffiths K. A non-invasive test for the pre-cancerous breast. *Eur. J. Cancer.* 1995;31A:1768–1772. doi: 10.1016/0959-8049(95)00379-W.

[42] Simpson HW, Griffiths K, McArdle C, Pauson AW, Hume P, Turkes A. The luteal heat cycle of the breast in disease. *Breast Cancer Res. Treat.* 1996;37:169–178. doi: 10.1007/BF01806498.

[43] Tannock IF. In vivo interaction of anti-cancer drugs with misonidazole or metronidazole: methotrexate, 5-fluorouracil and adriamycin. *Br J Cancer* 1980:42: 861-870; doi:10.1038/bjc.1980.334.

[44] Scheving LE. The dimension of time in biology and medicine--chronobiology. *Endeavour.* 1976: May;35(125):66-72.

[45] Haus E, Halberg F, Scheving LE, et al. International Cancer Research Workshop on "Chronotherapy of cancer - a critical evaluation" sponsored by the International Cancer Research Data Bank of the United States of America, and administered by the International Union Against Cancer. *Int. J. Chronobiol* 1979; 6(2):67-107.

[46] Scheving LE, Burns ER, Pauly JE, Halberg F, Haus E. Survival and Cure of Leukemic Mice after Circadian Optimization of Treatment with Cyclophosphamide and 1-ß-D-Arabinofuranosylcytosine. *Cancer Research* 1977:37, 3648-3655.

[47] Deka AC, Chatterjee B, Gupta BD, Balakrisnan C, Dutta TK. Temperature rhythm – an index of tumour regression and mucositis during the radiation treatment of oral cancers. *Indian J. Cancer* 1976:13 (1): 44-50.

[48] Scheving LE, Halberg F, Ehret CF. eds., Chronobiotechnology and Chronobiological Engineering, 1987. Martinus Nijhoff, The Netherlands.

[49] De Prins J, Cornelissen G. *Bull Cl.* Sc. 1971: 5e serie,LVII, 1243-1266.

[50] De Prins J, Cornelissen G. J, *Interdisciplinary Cycle Research.* 1975:6, 95-102.

[51] Malbecq W, De Prins J. Identification and modelling techniques for biological rhythms. In. rhythms in biology and other fields of application. M Cosnard, J. Demangeot, A. Le Breton (eds). 1983. New York. Springer. *Lecture Notes in Biomathematics* 49,345-353.

[52] Phillips MJ, Wilson DW., Simpson HW., Fahmy DR., Groom GV, Phillips MEA., Pierrepoint CG, Blamey RW, Halberg F, Griffiths K. Characterization of breast skin temperature rhythms of women in relation to menstrual status. *Acta endocr.* (Kbh.) 1981:96: 350- 360.

[53] Dunstan FDJ, Barham S, Kemp KW, Nix ABJ, Rowlands RJ, Wilson DW, Phillips MJ, Griffiths K. A feasibility study for early detection of breast cancer using breast skin temperature rhythms. *Statistician* 1982:31: 37-52.

[54] Nelson W, Tong YL, Lee JK, Halberg F. Methods for cosinor-rhythmometry. *Chronobiologia.* 1979 Oct-Dec;6(4):305-23.

[55] Wetherill GB. In: *Sampling inspection and quality control.* 2nd Edition. 1977. London, Methuen.

[56] Levine H, Cornelissen G, Halberg F, Bingham C. Self- measurement, automatic rhythmometry, transmeridian flights and aging. In: *Chronobiology: Principles and Applications to Shifts in Schedules*, L.E. Scheving, F. Halberg eds., 1980. Sijthoff and Noordhoff, Alphen aan den Rijn, The Netherlands, pp. 371- 392.

[57] De Prins J, Lechien JP. Cahiers du Centre d'Etudes de Recherche Operationnelle 1977:19, 357-364.

[58] Cornelissen G, Halberg F, Lee JK, Teslow T, Stebbings JE. Biological monitoring for environmental protection with rhythm-based chronodesms (tolerance intervals). *Abstract, Minn. Acad. Sci. 46th Ann.* Spring Meeting, Macalester College, St. Paul, Minnesota, 1978. April 28-29.

[59] Halberg F, Lee JK, Nelson WL. Time-qualified reference intervals chronodesms. *Experientia* (Basel) 1978:34: 713-716.

[60] Haus E, Nelson W, Halberg F, Lee JK, Lakatua D, Wallach LA, Goetz F, Barbosa J. Chronodesms (circadian merodesms and cosinor- desms) for human plasma cortisol and prolactin. *Abstract, Minn. Acad. Sci. 46th Ann. Spring Meeting,* Macalester College, St. Paul, Minnesota, 1978. April 28-29, p.19.

[61] Bulbrook RD, Greenwood FC, Hayward JL. Selection of breast-cancer patients for adrenalectomy or hypophysectomy by determination of urinary 17-hydroxycorticosteroids and aetiocholanolone. *Lancet.* 1960 May 28;1(7135):1154–1157.

[62] Bulbrook RD, Hayward JL, Spicer CC, Thomas BS. Abnormal excretion of urinary steroids by women with early breast cancer. *Lancet.* 1962 Dec 15;2(7268):1238–1240.

[63] Read GF, Wilson DW, Campbell FC, Holliday HW, Blamey RW Griffiths K. Salivary cortisol and dehydroepiandrosterone sulphate levels in postmenopausal women with primary breast cancer. *Eur. J. Cancer Clin. Oncol.* 1983:19, 477-483.

[64] Harper ME, Peeling WB, Cowley T, Brownsey BG, Phillips ME, Groom G, Fahmy DR, Griffiths K. Plasma steroid and protein hormone concentrations in patients with prostatic carcinoma, before and during oestrogen therapy. *Acta Endocrinol* (Copenh). 1976 Feb;81(2):409-26.

[65] Campbell IT, Walker RF, Riad-Fahmy D, Wilson DW. Griffiths K. Circadian rhythms of testosterone and cortisol in saliva: effects of activity-phase shifts and continuous daylight. *Chronobiologia* 1982:IX, 389-396.

[66] Walker RF, Read GF, Wilson DW, Riad-Fahmy D, Griffiths K. Chronobiology in laboratory medicine: principles and clinical applications illustrated from measurements of neutral steroids in saliva. In "*Chronobiology: Its Role in Clinical Medicine, General Biology, and Agriculture, Part A*". Eds. Hayes DK., Pauly JE., Reiter RJ. Publ. Wiley-Liss, Inc., 1990. New York, Toronto, Singapore, pp 105-117.

[67] Wilson D.W., Walker R.F., Read G.F., Griffiths K. Potential value of salivary steroids in chrono-epidemiological and endocrine-related studies. In *"Chronobiology: Its Role*

in Clinical Medicine, General Biology, and Agriculture, Part A". Eds. Hayes DK., Pauly JE., Reiter RJ.Publ. Wiley-Liss, Inc., 1990. New York/Toronto/Singapore, pp 1 19-130.

[68] Read GF, Walker RF, Wilson DW, K. Steroid analysis in saliva for the assessment of endocrine function. In *"Steroid Formation, Degradation, and Action in Peripheral Tissues. 1990:Vol. 595"*. Eds. Castagnetta L., d'Aquino S., Labrie F., Bradlow H.L. Publ. New York Academy of Sciences, New York, pp 260-274.

[69] Walker R.F., Wilson D.W., Read G.F., Riad-Fahmy D. (1980). Assessment of testicular function by the radioimmunoassay of testosterone in saliva. *Int. J. Androl.* 3, 105-120.

[70] Harper ME, Chaisiri P, Slade1, Peeling WB, Griffiths K. Hormonal Relationships, Receptors, and Tumour Markers, *Recent results in cancer research.* 1981. 78:44-59. Berlin, Springer-Verlag.

[71] Gibran, K. The Prophet. 1923. New York: Knopf.

[72] Griffiths K., Read G.F., Wilson D.W. (1983). Breast cancer research: a new approach through assays for steroids in saliva. *Thai J. Surg.* 4 49-58.

[73] Griffiths K., Read G.F., Wilson D.W., Fahmy D.R. (1983). Breast cancer research: a new approach through assays for steroids in saliva. *Rev. Endocr.-Rel. Cancer*, Suppl. 13, 45-53.

[74] Read GF, Fahmy DR, Wilson DW, Griffiths K. A new approach for breast cancer research: assays for steroids in saliva. In *Commentaries on Research in Breast Disease.* 1983:Volume 3. Eds. Bulbrook R.D., Taylor Jane D. Publ. Alan R. Liss Inc., New York, pp 61-92.

[75] Read GF, Bradley JA, Wilson DW, George WD, Griffiths K. Evaluation of luteal-phase salivary progesterone levels in women with benign breast disease or primary breast cancer. *Eur. J.Cancer Clin. Oncol.* 1985:21, 9-17.

[76] Read GF, Wilson DW, Bradley JA, George WD, Griffiths, K. The hormonal etiology of breast disease: a new approach using the determination of salivary progesterone. In: *Endocrinology of the Breast: Basic and Clinical Aspects.* (1986). Volume 464. Eds. Angeli A., Bradlow H.L., Dogliotti L. Publ. Annals of the New York Academy of Sciences, 1986: pp 573-575.

[77] Kumar S, Mansel RE, Wilson DW, Read GF, Truran PL, Hughes LE, Griffiths K. Daily salivary progesterone levels in cyclical mastalgia patients and their controls. *Br. J. Surg.* 1986:75, 260-263.

[78] Joyce BG, Wilson DW, Read GF, Riad-Fahmy D. An improved enzyme immunoassay for progesterone in human plasma. *Clin. Chem.* 1978:24, 2099-2102.

[79] Wilson DW, Read GF, Hughes IA, Walker RF, Griffiths K. Hormone rhythms and breast cancer chronoepidemiology: salivary progesterone concentrations in pre- and post-menarchal girls and in normal premenopausal women. *Chronobiol. Int.* 1984:1, 159-165.

[80] Read GF, Wilson DW, Hughes IA., Griffiths K. The use of salivary progesterone assays in the assessment of ovarian function in postmenarcheal girls. *J. Endocrinol.* 1984:102, 265-268.

[81] Read GF Hughes IA, Wilson DW, Griffiths K. Measurements of salivary progesterone concentrations as an index of corpus luteum function in adolescent girls. In " *Adolescence in Females*". Eds. Venturoli S., Flamigni C., Givens J.R. Publ. Year Book Medical Pub., Chicago, 1985. pp 471-473.

[82] Danutra V, Turkes A, Read GF, Wilson DW, Griffiths V, Jones R, Griffiths K. Progesterone concentrations in samples of saliva from adolescent girls living in Britain and Thailand, two countries where women are at widely differing risk of breast cancer. *J. Endocrinol.* 1989:121, 375-381.

[83] Wilson DW, Turkes A, Jones R, Danutra V, Read GF., Griffiths K. A comparison of menstrual cycle profiles of salivary progesterone in British and Thai adolescent girls. *Eur. J. Cancer* 21992:8A, 1162-1167.

[84] Walker RF, Wilson DW, Truran PL, Read GF, Richards G, Walker SM., Riad-Fahmy D. Characterization of profiles of salivary progesterone concentrations during the luteal phase of fertile and subfertile women. *Endocrinol.* 1985:104, 441-446.

[85] Campbell IT, Wilson DW, Walker RF, Griffiths K. The use of salivary steroids to monitor circadian rhythmicity on expeditions in the Arctic. *Chronobiol. Int.* 1985:2, 55-59.

[86] Salvador J, Wilson DW, Harris PE, Peters JR, Edwards C, Foord SM, Dieguez, Hall R, Scanlon M. Relationships between the circadian rhythms of TSH, prolactin and cortisol in surgically treated microprolactinoma patients. *Clin. Endocrinol.* 1985:22: 265-272.

[87] Francis SJ, Walker RF, Riad-Fahmy D, Hughes D, Murphy JF, Bray OP. Assessment of adrenal cortical activity in term newborn infants using salivary cortisol determinations. *J. Pediatr.* 1987:111:129-133.

[88] Hughes IA, Dyas J, Robinson J, Walker RF, Fahmy DR. Monitoring treatment in congenital adrenal hyperplasia: use of serial measurements of 17OH-progesterone in plasma, capillary blood, and saliva. *Ann. NY Acad. Sci.* 1985;458:193-202.

[89] Young MC, Robinson JA, Read GF, Riad-Fahmy D, Hughes IA. 170H-progesterone rhythms in congenital adrenal hyperplasia. *Arch. Dis. Child.* 1988 June; 63(6): 617–623.

[90] Butler GE, Walker RF, Walker RV, Teague P, Riad-Fahmy D, Ratcliffe SG. Salivary testosterone levels and the progress of puberty in the normal boy. *Clin. Endocrinol.* (Oxf). 1989 May;30(5):587-96.

[91] Barnes DM. Steroids may influence changes in mood. *Science,*13 June, 1986: 1344-1345.

[92] Halberg F, Cornelissen G, Spector NH, Sonkowsky RP, Otsuka K, Baciu I, Hriscu M, Schwartzkopff O, Bakken EE. Stress/strain/life revisited. Quantification by blood pressure chronomics: benetensive, transtensive or maletensive chrono-vasculo-neuro-immuno-modulation. *Biomed. Pharmacother.* 2003 Oct;57 Suppl 1:136s-163s.

[93] Cook NJ, Read G, Walker RF, Harris B, Riad-Fahmy D. Changes in adrenal and testicular activity monitored by salivary sampling in males throughout marathon runs. *Eur. J. Appl. Physiol.* 1986:55:634-638.

[94] Kuoppasalmi K, Naveri H, Harkonnen M, Adlercreutz H. Plasma cortisol, androstenedione, testosterone and luteinizing hormone in running exercise of different intensities. *Scand. J. Clin. Lab. Invest* 1980:40:403-409.

[95] Winget CM., Soliman MRI., Holley DC, Meylor JS. Chronobiology of Physical Performance and Sports Medicine, In: Touitou Y., Haus E., *Biologic Rhythms in Clinical and Laboratory Medicine*, 1994. Springer-Verlag, Berlin, Heidelberg, New York, pp 230-242.

[96] Haus E, Nicolau GY, Lakatua D, Sackett-Lundeen L. Reference values for chronopharmacology, Ann. Rev. *Chronopharmacology*, 1988:Vol. 4 pp 333-424.

[97] Cornelissen G, Halberg F, Schwartzkopff O, Katinas G, Johnson D, Otsuka K, Watanabe Y, Wang ZR, Wan CW, Perfetto F, Tarquini R, Maggioni C, Syutkina EV, Masalov A, Siegelova J, Zhao ZY, Singh RB, Singh RK, Delyukov A, Gorgo Y, Zaslavskaya RM, Gubin GD, Gubin DG, Kumagai Y, Uezono K, Wilson D, Weydahl A, Bakken E. Editor's foreword: What Gesell wished, Hellbrügge accomplished: Chronomics of child development. *Neuroendocrinol. Lett* 2003; 24 (Suppl 1): 14-24.

[98] Bulbrook RD, Hayward JL, Spicer CC. Relation between urinary androgen and corticoid excretion and subsequent breast cancer. *Lancet* 1971:2:395-398.

[99] Levine H, Thompson D, Shiotsuka R, Krzanowski M, Halberg F. Autorhythmometrically determined blood pressure ranges and rhythm of 12 presumably healthy men during an 18-day span. *Int. J. Chroobiol.* 1973:1: 337-338.

[100] Millar-Craig MW, Bishop CN, Raftery EB. Circadian variation of blood pressure. *Lancet.* 1978 Apr 15;1(8068):795–797.

[101] McIntyre, N. “E B Raftery”. *BMJ* 1995:311,1296.

[102] Halberg F, Haus E, Halberg E, Cornelissen G, Scarpelli P, Tarquini B, Cagnoni M, Wilson D, Griffiths K, Simpson H, Balestra E, Reale L. Chronobiologic challenges in social medicine: illustrative tasks in cardiology and oncology - Proc. 2nd Int. Conf. Medico-Social Aspects of Chronobiology, Florence, Halberg F., Reale L., Tarquini B., eds., 1984. Istituto Italiano di Medicina Sociale, Rome, 1986, pp. 13-42.

[103] Cugini P, Leone G, Di Palma L, Battisti P, Wilson DW, Kawasaski T, Tamura K, Halberg F, Cornelissen G. Construction of references standards for blood pressure 24-h patterns. *Statistician* 1990:39: 285-299.

[104] Breus TK, Halberg F, Cornelissen G. Effect of the solar activity on the physiological rhythms of human being. *Proceedings, International Symposium on Relations of Biological and Physicochemical Processes with Solar Activity and Other Environmental Factors*, Pushchino, Russia, September 27-October 1, 1993, pp. 29 (in Russian) and 30 (in English).

[105] Halberg Franz, Cornelissen G, Beaty L, Schwartzkopff O, Halberg Francine, Revilla M, Siegelova J, Chibisov S, Breus T. Aligned physiological-epidemiological-physical environmental monitoring of variability in Vernadsky’s noösphere: a chronousphere. *Proceedings, Space Weather Effects on Humans in Space and on Earth,* Moscow, 4-8 June 2012, p. 28–30.

[106] Halberg F, Powell D, Otsuka K, Watanabe Y, Beaty LA, Rosch P, Czaplicki J, Hillman D, Schwartzkopff O, Cornelissen G. Diagnosing vascular variability anomalies, not only MESOR-hypertension. *Am. J. Physiol. Heart Circ. Physiol.* 2013, 305:H279-H294.

[107] Drazen M. Grmek, Santorio Santorio, I njegovi aparati I instrumenti. 1952. Zagreb Institurt za medicinska istrazivanja Jugoslavenske academije.

[108] Virey JJ. In: Ephémérides de la vie humaine, ou Recherches sur la révolution journalière et la périodicité de ses phénomènes dans la santé et les maladies. 1814. Doctoral Thesis in Medicine. Val-de-Grâce, a military hospital, Paris.

[109] Graeber RC, Gatty R, Halberg F, Levine H. Human eating behavior: preferences, consumption patterns and biorhythms. NATICK/TR-78/022 Technical Reports, U.S. Army, 1978.

[110] Singh RB. About the Editor. *The Open Nutraceuticals Journal* 2011:4, pp25-30.

[111] De Meester F. Editorial: "Impact of Research Conducted by Dr Ram B Singh on Modern Concepts of Nutrition & Lifestyle in Mind-Body Diseases" *The Open Nutraceuticals Journal* 2011. 4, pp 31-33.

[112] Wilczyńska A., Łapiński Ł, Wilson DW, Siegelova J, Pella D, Wahli W, Galli C, Singh RB, Cornelissen G, Halberg F, De Meester F. 2010, Effect of one blood/tissue component (omega-6 status) on body/mind interactions (chronomics) as measured by changes in biomarkers of pre-metabolic syndromes. A lifespan study, 5th World Congress on Clinical Nutrition. El Sokhna Resort, Ain El Sokhna, Egypt.

[113] De Meester F, Wilson DW, Singh RB, Wilczyńska A, Cornelissen G, Halberg F. 2010, Omega-3 fatty acids and CVD: and the Chronomics Project, International Congress on Nutrition in Cardiovascular Diseases. Annamalai University, Annamalainagar, Chidambaran, India.

[114] Halberg F, Cornelissen G, Wilson, DW, Singh RB, Wilczyńska A., De Meester F. 2010, The TsimTsoum Institute (TTI) & World Council of Body & Mind (WCBM): Scope & Vision, 1st International Workshop of The TsimTsoum Institute. Evidence based Holistic Approach to Health. Krakow, Poland.

[115] De Meester F, Wilczynska A, Singh RB, Cornelissen G, Halberg, F., Wilson DW. 2010, The Chronomics Study: design and protocol, 11th International congress 'Health & Education Millennium'. Moscow, Russia., SOMVOZ, Russia, 16.

[116] Halberg F, Cornelissen G, Wilson DW, Hungin APS, De Meester F, Wilczynska A, Singh RB, Sanchez de la Pena, S. Chronobiological perspectives in health and disease: evolution, cardiovascular risk, diet, endocrinology, and premetabolic syndrome. Society of Biology, Northern Branch, Great North Museum; 2011. Newcastle-upon-Tyne, 26th March.

[117] De Meester F, Wilczyńska A, Singh RB, Wilson DW, Pella D, Fedačko J, Siegelová J, Fišer B, Galli C, Cornelissen G, Halberg F. Advances in pneumology, Venue: Ibis Hotel Bonn, Vorgebirgsstr. 33, 53119 Bonn, Germany, June 17-18, 2011. Host Dr. Rüdiger Siekmeier – Budesinstitut für Arzneimittel und Medizinprodukte (BfArM) – Bonn, Germany.

[118] Wilson DW, Hungin APS, Wilczynska A, De Meester F, Singh RB, Sánchez de la Peña S, Cornelissen G, Halberg F. Perspectives of mind and body mechanisms in gastroenterology and cardiovascular disease: understanding variability. 6th International Congress on Cardiovascular Diseases India Habitat Center, New Delhi, India. From Sept 12-14, 2011.

[119] Singh RB., De Meester F,Wilczynska A, Wilson DW, Hungin, APS, Singh, R, Cornélissen G, Halberg F. Chronobiological, clinical and dietary vistas in the care of the elderly: clinical and physiological monitoring and the Tsim Tsoum Concept. Singh, R.B., De Meester, F.,Wilczynska, A., Wilson, D.W., Hungin, A.P.S., Singh, R., Cornelissen, G., Halberg, F. 8th International Conference on Geriatric Care (GSICON). 2011. Guru Nanak Dev Hospital, Amritsar.

[120] Cornelissen G, Galli C, Halberg F, De Meester F, Risé P, Wilczynska-Kwiatek A, Singh RB, Guillaume F. Circadian time structure of fatty acids and vascular monitoring. *J. Appl. Biomed.* 2010;8:93–109.

[121] Wilson DW, Hungin APS, Howse JH, De Meester F, Singh RB, Wilczynska, A, Cornelissen G, Halberg, F. Role of hormones and neuropeptides in IBS and other GI

disorders: understanding variability and chrononutrition. *The Open Nutraceuticals Journal,* 2011.

[122] Singh RB, Singh AK, Singh RK, Kumar, A, Rastogi SS, Singh G., Kartikey K, Singh S, Fedacko J, Pella D, De Meester F, Wilczynska A, Wilson DW. Nutrition in chronomedicine; we are indebted to Franz Halberg. The *Open Nutraceuticals Journal* 2012. 5: 1876-3960/12, 45-46.

[123] Singh RB, De Meester F, Wilcynszka A, Cornelissen G, Wilson DW, Hungin, APS, Singh R, Halberg F. Chronobiologically interpreted clinical and physiological monitoring and dietary vistas in the care of the elderly. WHJ, 2011: v3(2).

[124] De Meester F, Wilczyńska A, Singh RB, Wilson DW, Pella D, Fedacko J, Siegelova J, Fiser B, Galli C, Cornelissen G, Halberg F. The Chronomics Trial: Report of a 3-month pilot study. In: *Non-Invasive Methods in Cardiology* 2011 (Masaryk University, Faculty of Medicine, Brno, Czech Republic) Symposium Proceedings (Halberg F, Kenner T, Siegelová J, eds), ISBN 978-80-210-5672-5, © 2011 Masarykova Univerzita, pp. 141-161. (https://www.med.muni.cz/dokumenty/pdf/Noninvasivemethods-in-cardiology-2011.pdf)

[125] Cornelissen G, DeMeester F, Wilczynska-Kwiatek A, Halberg F. Relative prominence of weekly-to-daily amplitude ratio of blood pressure changes with age. In: Proceedings of the World Forum – International Congress on Natural Cataclysms and Global Problems of the Modern Civilization, 2011. September 19-21, Istanbul, Turkey, pp. 284-7.

[126] De Meester F,Wilczynska A, Singh RB, Wilson DW, Hungin APS, Pella D, Fedacko J, Siegelova J, Fiser B, Galli C, Cornelissen G, Halberg F. Metabolism, Diet & Disease. Georgetown University Hotel & Conference Center, Washington DC, USA, May 29-31, 2012. Tissue inflammation and NCDs: dietary control, physical exercise and mind body interactions. BMC Proceedings 2012:6 (Suppl 3): P53 (http://www.biomedcentral.com/1753-6561/6?issue=S3).

[127] Singh RB, Cornelissen G, Sothern RB, Wilson DW, Revilla M, De Meester F, Chibisov S, Agarval R, Schwartzkopff O, Halberg F. In Memoriam, Salvador Sanchez Alfonso De La Pena, MD, PhD: A Comrade in the Chronosphere (November 14, 1948 - November 2, 2011) *World Heart J.* 2012;3(4):259-272.

[128] Cornelissen G, Halberg F, Halberg J, Schwartzkopff O, Cugini P. Remembering the father of chronobiology and chronomics: Franz Halberg, MD (5 July 1919-9 June 2013). *Clin. Ter.* 2013;164(4):I-VI.

In: Circadian Cardiology with Focus on Both Prevention ... ISBN: 978-1-63463-956-9
Editors: K. Hristova, A. Shehab, G. Cornelissen et al. © 2015 Nova Science Publishers, Inc.

Chapter 7

Proatherogenic Risk Factors and under-Nutrition among Adolescents in South East Asia: When to Eat and What to Eat?

Shabnam Omidvar*[*1], Sumit Karn[2], Shahla Shafiee[3], R. B. Singh[4], Miki Tokunaga[5], H. S. Buttar[6] and D. W. Wilson[7]

[1]Department of Nursing and Midwifery, Babol Medical Sciences University, Iran
[2]Nutrition Program, Child Health Division/SUAAHARA, Department of Health Services, Kathmandu, Nepal
[3]Department of Nursing and Midwifery, Zahedan Medical Sciences University, Iran
[4]Halberg Hospital and Research Institute, Moradabad, India
[5]Graduate School of Human Environment Science, Fukuoka Women's University, Japan
[6]Department of Pathology & Laboratory Medicine, University of Ottawa, Faculty of Medicine, Ottawa, Canada
[7]School of Medicine, Pharmacy, and Health, Durham University, Durham, UK

Abstract

Background: In majority of the lower middle income countries, adolescence is associated with a double burden of diseases related to either under-nutrition or over-nutrition. In countries with rapid economic development, proatherogenic risk factors are also developing concurrently, although diseases related to under-nutrition, such as infections are not fully controlled. In this review, we describe both problems.

Methods: Internet searches using different data bases and discussion with experts.

Results: Approximately, 19% of the global population which is about 1 in 5 persons in the world is comprised of adolescents. This represents roughly 1.2 billion people

* Correspondence: Dr. Shabnam Omidvar, Ph.D. Department of Nursing and Midwifery, Babol Medical Sciences University, Iran Email. omidvar50@gmail.com.

between the ages of 10–19 years, out of which 85% live in developing countries. Nearly two thirds of premature deaths and one third of the total disease burden in adults are associated with conditions or behaviors that began in youth, including tobacco use, lack of physical activity, unprotected sex or exposure to violence. Promoting healthy practices during adolescence, and efforts that better protect this age group from risks will ensure longer lives and would protect from non-communicable diseases (NCDs) in adult life. Reports from India, Bangladesh, Nepal and Myanmar indicated that 32%, 48%, 47% and 39% adolescents, respectively, are stunted in growth, whereas 53%, 67%, 36%, and 32% adolescents from these countries are thin and underweight. Diets of adolescents were deficient in all the essential nutrients, especially in caloric energy, iron, magnesium, potassium, vitamin A, vitamin D and calcium. Chronic dietary deficiency of magnesium, potassium, selenium, chromium, zinc, and antioxidants may predispose to obesity and insulin resistance during transition from under-nutrition to over-nutrition, resulting into metabolic syndrome which is a risk factor for cardiovascular diseases (CVDs) and other chronic diseases. Eating the same amount of calories in the morning and in the evening causes less increase in body weight than eating in the evening. Hence breakfast may have greater impact than dinner for the prevention of obesity. However, a functional food with smaller breakfast appears to be an important driver for the prevention of circadian rhythms of cardiovascular events among people predisposed to CVDs, as gauged by circadian patterns of adverse cardiovascular events.

Conclusions: South East Asian countries with lower income as well as upper and lower middle income countries are in a rapid transitional phase from poverty to affluence. Increase in carbohydrate and fat energy intake via chronotherapy, in the presence of micronutrient deficiency and sedentary behavior may prevent or cause obesity and the metabolic syndrome.

Keywords: South Asian adolescents, obesity, metabolic syndrome, fast food energy intake, over nutrition, micronutrient deficiency, chronotherapy, lifestyle

Introduction

Adolescence is a bridge between childhood and adulthood and is regarded as the most important stage in life when growth and development are accompanied by physical, physiological, behavioral and social changes. Adolescents (aged 10 to 19 years) have specific health and development needs, and many face challenges such as poverty, a lack of access to health education and health services, and detrimental environments which hinder their well being [1-3].

Interventions that address their needs can save lives and foster a new generation of productive adults who can help their communities' economic progress. Approximately, 1 in 5 persons in the world is an adolescent, representing nearly 1.2 billion people between the ages of 10–19 years, and 85% of them live in developing countries [1-3]. In other words, roughly 19% of the global adolescent population lives in developing countries [4].

The state of their health and well-being is important for their lives now and in the future, for this and the next generation. A huge proportion of the world's population (> 1.75 billion) is young, in the age range of 10 -24 years [2-4]. Nearly two thirds of premature deaths and one third of the total disease burden in adults are associated with conditions or behaviors that began in youth, including tobacco use, a lack of physical activity, unprotected sex indulgence or exposure to violence [4-6].

Promoting healthy practices and dietary habits during adolescence, and efforts that better protect this age group from these risk factors will ensure longer lives and would avoid non-communicable diseases in adult life (NCDs) [7-9].

NCDs have become a major cause of death and disability in both developed and developing countries of the world, causing 36 million deaths in 2008 [6]. On Sept 19, 2011, the UN took an important step towards modification of the future course of NCDs at the High-level Meeting on NCDs. By the year 2020, NCDs are projected to cause almost three-quarters as many deaths as communicable diseases, maternal, perinatal, and nutrition-related diseases, and by 2030 these risk factors may be responsible for the most common causes of death [6].

This review highlights the importance of nutritional status, and dietary patterns in relation to under-nutrition and cardiometabolic risks in adolescence with reference to South East Asia.

South East Asia Region (SEAR)

Eleven countries comprise the World Health Organization South-East Asia Region (WHO SEAR): Bangladesh, Bhutan, Democratic People's Republic of Korea (DPRK), India, Indonesia, the Maldives, Myanmar, Nepal, Sri Lanka, Thailand and Timor-Leste. The member countries of the WHO SEAR bear a disproportionate burden of disease, with 25% of the world's population (Table 1 and 2) and 30% of the global disease burden [5]. The adolescent population in the South East Asia Region is about 350 million, constituting 15%-26% of the population in the region's countries [3-5].

Adolescents form a diverse population segment because they are either in different stages of development, i.e., early, mid or late adolescence, they live in urban, rural or slum areas, or they may even be homeless [7-9].

Apart from affecting body weight, nutrition may be an important lifestyle factor influencing the timing of puberty [9]. Adolescents may be in different marital status as well as in various social groups with variable intake of fatty acids and variable diet and lifestyle factors [10-14].

Adolescents As a Nutritional Vulnerabie Group

Nutrition during adolescence plays an important role in the life cycle. Worldwide, there has been a great change in the nutritional status of adolescents during the past 2 decades because of global economic development and urbanization [7, 8]. Many health problems later in life can be improved or avoided by adopting healthy lifestyle habits during adolescence [9]. Adolescence is a unique intervention point in the life cycle as well. According to WHO reports, under-nutrition and anemia are a public health concern in the adolescent age group in South East Asia Region [7]. There is a trend of decreasing under-nutrition in developing countries but still under-nourishment is a huge and persistent problem in Asia [13, 14].

Reports from India, Bangladesh, Nepal and Myanmar indicated that 32%, 48%, 47% and 39% adolescents, respectively, are stunted, and 53%, 67%, 36%, and 32% adolescents, respectively, from these countries are thin and lean.

Table 1. Demographic, economic and health profiles for countries in the SEAR (2008)

Category	Indicators	Thailand	Bangladesh	Bhutan	India	Maldives	Nepal	Sri Lanka	Timor-Leste	DPRK	Myanmar	Indonesia	World	High income
Population	Total (million)	67	160	0.7	1140	0.3	28.8	20	1	24	50	227	6697	1069
growth	Annual growth (%)	0.5	1.4	–	1.3	–	1.7	0.7	3.3	0.3	2.1	.2	1.1	0.5
	465 years (% total)	7	4	5	5	4	4	7	3	9	5	6	7	15
Economy	GNI/ capita	3670	520	1900	1040	3640	400	1780	2460	–	–	1880	8654	39 687
	Annual growth GDP (%)	2.5	6.2	13.8	6.1	5.2	5.3	6	13.2	1.7	12.7	4.8	1.7	0.5
Health financing	Total expenditure on health (% GDP)	3.7	3.4	–	4.1	–	5.1	4.2	13.6	3.6	–	2.2	9.7	11.2
	General government expenditure on health (% total)	73.2	33.6	–	26.2	–	39.7	47.5	84.6	83.7	–	54.5	59.6	61.3
	Per capita total expenditure on health (USD)	136	15	–	40	–	20	68	58	22	–	42	806	4406
Development	HDI (Index, rank)	0.654 (92)	0.469 (129)	–	0.519 (119)	0.602 (107)	0.428 (138)	0.658 (91)	0.502 (120)	0.877 (12)	0.451 (132)	0.600 (108)	0.624	-

Source: The World Bank, World Development Indicators 2010.

Original data sources include WHO, UNICEF, UNFPA and WHO National Health Accounts (Health Financing data). GDP, gross domestic product; GNI, Gross National Income; US, Unites States; HDI, Human Development Index.

Table 2. Life expectancy and mortality rates, using international and national (in parentheses) data sources, in SEAR countries compared with global data

Country/region	Life expectancy (years)		Healthy life expectancy (years)	Neonatal mortality rate[a]		Infant mortality rate[a]		<5-year-old mortality rate[a]		Adult mortality[b]	Maternal mortality[c]	
Bangladesh	65	(65)	56	30	(41)	41	(53)	52	(88)	234	340	(380)
Bhutan	63	(66)	55	34		52	(40)	79	(61)	228	200	(255)
DPRK	70	(68)	59	18		26	(21)	33	(46)	167	250	(97)
India	65	(65)	56	34	(37)	50	(57)	66	(85)	212	230	(301)
Indonesia	68	(69)	60	19	(20)	30	(32)	39	(46)	190	240	(307)
Maldives	75	(73)	64	8	(8)	11	(12)	13	(16)	84	37	(72)
Myanmar	64	(60–64)	50	33	(40)	54	(50)	71	(67)	231	240	(380)
Nepal	67	(61)	55	26	(34)	39	(48)	48	(61)	196	380	(281)
Sri Lanka	71	(73)	63	9		13	(11)	16	(16)	182	39	(47)
Thailand	70	(69)	62	8	(10)	12	(24)	13	(28)	205	48	(14)
Timor-Leste	67	(58)	53	27	(32)	48	(98)	56	(140)	195	370	(420–800)
SEAR	65		57	31		45		59		209	240	
Global	68		59	24		42		60		176	260	
Low income	57		49	36		75		117		321	580	
Lower middle income	68		61	26		42		57		176	230	
Upper middle income	71		61	11		19		22		184	82	
High income	80		70	4		6		7		88	15	

[a]Per 1000 live births (Source: World Health Statistics 2011, WHO Regional Office for South-East Asia and 11 Health Questions about the 11 SEAR Countries, WHO 2007, in parentheses).

[b]Per 1000 live births (Source: World Health Statistics 2011, WHO Regional Office for South-East Asia and 11 Health Questions about the 11 SEAR Countries, WHO 2007, in parentheses).

[c]Per 100 000 live births (Source: World Health Statistics 2011, WHO Regional Office for South-East Asia and 11 Health Questions about the 11 SEAR Countries, WHO 2007, in parentheses).

Diets of adolescents were deficient in all the nutrients, especially in energy (calorie intake), iron, vitamin A, and calcium [15]. Individuals malnourished from a young age encounter serious consequences during adulthood as well as older age. Understanding the factors that shape adolescents' nutritional status is important to know their needs and to develop appropriate intervention strategies to improve the nutritional status of adolescents [16]. Globally, population-based data on adolescent nutritional status and related factors are limited, especially in developing countries [17]. Nutritional problems of adolescents, such as under-nutrition and nutrition-related chronic diseases, are mainly the result of dietary inadequacies, which may be related to a number of factors. Socio-cultural factors, such as peer influences, crazes for trendy foods, mood, body image, extreme changes in lifestyle, and food habits of adolescents in the recent past have affected both their nutrient intake and needs [18]. Consequently, lifestyle and eating patterns are major threats to adolescents' nutritional status [19]. The media, particularly television, also plays an enormous potential influence and can overshadow familial influences [20]. Product marketing and mass media and dietary behavior are interrelated and exert influences on food preferences, food purchases and children's food requests [20-22]. It appears that adolescent eating patterns are established through a complex process involving internal and external factors such as food preferences and availability, body weight perception, and parental and peer influences [23].

Physical and emotional changes that occur during adolescence also influence dietary and health-related behaviors [24, 25]. Adolescents have particular food choice and meal habits compared to young children and adults [20, 26] Adolescent's food habits, either individual or group consumption, reflect the complexity of biophysical and socio-cultural interactions [27]. Eating habits are associated with adolescent's need to express freedom from parental control and their effort to adopt what they perceive as adult preferences and lifestyle [25, 28-30].

The South East Asian region seems to have a double burden of diseases, including under-nutrition and related diseases in the rural population, as well as over-nutrition and related diseases such as obesity, metabolic syndrome and hypertension, which are emerging in adolescence [13, 14]. The Building Healthy Lifestyles Conference proposed that modifying lifestyles to enhance physical activity, and the Paleolithic diet can reduce obesity and cardiovascular disease [31, 32]. In the United States, the death and morbidity rate due to NCDs remains high, despite high income. Therefore, poverty itself does not appear to be the cause of 80% deaths due to NCDs in the developed countries [33-35].

Food Consumption Patterns in Adolescence

As adolescents become more independent, they spend more time with peers or friends. Peer relationships provide important role for developing identity, autonomy and intimacy. However, they can be the source of negative influences [36]. Social relationships that develop while eating meals with others build a sense of belonging, connectedness and security [37]. Today, nuclear families are the core of most shared or communal eating. The family meal can be an opportune time for socialization and role-modeling of cultural manners, table etiquette, and healthy eating habits [38, 39]. Research suggests a decreased frequency of family meals: from 89.2% in 1973 to 75.9% in 1994, 47% in 1998 and 58% in 2005 in various studies [38, 40-42]. Reports on whether socio-economic status is related to diet quality or food intake among adolescents are mixed, depending on the outcome variable, i.e., nutrient-level data or specific food groups. Studies looking at nutrient intake data have generally not found a relationship between household income and dietary intake in adolescents. Several studies have found that adolescents' fruit and vegetables consumption is related to socio-economic status. Therefore, lower income households are more likely to consume insufficient fruits and vegetables.

In a cross-sectional survey, in each State in India, 120 villages were selected from eight districts. The information on socio-demographic profile was collected in all the 20 households, while anthropometric data such as weight, height and clinical signs of nutritional deficiency were collected on all the available adolescents in the selected households [43]. The nutrient intakes were compared with recommended dietary allowances (RDA). Anthropometric and socio-economic information on 12,124 adolescent boys and girls and dietary information on 2,579 individuals in 1996-1997 was available for statistical analysis. The prevalence of under-nutrition (< median -2.SD of NCHS weight for age) is higher in boys (53.1%) than in girls (39.5%). The extent of stunting was higher (42.7%) among adolescents belonging to the underprivileged caste community. In the case of girls, the extent of underweight was considerably less in each age group than their male counterparts. About 70% of adolescents consumed more than 70% of RDA for caloric energy. Intakes of micronutrients such as vitamin A and riboflavin were woefully inadequate. The extent of under-nutrition was high among adolescents and was higher among boys than girls. Adolescent girls in the rural areas could be at greater risk of nutritional stress because of early marriage and early conception before completion of their physical growth.

A survey identified significant differences in weight- and height-for-age across provinces, between urban and rural children in Iran [44]. In 1997, an anthropometric assessment of male students aged 6-18 years was conducted in Isfahan, a centrally-located city in Iran, in a random cluster sample of 4,364 out of 200,000 male students, aged 6-18 years from all areas of the city. Results showed that the growth percentiles of 6-18-year male students of Isfahan in 1997 were comparable to the NCHS charts. These students were taller and heavier than their fellow-citizens 22 years earlier, their compatriots in rural areas of Isfahan (1997) and Rasht (1989). They were also taller than those who were living in Tehran (1994). A similar study was conducted at the same time on urban Isfahani female students aged 6-18 years. All the percentiles of their height and weight curves were comparable with those of the NCHS charts of USA.

A health screening in Indonesia was aimed to show the magnitude of nutritional problem in adolescent girls among 1,060 students, aged 10 – 12 years from 34 elementary schools in two sub-districts of a rural coastal area in Tangerang District [45]. Around 49.5% of the girls were underweight, 38.4% stunted and 26.7% wasted, which indicated that both acute and chronic malnutrition are the nutritional problems among adolescent girls in this population. Anemia was observed among 33.7% of the girls who also had lower weight, height and body mass index compared with non-anemic girls. A cross-sectional study was conducted among 384 female students, aged 10-16 years, to investigate the dietary pattern and nutritional status of adolescent girls attending schools in Dhaka city and to examine the association with various social factors [46]. The prevalence of under-nutrition among the participants assessed as stunting was 10% overall with younger girls being less stunted (2%) than older girls (16%), whereas 16% were thin with relatively more of the younger girls (21%) being thinner than those of the older girls (12%). A substantial proportion of girls did not consume eggs

(26%), milk (35%) or dark green leafy vegetables (20%) but a larger proportion consumed meat (50%) and fish (65%) at least four times a week. Only 9% and 17% of the girls, respectively, met the RDA for energy and protein and majority (77%) had lower fat consumption. Intakes of iron (77% of girls), calcium (79%), vitamin A (62%), vitamin C (67%), and riboflavin (96%) were lower than RDA. Cereals were the major source of energy (57%), thiamine (67%), niacin (63%) and iron (37%). Animal sources supplied 50% of dietary protein and cooking fats were the principal source of fat (67%) in the diet. Milk was the major contributor for riboflavin and preformed vitamin A (retinol). Leafy vegetables and fruits were the main sources of provitamin A (carotenes). Those girls from families with lower incomes and less educated parents had a dietary pattern which tended to be poor with regard to egg, milk, meat and fruit, with lower intakes of protein, fat and riboflavin (vitamin B2). These results indicate that the diets of these girls tended to be inadequate depending upon the social class, both for macronutrients and micronutrients, with significant health implications.

In Nepal, around 23% population is in the adolescent age group [47, 48]. A cross-sectional community-based study was carried out in Morang district to determine prevalence and distribution of anemia in terms of age, sex and location (urban and rural) among the adolescent population [47]. Three hundred and eight adolescents (127 living in urban areas, 181 rural; 151 male, 157 female) participated in the study. The overall prevalence of iron deficiency anemia in the adolescent population was 65.6% with the distribution of rural 62.4%, urban 70.0%, male 52.3% and female 78.3%. Sufficiency or deficiency of iron makes the living of adolescents different as it affects their growth requirement and cognitive performance. Iron reserves in females result in better reproductive outcome. In Nepal, iron deficiency anemia among male adolescent has not been documented as yet. This study reports more than half of them are found anemic, which warrants further investigation. In another survey, among 1,047 subjects, 589 (56.3%) had anemia which was comparable in both males and females [48].

Proatherogenic Risk Factors in Adolescents

Economic prosperity in the developed countries has been associated with increased intake of total and saturated fat and total energy resulting into obesity, which are important proatherogenic risk factors. The prevalence of overweight and obesity as well as weight dissatisfaction has been increasing worldwide [49-57]. An Australian study included 7,889 children from government, private and Catholic schools (82% response rate) in August-November 2006 [49]. The prevalence of obesity was 6.4% in males and 5.6% in females in primary school students ($P = 0.34$). More high school males were obese than females (7.7% vs. 5.7%, $P = 0.001$). Obesity was more prevalent among students from Pacific Islander backgrounds. Adolescents who were most likely to be obese were boys and girls of lower socio-economic status or Pacific Islander or Middle Eastern/Arabic background. In developed countries, obesity is also increasing among adolescents [49, 50]. Among adolescent girls in the National Health Interview Survey, consumption of foods high in fat was inversely related to socioeconomic status (SES) [49]. Breakfast skipping was associated with low family SES in adults and adolescent boys, but not in girls. Several earlier studies have found that breakfast skipping is particularly prevalent among individuals from low SES families [50].

One study to date has examined the frequency of family meals by socio-demographic characteristics. Findings indicate that more frequent family meals are associated with mother's employment status, high socioeconomic status, and race/ethnicity [51]. A fundamental component of a healthy diet is to eat a balanced breakfast, which is important for nutritional adequacy. A good breakfast consumed in the company of one's family can contribute to a healthy diet rich in carbohydrates and micronutrients [52, 53]. It is possible that adults and adolescents who skip breakfast tend to eat more for the rest of the day. Breakfast skipping was also associated with increased prevalence of obesity (Odds Ratio: OR=4.5; 95% confidence interval: 1.57 – 12.90) [54]. Adolescents who eat more frequently with their families consume fewer sweetened carbonated beverages and more fruits, vegetables, grains, and calcium-rich foods. If adolescents skip meals, they are more likely to grab fast food from a restaurant, vending machine, or convenience store. These foods are high in fat and sugar and tend to provide little nutritional value. In addition, eating too many fast foods can lead to weight gain and, in some cases, diabetes and CVDs. Eating outside the home has been identified as an important factor influencing adolescents' food choices for foods with high energy and low nutrient density.

In a longitudinal prospective investigation, 705 boys were initially enrolled in the study. The information about height and weight measures and pubertal stage by age 11.5 years was available for 401 boys [55]. Boys in the highest BMI trajectory (mean BMI z-score at age 11.5 years, 1.84) had a greater relative risk of being prepubertal compared with boys in the lowest BMI trajectory (mean BMI z-score at age 11.5 years, -0.76) (adjusted relative risk = 2.63; 95% confidence interval, 1.05-6.61; P = 0.04).The relationship between body fat and timing of pubertal onset is not the same in boys as it is in girls. Further studies are needed to better understand the physiological link between body fat and the timing of pubertal onset in both sexes.

In a cross-sectional survey, 17,817 adolescents from 405 randomly selected schools were included [56]. Of these subjects, 16.5% of the adolescents were overweight, almost twice that number (32.1%) were dissatisfied with their weight (i.e., dieting or perceiving a need to diet) [56]. One-fifth of the total number of adolescents (16.0% boys; 24.0% girls) were not overweight but were dissatisfied with their weight in association with better education of mothers or more affluence. In both genders, but especially among girls, weight dissatisfaction was positively associated with most of the outcome variables (body image, health complaints, risk behaviors, and television viewing) regardless of weight status, whereas weight status was associated with only a few of the outcome variables. Weight dissatisfaction, independent of weight status, is associated with body image, health complaints, risk behaviors and television viewing, and represents a potential health risk factor for adolescents. Preventive interventions should focus not only on weight status, but also on body weight dissatisfaction.

The role of diet quality and nutrient adequacy in the etiology of childhood obesity is poorly understood [58]. A cross-sectional study among 1,030 Hispanic subjects, aged 4 to 19 years, in Houston assessed dietary intake in low-SES and without overweight, enrolled in the Viva la Familia Study. Despite higher energy intake in children with overweight, food sources, diet quality, macro- and micro-nutrient composition were similar between non-overweight and overweight children. Mean nutrient intakes were adequate (70% to 98% probability) in the children without and with overweight, except for vitamins D and E, pantothenic acid, calcium, and potassium, for which z-scores cannot be interpreted given the uncertainty of their adequate intake levels. It seems that the diets of low-SES Hispanic children with and without overweight were adequate in most essential nutrients, other components of a healthful diet, which promote long-term health, were suboptimal. It seems that healthy eating in children has become an important issue for adult health apart from what to eat and when to eat, as suggested by Halberg [58-60].

Nutritional Problems and Deficiencies in Adolescence Iron Deficiency and Anemia

Iron-deficiency anemia is the most common nutritional deficiency in the world; it is undoubtedly the most widespread nutritional problem affecting girls and women. Anemia and under-nutrition are generally recognized as the greatest nutritional problem among adolescents particularly girls, and diets poor in iron are likely to be a major factor. It is highly probable that the dietary intake of these girls is low in energy and nutrients, considering the low socio-economic background [61, 62]. The prevalence of iron deficiency anemia among adolescent population in a recent study from Nepal was 56.3% with distribution of males and females 26.8% and 29.4%, respectively [61]. According to the report of WHO, the prevalence of anemia ranged from 45-90% in India; 22-90% in Bangladesh; 28% in Indonesia; 42% in Nepal; and 26% in Myanmar [61, 62].

In South-east Asian region, the low iron intake is further worsened by the fact that the bioavailability of iron consumed is low, since the diet is primarily cereal-based with little meat or vegetables.

Such diet has high concentration of inhibitors and low concentration of enhancers and lowers the bio-availability of dietary iron. This is evident from data for India where the prevalence of anemia is higher in Indian adolescents consuming a vegetarian diet (45.8%) as compared to those consuming a mixed diet, including animal protein (30%) [62]. Adolescents (age 10-19 years) are at high risk of iron deficiency and anemia due to accelerated increase in requirements for iron, poor dietary intake of iron, high rate of infection and worm infestation, as well as the social norm of early marriage and adolescent pregnancy. The highest prevalence is between the ages of 12-15 years when requirements are at their peak. More than 50% girls in this age group have been reported to be anemic [61, 62].

Muzammil et al. [64] reported that more than half of the adolescents were anemic: 37.4% and 55.5 % among boys and girls, respectively. In this study from Bangladesh, 23% of the girls were anemic (Hb <12 g/dl), which was similar to an estimate of 22% in an earlier study in urban adolescent schoolgirls.

A few studies carried out among adolescent girls in Nepal reported that prevalence ranges from 42-60% [64, 65]. Other micronutrient deficiencies coexist in anemic adolescent schoolgirls in rural areas [64]. Epidemiologic study of trace elements and magnesium and risk of coronary artery disease in the rural and urban populations have been reported from India [66-68]. These studies showed a deficiency of vitamin A, beta carotene, magnesium and potassium as well as of antioxidants. In another study, antioxidant vitamins A, E, C and beta carotene were inversely associated with risk of coronary artery disease [68].

Vitamin A Deficiency

Vitamin A deficiency exists as a public health nutrition problem among preschool-aged children in 118 developing countries around the globe, with the South-East Asian Region harboring the maximum number of cases [69].

A study from India revealed the total prevalence of vitamin A deficiency to be 1.3% and approximately double among adolescent boys (1.7 %) than adolescent girls (0.9 %) [69]. Vitamin A deficiency can coexist with other nutrient deficiencies.

Vitamin D Deficiency

Vitamin D inadequacy is a worldwide problem and affects developed as well as developing countries, subtropical and temperate regions, and populations of all ages. Vitamin D deficiency may be overlooked in Asian countries, perhaps on the assumption that vitamin D deficiency is unlikely to occur in regions with plentiful sunshine [70-73]. Vitamin D status in Southeast Asian countries has recently received more attention. Health examination surveys (N~2,500–7,000) have been performed in countries in this region. Vitamin D deficiency varied from 6–70% in South East Asia. The determinants for the variation of vitamin D status are skin pigmentation, aging, and the sun protection behaviors such as application of a sunscreen, religious, lifestyle and nutritional differences [73].

Advanced age is a known risk factor for vitamin D deficiency. Interestingly, elderly in countries such as Korea and Thailand, had higher 25-(OH)-D concentrations when compared with young people. Therefore, adolescents are at higher risk in this region [70-73]. Most of the studies concerning vitamin D status were generated from countries situated at temperate latitudes. It is less clear what the extent of vitamin D insufficiency is in countries situated in the tropics and how geographical regions within a country would affect vitamin D status [70-73].

Vitamin D status has been linked to the pathogenesis of hip fractures as well as other skeletal and non-skeletal disorders as well as several of the NCDs [72]. Reports from across the world indicate that hypovitaminosis D is widespread and is re-emerging as a major health problem globally. Let us take a global perspective of vitamin D status across different regions of the world in order to identify the common and significant determinants of hypovitaminosis D. The definition of vitamin D insufficiency and deficiency, as well as assay methodology for 25-(OH)-D vary among studies. However, serum 25-(OH)-D concentrations below 75 nmol/L are prevalent in every region studied whereas concentrations below 25 nmol/L are most common in regions such as South Asia and the Middle East. Older age, female sex, living at higher latitude, winter season, darker skin pigmentation, less sunlight exposure, dietary habits, and absence of vitamin D fortification are the main factors that are significantly associated with lower 25-(OH)-D concentrations. Another study consisted of 2,641 adults, aged 15 - 98 years, randomly selected from the Thai 4th National Health Examination Survey (2008-9) cohort [73]. Serum 25-hydroxyvitamin D was measured by liquid chromatography/tandem mass spectrometry. Data were expressed as mean ± SE. Subjects residing in Bangkok, the capital city of Thailand, had lower 25-(OH)-D concentrations than other parts of the country (Bangkok, central, northern, northeastern and southern regions: 64.8

± 0.7, 79.5 ± 1.1, 81.7 ± 1.2, 82.2 ± 0.8 and 78.3 ± 1.3 nmol/L, respectively; $P < 0.001$). Within each region, except for the northeastern part of the country, subjects living inside municipal areas had lower circulating 25-(OH)-D concentrations (central: 77.0 ± 20.9 nmol/L vs. 85.0 ± 22.1 nmol/L, $P < 0.001$; north: 79.3 ± 22.1 nmol/L vs. 86.8 ± 21.8 nmol/L, $P < 0.001$; northeast: 84.1 ± 23.3 nmol/L vs. 87.3 ± 20.9 nmol/L, $P = 0.001$; south: 76.6 ± 20.5 nmol/L vs. 85.2 ± 24.7 nmol/L, $P< 0.001$). Overall, the prevalence of vitamin D insufficiency was 64.6%, 46.7%, and 33.5% in Bangkok, municipal areas except Bangkok, and outside the municipal area in other parts of the country, respectively. In addition, the prevalence of vitamin D insufficiency according to geographical regions was 43.1%, 39.1%, 34.2% and 43.8% in the central, north, northeast and south, respectively. After controlling for covariates in multiple linear regression analysis, the results showed that low serum 25-(OH)-D concentrations were associated with being female, younger, and living in urban areas and Bangkok. Vitamin D insufficiency is common and varies across geographical regions in Thailand. Vitamin D deficiency is a re-emerging epidemic, especially in minority populations. Vitamin D is crucial not only for bone health but for proper brain development and functioning. Low concentrations of vitamin D are associated with depression, seasonal affective disorder, and schizophrenia in adults, but little is known about vitamin D and mental health in the pediatric population (74). In a case study, 104 adolescents presenting for acute mental health treatment over a 16-month span were assessed for vitamin D status and the relationship of 25-OH vitamin D concentrations to severity of illness, defined by the presence of psychotic features. Vitamin D deficiency (25-OH-D concentrations < 20 ng/ml) was present in 34%; vitamin D insufficiency (25-OH-D concentrations 20–30 ng/ml) was present in 38%, with a remaining 28% in the normal range. Adolescents with psychotic features had lower vitamin D concentrations (20.4 ng/ml vs. 24.7 ng/ml; $P = 0.04$). The association for vitamin D deficiency and psychotic features was substantial (OR = 3.5; 95% CI: 1.4-8.9; $P < 0.009$). Race was independently associated with vitamin D deficiency and independently associated with psychosis for those who were Asian or biracial vs. white (OR = 3.8; 95% CI: 1.1 - 13.4; $P < 0.04$). Race was no longer associated with psychosis when the results were adjusted for vitamin D concentration. Vitamin D deficiency and insufficiency are both highly prevalent in adolescents with severe mental illness, indicating that vitamin D might be a mediator of illness severity, result of illness severity, or both. Higher prevalence of vitamin D deficiency but no greater risk of psychosis in African Americans, if confirmed, may have special implications for health disparity and treatment outcome research. Low vitamin D concentrations were found to correlate with major depression and premenstrual mood symptoms in women, and with mood disorders and cognitive impairment in older adults. An increased prevalence of low bone mineral density (BMD) has been reported in patients with major depressive disorder (MDD), mostly women. In a further study, 89 premenopausal women with MDD and 44 healthy control women were enrolled in a prospective study of bone turnover [75]. The prevalence of low BMD, defined as a T-score of less than -1, was greater in women with MDD vs. controls at the femoral neck (17% vs. 2%; $P = 0.02$) and total hip (15% vs. 2%; $P = 0.03$) and tended to be greater at the lumbar spine (20% vs. 9%; $P = 0.14$). The mean ± SD BMD, expressed as g/cm^2, was lower in women with MDD at the femoral neck (0.849 ± 0.121 vs. 0.866 ± 0.094; $P = 0.05$) and at the lumbar spine (1.024 ± 0.117 vs. 1.043 ± 0.092; $P = 0.05$) and tended to be lower at the radius (0.696 ± 0.049 vs. 0.710 ± 0.055; $P = 0.07$). Women with MDD had increased mean concentrations of 24-hour proinflammatory cytokines and decreased concentrations of anti-inflammatory cytokines.

Low BMD is more prevalent in premenopausal women with MDD. The BMD deficits are of clinical significance and comparable in magnitude to those resulting from established risk factors for osteoporosis, such as smoking and reduced calcium intake. The possible contribution of immune or inflammatory imbalance to low BMD in premenopausal women with MDD needs further clarification. Two randomized controlled trials (RCT) have shown that raising vitamin D concentrations improved depression [75, 76]. Vitamin D deficiency is highly prevalent in this descriptive sample of acutely mentally ill adolescents, especially in African-American and Asian teens, and appears related to psychotic symptoms. This work is the first to report an association between psychosis and vitamin D deficiency in psychiatrically hospitalized adolescents [76]. Optimal vitamin D concentrations also may protect against several different cancers (breast, colon, pancreas, and prostate), and autoimmune disorders (lupus, multiple sclerosis, and Type I diabetes). Molecular mechanisms of vitamin D that are protective against cancer include reducing cellular oxidative stress [77, 78]. The importance of vitamin D to the CNS in both healthy and psychiatric populations is less well appreciated and is vastly under-studied compared to its known impact on bone health. Vitamin D receptors are present throughout the brain, and D-deficiency is associated with negative CNS effects in animal studies [79].

Malnutrition and Consequences in Adolescence

Adolescence is the longest duration in the entire growing age in humans and encompasses several physiological changes. These include growth of long bones, development of sex characteristics and growth of fat and muscle mass [43, 44]. Adolescence is an important growth and development stage which has implications for future nutrition status and food consumption habits. During this time an adolescent's body size will grow by 15 percent. According to the general consensus, rate of growth for example: increase in height and weight among boys and girls follow a similar pattern until adolescence, with a mean increase of weight around 2 kg/year and 1.5-3 cm in height. This brings about increased demand for energy and for a number of micronutrients such as iron and calcium. Health and nutrition professionals recognize that poor eating patterns may result in nutritional problems that can impair growth and development in growing stages. Furthermore, eating habits formed in adolescence continue into adulthood: consequently, poor dietary patterns among youth have important implications for health and well-being in adulthood. [23]. In males, the relationship between nutrition and pubertal timing is less clear than in females [55]. Large numbers of girls in Asia are malnourished and reach menarche at an average age of 13.2 years [80-83]. One of the characteristic of the adolescents is the rapid growth during this period, which needs adequate energy and nutrients to support growth and development as well as to maintain a good health. Chronic malnutrition in earlier years is responsible for stunting later in life and adverse consequences at adolescence period. Stunted girls tend to become stunted during adolescence and adulthood [45]. Adolescence occupies a crucial and important place in the life of human beings whereby transition is characterized by rapid rate of growth. Current observational studies suggest notable associations between dietary intakes and pubertal timing beyond contributions to an energy imbalance: children with the highest intakes of vegetable protein or animal protein experience pubertal onset up to 7 months later or 7 months earlier, respectively [9]. The water-soluble B vitamins and several other micronutrients such as iron, magnesium, zinc, coenzyme Q10 and other minerals and antioxidants have been implicated in immunoprotection and physical performance [84-86]. Owing to these reasons, nutrition has become an important issue in Asia [87].

This review outlines the state of knowledge and research gaps in the area of determinants of healthy eating among children and youth [58]. The article is structured around individual and collective determinants that affect healthy eating in children and youth. Healthy eating may be defined as "eating practices and behaviors that are consistent with improving, maintaining and/or enhancing health. Among collective factors, familial factors and the nature of foods available in the physical environment, including at home, schools and in fast-food establishments, stand out as significant influences on healthy eating in children and youth. The media, such as television, has an enormous potential influence on healthy eating, and can overshadow familial influences. Other factors identified include knowledge, attitudes and food preferences; only the latter have been identified as a strong determinant of healthy eating in both children and adolescents. The results of the review identified a significant body of literature in the area of determinants of healthy eating in children and youth; however, very little of this research has taken place in Canada. Only a few determinants, such as economic factors and food security, the content of media, nutritional messages, and the issue of flavors, neophobia and food preferences, have undergone some examination by researchers. Apart from these factors, when to eat and what to eat are other important factors in the pathogenesis of increased risk for CVDs [59, 60].

When to Eat and What to Eat?

Having an adequate breakfast has been related to both physical and cognitive function and better performance at school in children and adolescents. Breakfast skipping has increased during the last 30 years [2]. Regular breakfast eating has been associated with a health-conscious lifestyle [28-30, 42]. In middle income countries, breakfast eating may be encouraged because eating at breakfast causes less increase in body weight and hence less risk for obesity, compared to eating the same amount of calories at dinner [59]. The critical contributions made by Halberg may have far-reaching implications whether for feeding the under-nourished in the developing populations or to populations having over-nutrition and obesity in the developed countries [59]. Eating a smaller amount of calories at dinner could be more useful to those who have malnutrition and starvation or those who want to reduce body weight. However, in obese subjects, eating breakfast and small dinners can protect without gain in body weight. It is important to advise only small breakfast containing protective foods in the developed countries, because the majority of the population is overweight and such individuals are predisposed to circadian rhythms of cardiac events which are common during 6.00 AM to 12.00 noon [60].

In brief, economic growth and increasing economic stability in Asia over the last four decades have been accompanied by changing dietary and lifestyle habits, leading to significant changes in the food and nutrition issues among all age groups facing most Asian countries. Obesity and resulting NCDs, associated with excessive consumption of total and saturated fat and refined rapidly absorbable carbohydrates, are becoming increasingly apparent among adolescents [88]. In fact, Asia has a double burden of diseases and a disproportionate share of malnutrition as well as obesity. Underweight and stunting remain significant problems in many Asian communities, and micronutrient deficiencies of iron, iodine, and vitamin A continue to afflict large population groups in the lower income subgroups. There is an urgent need for effective data collection and analysis, which are essential to formulate and implement intervention programs and strategies to address both sides of the changing nutrition scenario in Asia. Implementation of Halberg's findings into practice is likely to make an important difference in the prevention of under-nutrition as well as obesity.

Conflict of interest has not been declared by the authors.

Acknowledgments

Acknowledgment is due to International College of Nutrition and International College of Cardiology for logistic support to prepare this article.

References

[1] W.H.O. Adolescent nutrition: A neglected dimension. updated July 2006. https://apps.who.int/nut/ado.htm

[2] Facts on Adolescent Health, http://www.who.int/features/factfiles /adolescent_health/facts/en/index.html (Adolescents an age of opportunity and challenge http://www.ft.lk/2011/03/05/adolescents-an-age-of-opportunity-and challenge/statistics of south east Asia).

[3] http://www.who.int/maternal_child_adolescent/topics/adolescence/adolescent_health20 14 /en/index.html

[4] W.H.O. Adolescent nutrition: A neglected dimension. updated July 2006. https://apps.who.int/nut/ado.htm

[5] Dhillon PK, Jeemon P, Arora NK, Mathur P, Maskey M, Prabhakaran D. Sukrina RD. Status of epidemiology in the WHO South-East Asia region: burden of disease, determinants of health and epidemiological research, workforce and training capacity. *Int. J. Epidemiol.* 2012; 41: 847–860 doi:10.1093/ije/dys046

[6] Sacco RL, Smith SC, Holmes D, Shurin S, Brawley O, Cazap E, Glass E, Komajda M, Koroshetz W, Mayer-Davis E, Mbanya JC, Sledge G, Varmus H. Accelerating progress on non-communicable diseases. *Lancet* 2013; 382:e3-e5. DOI:10.1016/S0140-6736(11)61477-3

[7] Strategic directions for improving Adolescent Health in South-East Asia Region. *World Health Organization* 2011 P: 4

[8] Freedman DS, Khan LK, Dietz WH, Srinivasan SR, Berenson GS. Relationship of childhood obesity to coronary heart disease risk factors in adulthood: The Bogalusa Heart Study. *Pediatrics* 2001: 108(35); 712-718.

[9] Cheng G, Buyken AE, Shi L, Karaolis-Danckert N, Kroke A, Wudy SA, Degen GH, Remer T. Beyond overweight: nutrition as an important lifestyle factor influencing timing of puberty. *Nutr. Rev*. 2012 Mar; 70(3): 133-152.

[10] De Meester F. Progress in lipid nutrition: the Columbus concept addressing chronic diseases. *World Rev. Nutr. Diet* 2009; 100: 110-121.

[11] Marmot M, Allen J, Bell R, Bloomer E, Goldblatt P; Consortium for the European Review of Social Determinants of Health and the Health Divide. WHO European review of social determinants of health and the health divide. *The Lancet* 2012; 380: 1011-1029.

[12] Singh RB, Sharma JP, Rasogi V, Niaz MA, Ghosh S, Beegom R, Janus ED. Social class and coronary disease in a rural population of north India. *Eur. Heart J.* 1997:18;588-95.

[13] Singh RB, Beegom R, Mehta AS, Niaz MA, De AK, Mitra RK, Haque M, Verma SP, Dube GP, Siddiqui HM, Wander GS, Janus ED, Postiglion A, Haque MS. Social class, coronary risk factors and undernutrition, a double burden of diseases, in women during transition, in five Indian cities. *Int. J. Cardiol* 1999:69;139-147.

[14] Agarwal KN, Saxena A, Bansal AK, Agarwal DK. Physical growth assessment in adolescence. Indian *Pediatrics* 2001; 38: 1217-1235.

[15] Report of the Regional Meeting Chandigarh, India 17-19 September 2002: Improvement of Nutritional Status of Adolescents. *World Health Organization*, Regional Office for South-East Asia, New Delhi, December 2002

[16] Abudayy A, Thoresenb M, Yehia Abedc Y, Holmboe-Ottesena G. Overweight, stunning, and anemia are public health problems among low socioeconomic groups in school adolescents (12-15 years) in the North Gaza strip. *Nutrition Research* 2007; 27: 762-771.

[17] Story M, Stang J. Adolescent growth and development (chapter 1) in guidelines for adolescent nutrition services. 2005. http://www.epi.umn.edu/let/pubs/adol_book.shtm

[18] National Nutrition Monitoring Bureau. Diet and nutritional status of rural population. Hyderabad: National institute of Nutrition, Indian Council of Medical Research, 2002.

[19] (WHO Library Cataloguing-in-Publication Data Nutrition in adolescence: issues and challenges for the health sector: issues in adolescent health and development.)

[20] Medeiros Rataichesck Fiates G, Amboni R, Teixeira E. Television use and food choices of children: qualitative approach. *Appetite* 2008; 50: 12-18.

[21] Ayala GX, Baquero B, Arredondo EM, Campbell N, Larios S, Elder JP. Association between family variables and Mexican American children's dietary behaviors. *Journal of Nutrition Education and Behavior* 2007; 39 (2): 62-69.

[22] Nielsen SJ, Siega-Riz AM, Popkin BM. Trends in energy intake in U.S. between 1977 and 1996: similar shifts seen across age groups. *Obes Res*. 2002; 10(5): 370-378.

[23] Videon TM, Manning CK. Influences on adolescent eating patterns: the importance of family meals. *J. Adolesc Health*. 2003; 32(5): 365-373.

[24] Yannakoulia M, Karayiannis D, Terzidou M, Kokkevi A, Sidossis LS . Nutrition-related habits of Greek adolescents. *European Journal of Clinical Nutrition* (2004) 58, 580–586

[25] Neumark-Sztainer D, Wall M, Story M, Fulkerson JA. Are family meal patterns associated with disordered eating behaviors among adolescents? *J Adolesc Health*. 2004 Nov;35(5):350-9.

[26] Wu T, Stoots JM, Florence JE, Floyd MR, Snider JB, Ward RD. Eating habits among adolescents in rural Southern Appalachia. *J Adolesc Health*. 2007 Jun;40(6):577-80.

[27] Moreno LA, Kersting M, de Henauw S, González-Gross M, Sichert-Hellert W, Matthys C, Mesana MI, Ross N. How to measure dietary intake and food habits in adolescence: the European perspective. *Int. J. Obes* (Lond). 2005 Sep;29 Suppl 2:S66-77.

[28] Yannakoulia M, Karayiannis D, Terzidou M, Kokkevi A, Sidossis LS. Nutrition-related habits of Greek adolescents. *European Journal of Clinical Nutrition* 2004; 58: 580-586.

[29] Chitra U, Reddy CR. The role of breakfast in nutrient intake of urban schoolchildren. *Public Health Nutrition* 2007: 10(1): 55-58.

[30] Jenkins S, Horner SD. Barriers that influence eating behaviors in adolescents. *Journal of Pediatric Nursing* 2005; 20(4): 258-267.

[31] Singh RB, De Meester F, Wilson DW. Globalization of wealth but not heart health. *World Heart J* 2013; 4: 301-307.

[32] Pettee KK, Ainsworth BE. The Building Healthy Lifestyles Conference: modifying lifestyles to enhance physical activity, diet, and reduce cardiovascular disease. *Am. J. Lifestyle Med* 2009; 3 (1 Suppl): 6s-10s.

[33] Frassetto LA, Schloetter M, Mietus-Synder M, et al. Metabolic and physiologic improvements from consuming a paleolithic, hunter gatherer type diet. *Eur. J. Clin. Nutr.* 2009; 63(8): 947-955.

[34] Singh RB, Anjum B, Takahashi T, Martirosyan D, Pella D, De Meester F, Wilson DW. Poverty is the absolute cause of deaths due to noncommunicable diseases. *World Heart J* 2012; 4: 221-236.

[35] Editorial. Wealth but not health in USA. *Lancet* 2013; 381: 177. doi:10.1016/S0140-6736(13)60069-0

[36] Updegraff KA, McHale SM, Whiteman SD, Thayer SM, Crouter AC. The nature and correlates of Mexican-American adolescents' time with parents and peers. *Child Dev* 2006; 77 (5): 1470-1486.

[37] Raine KD. Determinants of healthy eating in Canada: an overview and synthesis. *Can. J. Public Health* 2005 Jul-Aug; 96 Suppl 3: S8-14, S8-15.

[38] Currie C, Levin K, Todd J, the HBSC national team. Health behavior: in school-aged children. Findings from the 2006 HBSC survey in Scotland. WHO 2008.

[39] Taveras EM, Rifas-Shiman SL, Berkey CS, Rockett HRH, Field AE, Frazier L, Colditz GA, Gillman MW. Family dinner and adolescent overweight. *Obesity Research* 2005; 13 (5): 900-906.

[40] Videon TM, Manning CK. Influences on adolescent eating patterns: the importance of family meals. *J. Adolesc. Health.* 2003 May; 32 (5): 365-73.

[41] Nicklas TA, Morales M, Linares A, Yang SJ, Baranowski T, De Moor C, Berenson G. Children's meal patterns have changed over a 21-year period: the Bogalusa Heart Study. *J. Am. Diet Assoc.* 2004; 104 (5): 753-761.

[42] Gillman MW, Rifas-Shiman SL, Frazier AL, Rockett HR, Camargo CA Jr, Field AE, Berkey CS, Colditz GA. Family dinner and diet quality among older children and adolescents. *Arch. Fam. Med.* 2000 Mar; 9(3): 235-240.

[43] Venkaiah K, Damayanti K, Nayak MU, Vijayaraghavan K. Diet and nutritional status of rural adolescents in India. *Eur. J. Clin. Nutr.* 2002; 56(11): 1119-1125.

[44] Aminorroaya A, Amini M, Naghdi H, Zadeh AH. Growth charts of heights and weights of male children and adolescents of Isfahan, Iran. *J. Health Popul. Nutr.* 2003 Dec;21(4):341-6.

[45] Kurniawan YAI, Muslimatun S, Achadi EL, Sastroamidjojo S. Nutritional status of adolescent girls in rural coastal area of Tangerang District. *Maj. Kedokt Indon* 2007; 57 (3): 140-145.

[46] Ahmed F, Zareen M, Khan MR, Banu CP, Haq MN, Jackson AA. Dietary pattern, nutrient intake and growth of adolescent school girls in urban Bangladesh. *Public Health Nutr.* 1998; 1(2): 83-92.

[47] Baral KP and Onta SR. Prevalence of anemia amongst adolescents in Nepal: a community based study in rural and urban areas of Morang District. *Nepal. Med. Coll. J.* 2009; 11(3): 179-182.

[48] Sinha AK, Singh GM, Karki KK. Prevalence of anemia amongst adolescents in Biratnagar, Morang Dist., Nepal. *Int. J. Pharma. & Biol. Archives* 2012; 3(5): 1077-1081.

[49] O'Dea JA. Gender, ethnicity, culture and social class influences on childhood obesity among Australian schoolchildren: implications for treatment, prevention and community education. *Health Soc. Care Community* 2008; 16 (3): 282-290.

[50] Rashidi A, Mohammadpour-Ahranjani B, Karandish M, Vafa MR, Hajifaraji M, Ansari F, Shima Sadeghi S. Obese and female adolescents skip breakfast more than their non-obese and male peers. *Central European Journal of Medicine* 2007; 2 (4): 481-487.

[51] Neumark-Sztainer D, Hannan PJ, Story M, Croll J, Perry C. Family meal patterns: associations with sociodemographic characteristics and improved dietary intake among adolescents. *J. Am. Diet Assoc.* 2003; 103 (3): 317-322.

[52] Keski-Rahkonen A, Kaprio J, Rissanen A, Virkkunen M, Rose RJ. Breakfast skipping and health-compromising behaviors in adolescents and adults. *Eur. J. Clin. Nutr.* 2003 Jul;57(7):842-53.

[53] Sjöberg A, Hallberg L, Höglund D, Hulthén L. Meal pattern, food choice, nutrient intake and lifestyle factors in The Göteborg Adolescence Study. *Eur. J. Clin. Nutr.* 2003 Dec;57(12):1569-78.

[54] Ma YS, Bertone ER, Stanek EJ, Reed GW, Herbert JR, Cohen NL, Merriam PA, Ockene IS. Association between eating patterns and obesity in a free living US adult population. *Am. J. Epidemiol.* 2003; 158: 85-92.

[55] Lee JM, Kaciroti N, Appugliese D, Corwyn RF, Bradley RH, Lumeng JC. Body mass index and timing of pubertal initiation in boys. *Arch Pediatr Adolesc* Med. 2010; 164(2): 139-144. doi: 10.1001/archpediatrics.2009.258.

[56] Al Sabbah H, Vereecken C, Abdeen Z, Coats E, Maes L. Associations of overweight and of weight dissatisfaction among Palestinian adolescents: findings from the national study of Palestinian schoolchildren (HBSC-WBG2004). *J. Hum. Nutr. Diet.* 2009 Feb; 22(1): 40-49. doi: 10.1111/j.1365-277X.2008.00901.x. Epub 2008 Aug 27.

[57] Wilson TA, Adolph AL, Butte NF. Nutrient adequacy and diet quality in non-overweight and overweight Hispanic children of low socioeconomic status: the Viva la Familia Study. *J. Am. Diet Assoc.* 2009 Jun; 109(6):1012-21. doi: 10.1016/j.jada.2009.03.007.

[58] Taylor JP, Evers S, McKenna M. Determinants of healthy eating in children and youth. *Can. J. Public Health.* 2005 Jul-Aug; 96 Suppl 3: S20-26, S22-29.

[59] Cornelissen G. When you eat matters: 60 years of Franz Halberg's nutrition chronomics. *The Open Nutr. J.* 2012; 4: 16-44.

[60] Singh RB, Singh AK, Sharma JP, Singh RK, Kumar A, Rastogi SS, Singh Garima, Kartikey K, Singh S, Fedacko J, Pella D, De Meester F, Wilczynska A, Wilson DW. Nutrition in chronocardiology: we are indebted Professor Franz Halberg. *The Open Nutra J* 2012; 45-65.

[61] Sinha AK, Singh GM, Karki Karna KK. Prevalence of anemia amongst adolescents in Biratnagar, Morang Dist. Nepal. *Int. J. of Pharma. & Biological Archives* 2012; 3: 1077-1081

[62] Prevention of deficiency anaemia in adolescents. Role of weekly iron and folic acid supplementation prevention of iron deficiency anaemia in adolescents. WHO report. *Regional office for South-East Asia.* 2011 P 2-4.

[63] National Economic and Social Development Board. The eighth national economic and social development plan (1997-2001). Bangkok, 2001.

[64] Muzammil K, Kishore S, Semwal J. Common nutritional deficiencies of adolescents in Dehradun. *Indian J. Sci. Res.* 2010; 1: 77-80.

[65] Ahmed F, Khan MR, Banu CP, Qazi MR, Akhtaruzzaman M. The coexistence of other micronutrient deficiencies in anaemic adolescent schoolgirls in rural Bangladesh. *Eur. J. Clin. Nutr.* 2008; 62: 365-372.

[66] Singh RB, Rastogi V, Niaz MA, Sharma JP. Epidemiologic study of magnesium status and risk of hypertension in a rural population of north India. *Mag Res* 1996; 9: 173-181.

[67] Singh RB, Gupta UC, Mittal N, Niaz MA, Ghosh S, Rastogi V. Epidemiologic study of trace elements and magnesium and risk of coronary artery disease in the rural and urban populations. *J. Am. Coll. Nutr.* 1997; 16:62-67.

[68] Singh RB, Rastogi SS, Ghosh S, Niaz MA et al, Dietary intakes, plasma levels of antioxidant vitamins, and free radical damage, in relation to coronary artery disease in the elderly subjects. *Amer. J. Cardiol.* 1995,76:1233-38.

[69] West KP Jr. Extent of vitamin A deficiency among pre-school children and women of reproductive age. *J. Nutr.* 2002; 132, 2857S-2866S.

[70] Lu L, Yu Z, Pan A, Hu FB, Franco OH, Li H, et al. Plasma 25-hydroxyvitamin D concentration and metabolic syndrome among middle-aged and elderly Chinese individuals. *Diabetes Care* 2009; 32: 1278-1283.

[71] Choi HS, Oh HJ, Choi H, Choi WH, Kim JG, Kim KM, et al. Vitamin D insufficiency in Korea -- a greater threat to younger generation: the Korea National Health and Nutrition Examination Survey (KNHANES) 2008. *J. Clin. Endocrinol. Metab.* 2011; 96: 643-651.

[72] Mithal A, Wahl DA, Bonjour JP, Burckhardt P, Dawson-Hughes B, Eisman JA, et al.; IOF Committee of Scientific Advisors (CSA) Nutrition Working Group. Global vitamin D status and determinants of hypovitaminosis D. *Osteoporosis Int* 2009; 20: 1807-1820

[73] Chailurkit LO, Aekplakorn W, Ongphiphadhanakul B. Regional variation and determinants of vitamin D status in sunshine-abundant Thailand. *BMC Public Health* 2011; 11: 853.

[74] Gracious BL, Finucane TL, Campbell-Friedman M, Parkhurst MN, Messing S: 25-OH vitamin D deficiency associated with psychotic features in acutely mentally ill adolescents. In: *New Research Poster 3.3, 57th Annual Meeting, American Academy of Child and Adolescent Psychiatry.* New York, NY; 2010.

[75] Eskandari F, Martinez PE, Torvik S, Phillips TM, Sternberg EM, Mistry S, Ronsaville D, Wesley R, Toomey C, Sebring NG, Reynolds JC, Blackman MR, Calis KA, Gold PW, Cizza G; Premenopausal, Osteoporosis Women, Alendronate, Depression (POWER) Study Group. Low bone mass in premenopausal women with depression. *Arch Intern Med* 2007; 167 (21): 2329-2336.

[76] Wilkins C, Sheline Y, Roe C, Birge S, Morris J: Vitamin D deficiency is associated with low mood and worse cognitive performance in older adults. *Am. J. Geriatr Psychiatr* 2006, 14: 1032-1040.

[77] Garland C, Garland F, Gorham E, Lipkin M, Newmark H, Mohr S. The role of vitamin D in cancer prevention. *Am. J. Public Health* 2006; 96: 252-261.

[78] Bao BY, Ting HJ, Hsu JW, Lee YF: Protective role of 1α, 25- dihydroxyvitamin D3 against oxidative stress in nonmalignant human prostate epithelial cells. *Int. J. Cancer* 2008, 122: 2699-2706.

[79] McCann J, Ames B. Is there convincing biological or behavioral evidence linking vitamin D deficiency to brain dysfunction? *FASEB J.* 2008, 22: 982-1001.

[80] Lytle LA. Nutritional issues for adolescents. *J. Am. Diet Assoc.* 2002; 102 (Suppl 3): S8-12.

[81] Dicks L. The original rebels. *New Sci.* 2005; 185: 39-42.

[82] Kaplowitz PB. Link between body fat and the timing of puberty. *Pediatrics* 2008; 121 Suppl 3: 208-217.

[83] Towards adulthood: exploring the sexual and reproductive health of adolescents in South Asia/edited by Sarah Bott et al. WHO 2003.

[84] Williams MH. Dietary supplements and sports performance: minerals. *J. Int. Soc. Sports Nutr* 2005; 2: 43-49.

[85] Lukaski HC. Vitamin and mineral status: effects on physical performance. *Nutrition* 2004; 20: 632-644.

[86] Kreider RB, Almada AL, Antonio J, Broeder C, Earnest C, Greenwood M, Incledon T, Kalman DS, Kleiner SM, Leutholtz B, et al. ISSN exercise and sport nutrition review: research and recommendations. *J. Int. Soc. Sports Nutr.* 2004; 1: 1-44.

[87] Tee ES. Priority nutritional concerns in Asia. The United Nations University, *Food and Nutrition Bulletin,* vol. 23, No. 4 2002/

[88] Dans A, Ng Nawi, Varghese C, Tai ES, Firestone R, Bonita R. Health in South East Asia. The rise in chronic non-communicable diseases in Southeast Asia: time for action. *Lancet* 2011; published online Jan 25, DOI 10.1016/SO140.6736(10)61506-1.

In: Circadian Cardiology with Focus on Both Prevention ... ISBN: 978-1-63463-956-9
Editors: K. Hristova, A. Shehab, G. Cornelissen et al.

Chapter 8

Tides at Solid Earth: Biomedical Effects (An "Anatomy" of Tidal Waves)

Miroslav Mikulecký**[*]**
BIOCOS Group, University of Minnesota, Minneapolis, MN, US
and Department of Statistics and Biometry,
Neuroendocrinology Letters, Stockholm, Sweden

Dedication

This article is dedicated to the Blessed Memory of Professor Franz Halberg, MD, multi Dr.h.c., the father of chronobiology and circadian rhythms, the generous donor of his great wisdom and heart to all without respecting any borders, from India to Japan, China, Russia and anywhere in the world.

Abstract

Background. Lunisolar tides in the Earth crust run each second at any place on dry land. They not only move buildings up and down but also exert a so-far neglected influence on living matter. One recognition of that came from Gunter Klein, who explained the timing of the beginning of sudden nastic movements of certain plants' leaves. This has been a puzzle for botanists since the time of Alexander the Great: they start during the turning points of lunisolar gravity, i.e., with the highest or lowest tide.

Hypothesis. The present author, reviewer of Klein's original booklet (Springer USA), is offering and testing the hypothesis that Klein constituted a basis for a *general gravi-biologic law* which should be valid for any biomedical sudden event. *Methods.* The present contribution aims at testing this hypothesis on the starting times of one hundred

[*] Correspondence as well requests for data and other attachments should be addressed to the author's e-mail (statistics@nel.edu). Miroslav Mikulecký Sr. Studenohorska 10, SK 84103 Bratislava, Slovak Republic; Bratislava. Postal address: Studenohorska 10, SK 84103 Bratislava, Slovak Republic; Email: statistics@nel.edu.

tachyarrhythmia attacks, suffered by the present author. In contrast to Klein's original study, the statistical uncertainty of the studied phenomenon is also taken into account. The input data include time of the start of the attack of tachyarrhythmia up to minutes and geographic coordinates of the place. The data were evaluated on the basis of graphs of lunisolar gravity (uGal versus time) for the given location and time: the position of the attack on the tidal wave was expressed as relative wave height (0=bottom, 1=peak, with the scale from 0, 0.01, 0.02, 0.03 etc up to 0.97. 0.98, 0.99 and 1). All obtained hundred values were divided into 8 subgroups: the top, middle top, middle bottom and bottom, with each group being divided into subgroup at the ascending and descending side of the wave. The distribution of all such 100 values into these 8 subgroups was evaluated by chi-square test. Its significant outcome ($p < 0.0005$) means a significant heterogeneity. Similar testing showed a nonsignificant difference between the increasing and decreasing part of the wave. The overall heterogeneity is therefore explained only by significant accumulation of the cases at the bottom and top quarters of the tidal wave. The hypothesis is therefore upheld. Results. The author's tachyarrhythmia attacks happened significantly more often at the bottoms or peaks of the local lunisolar tidal waves. Conclusion. Similar effects of lunisolar gravity can be expected for many other biomedical events in plants, animals and man. Preliminary experience confirms so far this idea for human births and deaths. A new gravi-biologic law may emerge on the horizon.

Keywords: Tachyarrhythmia; attacks' timing; lunisolar gravity; tidal wave

General Perspective

More than 300 years ago, the English epidemiologist Richard Mead published a treatise *De Imperio Solis ac Lunae in Corpora humana, & Morbis inde oriundis* (On the Influence of the Sun and Moon upon Human Bodies and the Diseases Arising Therefrom; *en.wikipedia.org/wiki/Richard_Mead*).

The Hypothesis

It has recently been shown [1,2] and corroborated by data extension [3] that the nastic movements of bean leaves usually start suddenly in turning points of lunisolar gravity, when the latter culminates in acme or reaches its minimum in nadir. This Klein's „*turning point effect*“ could reveal a general rule, valid for living matter at the solid Earth: biomedical events would then tend to accumulate near the turning points -- acmes and nadirs -- of the lunisolar gravity waves.

Data

To test this hypothesis in man, the times of onset of one hundred attacks of atrial tachyfibrillation, suffered by the present author between December 11, 1980 and November 4, 1992, were processed. Their pathophysiologic basis is obviously connected with the partial (Wilson) right bundle branch block, acquired due to scarlatina in the subject's age of 8 years, in 1935. For each attack, the time of its start (year, month, day, hour, minute) was registered together with the geographic width, length and oversea height of the place where it happened. From these data, the actual curve of the dependence of the lunisolar tidal acceleration on time-- a *selenoheliogravigram* -- was constructed [4, 5]. An example is shown in Figure 1.

To quantify the position of an attack onset on the tidal wave, the relative height (*rh*) on the wave was considered separately for attacks in its increasing and decreasing arm. The wave bottom (nadir) corresponds with *rh* = 0, its top (acme) with *rh* = 1. 101 *rh* values will be evaluated from 0, 0.01, 0.02, ... , 0.98, 0.99, 1, with the corresponding values *n* (total *n* = 100) of attacks for each of the separate values of *rh*. All 100 *n vs. rh* values are given in the Table 1.

In Figure 2, an ideal wave was „anatomically" divided into bottom (B), middle bottom (Mb), middle top (Mt) and Top (T) segments, each being divided into increasing (j) and decreasing (d) subsegments, with corresponding numbers of attacks in each subsegment.

Methods

At first, the homogeneity of the counts in 8 subsegments was tested using χ^2 distribution (Figure 2). Significant (p<0.0005) heterogeneity was found.The difference of the counts of attacks during increasing and decreasing tide was tested, again using a χ^2 test of the difference between two proportions (44/82 versus 38/82). As this difference was nonsignificant, the overall heterogeneity is due only to counts' differences between the cells B versus Mb and between T versus Mt. The significance level of testing was set at α=0.05.

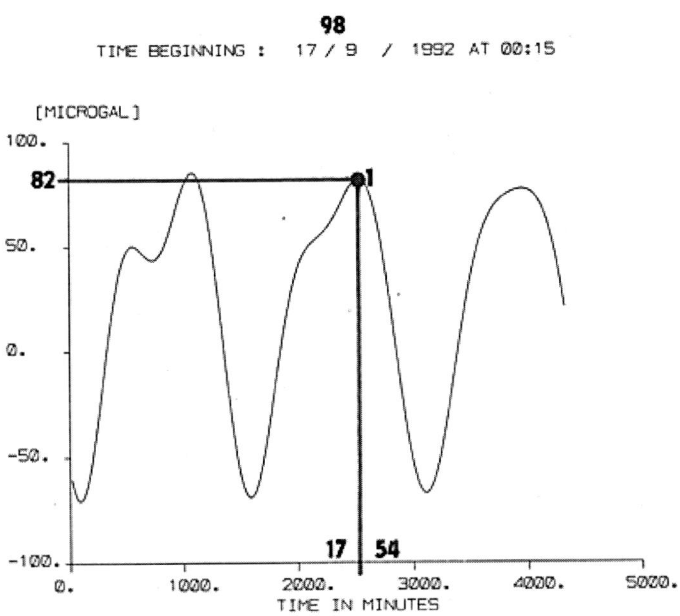

Figure 1. An attack location on the tidal wave. Attack # 98 exactly at the lunisolar gravity acme (rh = 1) started on 18 September 1982, 17:54 h in Bratislava (48°09' Northern latitude, 17°06' Eastern longitude, 130 m over sea).

Results

Accordingly, the attacks are significantly accumulated in the bottom and top segments, in agreement with the hypothesis, upheld now on the level of statistical significance $\alpha = 0.05$.

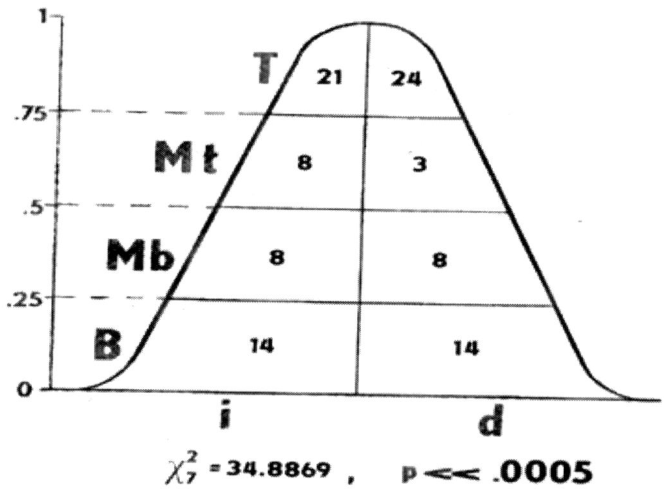

Figure 2. Schematic tidal wave divided into eight cells with the counts at the bottom (B), middle bottom (Mb), middle top (Mt) and top (T), each subdivided into increasing (i) and decreasing (d) arm. The significance of the heterogeneity of the counts in eight cells is shown below as $p<<0.0005$.

Table 1. Lunisolar tides and cardiac tachyarrhythmia attacks in man: observed data of the attack #1 to 100:
rh **= relative height of the tidal wave**
n **= number of cases given as ranking numbers of the attacks and as their sum**
+ = the ascending arm of the tidal wave
- = the descending arm of the tidal wave

rh	Ranking numbers of separate attacks	Number n of attacks
0.	4, 17, 19, 91	4
.01	15+, 92+	2
.02	20+, 54-	2
.03	-	-
.04	33+, 43-, 44-	3
.05	-	-
.06	52-, 76-	2
.07	-	-
.08	86-, 87-	2
.09	24+, 65+	2
.10	21+, 37+	2
.11	-	-
.12	82+	1
.13	-	-
.14	-	-
.15	93-, 99-	2

rh	Ranking numbers of separate attacks	Number n of attacks
.16	22+	1
.17	41+	1
.18	64-	1
.19	-	-
.20	-	-
.21	96+	1
.22	26+	1
.23	-	-
.24	59-	1
.25	60+	1
.26	-	-
.27	-	-
.28	-	-
.29	-	-
.30	10-, 69+	2
.31	-	-
.32	-	-
.33	-	-
.34	12+, 57-	2
.35	31-, 79-	2
.36	-	-
.37	-	-
.38	23-, 78-	2
.39	-	-
.40	11+	1
.41	-	-
.42	-	-
.43	-	-
.44	3+, 81+	2
.45	66-, 95+	2
.46	-	-
.47	18-	1
.48	67+	1
.49	-	-
.50	5-	1
.51	7+	1
.52	-	-
.53	-	-
.54	-	-
.55	-	-
.56	-	-
.57	-	-
.58	-	-
.59	34+	1
.60	-	-
.61	16-	1
.62	-	-
.63	-	-

Table 1. (Continued)

rh	Ranking numbers of separate attacks	Number n of attacks
.64	42+, 63-	2
.65	-	-
.66	39+	1
.67	-	-
.68	1+, 29+	2
.69	-	-
.70	-	-
.71	80+, 100+	2
.72	-	-
.73	-	
.74	-	
.75	-	-
.76	8+	1
.77	88+	1
.78	-	-
.79	-	-
.80	-	-
.81	14+	1
.82	74-	1
.83	47+	1
.84	-	-
.85	-	-
.86	28-, 56-	2
.87	-	-
.88	83-	1
.89	51+, 61+	2
.90	45-, 53+, 71+	3
.91	6-, 9-, 85-	3
.92	48-	1
.93	-	-
.94	36+, 77-	2
.95	73-, 84+, 94-	3
.96	13+, 25-	2
.97	30-, 55+, 70+	3
.98	68+	1
.99	2-, 62+, 75-	3
1. 27, 32, 35, 38, 40, 46, 49, 50, 58, 72, 84, 90,47, 98		14
		100

Summing up, there are 82 attacks belonging to 99 tidal wave places .01 to .99 and 18 attacks belonging to the places 0 and 1, i. e. to the bottom and to the peak of the wave. The difference 82/99 versus 18/2 is highly statistically significant.

Discussion

The tidal turning points were authentically, although without statistical confirmation, found (1, 2) to coincide with the start of nastic movements on the basis of Dr. Gunther Klein's long-lasting field observations and experiments with suddenly starting nastic movements of Phaseolus leaves -- a scientific question-mark for botanists since the times of Alexander the Great. Klein's published materials were extended and processed from the point of view of either turning point (3): from 22 turning points of lunar tidal force F, the exact coincidence with the turning points in leaf movement T happened, rigorously evaluated, in 12 of them, i.e., in 55%. This discovery was surprisingly exactly confirmed on older similar observed data [6], completed recently by lunisolar gravity curves [7]: this rediscovery on original Figure 1A, 1C, 1E, 1G and 2A shows that from altogether 37 turning points of lunisolar gravity, at least 11, i.e., 30%, does coincide exactly, rigorously evaluated again, with the turning points of leaves T. In our tachyarrhythmia, also after very rigorous evaluation, the exact coincidence (as e.g., on Figure 1) was found in 18 cases from 100 (18%).

Conclusion

The connection between turning points in the time course of lunisolar gravitation on one side and a biomedical event on the other was disclosed in the present contribution for human heart. There will be perhaps more sudden events in organisms living on the solid Earth -- plants or animals and man -- found „lunar connected" as to the time of their starting. This author has accumulated already some more encouraging experience with such biological sudden events as births or deaths. A Klein-Mikulecký's law of tidal effects on life at the dry Earth, beginning to appear on the horizon. may be analogous to that very well known and studied phenomena at seashores.

The author declares no competing financial interests.

References

[1] Klein, G. *Farewell to the Internal Clock. A Contribution in the Field of Chronobiology.* Springer, New York, USA. 2006: xxiii + 116 pp.

[2] Mikulecký, M Sen. *Farewell to the Internal Clock. A Contribution in the Field of Chronobiology.* Opinion. Biologia 2009; 64: 225-227.

[3] Barlow, P.W, Klingelé, E., +Klein, G., Mikulecký, M. sen. Leaf movements of bean plants and lunar gravity. *Plant Signaling & Behavior* 2008; 3: 1083-1090.

[4] Longman, M. Formulas for computing the tidal acceleration due to the Moon and the Sun. *J. Geophys. Res.* 1959; 64: 2351-2355.

[5] Klingelé, E. Computer programme „Etide" for computing the tidal acceleration. Eidgenossische Technische Hochschule Zuerich. (2012)..

[6] Kleinhoonte, A. Untersuchungen ueber die autonomen Bewegungen der Primärblätter von Canavalia ensiformis DC. Jahrbuecher fuer wissenschaftliche Botanik 1932; 75 : 679-725.

[7] Barlow, PW. Foreword. pp. vii-xx in: Klein G. *Farewell to the Internal Clock. A Contribution in the Field of Chronobiology.* Springer, New York, USA 2006: xxiii + 116 pp.

In: Circadian Cardiology with Focus on Both Prevention ... ISBN: 978-1-63463-956-9
Editors: K. Hristova, A. Shehab, G. Cornelissen et al. © 2015 Nova Science Publishers, Inc.

Chapter 9

Circadian Disruption and Vascular Variability Disorders (VVD): Mechanisms Linking Aging, Disease State and Arctic Shift-Work: Applications for Chronotherapy

In memory of Franz Halberg, the father of chronobiology: 1919-2013

D. G. Gubin*[1], G. Cornelissen[2], D. Weinert[3], A. S. Vetoshkin[4], L. I. Gapon[5], N. P. Shurkevich[5], F. A. Poshinov[5], N. V. Belozerova[5] and L. A. Danilova[1]

[1]Department of Biology, Medical Academy, Tyumen, Russia
[2]Halberg Chronobiology Center, University of Minnesota, Minneapolis, MN, US
[3]Institute of Zoology, Martin-Luther-University, Halle, Germany
[4]Medical Unit, "Gazprom Dobycha Yamburg" LLC, Yamburg, Russia
[5]Tyumen Cardiology Center – Branch of Research Institute of Cardiology, Siberian Branch of the Russian Academy of Medical Sciences, Tyumen, Russia

* Address for Correspondence: D. Gubin. Tyumen Medical Academy. Odesskaya, 52. 625023 Tyumen. dgubin@mail.ru.

Abstract

A fast growing body of evidence indicates that the circadian system is important for health. In turn, desynchrony has been associated with the development of numerous age-related diseases and the aging process. Alterations of the circadian variation occurring within the physiological range provide warning signs that, if acted upon, lead to truly primary prevention not across the board but targeted to the individual person. This is the case for instance for Vascular Variability Disorders. Particular attention is given to blood pressure and heart rate recorded in the elderly, in shift-workers in the Arctic, and to a personalized chronotherapeutic approach.

Keywords: Circadian rhythms, variability, heart rate, blood pressure, aging, hypertension, Arctic, chronotherapy, melatonin

Introduction

"For our body is like a clock; if one wheel be amiss, all the rest are disordered, the whole fabric suffers: with such admirable art and harmony is a man composed"
Robert Burton (1621)

"Just as you prescribe a dose for every drug, you must prescribe a timing for every drug"
Franz Halberg (1998)

Our organism is a multi-oscillatory complex system. Every single cell possesses the necessary genetic machinery to maintain circadian rhythms, yet needs to be synchronized and orchestrated with environmental time clues in order to fit evolutionary demands to predict and prepare for, rather than to respond to life events, a situation that prompted the partly endogenous nature of circadian rhythms. Recently we have learned much about the molecular mechanisms of circadian rhythms in our cells. Appreciation of the role of peripheral oscillators came with more knowledge. On the other hand, numerous overt physiologic rhythms, including clinically valuable ones, are dynamic functions that are coordinated by a vast net of mechanisms. These mechanisms themselves are quite often also dependent on oscillating models of the substrate -- enzyme -- receptor in a humoral framework, let alone in neuronal coordination. In such a complex multi-oscillatory system, the probability of internal desynchronization (circadian disruption) is high and may occur should single or more oscillators run out of phase with each other or abruptly change their amplitude [1]. This also means that numerous changes in variability of overt physiologic functions are inevitable and may have clinical relevance and emerge as Vascular Variability Anomalies (VVAs) or Vascular Variability Disorders (VVDs). Indeed, multiple ultradian and broader infradian rhythms are quite often demonstrated [2-4], though no obvious explanations for their evolutionary benefits are available. There are several cosmic and geomagnetic factors that may drive such rhythmicity; which may be clinically meaningful.

Not many dynamic functions in our organism, however, can be measured often and long enough to assess ultradian and infradian variability, and to look at interrelations among circadian rhythm parameters with adjacent spectral harmonics. Systolic (S) and diastolic (D)

blood pressure (BP) and heart rate (HR) are among the few variables whose dynamic analysis became routine after ambulatory blood pressure monitoring (ABPM) was introduced into clinical practice.

Analyses of dynamic variability in cardiovascular functions performed on databases of ABPM records usually reveal several rhythms with different frequencies. The relative spectral amplitudes and changes in phase stability under different physiologic conditions throughout ontogeny may help discriminate between health and certain diseases. The percentage rhythm (PR), the proportion of overall variance accounted for by a given periodic component fitted to the data, is a quite useful index to compare the relative prominence of this component among different variables that have been sampled in the same way (in terms of the length of observation, T, and the interval between consecutive measurements, Δt). Its advantage over the amplitude is that the PR does not depend on the variable's scale that restricts comparison of particular rhythms among different functions.

Herein, we suggest linking mechanisms that account for VVAs and VVDs that occur with increasing age, the presence of disease, and emerging under unique environmental factors in the Arctic. We study shift-workers engaged in monthly rotations to the station of Yamburg (68°N) and back to their home cities in the Russian Federation. We report results from several studies, united by their aim to show chronome modifications (frequency transformation, trends, noise) in cardiovascular dynamics under somewhat compromised circumstances (aging, loads of Arctic climate, disease state). We present a personalized approach to chronotherapy and show its benefits for health.

General Principles of Chronome Modifications with Age

In this first study, we sought general principles of age-dependent changes in the chronome (time structure) of cardiovascular variables and temperature, focusing on components in the middle frequency spectral domain extending from one cycle per week to 8 cycles per day. This frequency range covers ultradians with a period (τ) between 3 and 20 hours, circadians (τ of 24±4 hours), and infradians with a τ between 28 and 168 hours. Data were obtained by 7-day ABPM of 80 presumably normotensive subjects, 16-106 years of age, and temperature self-measurements from 335 clinically healthy subjects, 17-91 years of age, collected each day around 03:00, 08:00, 11:00, 14:00, 17:00, and 23:00 for 7 days. Participants providing an ABPM record were subdivided into 4 age groups: young adults (under 40 years, 28.5±1.5 years); adults (40-59 years, 46.8±0.8); seniors (60-74 years, 66.5±1.0); and elderly (over 75 years, 86.6±4.6). Participants providing temperature self-measurements were subdivided into the same 4 groups (young adults, 21.3±0.6 years; adults, 45.3±0.7 years; seniors, 67.0±0.6 years; elderly, 78.6±1.2 years). From SBP, DBP and HR measured by ABPM, mean arterial pressure (MAP), pulse pressure (PP), systolic volume (SV), minute blood flow (MBF) and peripheral vascular resistance (VR) were calculated for each sampling time.

Apart from expected age-dependent trends in MESOR and increased irregular variability, the most typical attribute of aging was a variance transposition from circadian to extracircadian frequencies: an increase in the amplitude and relative amplitude of ultradian and infradian components accompanied the circadian disruption. We defined this consistent pattern as ECD (for Extra-Circadian Dissemination) [5-9]. ECD was further classified into 4 types in terms of modifications in the (1) overall variability, gauged by standard deviation of the mean (SD), and in the (2) circadian and (3) pooled extracircadian variability, gauged by the PR and amplitude of the respective spectral components. The 4 types are: *I. True hypervariable ECD:* upward trends in certain frequencies of ultradian and infradian variability prevailing over downward trends in circadian amplitude, leading to an increase in overall variability with age (found for SBP, PP and VR); *II. True euvariable ECD:* upward trends in ultradian and infradian variability are counterbalanced by proportional downward trends in the circadian amplitude, resulting in no significant trend with age in overall variability (DBP, MAP, SV); *III. True hypovariable ECD:* downward trend in the circadian amplitude prevails over upward trends in ultradian and infradian variability, leading to a decrease in overall variability with age (temperature); *IV. Relative ECD:* downward trends in all frequency domains are found, predominantly in the circadian amplitude and PR, leading to a drastic decrease with age in overall variability (HR, MBF).

Changes in certain chronobiologic parameters of physiologic variables are heterochronous: a circadian phase advance (of about 90-120 min) develops earlier and was found to be typical for an age group of 60-74 years and, in some variables (SBP, DBP, MAP, VR) was then followed by either cycle-to-cycle and inter-individual phase scattering, or both, in groups of subjects older than 75 years. However, an even earlier decrease of cycle-to-cycle circadian phase stability can be found in some individuals (in an age group of 40-59 years, compared to an age group of young adults less than 40 years); such phase scattering is accompanied by a relative increase in particular harmonics of infradian spectra. Loss of cycle-to-cycle circadian phase stability of cardiovascular variables is a remarkable feature that can only be discerned in longitudinal surveys. Increase of ultradian variability emerges later, after 60 years of age. A noteworthy circadian amplitude decrease in separately analyzed 24-hour time series is the latest feature to develop and can be found only in some individuals after 75 years of age, also being not typical for every cardiovascular variable studied. Thus, in general, variability modification in spectral components of the cardiovascular system precedes age-dependent changes in circadian rhythm parameters per se, so such changes cannot be found unless longitudinal records are available and used for analysis (Figure 1). The most prominent and persistent modifications of cardiovascular spectra in terms of frequency transformation with age are an increased prominence of the 12-hour (circasemidian) component and an increase in 12-hour-to-24-hour amplitude ratio (Table 1). Other age-dependent changes found by us were previously described and published elsewhere [10-12; see 9 for review].

Vascular Variability Anomalies (VVA) and Disorders (VVD)

Some general principles underlying chronome alterations found in aging also apply in the development of cardiovascular diseases. Specific alterations of the circadian patterns of BP and HR have been referred to as VVAs and, if they persist, as VVDs [13]. These VVAs (VVDs), usually not routinely screened for in conventional practice, are associated with an increased cardiovascular disease risk [13]. To detect such VVAs (VVDs), around-the-clock monitoring for at least 7 days has been recommended [13-15]. Some (but not all) VVAs, such as MESOR-hypertension, Excessive PP (EPP, when PP>60 mmHg), ecphasia (odd timing of the circadian rhythm of BP but not of HR), ecfrequentia (statistically significant deviation from 24 hours of the circadian period of BP and/or HR), Deficient HR Variability (DHRV, when the standard deviation of HR is below 7.5 beats/min) [16] are similar to those seen during the aging process.

Table 1. Some age-dependent correlations in cardiovascular and body temperature variability

SBP	r*	P**
MESOR	0.582	<0.001
SD	0.389	<0.001
A12/A24	0.347	0.002
PR	-0.434	<0.001
DBP	r*	P**
MESOR	0.515	<0.001
SD	0.046	0.688
A12/A24	0.310	0.005
PR	-0.462	<0.001
MAP	r*	P**
MESOR	0.586	<0.001
SD	0.189	0.096
A12/A24	0.359	0.001
PR	-0.449	<0.001
PBP	r*	P**
MESOR	0.485	<0.001
SD	0.440	<0.001
A12/A24	0.225	0.043
PR	0.056	0.612
HR	r*	P**
MESOR	-0.147	0.194
SD	-0.565	<0.001
A12/A24	0.112	0.325
PR	-0.323	0.004
SV	r*	P**
MESOR	-0.938	<0.001
SD	0.163	0.152
A12/A24	0.296	0.008
PR	-0.286	0.011
MBF	r*	P**
MESOR	-0.914	<0.001
SD	-0.691	<0.001
A12/A24	0.299	0.007
PR	-0.530	<0.001
Temperature	r*	P**
MESOR	-0.189	0.004
SD	-0.210	0.001
A12/A24	0.282	<0.001
PR	-0.250	<0.001

Notes: A12/A24 – 12-hour to 24-hour component amplitude ratio; SD – standard deviation of 168-ABPM mean as measure of overall intra-individual variability; PR – 24-hour percentage rhythm calculated on 168-hour long time-series;*r – Pearson's correlation coefficient with age.**P – P-value for age-effect.

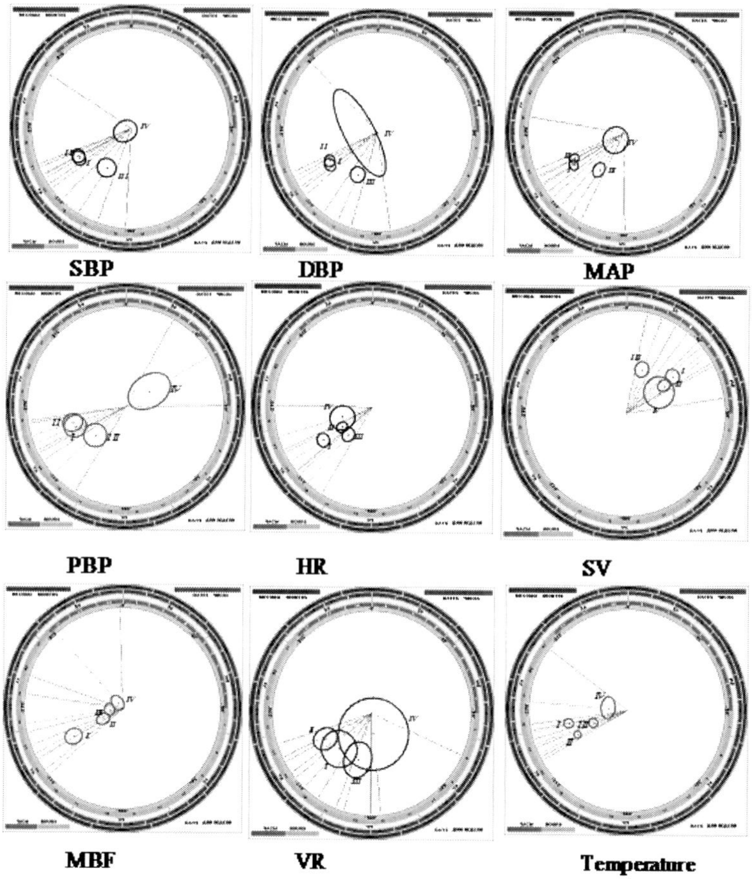

Figure 1. Population-mean cosinor analysis reveals phase advance (in group III, 60-74 years of age) and later on phase scatter of circadian acrophase, accompanied by reduction in circadian amplitude is typical for cardiovascular variables and temperature with age. Note that in group IV (>75 years of age) synchronized circadian pattern of numerous cardiovascular variables is disrupted so that significant rhythm is no more detected.

Of course, the likelihood of morbidity and mortality also increases with age! For this reason, VVAs (VVDs) are defined, whenever possible, in the light of reference values qualified by gender and age.

In our comparison of 24-hour ABPM data from normotensive (N=27, age: 40.7±1.5 years) vs. mild hypertensive (N=37, age: 39.4±1.2) male adults, some VVAs similar to that of elderly subjects has been found [17, 18]. A tendency for a true hypervariable ECD of DBP, MAP and SV (increased ultradian-to-circadian variability) has been found in hypertensive patients [17, 19, 20]. Increased individual variability of circadian rhythm characteristics of VR has led to the loss of a statistically significant circadian rhythm on a group basis, assessed by means of the population-mean cosinor. In hypertensive patients, the 24-hour acrophase of PP was delayed relatively to that of SBP, DBP and MAP which remained synchronized. Signs of intrinsic desynchronization among certain hemodynamic functions are thus already manifested in a mild hypertensive disease state. Negative correlations have also been found between increases in SBP MESOR (P=0.026) and DBP MESOR (P=0.002) and decreases in their PR, accompanied by a negative correlation between the SBP MESOR and the SV PR (P=0.036) [17].

Vascular Variability and Its Alterations in Normotensive and Hypertensive Arctic Shift-Workers

Photoperiodic and geomagnetic features of the Arctic are also factors of compromised circadian variation, especially for subjects engaged in monthly shift-work rotations. Not only in hypertensive but also in normotensive populations, numerous novel BP and HR variability patterns, not typical for residents of lower latitudes, are often found [18, 21-24].

Circadian profiles and chronobiologic parameters of BP and HR have been prospectively analyzed in 373 hypertensive and 188 presumably normotensive male shift-workers in the Arctic (Tyumen Far North, Yamburg, 68°N), who provided 5 to 6 24-hour ABPM records in consecutive seasons. All subjects were engaged in monthly shift-work rotations (1 month at work, 1 month back in their home city). Vascular variability in both normotensive and hypertensive Arctic shift-workers differs markedly from that of residents of lower latitudes. Elevated MESORs of DBP and HR with no increase in SBP-MESOR are in keeping with a consistently lowered PP (Table 2 and Figure 2). Drastic changes in circadian rhythm parameters and overall variability (diminished 24-hour amplitude and percentage rhythm, PR; increased overall and ultradian variability, elevated nocturnal values, scattered acrophases) have been found in Arctic shift-workers by comparison with peers in Tyumen (57° N). Such changes were most prominent in HR, which has a more solid endogenous circadian component that is not much influenced by hypertension per se, as compared to BP.

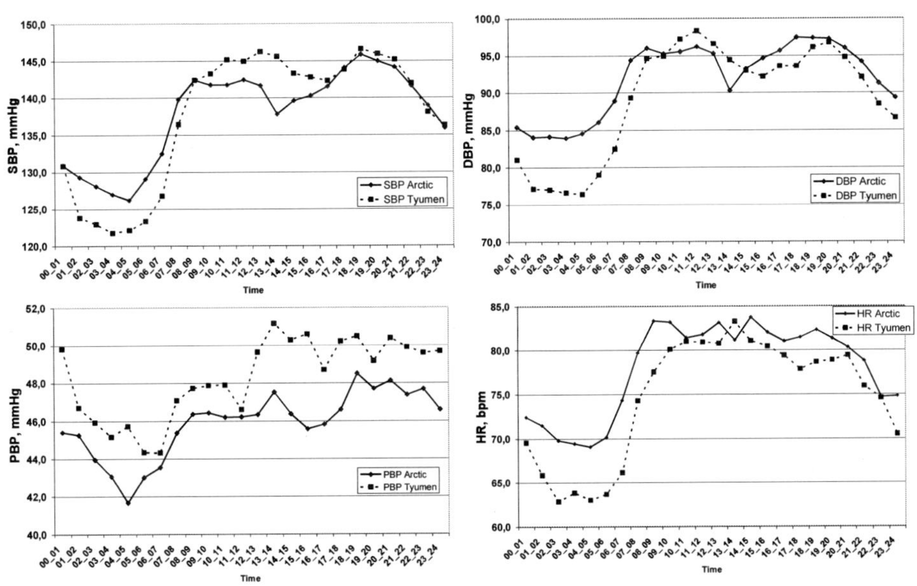

Figure 2. 24-hour dynamics of SBP, DBP, PBP and HR in hypertensive shift-workers of Arctic and in hypertensive Tyumen residents.

Observations of profoundly compromised HR circadian amplitude and PR, not only in hypertensive but also in normotensive subjects, indicate that climatic and photoperiodic features of the Arctic may leave their specific signature on the circadian system.

Table 2. Variability modifications of SBP, DBP and HR in hypertensive shift-workers of Arctic vs. hypertensive Tyumen residents

Location	Tyumen (57° N) N=144	Yamal (68° N) N=373	P
PR (SBP) (%)	33.3±16.5	19.9±15.4	<0.001
PR (DBP) (%)	31.2±17.5	18.7±15.4	<0.001
PR (HR) (%)	35.7±17.5	19.8±15.5	<0.001
MESOR (SBP) (mmHg)	138.3±10.8	138.6±8.8	0.778
MESOR (DBP) (mmHg)	89.9±8.9	92.7±7.9	0.013
MESOR (HR) (beats/min)	74.8±7.9	77.8±9.2	0.014
24-h Amplitude (SBP) (mmHg)	12.9±5.4	8.8±4.7	<0.001
24-h Amplitude (DBP) (mmHg)	10.5±4.4	6.7±3.7	<0.001
24-h Amplitude (HR) (beats/min)	10.3±4.5	7.4±3.9	<0.001
*Best-fit period			
τ=24h (SBP) (%)	70.8	41.3	<0.001
τ =24h (DBP) (%)	73.6	43.4	<0.001
τ =24h (HR) (%)	81.3	60.9	0.002
τ =12 h (SBP) (%)	14.6	20.9	0.103
τ =12 h (DBP) (%)	14.6	19.6	0.187
τ =12 h (HR) (%)	4.6	13.9	0.038
τ ≤8h (SBP) (%)	14.6	37.8	<0.001
τ ≤8h (DBP) (%)	11.8	37.0	<0.001
τ ≤8h (HR) (%)	14.1	25.2	0.059

* Best-fit period – percentage of subjects who have best-fit period of certain τ (as indicated); taken from linear least-squares analysis of circadian, τ =24 hours and 8 consecutive ultradian harmonics; subjects who had best-fit period, τ ranging from 8 to 3 hours (3rd to 8th harmonics of circadian) are united.

The most prominent feature in variability modifications of BP and HR in Arctic shift-workers is a PR reduction, almost two-fold for HR. PR decreased to an even greater extent than the 24-hour amplitude. This can be accounted for by a lesser decline in nighttime values combined with an increase in short-term ultradian variability. While the MESOR of SBP was not elevated, the higher MESORs of DBP and HR were again mainly due to higher nighttime values. The circadian rhythm was the best-fitting component in statistically significantly fewer subjects (Table 2). A variance transposition toward the 12-hour component (for HR) and to periods shorter than 8 hours (for SBP and DBP) was found (Table 2). A rise in harmonics with periods shorter than 6 hours provided this variance transposition of BP.

Comparative analyses of BP and HR variability patterns between the Tyumen population and the normotensive Arctic shift-workers revealed very similar results. No statistically significant difference in the 24-hour mean BP acrophase was found, either within normotensive vs. hypertensive groups or among them. It must be underlined, however, that the further progression of ECD can be observed in comparison of BP and HR variability in normotensive vs. hypertensive Arctic shift-workers. In hypertensive patients, the 24-hour PR and amplitude are statistically significantly decreased, whereas ultradian variability is even larger. The percentage of persons for whom the 24-hour rhythm is the best-fitting component also tends to decline, though this difference does not reach statistical significance.

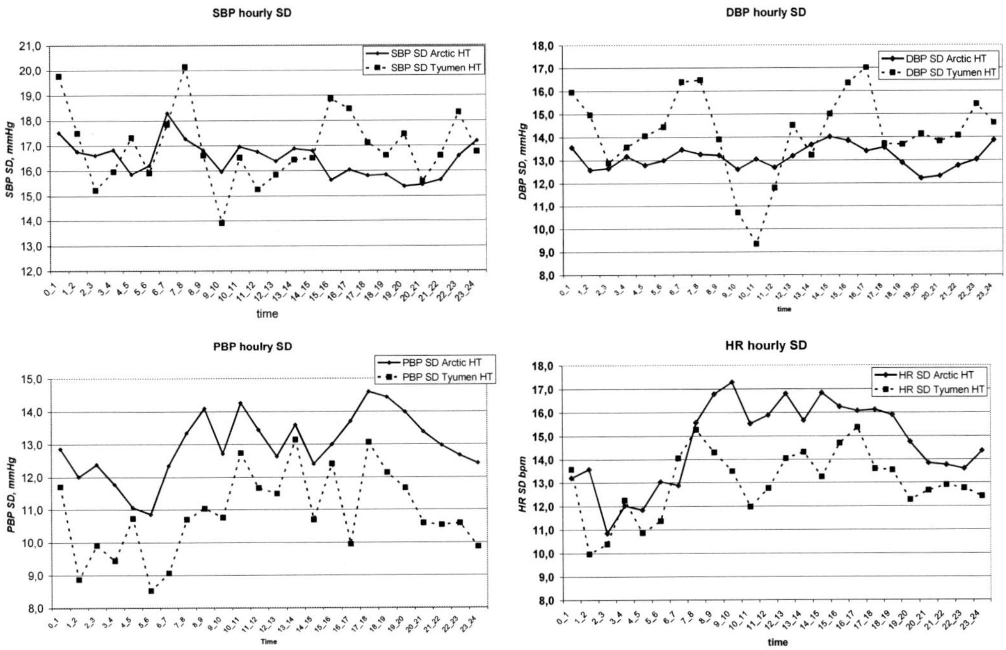

Figure 3. Differences in 24-hour patterns of standard deviation (SD) of hourly means of SBP, DBP, PBP and HR in hypertensive (HT) patients between Tyumen residents and Arctic shift-workers, as a measure of inter-individual variability.

Another impressive feature of BP and HR variability is depicted in Figure 3, showing statistically significantly elevated daytime variability of HR within 1-hour intervals (gauged by hourly SDs). The presence of circadian and ultradian variability in hourly HR-SDs is easily discerned by the naked eye. For SBP and DBP, however, hourly SDs in Arctic hypertensive patients reveals a more rigid pattern throughout the whole 24-hour record. The difference in pattern is obvious for DBP: whereas the hypertensive patients in Tyumen have a prominent 12-hour component in hourly DBP-SDs, it is absent in the Arctic group.

Despite the lower PP itself, Arctic shift-workers have higher hourly PP-SD values throughout the entire 24-hour record. Increased short-term variability of PP in the Arctic hypertensive shift-workers also indicates complex disturbances in the coordination of their vascular system.

Additional data analyses for the group of normotensive Arctic shift workers (N=129) were performed by sphygmochron [25] to identify VVAs. By comparison with clinically healthy peers matched by gender and age, the following VVAs were identified. There were 8 records from 6 subjects with systolic MESOR-hypertension, and 25 records from 18 subjects with diastolic MESOR-hypertension. Systolic and diastolic MESOR-hypotension was found in 3 records from 3 different workers and in 1 record from a single worker, respectively. EPP was found in only one record. Systolic CHAT (circadian-hyper-amplitude-tension, a condition consisting of an excessive circadian amplitude of BP) occurred in 5 records from 5 different subjects, and diastolic CHAT in 5 records from 4 different subjects, indicating that

CHAT may have been transient, perhaps related to the work load, rather than persistent. Systolic and diastolic ecphasia found in 37 and 30 records, respectively, may have been related to shift-work. A higher-than-expected number of records (123/659 or 18.7%) from 75 workers have a pulse pressure product (PPP) above 100 mmHg.beats/min.%, while in another 207 records, the PPP is between 90 and 100, a level already associated with a statistically significant increase in cardiovascular disease risk in one outcome study [16]. DHRV is also detected in a larger-than-expected number of records, namely in 95 records (14.4%) from 62 subjects (48.1%, at least part of the time).

For SBP, DBP and HR, the standard deviation (SD), MESOR (M), 24-hour amplitude (A), and corresponding PR of each subject's record were expressed as a percentage of all profiles [23]. These newly formed time series were analyzed by least squares spectra in the frequency range of 1 cycle in 5 years to 15 cycles per year, complemented by the extended nonlinear cosinor to estimate the period with its 95% confidence interval (CI), using Marquardt's algorithm [26]. As subjects came from different locations varying between 2611 and 3458 miles from the geomagnetic North Pole, circadian rhythm parameters were compared among the different sites by 1-way analysis of variance (ANOVA) and linear regression analysis.

Table 3. Para-annual components of vascular variability observed in normotensive Arctic shift-workers of Yamburg station (68N)

Endpoint	Para-Annual Peak			Overall largest peak			Period = 1.32y		Period = 1y		A-Ratios		
	Period	Amplitude		Period	Amplitude		Amplitude		Amplitude		ParaA/1y	Largest/1y	1.32y/1.0y
SBP-SD	1.786	4.916	ss	1.786	4.916	ss	1.788	ns	1.379	ns	356.490	356.490	129.659
SBP-PR	0.532	8.070	ss	0.142	8.184	ss	4.129	ns	3.446	ns	234.185	237.493	119.820
SBP-M	2.083	0.954	ss	2.083	0.954	ss	0.529	ss	0.278	ns	343.165	343.165	190.288
SBP-A	0.781	4.638	ns	0.154	5.483	ss	1.739	ns	2.390	ns	194.059	229.414	72.762
MAP-SD	1.923	4.731	ss	1.923	4.731	ss	1.711	ns	2.492	ns	189.848	189.848	68.660
MAP-PR	0.532	12.350	ss	0.532	12.350	ss	6.606	ns	4.404	ns	280.427	280.427	150.000
MAP-M	1.923	0.561	ss	0.305	0.663	ss	0.345	ns	0.333	ns	168.468	199.099	103.604
MAP-A	0.833	6.126	ss	0.163	6.322	ss	5.695	bs	1.824	ns	335.855	346.601	312.226
DBP-SD	1.786	4.847	ss	1.786	4.847	ss	2.508	bs	1.950	ns	248.564	248.564	128.615
DBP-PR	1.316	12.714	ss	0.142	14.318	ss	12.714	ss	4.517	ns	281.470	316.980	281.470
DBP-M	1.087	0.486	ns	0.152	0.857	ss	0.396	ns	0.465	ns	104.516	184.301	85.161
DBP-A	1.316	9.013	ss	1.316	9.013	ss	9.013	ss	1.130	ns	797.611	797.611	797.611
HR-SD	3.125	6.158	ss	0.263	6.503	ss	1.183	ns	1.194	ns	515.745	544.640	99.079
HR-PR	2.778	19.323	ss	2.778	19.323	ss	1.703	ns	4.064	ns	475.468	475.468	41.905
HR-M	3.125	2.734	ss	3.125	2.734	ss	0.338	ns	0.427	ns	640.281	640.281	79.157
HR-A	2.273	8.085	ss	0.212	8.291	ss	0.949	ns	0.427	ns	1893.443	1941.686	222.248

SBP: Systolic Blood Pressure; MAP: Mean Arterial Pressure; DBP: Diastolic Blood Pressure; HR: Heart Rate; SD: Standard Deviation; M: MESOR; A: 24-hour Amplitude; PR: Percentage Rhythm of 24-hour component; y: year; ss: Statistically Significant ($P<0.05$); ns: Not statistically significant ($P>0.05$); P-values from zero-amplitude (no rhythm) test, not corrected for multiple testing. Amplitude ratios are expressed in %; Periods are in years.

Least squares spectra of longitudinal series of pooled relative endpoints reveal the presence of peaks corresponding to periods longer than 1 year, whereas no spectral peak is observed at precisely 1 year. Table 3 summarizes the results by listing the largest peak detected in the para-annual region of the spectrum, and the overall largest peak, whether it was anticipated or not. Amplitude ratios (in %) in Table 3 also compare these amplitudes and that of an anticipated 1.32-year component (corresponding to a signature of the solar wind, found in all BP and HR longitudinal records available for analysis thus far [27, 28]) to the amplitude of the calendar year. The about 1.32-year (transyear) tends to be more prominent than the calendar year (Table 3). In the case of DBP, nonlinear analyses confirm the presence of a statistically significant transyear for the overall SD, and for the 24-hour amplitude and its corresponding PR. Period estimates [and their CIs] are for DBP-SD 1.75 [1.45, 2.04] years, for DBP-A 1.30 [1.08, 1.53], and for DBP-PR 1.31 [1.07, 1.54] years, with corresponding amplitudes (in %) of 4.9 [1.7, 8.1], 9.0 [1.2, 16.8], and 12.7 [1.2, 24.2], respectively.

One lesson learned from this study was the dominance of para-annual components over the calendar-yearly variation associated with the changing of the seasons in a region with large fluctuations in both temperature and day length between winter and summer. Of particular interest is the detection of a transyear modulating the circadian amplitude of DBP, in keeping with results documented in longitudinal (rather than hybrid) records [27, 28]. Another important finding is the larger-than-expected number of abnormalities in the pulse pressure product and heart rate variability of this population. Whether these VVAs are associated with the monthly rotational shift work these subjects are engaged in, or whether they relate to their geographic location as it may be affected by terrestrial and space weather, remains to be determined. In any event, the identification of VVAs in 5 to 6 records in consecutive seasons from 129 shift-workers has laid the foundation for determining their ability to predict actual adverse cardiovascular outcomes, should it be possible to follow-up on their health status, preferably with repeated weeklong monitoring, as recommended in a consensus document [13].

Benefits of Personalized Chronotherapy of Hypertension in Arctic Shift-Workers

Timing of treatment is crucial and basic to reach its aim. Thus, chronotherapy of cardiovascular diseases in general, and hypertension in particular, is a very promising approach capable of increasing efficiency while decreasing dosage and side effects. Its advantages have been described by many researchers [22, 24, 29-37]. As opposed to recommendations made across-the-board for all patients [e.g., 30, 31], we recommend an individualized approach used in only a few studies thus far [37]. Intuitively, a chronotherapeutic regimen should be best applied in the light of results from the chronodiagnosis, an approach which has been called chronotheranostics [38].

To verify implied advantages of a personalized chronotherapeutic approach based on individual variability, we analyzed and compared outcomes of non-individualized, standard therapy and individualized, chronobiologic therapy of hypertension in Arctic shift-workers and in Tyumen residents [23].

Here we discuss the results of an Arctic study in which 424 men (of the 538, aged 30-59 years, initially enrolled) participated in all 5 consecutive ABPM sessions, taken in each season at about 3-month intervals during the entire 12-month survey and completed individual treatment protocols. Treatment was with one of 3 groups of drugs: calcium channel blocker, amlodipine (5-10 mg/day); beta-blocker, bisoprolol (5-10 mg/day), or angiotensin-converting enzyme (ACE) inhibitor, lisinopril (5 mg/day). Data from the 130 Arctic shift-workers with no anamnesis of hypertension (presumably normotensive) served to seek predictors of development of hypertension within 1 year. The remaining 294 men were found to have high BP and entered the intervention arm of the study.

In order to implement an individualized chronoselective approach, all hypertensive patients were randomized into 2 groups, and then one group was randomly assigned to standard therapy (group "S"; 120 men) and the other group to chronotherapy (group "CH"; 174 men). In group "S", therapy was based solely on the 24-hour ABPM results: those who had "dipper" or "over-dipper" profiles were treated in the morning (preventive therapy, 8-10 a.m.), those who had either "non-dipper" or "night-peaker" profiles were treated in the evening (imposition therapy, 8-10 p.m.). In group "CH", timing and number of dosages were prescribed individually, after consideration not only of the circadian amplitude, acrophase and bathyphase (phase of predicted minimum), but also of the 12-hour and 8-hour characteristics. To compare and unify results, patients in group "CH" were considered as (1) eurhythmic (or EU-HT) if MESOR-hypertension was not accompanied by a diminished 24-hour amplitude or acrophase alteration and the circadian rhythm was still the best-fitting component for SBP; (2) aperiodic (or ISO-HT) if the 24-hour BP amplitude was diminished and the 24-hour component was not statistically significant, usually accompanied by the predominance of ultradian components with periods shorter than 8 hours; or (3) ecphasic (or ALLO-HT) if either the 24-hour SBP acrophase was statistically significantly shifted, or the 12-hour component was predominant. Then, for EU-HT patients, preventive, morning therapy was used; for ISO-HT patients, rhythm imposition, evening therapy was used, and for ALLO-HT patients, a combined therapeutic approach was used. The individualized 12-month chronotherapeuty of hypertension in Arctic shift-workers led to a great increase in treatment compliance. This led in turn to a 2-fold gain in treatment efficiency and enabled us to reach target BP values in 68% of the patients vs. 34% in the "S" group. In several publications [22-24, 39], we have shown numerous advantages of personalized chronotherapy. Inter alia, in the "CH" group, 14-day treatment with the angiotensin-converting enzyme inhibitor lisinopril exceeds, in terms of its efficiency, the standard approach applied in the "S" group. The major factor was awareness and taking into consideration the 12-hour component, should it dominate in BP spectra. In subjects with a predominant 24-hour rhythm, the efficiency of chronotherapy was similar in both groups. Each of the 3 drugs studied was more effective in leveling an excessive 24-hour BP-SD and normalizing elevated nighttime BP values in the "CH" group. The efficiency of mono-drug therapy was also improved. The beta-blocker, bisoprol, was relatively less effective in correcting intrinsic desynchronization, mainly due to decreasing daytime HR if taken in the morning. The ability of beta-blockers to suppress the production of melatonin when taken in the evening might be one of several possible causes. On the other hand, a preventive strategy of chronotherapy with bisoprolol can be most effective in subjects with CHAT and/or tachycardia.

Prospectively, 21.3% of the normotensive shift-workers became hypertensive within 12 months. Logistic regression revealed high prognostic value of certain chronome alterations (SBP phase shift and lowered 24-hour amplitude of SBP) in predicting hypertension (74.6%, $\chi^2 = 33.3$, $P = 0.009$). Subjects with "ALLO-normotension" and "ISO-normotension") were found to have a statistically significantly larger left atrium diameter (38.0±3.6 and 38.8±3.9 vs. 34.9±5.7 mm, P=0.003 and P=0.004, respectively) and a non-significantly higher left ventricle mass. While the above results were obtained in relation to alterations of SBP variability, similar results hold for DBP.

Our chronobiologic approach was also superior to the standard ABPM approach in predicting the development of hypertension in terms of its diagnostic sensitivity and prognostic value of a negative result (Table 4).

Table 4. Advantages of chronobiologic vs. standard ABPM data analysis and interpretation in predicting of hypertension in Arctic shift-workers

Index	Formula	S	CH	P
Diagnostic sensitivity, DSS	a/(a+c)	94.6	98.3	0.016
Diagnostic specificity, DSP	d/(d+b)	84.6	90.0	0.192
Diagnostic efficiency, DE	(DS+DSP)/2	89.6	94.1	0.791
Prognostic value of positive result	a/(a+в)	93.3	95.7	0.198
Prognostic value of negative result	c/(c+d)	12.7	4.1	0.016

Diagnostic sensitivity (DS), diagnostic specificity (DSP) and diagnostic efficiency (DE); prognostic value of positive result (PPR) and prognostic value of negative result (NPR) were assessed as: DS=a/(a+c)×100%, DSP = d/(d+b)×100%, DE = (DS+DSP)/2, PPR=a/(a+b)×100%, NPR = d/(c+d)×100%, where a – true positive result, b – false positive result, c – false negative result, d – true negative result.

Despite a few cases with negative results, our chronobiologic approach shows a high prognostic value. An inter-rater agreement between two diagnostic methods is also high (92.7%), and confirmed by a kappa index of 0.46±0.08.

Chronotherapy and Chronobiotics in Correction of Compromised Time Structure

We studied 97 residents of the Tyumen Elderly Veterans Home of both genders from 63 to 91 years of age, mean age 80.2±1.2 years, living on a self-chosen sleep-wake regimen to suit their personal convenience. This study aimed at evaluating age-dependent changes of cardiovascular circadian rhythms and at investigating whether any chronome alterations could be reversed by a single daily ingestion of melatonin [40, 41]. The experiment lasted three weeks. After one reference week, 63 of the subjects were randomly assigned to receive 1.5 mg melatonin (Melaxen^{TM}) each day at 22:30 for two weeks (group “M”). The other 34 subjects were placebo-treated (group “P”). SBP, DBP and HR were measured using semi-automated devices at 03:00, 08:00, 11:00, 14:00, 17:00 and 23:00 each day during the first and the third weeks. Specially trained personnel took the measurements, avoiding disturbing the subjects' sleep. Neither participants nor experimenters were aware of the group to which subjects had been assigned (double-blind study). As effects of treatment were expected mainly in group M, this group was made twice as large as the P group.

Rhythm characteristics were estimated by cosinor [42]. Each record of 49 values (7 time points each on 7 consecutive days) was analyzed. The method involved the least squares fit of single cosine curves with frequencies in the range of 1 to 21 cycles per week, corresponding to the 7 consecutive harmonics of a weekly (168-hour) cycle and 3 consecutive harmonics of circadian (24-hour) rhythm, with corresponding periods of 24, 12 and 8 hours. For each component, estimates of the amplitude and acrophase were obtained as measures of the extent and timing of predictable change within a cycle. In addition, estimates of the MESOR and standard deviation, a gauge of overall variability, were obtained.

HR preserved its 24-hour rhythmicity. Circadian HR rhythms were still monophasic but with an earlier increase in the morning as compared to younger cohorts. The maximum was reached already at 11:00. Thereafter, HR remained high with an only moderate reduction. After 20:00 it decreased more steeply and reached its lowest value around 03:00. Circadian rhythms of SBP and DBP had a biphasic waveform. Peaks occurred around 08:00 and 17:00 and troughs between 11:00-14:00 and around 03:00. The percentage rhythm of the 12-hour component was increased, as was overall SBP variability, thus confirming a previously described tendency for *hypervariable ECD*, found in the ABPM aging study [10, 11, 17].

During melatonin treatment, the increase of HR was delayed. As a consequence, the maximum was observed only between 14:00 and 20:00. The nocturnal minimum, although occurring around the same time as before (03:00), was deeper. SBP and DBP values were also reduced during melatonin treatment, in keeping with earlier reports [43]. The hypotensive effect of melatonin was most pronounced between 03:00 and 8:00, the time of the highest risk of adverse cardiovascular events. The SBP decrease was the bigger the higher SBP was before treatment. Melatonin was thus shown to be of great value for the treatment of hypertension.

Melatonin had no effect on the percent rhythm and 24-hour amplitude of BP (Table 5). The 24-hour PR and amplitude of SBP were statistically significantly higher than those of the 12-hour component. In the case of DBP, however, the 12-hour amplitude (but not the PR) was more pronounced than that of the 24-hour component before treatment. Thereafter, the 24-hour amplitude became more pronounced by comparison with that of the 12-hour component. During treatment, the 12-hour amplitudes of SBP and the DBP were statistically significantly reduced, as was the percent rhythm of the 12-hour DBP component. Thus, melatonin treatment mostly reduced the 12-hour variability of both SBP and DBP, but had no statistically significant effect upon the 24-hour amplitude. The standard deviation, an estimate of the overall BP variability, was reduced during melatonin treatment for SBP (14.25 ± 0.59 vs. 15.77 ± 0.52). For DBP the difference was not significant (10.86 ± 0.53 vs. 11.22 ± 0.46).

Table 5. SBP and DBP circadian (24-hour) and circasemidian (12-hour) components before and during melatonin treatment. Results of Cosinor analyses (Mean values ± standard errors)

Circadian function	Rhythm component	Before Treatment		During Treatment	
		PR	Amplitude	PR	Amplitude
SBP (mm Hg)	24 h	15.5 ± 1.8	7.9 ± 0.5	15.2 ± 2.1	6.9 ± 0.5
	12 h	11.1 ± 1.5	6.5 ± 0.5	10.4 ± 1.4	5.6 ± 0.3*
DBP (mm Hg)	24 h	10.4 ± 1.2	4.5 ± 0.4	10.7 ± 1.5	4.3 ± 0.3
	12 h	12.1 ± 1.4	5.1 ± 0.4	9.3 ± 1.1*	4.1 ± 0.3*

PR – Percent Rhythm: Percentage of overall variability accounted for the rhythm of the given period length

* - indicate significant differences of amplitude or percent rhythm values before and during melatonin treatment.

Bold numbers indicate significant differences between 24- and 12-hour components.

In the case of HR, the standard deviation remained unchanged (7.57 ± 0.24 vs. 7.56 ± 0.28). Melatonin was thus shown to diminish manifestations of ECD and ecphasia. The latter can be extrapolated from the population-mean cosinor analysis. For this purpose, we added results of temperature variability estimated simultaneously, as described elsewhere [44]. On a population level, intrinsic desynchronization is typical for this age group (Figure 4). It also shows that melatonin is a potent resynchronizer of compromised circadian order manifested as intrinsic phase misalignment among body temperature and hemodynamic functions in the elderly (particularly between SBP and HR).

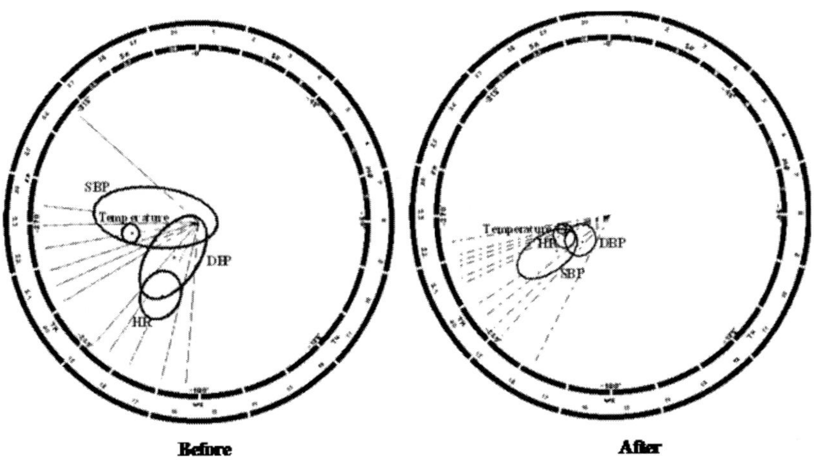

Figure 4. Melatonin treatment compensates for intrinsic circadian disruption in the elderly by synchronizing scattered phases of BP, HR and temperature.

Such an effect, however, is not immediate but is gained over at least one week. The delayed effect might be accounted for by a diversity of interactive feedback mechanisms of melatonin with the central and peripheral oscillatory molecular machinery, and also by the involvement of temperature circadian rhythm-mediated synchronization. Indeed, it was previously shown by others that melatonin lacks an immediate effect on core clock genes in the suprachiasmatic nuclei (SCN), but has a delayed effect, which implicates mostly post-translational mechanisms [45, 46]. Post-translational mechanisms of melatonin action on a cell are very diverse [47], and likely vary from tissue to tissue. Also, hypothetically, delayed melatonin effects on the SCN may in part be a byproduct of feedback mechanisms: melatonin helps synchronize the circadian temperature rhythm, which, in turn, with a lag, helps the SCN reset disturbed rhythms [48].

Interestingly, a melatonin effect likely depends on a pre-existing degree of circadian disruption: its synchronizing properties are correlated with phase shifts and cycle-to-cycle phase scatters, as shown earlier for temperature [44]. Circadian rhythms in BP, HR and temperature all became statistically significant after melatonin treatment, but not after placebo treatment, ruling out bias from the social routine (e.g., the presence of personnel taking measurements).

The effect of melatonin in compensating for age-dependent circadian disruption may be derived from both direct physiologic and complex chronobiotic properties. A direct physiologic effect, particularly on the thermoregulatory and the cardiovascular systems, initiates distal cutaneous vasodilation, thus promoting heat loss and decrease of core body temperature. The best investigated physiologic effect is the hypothermic effect of melatonin [44]. A temperature rhythm may have been an important secondary synchronizer, noting recent data of its importance for the synchronization of the SCN [48]. In addition, melatonin induces sleepiness and may even be involved in the coordination of the sleep-wake cycle and of the circadian rhythm in body temperature [49-51]. At the same time, melatonin may have an effect on multiple cardiovascular functions via receptors in the peripheral vasculature and in the heart itself. Thus, the administration of exogenous melatonin has been found to decrease BP in humans and animals [43]. In a double-blind crossover study, aimed at investigating the effect of repeated melatonin intake on the circadian BP rhythm in untreated hypertensive patients, it was shown that daily doses of melatonin for 3 weeks, taken 1 hour before sleep (but not a single dose) lowered nighttime SBP and DBP [52]. According to the authors, this suggests that melatonin did not act as a direct vasodilating drug, rather pointing to an important role for the SCN in the BP-lowering effect. A reduction of nighttime BP (even a 2-3 mmHg drop in SBP) may be clinically relevant [53]. A single melatonin dose given daily at 22:30 has its maximum SBP-lowering effect between 03:00 and 08:00, the time of the highest risk of adverse cardiovascular events such as heart attacks and strokes [54-56]. The later morning increase in HR during melatonin treatment is an additional beneficial effect. In the study of Scheer et al. [52], a repeated dose of melatonin (2.5 mg, 1 hour before sleep daily for 3 weeks) reduced SBP and DBP during sleep by 6 and 4 mmHg, respectively. In our study, we found an even more pronounced reduction in nighttime SBP and DBP with a smaller dose (1.5 mg vs. 2.5mg) of melatonin, taken repeatedly for a shorter span (2 weeks vs. 3 weeks). The age of our subjects may account for the difference, as melatonin is known to decrease with age [57].

Discussion

In the last few years a rapidly growing body of knowledge has shown that circadian desynchronization, or circadian disruption, has its own impact on the development of numerous age-associated diseases and in the aging process per se [58-63]. It has become increasingly obvious that health and disease are increasingly dependent on the proper internal synchrony in the vast network of intrinsic self-sustained molecular, tissue-specific and overt physiologic oscillations. Circadian desynchronism (disturbed acrophase and amplitude of input molecular and cellular processes) impairs the organism's ability to prepare for and to react to certain demands recurring at more or less regular interval, and affects repair mechanisms, leading to damage accumulation and apoptosis [47]. Different scholars of chronobiology suggested original, yet close, concepts for the interpretation of temporal disorders such as the "spinning top hypothesis" to account for uniform changes in circadian amplitude and acrophase during ontogeny in vertebrates, suggested by Gubin GD [64]. The concept of chronodisruption, proposed in 2003, addressed putative links between shift-work, transmeridian travel, and human cancers with similar causal interpretations [65]. Terms such as allochronism and dyschronism were coined to discriminate between the absence vs. the presence of shift-work intolerance and illness in shift-workers experiencing similar internal desynchronization [66]. Classification of desynchronizations related to different factors and mechanisms, and to principles of their chronotherapy has also been proposed [67]. Herein, we define our concept of changes in spectra with a deeper insight into the structure of variability modifications and its putative origins. Compromised orchestration of harmonic cellular circadian oscillations results in the manifestation of non-circadian variations of physiologic and biochemical variables (overt output rhythmicity). The ECD, a lower ratio of circadian amplitude to less regular ultradian or infradian amplitudes, is a typical feature of gain in chaotic variability and loss of circadian order that must accompany senescence, shift-work, high latitude residence and a number of diseases.

Hypertension, which is discussed herein as an example, remains a major cause of incapacity and mortality today. Its incidence increases with advancing age [68-70]. Also, a certain degree of internal and external desynchronization among different circadian rhythms and in relation with their periodic environment is observed in elderly subjects. Decreased circadian amplitudes are typical, while the percentage of ultradian components may be elevated. This is particularly true for the circadian BP and HR variation [10, 71-73]. Both phenomena are not independent from each other. Thus, a diminished temporal, particular circadian order might have adverse consequences on the health of the elderly [74]. On the other hand, hypertension was shown to have direct, adverse effects on the SCN (for review, see [75]).

Two peaks in the circadian pattern of SBP are generally observed in the adult population. However, the percentage of ultradian components increases with increasing age [76, 77] and in hypertension, correlating with disease progression [22]. Blunted circadian and modified ultradian amplitudes were found in children with chronic renal failure, and ultradian but not circadian rhythms were associated with renal function and proteinuria [78], and ultradian and circadian rhythm integrity were lost in hypertension [79]. The nocturnal fall in SBP also declines in the elderly; hence the night-time SBP remains elevated [80-83]. A shift in prominence from the circadian domain to higher frequencies was found for BP but not for HR in centenarians [72].

In general, BP behavior is more irregular in old age. From a chronobiologic point of view, this means a higher overall variability, usually apparent in the standard deviation (SD) of the mean, accompanied by a less predictable circadian pattern. As shown earlier by means of ABPM [10, 11, 71], the SBP-SD correlates positively with age ($r=0.389$, $P<0.001$), whereas DBP-SD does not ($r=0.046$, $P=0.688$). Similar results were found with self-measurements [10]. As high SD values are correlated with many adverse events, i.e., with left ventricular hypertrophy, the observation that melatonin reduces SBP-SD is clinically valuable. But can we learn more from studies of chronome alterations about the nature of transformed variability? We start with very intimate mechanisms of endogenous circadian rhythmicity within a cell and then extrapolate some of it to a systemic level.

In a recent review [8], we discussed the current molecular factors involved in the modulation of circadian rhythms and the variability of overt physiologic functions in mammals (Table 6).

Mutations and polymorphisms of well-known genomic factors such as core clock genes (*BMal*, *Clock*, *Per1, 2, 3* and *Cry 1,2*) impact circadian rhythm parameters, as reviewed elsewhere [cf. 84]. In addition, many other factors (epigenetic, post-transcriptional, and post-translational) influence circadian parameters. Even mild deviations in the circadian period (ecphasia) in one or a few intrinsic (endogenous) mechanisms may lead to numerous perturbations in downstream variability of overt physiologic (phenotypic) functions.

Table 6. General intrinsic and extrinsic factors involved in the manifestation of ECD and other variability modifications

Intrinsic/Endogenous	Extrinsic/Exogenous
Core clock gene polymorphism	LAN
SIRT1 polymorphism, etc. (modulator factors)	Decrease in physical activity
Loss of synchrony among SCN neurons	Suboptimal food schedule
Decrease in nighttime melatonin production	Intoxications
Sleep problems (primary)	Light perception deteriorations
Loss of temperature rhythm – induced synchronization/metabolic factors	

Phase shifts or phase instability (development of ECD) become inevitable [8, 9]. Over two decades ago, Gennady D. Gubin proposed that the circadian amplitude at a certain age can be used as a good estimate of "biological age" and as a predictor of longevity [85]. Accordingly, in keeping with the ECD concept, higher circadian amplitudes rely on stable periods of upstream input rhythms. In recent studies, somewhat larger deviations from periods of exactly 24 hours in endogenous activity rhythms were indeed shown to be associated with a statistically significantly reduced lifespan in mice [86, 87], and vice versa. While having preserved high amplitudes of core clock genes expression in the SCN due to endogenously high production of SIRT1 is associated with faster resynchronization after a phase shift (induced jet lag) in aged mice, these animals also have a larger circadian amplitude and a less fragmentized (i.e., less ultradian) activity rhythm [88]. These results are also in keeping with a larger circadian amplitude of core temperature associated with a prolonged lifespan achieved by means of restricted feeding, as original shown by Franz Halberg [see 89 for review].

On the cellular level, modeling studies showed that even subtle fluctuations in the number of factors in post-transcriptional machinery, namely miRNAs, can modulate the amplitude, acrophase and period of core clock genes expression [90, 91]. Interestingly, multiple ultradian rhythms have been found in transcriptomes of different peripheral tissues that were undetectable in disassociated cells. Moreover, harmonics were lost in the liver when mice are subjected to restricted feeding, suggesting that at least one component of circadian harmonics is driven by feeding [92]. These authors also discuss how different factors, such as feeding and endoplasmic reticulum stress underlying circadian variation may yield ultradian oscillations in tissues. On the other hand, the expression of the number of genes involved in liver lipid metabolism has a predominant 12-hour rhythmicity [93, 94]. Restricted feeding can also lead to temporary uncoupling and phase changes up to 12 hours in gene expression rhythms of peripheral circadian oscillators from that of the central oscillator in the SCN. It takes about 168 hours for peripheral rhythms in different tissues to be resynchronized [95]. Finally, in animals bearing mutations capable of eliminating circadian rhythmicity, such as *Clock*$^{-/-}$; *Npas2*$^{m/m}$; *Bmal*$^{-/-}$; *Per1/Per2* and *Cry1,2*, a double knockout not only abolishes the 24-hour rhythm, but leads to amplification of harmonics with periods between 4 and 6 hours [cf. 96 for review].

Overt phenotypic rhythmicity, observed systemically, is a result of multiple peripheral dynamic functions, which are more or less fine-tuned in extent by governing central oscillators. Systemic desynchronization is complex by its very nature. Every physiologic function is modulated throughout the whole upstream "homeostatic" framework, which is not actually "static", but essentially "kinetic". To comprehend the genesis of chronome alterations one should be aware of the acrophase and amplitude of circadian rhythms in catecholamines, adrenoreceptors, aldosterone, natriuretic peptide and prostaglandins, and not discount the variability in blood flow, vascular resistance, among others [see 97 for review].

Some infradian components may just be beat-harmonics, emerged as an outcome of superimposed rhythmicity of upstream factors with periods close to, yet deviant from precisely 24 hours [9]. Some receptors may become more susceptible to secondary extrinsic time clues. For instance, certain cosmic, solar, geomagnetic or social cycles may then yield physiologic or biochemical cycles of amplified amplitudes responding thereto. Thus it is likely that irregular infradian variability will "seek" for such influencers to become synchronized (at least in period, if not in phase).

The leveling of leading endogenous signal may potentially be hazardous for health. Underlying reasons may be that "time windows" may either be interrupted or not opened and/or because a given substrate has not been produced in enough quantity or for a long enough span, so that the time window for a physiologic effect to be gained was too shot. For example, the concentration of receptors may not be high enough yet, while the enzyme is still too active to enable a substrate to enter a certain cellular compartment. In circumstances of weakened synchronization to major time clues, a variability of opportunistic factors emerges (i.e., magnetic, cosmic, and solar). While this topic is still very poorly understood and well-studied molecular mechanisms are lacking, pioneering investigations by Franz Halberg and the International BIOCOS initiative started the mapping of long cycles and showed their presence in human physiology and pathology.

Chronome alterations in BP and HR during the aging process should also be viewed in the light of peripheral hormonal factors that are also associated with changes in circadian rhythm parameters without necessarily affecting the central oscillatory machinery. For instance, the emergence in aged subjects of a newly discovered serum thermolabile chemical factor reportedly modulated periods in fibroblasts and shifted the circadian acrophase through a process that seemingly does not involve central mechanisms [98].

Notwithstanding relatively intact genetic mechanisms underlying a circadian system that is often preserved in the elderly, in many diseases and in shift-work, it has to operate in circumstances when the driving force of either external time clues, central or peripheral structures are weakened or compromised. Thus, extrinsic, exogenous factors, should they be present, must also contribute to chronome alterations in overt physiologic or biochemical functions, thereby affecting circadian rhythm parameters, usually with an increase in ultradian or infradian variability (Table 6).

Franz Halberg once said "nothing is more important than Δt and capital T". Indeed, the longer the time series, the more precisely can the circadian period be estimated to document any deviation from 24 hours ("ecfrequentia"), and more precisely can emerging infradian components be assessed. The shorter the Δt, the better can ultradian components be examined to help discern underlying intrinsic and extrinsic mechanisms associated with discrepancies between the more endogenously driven HR and temperature and the less so BP and other more exogenously driven and labile cardiovascular and other functions.

Because of extensive genetic polymorphism in factors involved in modulation of overt physiologic functions, an individualized response to a variety of intrinsic and extrinsic disturbances is a sine qua non, and thus, a personalized chronobiologic approach to the treatment of VVDs (including MESOR-hypertension) and other chronome alterations is strongly recommended.

The timing of variables with a prominent circadian rhythm such as rest-activity, sleepiness, metabolism, alertness and performance under usual conditions is such that each is optimal during a particular span of the day. Alertness and performance reach their nadir at night during peak sleep propensity and fatigue [99, 100] and near the low point of core body temperature and the peak of melatonin secretion. In the Arctic and Antarctic, people are deprived of natural sunlight in winter and have continuous daylight in summer. Light of sufficient intensity and suitable spectral composition is a major factor necessary for maintaining a 24-hour synchronized circadian system in humans. At high latitudes, with little or even no sunlight in winter, delayed sleep and circadian acrophase has frequently been reported, with overt free-running in some cases. Seasonal affective disorder is also linked to photoperiodic factors and is typical at higher latitudes. Thus, the status of the circadian system is of interest [101]. Factors capable of compromising a proper time structure of biological functions and resulting in circadian disruption have been shown to be hazardous for human well-being [65, 102].

Circulating melatonin also declines with age in humans, and in circumstances of light at night (LAN), such as a sojourn in the Arctic summer. Changes in the timing of the circadian melatonin rhythm, both with respect to the day-night cycle and to other circadian rhythms, have been reported [57, 103-106]. Taken together, these changes with age may have adverse consequences for the synchronization of the circadian system, in general, and in relation to sleep, body temperature and BP, in particular. Many adverse effects of circadian disruption as a function of age, in the presence of disease, and at high latitudes such as the Arctic region may be overcome with melatonin treatment, also best optimized for each person.

Epilogue

Herein, we illustrated how a chronobiologic approach can be beneficial, whether for improving the quality of life of the elderly, for the treatment of overt disease or prophylactically, or for alleviating negative environmental effects in high latitude regions, with particular applications to shift work that presents a special challenge in the Arctic with very long summer days and very long winter nights. The father of chronobiology, the late Franz Halberg, was a pioneer in these studies, and we are proud to dedicate this paper to his memory.

References

[1] Aschoff J, Wever R. Human circadian rhythms: a multioscillatory system. *Fed. Proc.* 1976; 35: 2326–2332.

[2] Halberg F, Cornelissen G, Regal P, Otsuka K, Wang Z, Katinas GS, Siegelova J, Homolka P, Prikryl P, Chibisov SM, Holley DC, Wendt HW, Bingham C, Palm SL, Sonkowsky RP, Sothern RB, Pales E, Mikulecky M, Tarquini R, Perfetto F, Salti R, Maggioni C, Jozsa R, Konradov AA, Kharlitskaya EV, Revillam M, Wan C, Herold M,

Syutkina EV, Masalov AV, Faraone P, Singh RB, Singh RK, Kumar A, Singhs R, Sundaram S, Sarabandi T, Pantaleoni G, Watanabe Y, Kumagai Y, Gubin D, Uezono K, Olah A, Borer K, Kanabrockia EA, Bathina S, Haus E, Hillman D, Schwartzkopff O, Bakken EE, Zeman M. Chronoastrobiology: proposal, nine conferences, heliogeomagnetics, transyears, near-weeks, near-decades, phylogenetic and ontogenetic memories. *Biomed. & Pharmacother.* 2004;58(Suppl 1): S150-S187.

[3] Chibisov SM, Cornelissen G, Halberg F. Magnetic storm effect on the circulation of rabbits. *Biomed & Pharmacother* 2004; 58 (Suppl 1): S15-S19.

[4] Gubin D, Cornelissen G, Halberg F, Gubin GD, Turti T, Syutkina EV, Grigoriev AE, Mitish M, Yatsyk GV, Ikonomovz O, Stoynev A, Madjirova N, Siegelova J, Fiser B, Dusek J. Half-weekly and weekly blood pressure patterns in late human ontogeny. *Scripta medica* (Brno)1997; 70: 207-216.

[5] Gubin D, Gubin G. A circadian to extra-circadian variance transposition of human body temperature with advancing age. *Chronobiol. Int.* 1999; 16: 41.

[6] Gubin D, Gubin G, Halberg F, Cornelissen G. Index of extra-circadian variability (IECV) of systolic blood pressure (SBP) in human aging: Results from 168-hour ambulatory blood pressure monitoring. *Chronobiol. Int.* 1999; 16: 42.

[7] Gubin DG, Gubin GD, Komarov PI. [Classification of forms of rhythmical spectra disorders] (Classifikaziya variantov narushenii spectra bioritmov phiziologicheskykh pokazatelei). *Advances in Current Natural Sciences* 2004; 11: 47-48. (In Russian).

[8] Gubin DG. [Molecular Basis of Circadian Rhythms and Principles of Circadian Disruption. *Usp. Fiziol. Nauk.* 2013; 44 (4): 65-87. (In Russian).

[9] Agadjanyan NA, Gubin DG. [Desynchronization: Mechanisms of Development from Molecular to Systemic Levels]. *Usp. Fiziol. Nauk.* 2004; 35(2): 57-72. (In Russian).

[10] Gubin D, Cornelissen G, Halberg F, Gubin G, Uezono K, Kawasaki T. *The human blood pressure chronome: A biological gauge of aging*. In Vivo 1997; 11: 485-494.

[11] Gubin D, Ikonomov O, Stoynev A, Cornelissen G, Halberg F. [Human blood pressure chronomes with respect to aging]. Nauchny Vestnik Tyum GU. *Biology* 1997; 2: 55-58.

[12] Gubin DG, Gubin GD, Komarov PI. [Aging viewed through the prism of time order of living systems]. (Starenyie v svete vremennoi organizazii biologicheskykh system). *Adv. Gerontol.* (Uspehi gerontologii) 1998; 2: 67-73. (In Russian).

[13] Halberg F, Cornelissen G, Otsuka K, Siegelova J, Fiser B, Dusek J, Homolka P, Sanchez de la Pena S, Singh RB, BIOCOS project. Extended consensus on need and means to detect vascular variability disorders (VVDs) and vascular variability syndromes (VVSs). *World Heart J.* 2010; 2(4): 279-305.

[14] Cornelissen G, Halberg F, Bakken EE, Singh RB, Otsuka K, Tomlinson B, Delcourt A, Toussaint G, Bathina S, Schwartzkopff O, Wang ZR, Tarquini R, Perfetto F, Pantaleoni GC, Jozsa R, Delmore PA, Nolley E. 100 or 30 years after Janeway or Bartter, Healthwatch helps avoid "flying blind". *Biomed. & Pharmacother* .2004; 58 (Suppl 1): S69-S86.

[15] Gubin D, Vetoshkin A, Beaty L, Gumarova L, Halberg F, Cornelissen G. Chronobiologically-interpreted ambulatory blood pressure monitoring (C-ABPM) of shift-workers in Siberia. Proceedings of Scientific and Practical Conference "*Integration of Science and Practice: Results, Achievements and Prospects*", dedicated to 50th Jubilee of Tyumen State Medical Academy 2013: 15-17.

[16] Cornelissen G, Siegelova J, Watanabe Y, Otsuka K, Halberg F. Chronobiologically-interpreted ABPM reveals another vascular variability anomaly (VVA): excessive pulse pressure product (PPP) updated conference report. *World Heart J.* 2012; 4 (4): 237-245.

[17] Gubin D. [Some general principles of dynamic modifications in chronoinfrastructure of physiologic variables in human ontogeny]. Doctoral Dissertation. Moscow. Russian People Friendship University 2002; 368 pp. (in Russian).

[18] Gapon LI, Shurkevich NP, Vetoshkin AS, Gubin D. [The rhythms of arterial pressure and heart rate in individuals with arterial hypertension under the conditions of Far North]. Klinicheskaia meditsina 2006; 84(2): 39-44 (in Russian).

[19] Cornelissen G. Instrumentation and data analysis methods needed for blood pressure monitoring in chronobiology. In: Scheving LE, Halberg F, Ehret CF. (Eds.) *Chronobiotechnology and Chronobiological Engineering.* Dordrecht, The Netherlands: Martinus Nijhoff; 1987; pp. 241-261.

[20] Halberg F, Cornelissen G, Halberg E, Halberg J, Delmore P, Shinoda M, Bakken E. Chronobiology of human blood pressure. *Medtronic Continuing Medical Education Seminars,* 4th ed. Minneapolis: Medtronic Inc.; 1988. 242 pp.

[21] Gapon LI, Milailova IM, Shurkevich NP, Gubin DG. Khronostruktura arterial'nogo davleniya i chastoty serdechnykh sokrashchenii v zavisimosti ot sezonnogo ritma u bol'nykh arterial'noi gipertenziei v Khanty-Mansiiskom okruge [Chronostructure of arterial pressure and cardiac rate depending on seasonal rhythm in patients with arterial hypertension in Khanty-Mansi Okrug]. *Vestnik aritmologii* 2003; 31:. 32-36. [in Russian].

[22] Gapon LI, Shurkevich NP, Vetoshkin AS, Gubin DG. Hypertension in Tyumen North. Desynchronization and hyperreactivity as factors of disease. [Arterialnaya hipertoniya v usloviyah Tumenskogo Severa. Desynchronos kak factor formirovaniya bolezni). Moscow. *Medicinskaya kniga.* 2009; 208 pp.

[23] Gubin DG, Vetoshkin AS, Shurkevich NP, Gapon LI, Belozerova NV, Poshinov FA. [Chronobiological and standard analysis of 24-hour blood pressure and heart rate monitoring data in subjects engaged in monthly rotational shift-work in Yamal (Yamburg, Tyumen Region)]. *Advances in Current Natural Sciences* 2013; 6: 27-33. (In Russian).

[24] Vetoshkin AS, Gubin DG, Gapon LI, Poshinov FA, Belozerova NV. [Benefits of chronotherapy in treatment of hypertension in shift-workers of Arctic] Preimuschestva hronoterapii v lechenii arterialnoy gipertenzii (AG) u vahtovikov zaoplatyia. *Vestnik of Russian People Friendship University.* Medicine 2012; 7: 72-73. (In Russian).

[25] Cornelissen G, Halberg F. Chronomedicine. In: Armitage P, Colton T. (Eds.) *Encyclopedia of Biostatistics,* 2nd ed. Chichester, UK: John Wiley & Sons Ltd; 2005. p. 796-812.

[26] Marquardt DW. An algorithm for least-squares estimation of nonlinear parameters. *J. Soc. Indust. Appl. Math.* 1963; 11: 431-441.

[27] Cornelissen G, Masalov A, Halberg F, Richardson JD, Katinas GS, Sothern RB, Watanabe Y, Syutkina EV, Wendt HW, Bakken EE, Romanov Y. Multiple resonances among time structures, chronomes, around and in us. Is an about 1.3-year periodicity in solar wind built into the human cardiovascular chronome? *Human Physiology* 2004; 30 (2): 86-92.

[28] Cornelissen G, Halberg F, Rostagno C, Otsuka K. A chronomic approach to cardiac arrhythmia and sudden cardiac death. *The Autonomic Nervous System* 2007; 44: 251-254.

[29] Halberg F, Cornelissen G, Otsuka K, Watanabe Y, Wood MA, Lambert CR, Zaslavskaya R, Gubin D, Petukhova EY, Delmore P, Bakken E. Rewards in practice from chrono-meta-analyses 'recycling' heart rate, ectopy, ischemia and blood pressure information. *J. Med. Eng. Technol.* 1997; 21(5): 174-184.

[30] Hermida RC, Ayala DE, Mojon A, Fernandez JR. Influence of circadian time of hypertension treatment on cardiovascular risk: results of the MAPEC study. *Chronobiol. Int.* 2010; 27(1): 629-651.

[31] Hermida RC, Ayala DE, Mojón A, Fernández JR. Bedtime dosing of antihypertensive medications reduces cardiovascular risk in CKD. *J. Am. Soc. Nephrol.* 2011; 22: 2313-2321.

[32] Takeda N, Maemura K. Cardiovascular disease, chronopharmacotherapy, and the molecular clock. *Adv Drug Deliv Rev* 2010; 62: 956-966.

[33] Takeda N, Maemura K. Circadian clock and cardiovascular disease. *Journal of Cardiology*. 2011; 57: 249-256.

[34] Maury E, Ramsey KM, Bass J. Circadian rhythms and metabolic syndrome: from experimental genetics to human disease. *Circ Res* 2010; 106 (3): 447–462.

[35] Sivagnanam G. Chronotherapy in hypertension. *J. Pharmacol Pharmacother.* 2012; 3: 209.

[36] Campos LA, Cipolla-Neto J, Amaral FG, Michelini LC, Bader M, Baltatu OC. The angiotensin-melatonin axis. *International Journal of Hypertension* 2013; Article ID 521783, 7 pages, http://dx.doi.org/10.1155/2013/521783.

[37] Watanabe Y, Halberg F, Otsuka K, Cornelissen G. Toward a personalized chronotherapy of high blood pressure and a circadian overswing. Clin Exp Hypertens 2013; 35 (4): 257-266. doi: 10.3109/10641963.2013.780073. PMID: 23541237.

[38] Cornelissen G, Halberg F. Treatment with open eyes: markers-guided chronotheranostics. In: Youan BC. (Ed.) *Chronopharmaceutics: Science and Technology for Biological Rhythm-Guided Therapy and Prevention of Diseases*. Hoboken, NJ: Wiley; 2009; pp. 257-323.

[39] Shurcevich NP, Vetoshkin AS, Poshinov FA, Belozerova NV, Gapon LI, Gubin DG. [Effect of chronotherapy on the clinical course and way of life of patients with hypertension in a Polar watch]. *Uralskyi Med. Journal.* 2013; 106(3): 41-48.

[40] Gubin DG, Gubin GD, Zadumina EV, Alexeev AA, Golubkov AA. [Efficiency of exogenous melatonin in correction of age-dependent desynchronization in temperature and cardio-vascular homeostasis]. *Proceedings of III Russian Congress for Pathological Physiology "Dysregulational Pathology of Organs and Systems"*. Moscow. 2004; 165 pp. (In Russian).

[41] Gubin D, Gubin G, Weinert D, Gapon LI. The effect of single daily melatonin dosing on cardiovascular rhythms in the elderly. *Adv. Gerontol. 2014.* In preparation.

[42] Bingham C, Arbogast B, Cornelissen-Guillaume G, Lee JK, Halberg F. Inferential statistical methods for estimating and comparing cosinor parameters. *Chronobiologia* 1982; 9: 397-439.

[43] Zaslavskaya RM, Makarova LA, Shakarova AN, Komarov F, Wang ZR, Wan CM, Katinas G, Cornelissen G, Halberg F. Individualized time series-based assessment of

melatonin effects on blood pressure: model for pediatricians. *Neuroendocrinol. Lett.* 2003; 24 (Suppl 1): 238-246.
[44] Gubin DG, Gubin GD, Waterhouse J, Weinert D. The circadian body temperature rhythm in the elderly. Effect of single daily melatonin dosing. *Chronobiol. Int.* 2006; 23: 639-658.
[45] Poirel VJ, Boggio V, Dardente H, Pevet P, Masson-Pevet M, Gauer F. Contrary to other non-photic cues. Acute melatonin injection does not induce immediate changes of clock gene mRNA expression in the rat suprachiasmatic nuclei. *Neuroscience* 2003; 120: 745–755.
[46] Agez L, Laurent V, Guerrero HY, Pévet P, Masson-Pévet M, Gauer F. Endogenous melatonin provides an effective circadian message to both the suprachiasmatic nuclei and the pars tuberalis of the rat. *J. Pineal. Res.* 2009; 46: 95–105.
[47] Di Bella G, Mascia F, Gualano L, Di Bella L. Melatonin anticancer effects. *Int. J. Mol. Sci.* 2013; 14: 2410–2430.
[48] Buhr ED, Yoo SH, Takahashi JS. Temperature as a universal resetting cue for mammalian circadian oscillators. *Science* 2010; 330: 379–385.
[49] Cajochen C, Kräuchi K, Wirz-Justice A. Role of melatonin in the regulation of human circadian rhythms and sleep. *J. Neuroendocrinol.* 2003; 15: 432-437.
[50] Kräuchi K. How is the circadian rhythm of core body temperature regulated? *Clin. Autonom. Res.* 2002; 12: 147-149.
[51] van Someren EJW. Sleep propensity is modulated by circadian and behavior-induced changes in cutaneous temperature. *J. Therm. Biol.* 2006; 29: 437-444.
[52] Scheer FA, Van Montfrans GA, van Someren EJ, Mairuhu G, Buijs RM. Daily nighttime melatonin reduces blood pressure in male patients with essential hypertension. *Hypertension* 2004; 43: 192-197.
[53] Staessen JA, Wang JG, Thijs L. Cardiovascular protection and blood pressure reduction: a meta-analysis. *Lancet* 2001; 358: 1305-1315.
[54] Stoynev AG, Ikonomov OC, Minkova NK, Zacharieva SZ, Stoyanovsky VG. Circadian rhythms of arterial pressure: basic regulatory mechanisms and clinical value. *Acta Physiol. Pharmacol. Bulg.* 1999; 24: 43-51.
[55] Cornelissen G, Tamura K, Tarquini B, Germanò G, Fersini C, Rostagno C, Zaslavskaya RM, Carandente O, Carandente F, Halberg F. Differences in some circadian patterns of cardiac arrhythmia, myocardial infarction and other adverse vascular events. *Chronobiologia* 1994; 21: 79-88.
[56] Singh RB, Cornelissen G, Weydahl A, Schwartzkopff O, Katinas G, Otsuka K, Watanabe Y, Yano S, Mori H, Ichimaru Y, Mitsutake G, Pella D, Fanghong L, Zhao Z, Rao RS, Gvozdjakova A, Halberg F. Circadian heart rate and blood pressure variability considered for research and patient care. *Int. J. Cardiol.* 2003; 87: 9-28.
[57] Cornelissen G, Halberg F, Burioka N, Perfetto F, Tarquini R, Bakken EE. Do plasma melatonin concentrations decline with age? *Am. J. Med.* 2000; 109: 343-344.
[58] Haus E., Smolensky M. Biological clocks and shift work: circadian dysregulation and potential long-term effects. *Cancer Causes Control.* 2006; 17: 489-500.
[59] Takahashi JS, Hong HK, Ko CH, McDearmon EL. The genetics of mammalian circadian order and disorder: implications for physiology and disease. *Nat. Rev. Genet.* 2008; 9: 764-775.

[60] Wood PA, Yang X, Hrushesky WJ: Clock genes and cancer. *Integr. Cancer Ther.* 2009; 8: 303-308.

[61] Stevens RG. Working against our endogenous circadian clock: breast cancer and electric lighting in the modern world. *Mutat. Res. 2009*; 680: 106-108.

[62] Menet JS, Rosbash M. When brain clocks lose track of time: cause or consequence of neuropsychiatric disorders. *Current Opinion in Neurobiology* 2011; 21: 1–9.

[63] Bednarova K, Kodrik D, Krishnan N. Nature's timepiece—molecular coordination of metabolism and its impact on aging. *Int. J. Mol. Sci.* 2013; 14: 3026-3049.

[64] Gubin GD, Gerlovin ASh. Sutochnye Ritmy Biologicheskikh Protsessov v Ontogeneze Pozvonochnykh. [Circadian Rhythms of Biological Processes in Ontogeny of Vertebrata]. *Nauka*. Novosibirsk, 1980; 276 pp. (In Russian).

[65] Erren TC, Reiter RJ. Defining chronodisruption. *J Pineal Res* 2009; 46: 245–247.

[66] Reinberg AE, Ashkenazi I, Smolensky MH. Euchronism. allochronism. and dyschronism: is internal desynchronization of human circadian rhythms a sign of illness? *Chronobiol. Int.* 2007; 24: 553–588.

[67] Gubin GD, Gubin DG. [Classification of desynchronosis due to its causal factors and mechanisms. Two principles of chronotherapy] (Classifikaciya desynchronozov po prichinnomu factoru I mehanismam razvitiya. Dva principa terapii desinhronoza). *Fundamental Research* 2004; 1: 50. (In Russian).

[68] Bild DE, Fitzpatrick A, Fried LP, Wong ND, Haan MN, Lyles M, Bovill E, Polak JF, Schulz R. Age-related trends in cardiovascular morbidity and physical functioning in the elderly: the Cardiovascular Health Study. *J Am Geriatr Soc* 1993; 41: 1047-1056.

[69] O'Sullivan C, Duggan J, Atkins N, O'Brien E. Twenty-four-hour ambulatory blood pressure in community-dwelling elderly men and women, aged 60-102 years. *J. Hypertens.* 2003; 21: 1641-1647.

[70] Trenkwalder P, Ruland D, Stender M, Gebhard J, Trenkwalder C, Lydtin H, Hense HW. Prevalence, awareness, treatment and control of hypertension in a population over the age of 65 years: results from the Starnberg Study on Epidemiology of Parkinsonism and Hypertension in the Elderly (STEPHY). *J. Hypertens.* 1994; 12: 709-716.

[71] Gubin DG, Gubin GD. Some general effects of aging upon circadian parameters of cardiovascular variables assessed longitudinally by ambulatory monitoring. *Chronobiol. Int.* 2001; 18: 1106-1107.

[72] Ikonomov O, Stoynev G, Cornelissen G, Stoynev A, Hillman D, Madjirova N, Kane R, Halberg F. The blood pressure and heart rate chronome of centenarians. *Chronobiologia* 1991; 18: 167-179.

[73] Otsuka K, Watanabe H, Cornelissen G, Shinoda M, Uezono K, Kawasaki T, Halberg F. Gender, age and circadian blood-pressure variation of apparently healthy rural vs metropolitan Japanese. *Chronobiologia* 1990; 17: 253-265.

[74] Weinert D. Age-dependent changes of the circadian system. *Chronobiol. Int.* 2000; 17: 261-283.

[75] Scheer FA, Kalsbeek A, Buijs RM. Cardiovascular control by the suprachiasmatic nucleus: neural and neuroendocrine mechanisms in human and rat. *Biol. Chem.* 2003; 384: 697-709.

[76] Benton LA, Berry SJ, Yates FE. Ultradian oscillations in human blood pressure: effects of age. In: Temporal Disorder in Human Oscillatory Systems. *Springer Series in Synergetics* 1987; 36: 141-149.

[77] Kawamura H, Ozawa Y, Jumabay M, Mitsubayashi H, Izumi Y, Mahmut M, Ming MY, Aisa M, Cheng ZH, Wang SZ. Time-series analysis of systolic blood pressure variation in thirty-three Uygur centenarians in China. *Hypertens. Res.* 2003; 26: 597-601.
[78] Wühl E, Hadstein C, Mehls O, Schaefer F. and the ESCAPE Trial Group. Ultradian but not circadian blood pressure rhythms correlate with renal dysfunction in children with chronic renal failure. *JASN* 2005; 16(3): 747-754.
[79] Perez-Lloret S, Aguirre AG, Cardinali DP, Toblli JE. Disruption of ultradian and circadian rhythms of blood pressure in nondipper hypertensive patients. *Hypertension* 2004; 44(3): 11-15.
[80] Fotherby MD, Potter JF. Twenty-four-hour ambulatory blood pressure in old and very old subjects. *J. Hypertens.* 1995; 13: 1742-1746.
[81] Kobrin I, Oigman W, Kumar A, Ventura HO, Messerli FH, Frohlich ED, Dunn FG. Diurnal variation of blood pressure in elderly patients with essential hypertension. *J. Am. Geriatr. Soc.* 1984; 32: 896-899.
[82] Ohya Y, Ohtsubo T, Tsuchihashi T, Eto K, Sadanaga T, Nagao T, Abe I, Fujishima M. Altered diurnal variation of blood pressure in elderly subjects with decreased activity of daily living and impaired cognitive function. *Hypertens. Res.* 2001; 24: 655-661.
[83] Suzuki Y, Kuwajima I, Aono T, Kanemaru A, Nishinaga M, Shibata H, Ozawa T. Prognostic value of nighttime blood pressure in the elderly: a prospective study of 24-hour blood pressure. *Hypertens Res.* 2000; 23: 323-330.
[84] Vitaterna MH, Takahashi JS, Turek FW. Overview of circadian rhythms. *Alcohol Research & Health* 2001; 25: 85-92.
[85] Gubin GD, Weinert D. [Biorhythms and age] Bioritmi i vozrast. *Usp. Fiziol. Nauk* 1991; 22: 77-96.
[86] Libert S, Bonkowski MS, Pointer K, Pletcher SD, Guarente L. Deviation of innate circadian period from 24 h reduces longevity in mice. *Aging Cell* 2012; 11(5): 794-800.
[87] Wyse CA, Coogan AN. Impact of aging on diurnal expression patterns of CLOCK and BMAL1 in the mouse brain. *Brain Res.* 2010; 1337: 21–31.
[88] Chang HC, Guarente L. SIRT1 mediates central circadian control in the SCN by a mechanism that decays with aging. *Cell* 2013; 153: 1448-1460.
[89] Cornelissen G. When you eat matters: 60 years of Franz Halberg's nutrition chronomics. The Open *Nutraceuticals J.* 2012; 5 (Suppl 1-M1): 16-44.
[90] Zhou W, Li Y, Wang X, Wu L, Wang Y. MiR-206-mediated mechanism of the mammalian circadian clock. *BMC Systems Biology* 2011; 5: 141.
[91] Nandi A, Vaz C, Bhattacharaya A, Ramaswamy R. miRNA-regulated dynamics in circadian oscillator models. *BMC Systems Biology* 2009; 3: 45.
[92] Hughes ME, DiTacchio L, Hayes R, Vollmers C, Pulivarthy S, Baggs JE, Panda S, Hogenesch JB. Harmonics of circadian gene transcription in mammals. *PLoS Genet.* 2009; 5. e1000442.
[93] Gachon F, Bonnefont X. Circadian clock-coordinated hepatic lipid metabolism: only transcriptional regulation? *Aging* 2010; 2:101–106.
[94] Cretenet G, Le Clech M, Gachon F. Circadian clock-coordinated 12 hr period rhythmic activation of the ire1α pathway controls lipid metabolism in mouse liver. *Cell Metab.* 2010; 11: 47–57.

[95] Damiola F, Le Minh N, Preitner N, Kornmann B, Fleury-Olela F, Schibler U. Restricted feeding uncouples circadian oscillators in peripheral tissues from the central pacemaker in the suprachiasmatic nucleus. *Genes Dev.* 2000; 14(23): 2950-2961.

[96] Yu E, Weaver DR. Disrupting the circadian clock: Gene-specific effects on aging cancer and other phenotypes. *Aging* 2011; 3: 479-493.

[97] Gubin DG, Gubin GD, Gapon LI. [Advantages of chronobiologic time-specified reference limits for ambulatory blood pressure monitoring data analysis]. *Vestnik aritmologii* 2000; 16: 84-94. (In Russian).

[98] Pagani L, Schmitt K, Meier F, Izakovic C, Roemer K, Viola A, Cajochen C, Wirz-Justice A, Brown SA, Eckert A. Serum factors in older individuals change cellular clock properties. *Proc. Natl. Acad .Sci.*2011; 108: 7218-7223.

[99] Rajaratnam SM, Arendt J. Health in a 24-h society. *Lancet* 2001; 358: 999–1005.

[100] Arendt J. Biological rhythms during residence in Polar Regions. *Chronobiol Int* 2012; 29(4): 379-394.

[101] Arendt J. Shift work: coping with the biological clock. *Occup Med* (Lond). 2010; 60: 10–20.

[102] Blask D, Brainard G, Gibbons R, Lockley S, Stevens R, Motta M. Light Pollution: Adverse Health Effects of Nighttime Lighting. *Report 4 of the Council on Science and Public Health,* American Medical Association 2012; 25 pp.

[103] Skene DJ, Swaab DF. Melatonin rhythmicity: effect of age and Alzheimer's disease. *Exp. Gerontol.* 2003; 38: 199-206.

[104] Touitou Y. Human aging and melatonin. Clinical relevance. *Exp. Gerontol.* 2001; 36: 1083-1100.

[105] Youngstedt SD, Kripke DF, Elliott JA, Klauber MR. Circadian abnormalities in older adults. *J. Pineal. Res.* 2001; 31: 264-272.

[106] Zhao ZY, Xie Y, Fu YR, Bogdan A, Touitou Y. Aging and the circadian rhythm of melatonin: a cross-sectional study of Chinese subjects 30-110 yr of age. *Chronobiol. Int.* 2002; 19: 1171-1182.

In: Circadian Cardiology with Focus on Both Prevention ... ISBN: 978-1-63463-956-9
Editors: K. Hristova, A. Shehab, G. Cornelissen et al.

Chapter 10

Personalized Circadian Timing of Exercise

R. B. Singh [1,*]***, Franz Halberg*** [2]***,***
Germaine Cornelissen [2]***, Jarmila Siegelova*** [3]***, Krasimira Hristova*** [4]***,***
Eri Toda [5]***, Toru Takahashi*** [6]***, Jan Fedacko*** [7]
and Kuniaki Otsuka [8]

[1]Halberg Hospital and Research Institute, Moradabad, India;
[2]Halberg Chronobiology Center, University of Minnesota, Minneapolis, MN, US,
[3]Masaryk University, Brno, Czech Republic
[4]Department of cardiac Imaging, National Heart Hospital, Sofia, Bulgaria
[5]Department of Cardiology, Tokai University Hachioji Hospital, Tokyo, Japan
[6]Department of Nutrition, Graduate School of Human Environment Science, Fukuoka Women's University, Japan
[7]PJ Safaric University, Kosice, Slovakia
[8]Department of Internal Medicine, Tokyo Women's Medical University, Tokyo, Japan

Abstract

That exercise effects are circadian stage- dependent and that exercise at the wrong circadian stages can induce a Vascular Variability Disorder (VVD) has been documented earlier. Herein we show how statistically significant results can be obtained with a simple individualized design, that can be self-applied by everybody to optimize a desired effect by Chronobiologically-interpreted Ambulatory Blood Pressure (BP) and heart rate (HR) Monitoring C-ABPM). A 68-year old internist cardiologist monitored himself at half-hourly intervals, with interruptions, for 3- to7-day sessions, exercise timing being kept the same within a given session and changed from one session to another. Exercise training in the morning was associated with lowest BP and HR MESORs (Midline

* Correspondence: Prof. Dr. R.B. Singh, MD Halberg Hospital and Research Institute, Civil Lines, Moradabad-10(UP) 244001, India. Email: rbs@tsimtsoum.net, drkk@dataone.in.

Estimating Statistic Of Rhythm, a rhythm-adjusted mean) as compared to exercise done at mid-day, late afternoon or in the evening. We review some of the literature, speculate about its meaning, yet do not stray beyond a single case in our conclusions for everybody: it seems possible and desirable to personalize exercise at a time of pertinence rather than convenience. One shoe or one timing does not fit all.

Keywords: Timing, physical activity, circadian, heart rate, blood pressure

Introduction

The International College of Cardiology; American Heart Association Council on Nutrition, Physical Activity, and Metabolism; American Heart Association Council on Clinical Cardiology; and American College of Sports Medicine call on all populations of the world, and their employers to create a culture of physical activity and health for the prevention of cardiovascular diseases (CVDs) and diabetes mellitus [1-3]. Usually there is no prescribed time for exercise. Regular exercise increases life expectancy, quality of life and work capability and productivity. In a majority of populations, some individuals work around the clock, and many of them exercise when convenient. An acute bout of exercise might increase the risk of an adverse clinical event, such as sudden cardiac death (SCD) or acute coronary syndrome (ACS), particularly in vulnerable subjects (3, 4, cf. 5), but when this or other harm is likely to occur is usually not considered.

Differential long-term versus short-term responses in heart rate variability to exercise may be a contributory factor. Failure to achieve target HR and a smaller increase in HR with exercise are predictive of overall mortality and incident coronary heart disease [6]. Exercise training in patients at vascular disease risk is reportedly beneficial [7]. HR variability (HRV), diminished with aging, is augmented after exercise training [8]. HRV is higher in athletes than in healthy sedentary controls [9, 10], as it is in physically active individuals as compared to sedentary and smoker counterparts [11], whereas it is reduced in association with obesity [12]. During exercise [13-21] and during recovery from exercise [22], however, HRV is reportedly reduced. There may thus be a dual effect of exercise on HRV, an acute effect resulting in decreased HRV, and a chronic effect resulting in increased HRV [23, 24].

There is evidence that the incidence of cardiovascular events peaks in the morning around 09:00 to 11:00 [25, 26]. This evidence, however, is statistical. Since there is variability even among groups, i.e., among the populations investigated, whether one focuses on cardiovascular or overall mortality (Figure 1A) or beyond [26], it appears that timing not only of death but of all interventions for sustaining health is best individualized (Figure 1B).

In one subject, based on 7-day C-ABPM, a span of evening exercise was associated with the presence of a circadian overswing (CHAT, brief for Circadian Hyper-Amplitude Tension), not found when the same subject exercised in the morning [5]. This may not be trivial, as CHAT has been associated with a larger increase in adverse vascular outcomes than MESOR-hypertension in a prospective outcome study (Figure 2) [27]. As noted above, effects of exercise on HRV may be opposite, whether short-term or long-term effects were examined: acute effects were reported as a transient lowering of HRV, yet in the long-run, HRV was higher in association with exercise [3, 4]. Some cardiac patients prefer to schedule leisure-time physical activity before lunch or in the evening [28-30].

The question arises as to whether exercise at certain about-daily rhythm stages (the clock-hours may be different in shift-workers) may be inherently more risky than physical activity performed at other times (not of day but of our built-in circadian system). Epidemiological studies suggest that morning exercise is just as safe as afternoon exercise for cardiac patients enrolled in a supervised rehabilitation program [31]. It is not clear whether circadian time alters the risk of a cardiac event, occurring during spontaneous physical activity performed by individuals with established risk factors for heart disease. Regarding the safety and usefulness of giving priority for morning, evening or daytime exercise, there may be various reasons [31], yet the individualized timing is one of them [5].

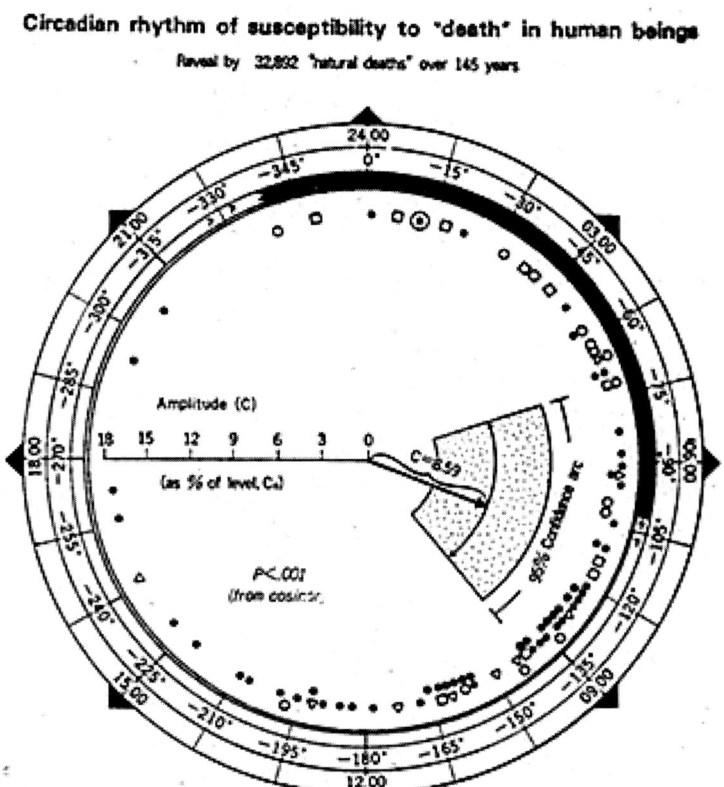

Mortality category	No. of cases	No. of (C, φ)	Symbol for acrophase
Pediatric	21,673	12	○
Postsurgical	500	1	⊙
Cardiovascular	7,644	9	△
Pulmonary	9,357	12	□
Cause unspecified	393,718	54	•
Cosinor of above	432,892	88	

Figure 1A. Polar plot summarizes the timing of individual acrophases for 88 time series representing nearly 433,000 deaths. Almost 65% of the acrophases occur between 04:00 and 11:00. Greatest overall sample mortality occurs at 07:12. Inner shaded portion indicates suspected nocturnal rest span of sample prior to death. From (25). © Halberg.

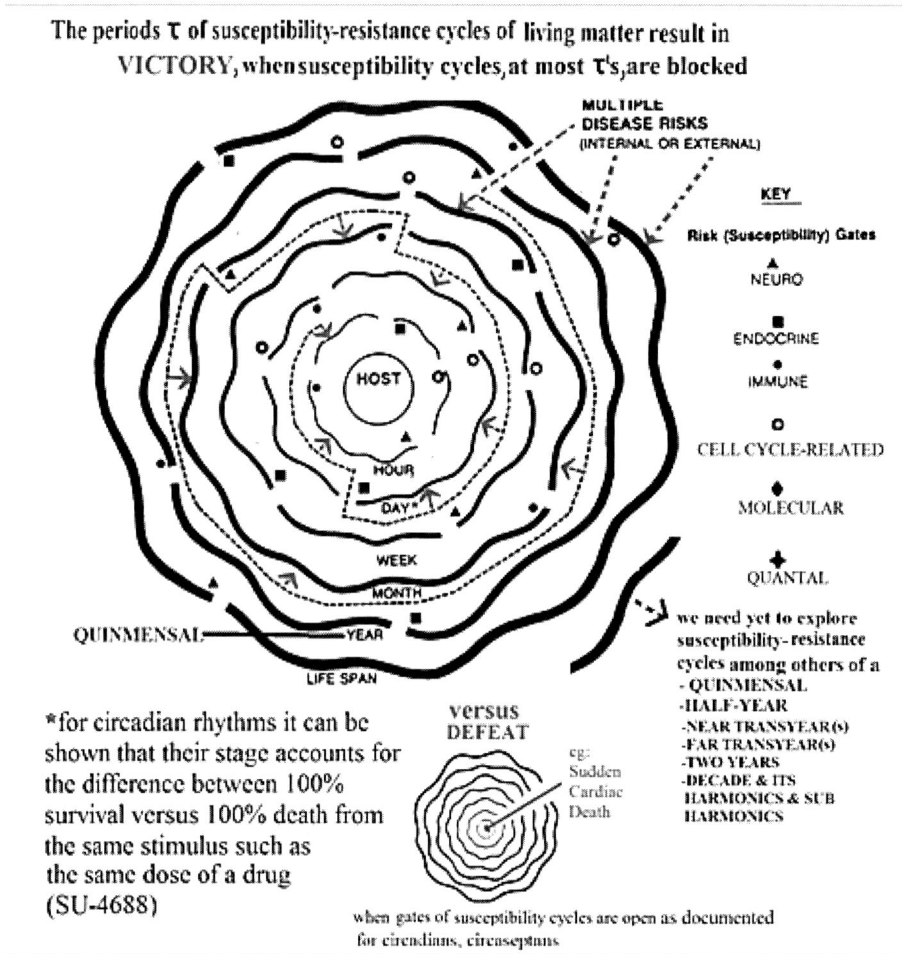

Figure 1B. Susceptibility resistance cycles occur in a broad spectrum of extra-circadian as well as circadian rhythms. We die only once, and not necessarily after awakening, even in population studies (see scatter in Figure 1A). © Halberg.

While exercise is an easily accessible and inexpensive approach to improve cardiovascular health, control weight gain, and increase survival chances after a morbid event such as a myocardial infarction [1-3], it is sometimes associated with untoward effects in healthy [5] and perhaps more so in vulnerable subjects. As noted above, a contributory factor may be HRV [4-24, 28].

Circadian rhythms of increased oxidative stress, platelet aggregation, deficiency of magnesium, antioxidants, vitamin A,E,C, CoQ10, increased lipoprotein (a), catecholamines and cortisol have been reported to peak in the second quarter of the day (that is between 06:00-12:00) [32-35]. There may be circadian variations in the circulatory responses to exercise among subjects not having exercise training [36, 37]. One may speculate that exercise training in the early morning (between 05:00 and 07:00) may prepare our body and cause adaptation to fight the circadian biological stress impinging upon our body after awakening, in the second quarter of one habitual day. As part of a healthy lifestyle or for rehabilitation purposes, exercise need not be practiced at times of convenience rather than pertinence.

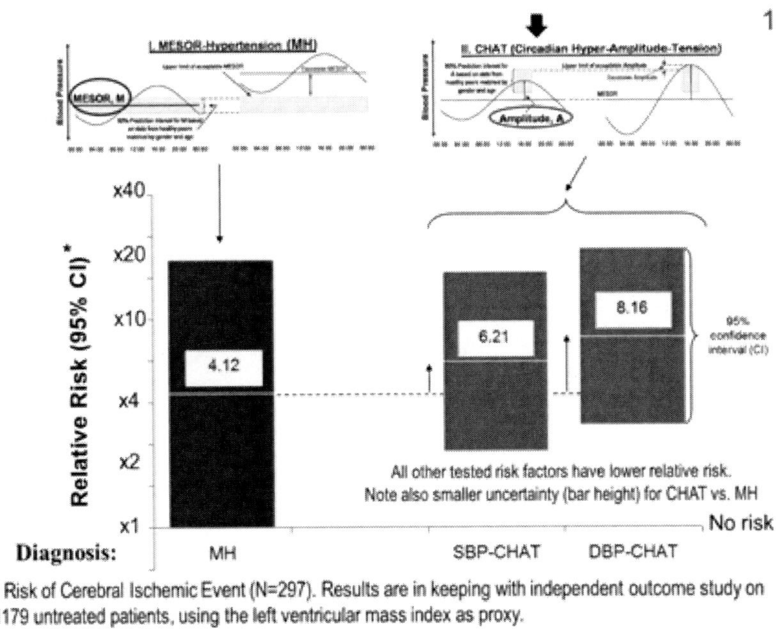

Figure 2. High blood pressure (BP) is a known risk factor for adverse cardiovascular outcomes. In a 6-year prospective study of 297 patients, MESOR-hypertension (MESOR, a rhythm-adjusted mean, is an acronym for Midline Estimating Statistic Of Rhythm) was associated with a relative risk of 4.12 (left). In other words, patients with an elevated BP were 4.12 times more likely to suffer an adverse event than normotensive individuals. The uncertainty of this relative risk, gauged by its 95% confidence interval (height of bar) is barely above the no-risk line. By comparison, an abnormal circadian pattern of BP characterized by an excessive circadian swing (CHAT, short for Circadian Hyper-Amplitude-Tension) is associated with an even larger relative risk of 6.21 (systolic CHAT) or 8.16 (diastolic CHAT), their 95% confidence intervals being well above the no-risk line. The diagnosis of CHAT and of other Vascular Variability Disorders (VVDs) requires the around-the-clock monitoring of BP and heart rate, preferably for a minimum of 7 days at the outset. It cannot be obtained from single measurements typically taken in the physician's office as is current practice today. CHAT must not be induced by exercise. © Halberg.

To examine whether the timing of exercise can be optimized on an individual basis, this study examines the effects of exercise done during different times of the 24-hour day by the senior author (RB), an internist-cardiologist who has been regularly doing exercise twice in a day for the last few years.

Case Report and Methods

A brief description of this subject's results has been published [38]. Here, in the context of the literature [39], we describe diet and lifestyle factors in this case in more detail and with a review of prior work. RB is from social class 2, physician by occupation, working about 10-12 hours a day and has regular moderate spare time activity without any occupational physical activity. This 68-year old man believes in and recommends self-experimentation, in this case applicable to everybody to study the effects of timed exercise on BP and HR, which he implemented during a 4-week span. RB was on a schedule of sleep from about 22:30 to about 06:00, alternating with wakefulness, with an afternoon nap from 14:00 to 16:00. According to a 7-day food intake diary, the consumption of fruits, vegetables and nuts was 420 g/day, whole grains, pulses, porridge, wheat and rice about 310 g/day, milk 100 ml/day, mustard oil 25 g/day, Indian clarified butter 10 g/day, sugar 15g/day, and salt 6.5g/day. He also took 2-3 eggs and 1-2 servings of chicken in a week. Alcohol intake was 5-6 drinks per week on 2-3 days in a week without tobacco use. RB had been doing regular physical activity for 25-30 minutes daily in the morning at about 6:30 hours, as well as in the evening around 18:30 for the last several years. He has a history of chronic bronchitis mostly in the winter. He also has had "hypertension" (a 160/100 mmHg spotcheck) and MESOR-hypertension of about 4-week duration in 2010, during his visit to the USA, followed by travel to Russia and Poland for about 15 days.

At another physical examination his BP happened to be 130/88 mmHg and his pulse 78 beats/min. His waist circumference is 91 cm, and his body mass index 25 kg/m^2. Systemic examination revealed no cardiovascular, respiratory, gastrointestinal, or nervous system abnormality. He further measured his BP and HR around the clock at 30-min intervals, with interruptions, for spans of 3-7 days with an ambulatory monitor (TM-2421, A&D, Tokyo, Japan) during which he exercised at different times (around 06:00, 09:00, 12:00, 15:30, and 18:00). Each day's data as well as each record as a whole was analyzed by cosinor (40, 41, cf. 27, 42-44) to obtain estimates of the MESOR (M, a rhythm-adjusted mean), and 24-hour amplitude and acrophase (phase of fitted model's maximum). The cosinor method is illustrated in Figure 3.

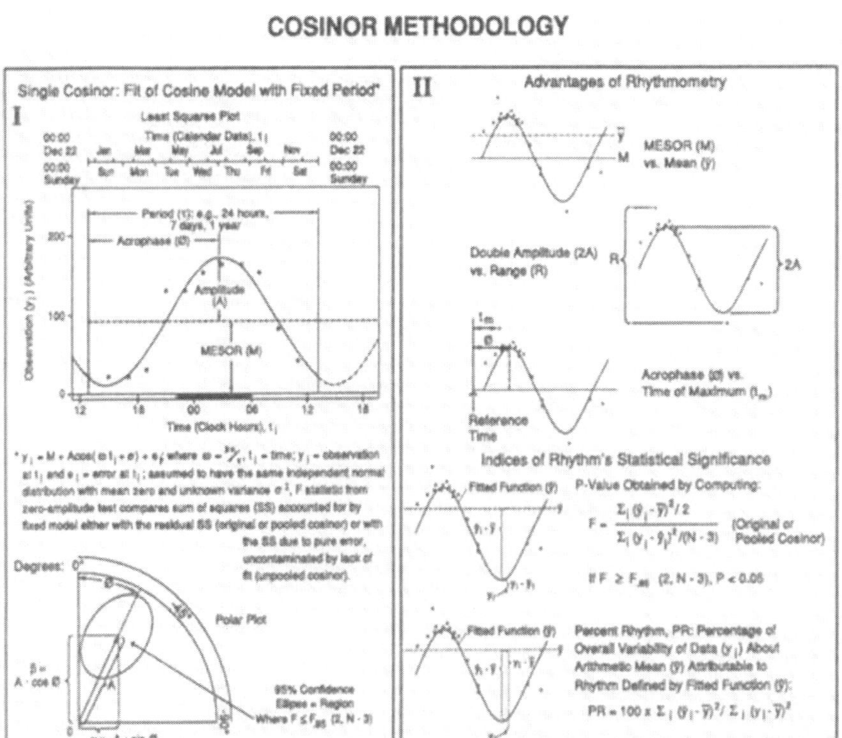

Figure 3. Parameters from cosine fitting, including the amplitude discussed in Figures 3, 5 and 6, and the MESORs in Figure 4. © Halberg.

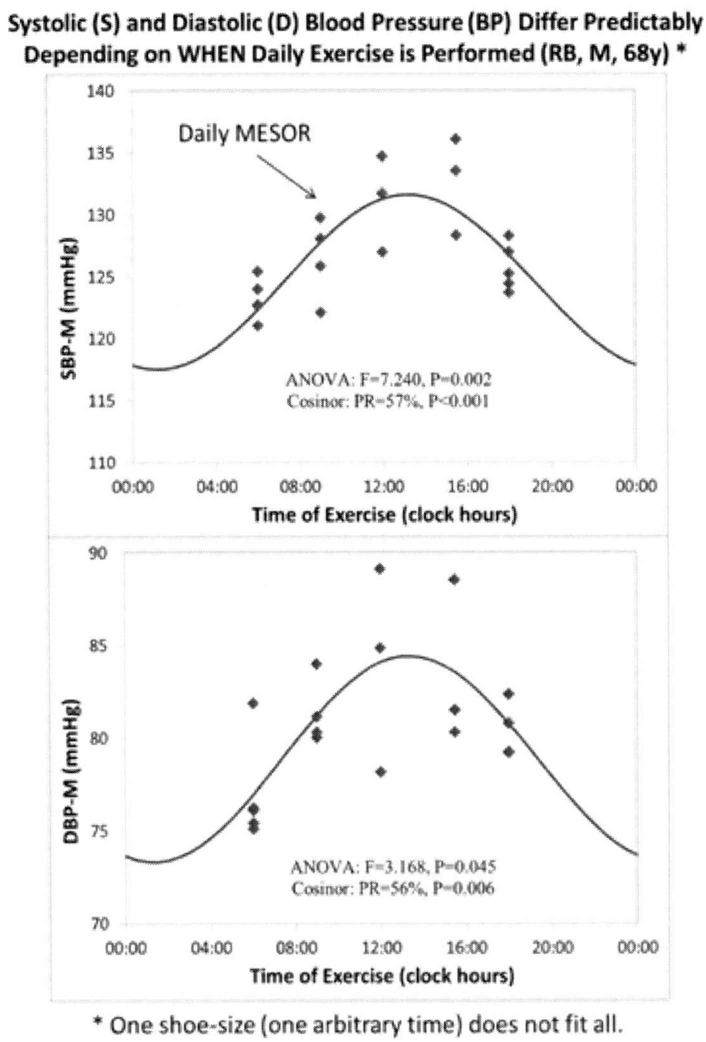

Figure 4. The same subject exercised at different times after awakening in a schedule of sleep/rest from 22:30 to 06:00, demonstrating a circadian rhythm in response to exercise validated by both cosinor and 1-way analysis of variance (ANOVA). © Halberg.

Results of these intermediate computations were assigned to the clock hour of exercise during the given span to yield new time series amenable to further cosinor analysis, complementing a one-way analysis of variance, the main effect being the timing of exercise.

Results

Systolic (S) and diastolic (D) BP values were predictably higher during the span when exercise was performed daily around noon as compared to spans when it was done earlier or later in the day, Figure 4.

On the average, daily MESORs were 9.5 or 7.1 mmHg higher when exercise was done around 15:30 (SBP) or noon (DBP) by comparison to 06:00 in the morning (Table 1).The circadian amplitude of HR was also predictably larger (by a factor of about 2) when exercise was performed around noon (P = 0.042 by cosinor).

There is evidence that the same daily dose of the same medication administered to the same patient has different effects on the circadian pattern of BP depending on when it is administered (45, cf. 27, 42-44). A circadian stage-dependent effect of a given treatment has been observed with both pharmacologic [45] and non-pharmacologic [46, 47] interventions, including exercise [48], Figure 5. Whereas an elevated BP is a known cardiovascular disease risk factor, other alterations of the circadian patterns of BP and HR have also been related to an increased cardiovascular disease risk, including too large a circadian amplitude of BP [27]

Table 1. Circadian Characteristics of Blood Pressure and Heart Rate on Days with Exercise at Different Times*

Start date (2011)	T (hours)	Time of Exercise	SBP-M	SBP-A	SBP-ϕ	DBP-M	DBP-A	DBP-ϕ	HR-M	HR-A	HR-ϕ	HR-SD
10/21	167	06:00	122.3	2.92	13:36	76.2	3.13	10:28	88.0	3.99	15:48	9.95
10/31	128	09:00	125.8	6.24	15:32	80.9	7.18	14:24	80.4	6.56	15:52	10.22
11/12	73	12:00	130.6	6.65	14:32	83.5	5.56	14:16	85.4	2.56	02:12	11.87
11/15	90	15:30	132.3	6.03	16:44	82.6	8.06	15:16	84.0	4.97	15:16	14.83
11/6	143	18:00	124.7	7.10	14:08	79.6	4.54	13:12	83.7	2.44	00:12	10.48
P-value (cosinor)			<0.001			0.006				0.042		0.098
P-value (ANOVA)			0.002			0.045				0.105		0.055

* Exercise performed daily; Data during days when exercise was performed at the same time were analyzed to yield estimates of the MESOR (M), 24-hour amplitude (A) and acrophase (ϕ) of systolic (S) and diastolic (D) blood pressure (BP) and heart rate (HR), and of the standard deviation (SD) of HR. For assessing any circadian stage-dependent effects of exercise on these parameters, data were analyzed for each day separately, and daily estimates were assigned to the time of exercise.

T: Duration of monitoring; M and A of SBP and DBP in mmHg; M and A of HR in beats/min; □ in hh:mm.

Discussion

As seen in Figure 5, the immediate response of BP and HR to graded exercise was also circadian stage-dependent in an earlier study on four marathon runners [48]. Exercise by a 46-year old man in the Czech Republic was associated with an increase in the circadian amplitude of BP that was more pronounced when exercise was done in the evening than in the morning, in his case leading to an abnormal circadian pattern carrying an increased cardiovascular disease risk [5]. Depending on whether the circadian pattern of BP tends to have too large an amplitude at the outset or not, exercise should be timed to reduce both long-term and short-term BP without bringing about VVDs that may be harmful and associated with a higher risk of a hard ischemic event than MESOR-hypertension, Figure 2 C-ABPM can serve for the optimization of the timing of exercise on an individual basis [27].

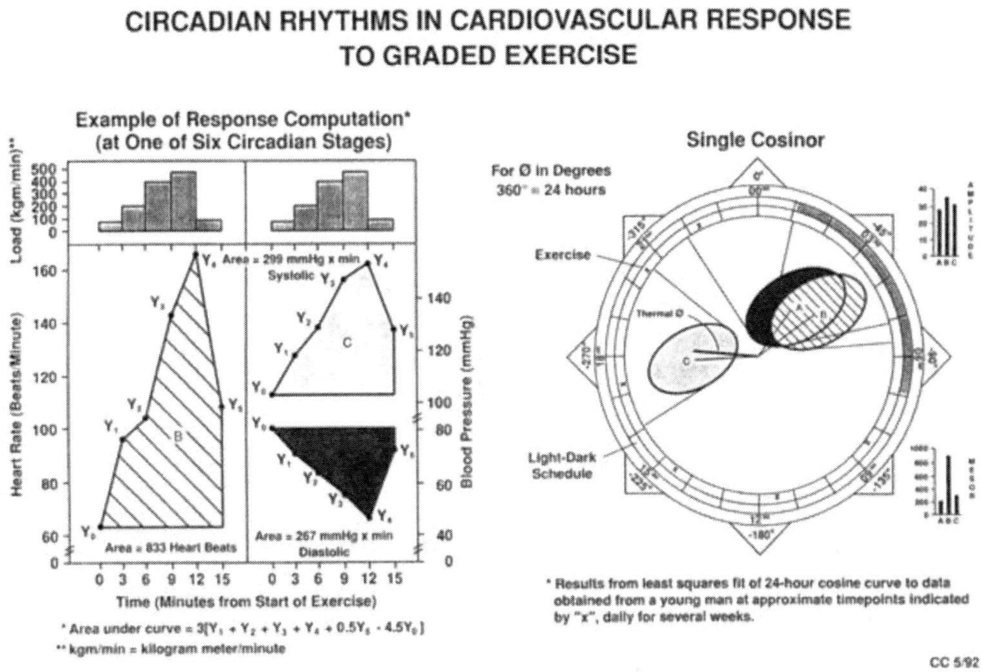

Figure 5. Changes in blood pressure (BP) and heart rate (HR) in response to graded exercise vary according to the circadian stage at which exercise is performed. A young, healthy subject, on an approximately isocaloric food ration after each exercise session, exercised for a 15-minute span about every 3 hours during wakefulness for about one month. The subject exercised with individually fixed loads, starting with a 3-min light load warm-up, followed by 3-minute increments of loads to obtain 85% of maximal HR and ending with a 3-min deceleration by pedaling without a load. At the end of each 3-min span, the BP and HR were self-measured. For this subject, the largest responses to exercise were observed during the rest span for diastolic BP and HR, and during the late afternoon for systolic BP, when oral temperature was also highest (not shown). © Halberg.

High concentrations of homocysteine are associated with increased cardiovascular disease risk, by causing endothelial dysfunction and platelet aggregation. Exercise in the evening increases homocysteine more than exercise in the morning [37]. In view of this finding, exercise may be preferred in the morning compared to midday and evening, an observation also confirmed in our study in relation to its effects on RB' BP. There may be different conditions regarding the safety and usefulness of giving priority for morning, evening or daytime exercise in different individuals. Clinical studies indicate that acute ischemic events show a circadian rhythm in their incidence which correlates with increased activity of the sympathetic nervous system and increased platelet aggregation during early morning, underlying results in Figure 1. Increased concentrations of plasma catecholamine and cortisol following sympathetic nervous system activation can result in increased platelet aggregation (possibly by activating platelet α2-adrenergic receptors), oxidative stress and deficiency of antioxidants and magnesium which are arrhythmogenic [32-35]. Further evidence indicates that acute vigorous physical activity or exercise may trigger a coronary syndrome and induce thrombogenesis [49].

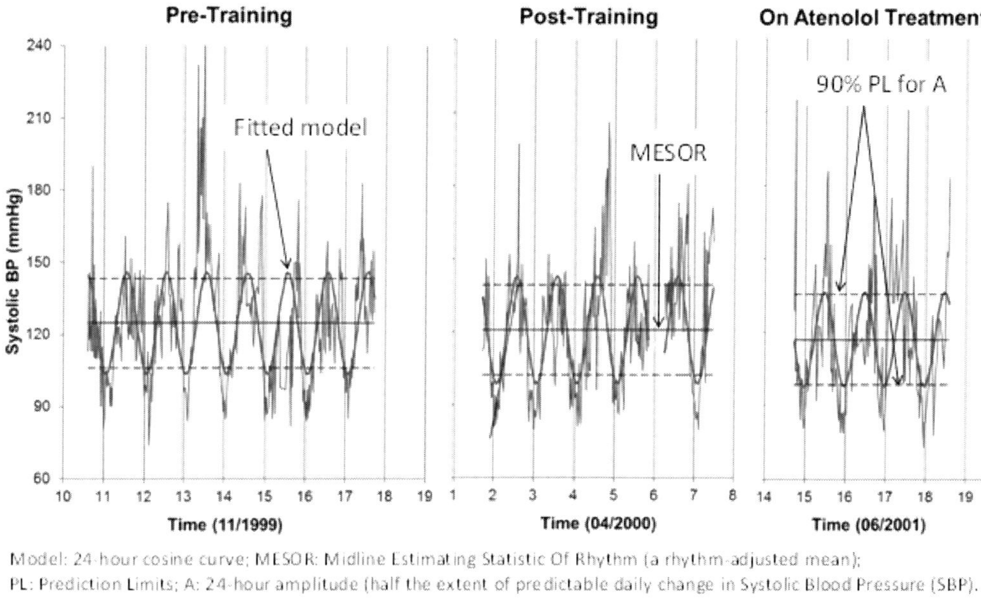

Figure 6. Systolic CHAT seen before (left), and after (center), 15 weeks of aerobic training in a MESOR-normotensive physically fit postmenopausal African American woman, and its relative abatement after 12 weeks of atenolol treatment that was initiated eleven months after discontinuation of training (right). Middle horizontal line indicates MESOR (Midline Estimating Statistic Of Rhythm), and upper and lower lines are chronodesms or time-specified 90% prediction limits derived from clinically healthy peers matched to this subject's gender and age. Mean BP, HR, and HR variability were normal at the outset. Training had no effect on circadian BP variation, which exceeded gender- and age-specified limits, indicative of CHAT (Circadian Hyper-Amplitude-Tension). A 12-week treatment with the β1-adrenoceptor blocker atenolol, 14 months post-training, normalized CHAT and HR variability and further reduced BP and HR MESORs. These results do not prove but are in keeping with β1-adrenergic mediation of CHAT and its refractoriness to exercise training in one case. © Halberg.

An interaction of circadian rhythm and exercise appears to be important in the pathogenesis of atherothrombosis, because abnormalities in endogenous coagulation and fibrinolysis may contribute in the development of an acute cardiovascular event [4, 28, 32-35]. Circadian acrophases of fibrinogen and platelet rhythms have been reported in the afternoon and evening but no study was found in a review that related to the influence of exercise during morning and evening on fibrinogen and platelets [50]. There is some evidence indicating increased concentration of fibrinogen and decreased platelet counts following an acute single bout of high intensity exercise [51].

In an interesting experiment, participants exercised at various times of the day and the change in phase of various circadian rhythms was measured with a view of exercise-induced phase shifting [52]. The substantial levels of activity needed to obtain phase shifts may not be attainable by the majority of people. A lack of agreement with the phase-shifting effects of bright light was interpreted to indicate that exercise is not exerting its effects via photic entrainment pathways. An alternative explanation may involve exercise-induced hyperthermia. Exercise may have a different phase response curve to light in diurnally versus nocturnally active species. Nocturnal exercise can induce phase delays in the onset of melatonin.

Eurythmy therapy (EYT) is an anthroposophical movement therapy aimed at reducing fatigue symptoms and strain [53]. In a study examining whether the improvement in fatigue symptoms was accompanied by changes in the circadian rhythm of HRV, 25 women performed 10 hours of EYT over six weeks. Electrocardiograms (ECGs) were recorded before and after the EYT trial. HRV was quantified by parameters in the frequency and time domains and the nonlinear parameters of symbolic dynamics. The day-night contrast with predominance of vagal activity at night became more pronounced after the EYT training, and with decreased ultra-low and very low frequencies, the HRV shows changes interpreted as calmer sleep. During the night, the complexity of the HRV is increased as indicated by nonlinear parameters. The study reported "improvements" in HRV in terms of greater day-night contrast associated with an increase of vagal activity and calmer and more complex HRV patterns during sleep.

During human evolution, early man used to go for hunting near the river banks where animals come at sunrise to drink water [4]. Early man had marked excitement during running, climbing and maybe horse riding, all for hunting, causing a marked increase in mental and physical strain and BMR with an increased release of catecholamines and cortisol and excessive sweating. These physiological and biochemical changes may have been responsible for the origin and the evolution of the circadian rhythm of increased cardiovascular biomarkers and cardiovascular events in the second quarter of the day [4, 28, 32-35, 54-57]. One may speculate, extrapolating from a single case, that morning time exercise may help the adaptation of our cardiovascular responses, associated with a lower BP and HR which may be protective against any increase of cardiovascular events. In one study, among 209 adults, increased HR and BP MESORs were observed in the second quarter of the day [54]. Fruit and vegetable intake were inversely associated with MESOR hypertension [55]. In another study measuring HR and BP at half-hour intervals for 7 days, a larger circadian amplitude of HR was associated with active prayer in Hindu Indians in Asia [56].

In brief, exercise training undergone in the morning showed lowest cardiovascular responses in BP and HR MESORs, indicating, if and only if one could extrapolate beyond results of a single study (obviously one cannot do so), that the morning may be the best to have least adverse effects of the exercise. In addition, in a case of CHAT, Figure 6, fifteen weeks of aerobic training had no effect on this vascular variability disorder [58]. This study also discusses the potential cardiovascular complications of exercise as well as circadian rhythms of cardiovascular events that occur in the second quarter of the day. It is possible that screening patients before participation in exercise, excluding high-risk patients from certain activities, promptly evaluating possible prodromal symptoms, training fitness personnel for emergencies, and encouraging patients to avoid high-risk activities, and do exercise in the morning, appear prudent but this possibility has not been systematically evaluated. The only firm conclusion is that like all other measures, exercise is not only circadian stage-dependent [59], but it can also be personalized. The design of the present study shows that this can be done relatively promptly and simply. More generally interested readers can be referred to yet other publications (60-65).

Acknowledgments are due to International College of Nutrition for logistic support.
Conflict of interest has not been declared by the authors.

References

[1] Singh RB, and Indian Consensus Group. Prevalence and prevention of hypertension, diabetes mellitus, and coronary artery disease in India: A scientific statement of the International College of Nutrition and International College of Cardiology. *World Heart J*. 2010; 2: 31-44.

[2] Singh RB, Kumar A, Neki NS, Pella D, Rastogi SS, Basu TK, Acharya SN, Juneja L, Toru T, Otsuka K, De Meester F and Wilson DW. Diet and Lifestyle Guidelines and Desirable Levels of Risk Factors for Prevention of Cardiovascular Disease and Diabetes among Elderly Subjects. A Revised Scientific Statement of the International College of Cardiology and International College of Nutrition-2011.*World Heart J*. 2011; 3: 305-18.

[3] American Heart Association Council on Nutrition, Physical Activity, and Metabolism; American Heart Association Council on Clinical Cardiology; American College of Sports Medicine. Exercise and acute cardiovascular events placing the risks into perspective: a scientific statement from the American Heart Association Council on Nutrition, Physical Activity, and Metabolism and the Council on Clinical Cardiology. *Circulation*. 2007; 115: 2358-68.

[4] Singh RB, Gupta A, Pella D, Kumar A. Circadian Cardiology. *World Heart J*. 2008; 1: 101-125.

[5] Homolka P, Cornelissen G, Homolka A, Siegelova J, Halberg F. Exercise-associated transient circadian hypertension (CHAT)? Abstract, III International Conference, Civilization diseases in the spirit of V.I. Vernadsky, People's Friendship University of Russia, Moscow, Oct. 10-12, 2005, p. 419-421.

[6] Lauer MS, Okin PM, Larson MG, Evans JC, Levy D. Impaired heart rate response to graded exercise. Prognostic implications of chronotropic incompetence in the Framingham Heart Study. *Circulation* 1996; 93: 1520-1526.

[7] Malfatto G, Facchini M, Bragato R, Branzi G, Sala L, Leonetti G. Short- and long-term effects of exercise training on the tonic autonomic modulation of heart rate variability after myocardial infarction. *Eur. Heart J*. 1996; 17: 532-538.

[8] De Meersman RE. Heart rate variability and aerobic fitness. Am Heart J 1993; 125: 726-731.

[9] Costa O, Freitas J, Puig J, Carvalho MJ, Freitas A, Ramos J, Puga N, Lomba I, Fernandes P, de Freitas F. Spectrum analysis of the variability of heart rate in athletes. *Revista Portuguesa de Cardiologia* 1991; 10: 23-28.

[10] Puig J, Freitas J, Carvalho MJ, Puga N, Ramos J, Fernades P, Costa O, de Freitas AF. Spectral analysis of heart rate variability in athletes. *J. Sports Med & Physical Fitness* 1993; 33: 44-48.

[11] Gallagher D, Terenzi T, de Meersman R. Heart rate variability in smokers, sedentary and aerobically fit individuals. *Clinical Autonomic Research* 1992; 2: 383-387.

[12] Zahorska-Markiewicz B, Kuagowska E, Kucio C, Klin M. Heart rate variability in obesity. *Int. J. Obesity & Related Metabolic Disorders* 1993; 17: 21-23.

[13] Bernardi L, Salvucci F, Suardi R, Solda PL, Calciati A, Perlini S, Falcone C, Ricciardi L. Evidence for an intrinsic mechanism regulating heart rate variability in the

transplanted and the intact heart during submaximal dynamic exercise? *Cardiovascular Res.* 1990; 24: 969-981.

[14] Kamath MV, Fallen EL, McKelvie R. Effects of steady-state exercise on the power spectrum of heart rate variability. *Medicine & Science in Sports & Exercise* 1991; 23: 428-434.

[15] Yamamoto Y, Hughson RL, Peterson JC. Autonomic control of heart rate during exercise studied by heart rate variability spectral analysis. *J. Appl. Physiol.* 1991; 71: 1136-1142.

[16] Nakamura Y, Yamamoto Y, Muraoka I. Autonomic control of heart rate during physical exercise and fractal dimension of heart rate variability. *J. Appl. Physiol.* 1993; 74: 875-881.

[17] Perini R, Orizio C, Milesi S, Biancardi L, Baselli G, Veicsteinas A. Body position affects the power spectrum of heart rate variability during dynamic exercise. *Eur. J. Appl. Physiol. Occupational Physiol.* 1993; 66: 207-213.

[18] Breuer HW, Skyschally A, Schulz R, Martin C, Wehr M, Heusch G. Heart rate variability and circulating catecholamine concentrations during steady state exercise in healthy volunteers. *Br. Heart J.* 1993; 70: 144-149.

[19] Saito M, Nakamura Y. Cardiac autonomic control and muscle sympathetic nerve activity during dynamic exercise. *Jpn. J. Physiol.* 1995; 45: 961-977.

[20] Hagerman I, Berglund M, Lorin M, Nowak J, Sylven C. Chaos-related deterministic regulation of heart rate variability in time and frequency domains: effects of autonomic blockade and exercise. *Cardiovasc. Res.* 1996; 31: 410-418.

[21] Shin K, Minamitani H, Onishi S, Yamazaki H, Lee M. The power spectral analysis of heart rate variability in athleyes during dynamic exercise: part I. *Clin. Cardiol.* 1995; 18: 583-586.

[22] Weydahl A, Sothern RB, Cornelissen G. Heart rate variability during recovery from 30-minute exercise is greatest in the morning. Abstract, 25th Meeting, International Society for Chronobiology, Kemer-Antalya, Turkey, October 9-13, 2001. *Chronobiol. Int.* 2001; 18: 1217.

[23] Weydahl A., Cornelissen G., Halberg F., Siegelova J., Kumagai Y., Otsuka K. Chronobiologic optimization of exercise physiology and practice guided by heart rate variability. *Scripta medica* (Brno) 71: 215-230, 1998.

[24] Weydahl A, Sothern RB, Cornelissen G. Non-linear relation of heart rate variability during exercise recovery with local geomagnetic activity. *Biomed. Pharmacother.* 2002; 56 (Suppl 2): 298s-300s.

[25] Smolensky M, Halberg F, Sargent F II. Chronobiology of the life sequence. In: Itoh S, Ogata K, Yoshimura H. (Eds.) Advances in Climatic Physiology. Tokyo: *Igaku Shoin* Ltd.; 1972. p. 281-318.

[26] Cornelissen G, Tamura K, Tarquini B, Germanò G, Fersini C, Rostagno C, Zaslavskaya RM, Carandente O, Carandente F, Halberg F. Differences in some circadian patterns of cardiac arrhythmia, myocardial infarction and other adverse vascular events. *Chronobiologia* 1994; 21: 79-88.

[27] Halberg F, Cornelissen G, Otsuka K, Siegelova J, Fiser B, Dusek J, Homolka P, Sanchez S, Singh RB, BIOCOS Project. Extended consensus on need and means to detect Vascular Variability Disorders (VVDs) and Vascular Variability Syndromes (VVSs). *World Heart J.* 2010; 2(4): 279–305.

[28] Singh RB, Pella D, Sharma JP, Rastogi S, Kartikey K, Goel VK, Sharma R, Neki NS, Kumar A, Otsuka K. Increased concentrations of lipoprotein(a), circadian rhythms, and metabolic reactions, evoked by acute myocardial infarctions, associated with acute reactions in relation to large breakfast. *Biomed. Pharmacother.* 2004; 38 (suppl): 116-123.

[29] Singh RB, Pella D, Neki NS, Chandel JP, Rastogi S, Mori H, Otsuka K. Mechanism of acute myocardial infarction study. Biomed Pharmacother 2004; 58 (Suppl): 111-115.

[30] *Singh RB, Singh AK, Sharma JP, Singh RK, Kumar A, Rastogi SS, Singh G, Kartikey K, Singh S, Fedacko J, Pella D, De Meester F, Wilkzynska A* and *Wilson DW.* Nutrition in Chronocardiology: We are Indebted Professor Franz Halberg. *The Open. Nutra J.* 2012; 5: 45-65.

[31] Atkinson G, Drust B, George K, Reilly T, Waterhouse J. Chronobiological considerations for exercise and heart disease. *Sports Med.* 2006; 36: 487-500.

[32] Dubnov G, Pella D, Singh RB. The effect of an alpha-linolenic acid rich diet on the circadian rhythm of cardiac events. *World Heart J.* 2008; 1: 49-56.

[33] Singh RB, Fedacko J, Sharma JP, Vargova V, Sharma, Moshiri M, De Meester F, Otsuka K. Association of inflammation, heavy meals, magnesium, nitrite, and coenzyme Q10 deficiency and circadian rhythms with risk of acute coronary syndromes. *World Heart J.* 2010; 2: 219-228.

[34] Singh RB, Weydahl A, Otsuka K, Watanabe Y, Yano S, Mori H, Ichimaru Y, Mitsutake G, Sato Y, Fanghong L, Zhao ZY, Kartik C, Gvozdjakova A. Can nutrition influence circadian rhythm and heart rate variability? *Biomed. Pharmacother.* 2001; 55: 115-124.

[35] Singh RB, Niaz MA, Cornelissen G, Otsuka K, Siegelova J, Fiser B, Halberg F. Circadian rhythmicity of circulating vitamin concentration. *Scripta Medica* (Brno) 2001; 74: 93-96.

[36] Atkinson G, Jones H, Ainslie PN. Circadian variation in the circulatory responses to exercise: relevance to the morning peaks in strokes and cardiac events. *Eur. J. Appl. Physiol.* 2010; 108: 15-29.

[37] Saboorisarein M, Yazdanpoor F, Koushkie Jahromi M. The Influence of acute morning and evening exercise on homocysteine, fibrinogen and platelet. *Int. J. Cardiovasc. Res.* 2012, 1: 4. doi:http://dx.doi.org/10.4172/2324-8602.1000109.

[38] Singh RB, Halberg F, Siegelova J, Cornelissen G. What is the best time to exercise? *Noninvasive Methods in Cardiology.* 2012; 163-64.

[39] Borer KT, The autonomic nervous system in exercise: An integrative view. In C Tipton (Ed.) History of exercise physiology. Champaign: Human Kinetics, 2013.

[40] Halberg F. Chronobiology: methodological problems. *Acta med. Rom.* 1980; 18: 399-440.

[41] Cornelissen G, Halberg F. Chronomedicine. In: Armitage P, Colton T. (Eds.) Encyclopedia of Biostatistics, 2nd ed. Chichester, UK: John Wiley & Sons Ltd; 2005. pp. 796-812.

[42] Cornelissen G, Otsuka K, Halberg F. Blood pressure and heart rate chronome mapping: a complement to the human genome initiative. In: Otsuka K, Cornélissen G, Halberg F. (Eds.) Chronocardiology and Chronomedicine: *Humans in Time and Cosmos.* Tokyo: Life Science Publishing; 1993. p. 16-48.

[43] Youan BC (Ed.) Chronopharmaceutics: Science and Technology for Biological Rhythm-Guided Therapy and Prevention of Diseases. Hoboken, NJ: Wiley; 2009 (see pp. xvii–xxxii, 217-255, and 257-323).

[44] Halberg F, Cornelissen G, Wilson D, Singh RB, De Meester F, Watanabe Y, Otsuka K, Khalilov E. Chronobiology and chronomics: detecting and applying the cycles of nature. *Biologist* 2009; 56 (4): 209-214.

[45] Watanabe Y, Halberg F, Otsuka K, Cornelissen G. Toward a personalized chronotherapy of high blood pressure and a circadian overswing. *Clin. Exp. Hypertens.* 2013; in press.

[46] Watanabe Y, Cornelissen G, Watanabe M, Watanabe F, Otsuka K, Ohkawa S-i, Kikuchi T, Halberg F. Effects of autogenic training and antihypertensive agents on circadian and circaseptan variation of blood pressure. *Clin. Exp. Hypertens.* 2003; 25: 405-412.

[47] Lee MS, Lee JS, Lee JY, Cornelissen G, Otsuka K, Halberg F. About 7-day (circaseptan) and circadian changes in cold pressor test (CPT). *Biomed Pharmacother.* 2003; 57 (Suppl 1): 39s-44s.

[48] Levine H, Saltzman W, Yankaskas J, Halberg F. Circadian state dependent effect of exercise upon blood pressure in clinically healthy men. *Chronobiologia* 1977; 4: 129-130.

[49] Thrall G, lane D, Caroll D, Lip GY. A systematic review of the effects of acute psychological stress and physical activity on haemorheology, coagulation, fibrinolysis and platelet reactivity: Implications for the pathogenesis of acute coronary syndromes. *Thromb. Res.* 2007; 120: 819-847.

[50] Berger J. Chronohaematology. *J. Appl. Biomed.* 2004; 2: 179-185.

[51] van den Burg PJ, Hospers JE, van Vliet M, Mosterd WL, Bouma BN, Huisveld IA.Changes in haemostatic factors and activation products after exercise in healthy subjects with different ages. *Thromb. Haemost.* 1995; 74: 1457-1464.

[52] Atkinson G, Edwards B, Reilly T, Waterhouse J. Exercise as a synchroniser of human circadian rhythms: an update and discussion of the methodological problems. *Euro J. Applied Physiol.* 2007; 99: 331-341.

[53] Seifert G, Kanitz JL, Pretzer K, Henze G, Witt K, Reulecke S, and Voss A. Improvement of circadian rhythm of heart Rate variability by Eurythmy Therapy Training. Evidence-Based Complementary and Alternative Medicine 2013 (2013), Article ID 564340, 9 pageshttp://dx.doi.org/10.1155/2013/564340.

[54] Singh RB, Cornelissen G, Siegelova J, Halberg F. About half weekly pattern of blood pressure and heart rate in men and women of India. *Scripta Medica* (Brno) 2002; 75: 125-128.

[55] *Singh RB, Cornelissen G, Otsuka K, Juneja LR, Halberg F.* Coronary Risk Factors and Ambulatory Blood Pressure and Heart Rate in Asian Indians. *The Open Nutra J.* 2012; 5: 79-80.

[56] Singh RB, Cornelissen G, Kumar A, Bathina S, Halberg F. Larger Circadian Amplitude of Heart Rate Associated with Active Prayer in Hindu Indians in Asia. *World Heart J.* 2009; 1: 219-222.

[57] Thrall G, Lane D, Caroll D, Lip GY. A systematic review of the effects of acute psychological stress and physical activity on haemorheology, coagulation, fibrinolysis

and platelet reactivity: Implications for the pathogenesis of acute coronary syndromes. *Thromb Res.* 2007; 120: 819–847.

[58] Borer KT, Cornelissen G, Halberg F, Brook R, Rajagopalan S, Fay W. Circadian blood pressure overswinging in a physically fit, normotensive African American woman. *American Journal of Hypertension* 2002; 15: 827-830.

[59] Weydahl A, Halberg F, Serfass RC. Circadian cardiovascular and thermal modulation, ultradian entrainment or resonance with exercise at 1.2-, 1.7- and 2.3-h intervals. *Chronobiologia* 13: 260-262, 1986.

[60] Reilly T. Human circadian rhythms and exercise. *Critical Rev. Biomed. Engineering* 1990; 18: 165-180.

[61] Reilly T, Atkinson G, Waterhouse J. Travel fatigue and jet-lag. *Journal of Sports Sciences* 1997; 15 (3), 365-369.

[62] Reilly T, Brooks GA. Investigation of circadian rhythms in metabolic responses to exercise. *Ergonomics* 1982; 25 (11): 1093-1107.

[63] Reilly T, Baxter C. Influence of time of day on reactions to cycling at a fixed high intensity. *Brit. J. Sports Med.* 1983; 17 (2): 128-130.

[64] Reilly T, Garrett R. Investigation of diurnal variation in sustained exercise performance. *Ergonomics* 1998; 41 (8): 1085-1094.

[65] Reilly T, Waterhouse J. Sports performance: is there evidence that the body clock plays a role? *Eur. J. Appl. Physiol.* 2009; 106: 321-332.

In: Circadian Cardiology with Focus on Both Prevention ... ISBN: 978-1-63463-956-9
Editors: K. Hristova, A. Shehab, G. Cornelissen et al. © 2015 Nova Science Publishers, Inc.

Chapter 11

The Origin of Life and the Evolution of Circadian System and the Role of Circadian Rhythms in Cardiovascular Diseases

***D. W. Wilson*[1,*], *F. De Meester*[2], *R. B. Singh*[3] *and Elena Gerasimova*[4]**
[1]School of Medicine, Pharmacy and Health, Durham University, UK
[2]The Tsim Tsoum Institute, Krakow, Poland
[3]Halberg Hospital and Research Institute, Moradabad, India
[4]State Medical Academy of NN Burdenko in Voronezh, Russia

Abstract

Background. Following the emergence of cosmic space-time, fundamental particles, mass, gravity and radiation that led to the formation of Earth, the origin of life occurred probably with the emergence of the protocell. The functions of this protocell were under influence of geophysical forces and solar periodicities resulting in circadian and other frequency rhythms in biological systems integrated with the external environment i.a. comprising nutrients.

Methods. Internet search and discussion with colleagues.

Results. The discourse in space-time, fundamental particles, gravity, mass and radiation allows progression towards periodic events in parts of the observable universe and of relevance to this paper, in life itself.

Conclusion. Diet and lifestyle factors and geophysical cycles may trigger the circadian system and adversely affect cardiovascular events.

* Corresponding author: Dr D. W. Wilson. School of Medicine, Pharmacy and Health, Durham University, Durham, UK. E-mail: d.w.wilson@durham.ac.uk.

Space-Time, Fundamental Particles, Mass and Radiation

The current age of the current observable universe, as defined by the Lambda (Hubble constant)-cold dark matter (Lambda-CDM) model [1] which is the time elapsed since the Big Bang (from an infinitely dense hot uniform primordial singularity state at a point of 'no size'.), is estimated to be 13.798 (SD; 0.037) × 10^9 years [2].

Subject to general cosmological belief, and associated ideas, it was the Big Bang (more appropriately expansion) which marked the beginning of four-dimensional time-space: the question as to cause (which implicates time) is meaningless. The Standard Model represents fundamental particles [3] (and their respective antimatter particle [4]) that make up all matter, and their interaction with fundamental forces, each force is carried by a force carrier particle. There are 12 fundamental particles [5] (Fermions: 6 leptons and 6 quarks) and 5 force carrier particles (4 Gauge bosons, and the recently- found Higgs boson).

It was Peter Higgs [6] together with others [7, 8] who proposed the Higgs boson's existence in 1964 to explain what dictated the shape and size of matter. To date, with an extremely high probability it is the Large Hadron Collider's most notable success [9] and current technical upgrades expect to find more force carrier particles that change the Standard Model to account for more than about 4% of observable universe. The contribution made by Satyendra Nath Bose of Kolkatta, India, should be considered fundamental to the project and he is an historic father of the project. After all, the boson is named in his honor for work he did in the 1920s [10] with Albert Einstein in defining one of two basic classes of subatomic particles: unfortunately Bose was not nominated for the Nobel Prize and consequently could not be considered for the coveted prize in particle physics.

We conveniently avoid cosmogamy (Nobel Laureate in physics, 1964; [11]) and choose arbitrary time spans of note, with general comments in parentheses [12] subject to some in exactitude, viz. the first fraction of a millisecond (speculative theories about the Planck epoch (10^{-43}s) symmetry breaking of strong and weak electromagnetic interactions, thermalization, vacuum energy and cosmic inflation (10^{-32}s), disappearance of some exotic particles, quark-gluon plasma, temperature drop to 10^{13}K (1GeV, etc.), there after within an hour(q.v. within 3 minutes, Steve Weinberg, Nobel Laureate in Physics 1979 [13]) (There were electrons, neutrinos, photons, and quarks combined to form neutrons and protons, neutrons were interacting to form protons, at 10^{11} K protons dominated; no nuclei existed initially later nucleosynthesis [14] of hydrogen and deuterium took place, continuation of hadron epoch and lepton epoch.), then the next three hundred thousand years (fusion reactions almost completed, energy density of radiation> matter, plasma of separate nuclei and electrons), and finally first neutral atoms), the next 10 million years (<13.6 eV, hydrogen and other atoms formed, age of decoupling at c3000K; origin of microwave background at present at 2.7K)) and thereafter to the present day (gravitational condensation, galaxy/star formation 10^7 to 10^9 years, etc.). Indeed it is tempting to conveniently think of matter as frozen energy but it is an inexactitude and not discussed further.

Particles interact with the Higgs field to yield mass [15] but it is gravity which led to our observable universe and gravity is not in the Standard Model.

This brief discourse in space-time, fundamental particles, gravity, mass and radiation allows progression towards periodic events in parts of the observable universe and of relevance to this paper, is life itself here on Earth (Virgo supercluster/Virgo cluster/Local Group/Milky Way/Sun -110 million light years diameter of VSC) [16], specifically circadian rhythms, and how these appear in selected examples of cardiovascular disease.

Periodicity within the Universe

Leaving aside the myths and anthropomorphism (e.g., 17), focus is placed on the scientific method for understanding periodicity in the cosmos [18]. The Ptolemaic (modified geocentric building on that of Aristotle – i.e., 'equant') astronomic model of the cosmos was superseded by Copernicus' heliocentric model, though earlier Persian [19] and Greek astronomers (e.g., Aristachus) may have had some influence on Copernicus [20]: as Earth movement was recognized. Tycho Brahe [21], an anti-Aristotlean with respect to geocentricism, made many precise astronomical and planetary observation which his erstwhile guest Johannes Kepler used to develop his theories on motion and gravitation in his *Astronomia nova* [22] late 1602 Kepler reformulated planetary motion in terms of geometry: planets sweep out equal areas in equal times—Kepler's second law of planetary motion and later his third law, which shows a constant proportionality between the cube of the semi-major axis of a planet's orbit and the square of the time of its orbital period, *Harmonices Mundi* [23].

Newton subsequently developed Newtonian mechanics and the law of gravitation [24] and over 3 centuries later Einstein introduced his special (and general) theory of relativity [25, 26] in 4-dimensional space-time. Actually Earth's has precessional motion around its axis [27], and similarly those of planets around the Sun and so orbits are not truly elliptical but can affect climate and therefore the biosphere. Gravitational coupling between the Moon and the bulge nearest the Moon acts as a torque on the Earth's rotation, taking kinetic energy from the Earth's spin whilst at the same time increasing the angular momentum of the moon and placing it in a higher orbit with a longer period: also the expansion of the universe means there is a change in the matter surrounding Sun-Earth-Moon [28].

Thus in consideration of the biological evolution [29] the rotation of the Earth has changed, irrespective of model, in Silurian times it was only *c*21h instead of the present of *c*24h. Planets have a well-known periodic motion around the Sun; the Milky Way has of order $2\text{-}4 \times 10^{11}$ stars and an estimated mass of 10^{11} Suns.

The rotational velocity of stars in the Milky Way (G) is surprising [30] in that unlike planets in our solar system around the Sun the velocity is approximately constant over a substantial range of radii from the galactic centre and this is believed to be due to the presence of dark matter. Consideration of clusters and superclusters of galaxies is beyond the scope of this paper but every aspect of our expanding universe is dynamic.

Most of the galaxies in the observable universe appear to recede from us with high velocities and exhibit a Doppler red shift which is an expression of the expansion of the space-time volume which has been explained by the expansion model of the universe. This brief discourse on space-time, matter, radiation (electromagnetic forces), and mass forms a backdrop to chronomics, a science developed by Professor Halberg, University of Minnesota [31].

The Origin of Life?

Despite theories and deities, who knows the origin of life [32-34]? However, modern organisms have characteristics of: - interdependence of nucleic acids with its information storage [36] and metabolism; replication; cellular units; existing within a range of environmental conditions. Origins of life may have begun in a primordial anaerobic oceanic soup [37], as anaerobic heterotrophs, sometime after the Hadean era [38] when condensation of water vapour was unlikely, and condensation occurred, viz. the Palaeoarchean era (last universal ancestor (LUA) (c3600-3200 Ma), preceded by abiogenesis and perhaps the early forms of life in the Eoarchean era (c3900-3500 Ma); though extraterrestrial origins through meteoric bombardment, e.g., the Murchison meteorite [39, 40], and delivery of complex organic molecules necessary for life are a possibility. Standing back to examine life, what do we see? We see an enclosure with complex chemical walls, and chemicals within this enclosure which is capable of self-organization, maintaining its integrity, fulfilling basic functions such its energy needs, and furthermore is self-replicating [35]. The origins of life are murky but a reasonable approach seems to be biopoiesis which is a step-wise approach to trying to evaluate this great mystery; thus we can assume from the literature that monomers of amino acids were readily formed, and intermediates in the plastics industry are known as living molecules [41] but how does one jump to 'an enclosure with extremely complex chemical walls' –Fox's thermal synthesis theory of amino acids leading to thermally produced protein and thence to highly self-ordered proteinoid microspheres (e.g., using scanning electron microscopy), possibly formed in hot water, the hot-spring scenario, and the first protocell [42, 43] is plausible but not proven, of course. It seems an increased order of difficulty to incorporate nucleic acids into this capsule-like structure, perhaps by replicating clays, autotrophic feeding, or even the naked gene [44, 45], etc. *Non liquet; non omnia possumus omnes.* However, the experiments of Belousov-Zhabotinsky [46, 47] indicated that for oscillatory systems to occur in a medium it did not have to be far from its thermodynamic equilibrium in contrary to erstwhile belief. Prigogine, Nobel Prize in chemistry 1977 discussed non-equilibrium thermo-dynamics, particularly the theory of dissipative structures which had two linking elements, viz. "function as expressed by the chemical equations, and the space-time structure, which results from the instabilities, and the fluctuations, which trigger the instabilities [48]. The interplay between these three aspects leads to most unexpected phenomena, including "order through fluctuations" and is a consequence of the Second Law of Thermodynamics: from which, for cyclic processes, the Kelvin-Planck and Clausius statements apply [49]. Chevreul (fatty acids), Wohler (Urea synthesis – possibly myth in the vitalism debate) and Kolbe (acetic acid) experiments largely dismissed the concepts of vitalism but it was Alexander Leonidovich Chizhevsky who described life on the Earth as "an echo of the Sun". "Peut-être meme nos sentiments et nos pensées ne sont-ils qu'unfaible écho de ces vibrations du cosmos. Involontairement, une antique idée nous vient à l'esprit: notre connaissance des phénomènes de la nature ne serait pas autre chose qu'un écho, reçu par nos organes, des processus réels de l'univers" [50, 51].

Chizhevsky was deservedly offered a Nobel Prize in 1939 but was forced to turn it down by the Stalinist government. In recent decades, Halberg at the University of Minnesota has focused i.a. on evidence that fluctuations in solar magnetism can influence gross clinical phenomena such as rates of strokes and heart attacks, and related cardiovascular variables such as blood pressure and heart rate, which opens the door for broad studies of clinical aspects of such work [31]. Even the scriptures (Ecclesiastes 3, 1-8) and popular poets recognize the ubiquity of rhythmic processes, such as that by Kahlil Gibran [52], Phardrous, the Greek struck his foot upon a stone… "Oh dead thing in my path!". Almustafa, the chosen and beloved, said, "Why say you: "O dead thing"? All things live and glow in the knowledge of the day and the majesty of the night. You and the stone are one. There is a difference only in heart-beats. Your heart beats a little faster, does it my friend?"…"Its rhythm may be another rhythm". The protocell now has to develop and evolve and to do this it needs nutrients and self-sustaining energy mechanisms for reproduction and existence in often a harsh environment which is now discussed.

The Origin of Nutrients in the Protocells

There must be several steps or dissipated energy structures which developed from the protocell, such as passive and dynamic access to energy, notably nutrients used in physicochemical photo-physiochemical processes leading to the early phylogenic tree of life comprising bacteria, archaea and eukaryote that is now recognized as a living cell. This energy would include radiation (no ozone layer), geothermic, electric discharge, meteorite impacts, and radioactive, etc., heat. Heterotrophy and autotrophy are two forms of metabolism, the former utilize complex organic molecules as food for energy and carbon for biosynthesis, and the latter use light, inorganic molecules and usually, though not exclusively, carbon dioxide for energy and biosynthesis. Both autotrophs and heterotrophs metabolize, grow, mutate, reproduce, and die. We may assume that the early life forms were bacteria with RNA [45] possibly performing an informational role and catalytic activity, rather than DNA today, for which Altman and Cech received the Nobel Prize in chemistry in 1989 [44], and existing in a cell with no nucleus, mitochondrion or chloroplasts. Thus the prior steps were possibly the evolution of RNA, protein, and later DNA within the cell, the specialization of proteins into enzymes, cytoplasm, and the related subsystems, and the formation of a protective semipermeable membrane. Cyanobacterial photo-autotrophs using water as an electron donor and fixing carbon by the Calvin cycle are found in fossil records c3800Ma, supplying oxygen to heterotrophs. It is likely that the first life forms, as ancestral prokaryotes were chemo-autotrophs in hot sulphurous aqueous environments.

Unicellular eukaryotes have been identified in the Precambrian fossil record [53, 54] evolved perhaps from a prokaryote ancestor which had e.g., independent mitochondrial free-living guests (or chloroplasts) in the cytoplasm which became, though loss of independence, an endosymbiotically organism [55]: endosymbiosis is well documented for a variety of organism, e.g., the giant amoeba aerobe *Pelomyxa palustris* and aerobic bacteria [35]. It is unclear whether the origin of the nucleus preceded or came after mitochondria, the inner membrane of the latter where respiratory electron transport takes place being the outer layer of the symbiont. It is possible that the nucleus was first formed autogeneously by prokaryote cells and symbiosis with other organelles took place later. The reproductive processes involving sex [56], ploidy and meiosis are beyond the remit of this paper but the late Mesoproterozoic era is a time when sexual evolution occurred.

Finally, diversification of eukaryotes took place, and multicellular organisms with clusters of eukaryote cells with different functions, and algae have been found c1000Ma in the Hunting Formation in arctic Canada [57, 58] and molecular evidence points to a common lineage perhaps with *Bangiomorpha pubescensa* found in the late Mesoproterozoic [57]. The evolution of species through adaptation of the organism to its environment is well documented, and of relevance is the mammalian order [59], a clade of warm-blooded amniotes, whose adaptive radiation seemingly began about 65Ma. Studies of mtDNA have helped to establish that chimpanzees (*Pan spp*., including *P. troglodytes* and *P. paniscus*) [60], with which we shared a common ancestor c5 MYBP thence c7MYBP when *Homo*, with chimpanzees, shared a common ancestry with gorillas and it is interesting to note that the nuclear genome sequences of chimps and humans >98% in similarity. In 1.5-2 Ma hominoids migrated from Africa and occupied parts of southern Europe, Egypt, India and China. An evolutionary discourse on human, technical and cultural evolution is beyond the scope of this paper [61-63]. So from the aforesaid on the origins of life and archaeological records of hominoids it is pertinent to focus on present day circadian rhythms defined by Halberg as having a cycle length of about 24 h (> 20 to < 28 h) whether they are synchronized with a 24-h periodic surrounding or not [64-67].

The Origin of Circadian System

The suprachiasmatic nuclei (SCN) is under influence of daily alternation between light and darkness directly via the eyes and results in plasma melatonin concentrations secreted by the pineal gland, which is a window to both light and geophysical forces. In mammals, the adrenal and the pineal-hypothalamic-pituitary network coordinates a circadian cell cycle, with peripheral timing mechanisms which are being documented in molecular biologic terms at about 24-hour (circadian) and higher (ultradian) frequencies, with coordination, [61-65]. In everyday life, the SCN and clock gene contribute to the coordination of the circadian rhythms' phase and amplitude, (Figure 1) [62-65]. There is emergence of cardiovascular events, when our neuroendocrine time structures (chronomes) are not able to cope with the adverse effects of stimuli from within or from without, acting, e.g., via the sympathetic nervous system [62, 63]. Circadian rhythms of various biomarkers of health have been observed after its (circadian) discovery by Franz Halberg, which are under control of circadian clocks present in all body cells [64-68]. The evolution of circadian clocks has occurred to synchronize the physiology, metabolism and behaviour to the 24-h geophysical cycles of the Earth [62, 63].

The words circadian and chronobiology were used for the first time by Franz Halberg in 1950 and officially introduced to a nomenclature committee in Stockholm for the first time in 1955 [64, 65]. Chronobiology developed globally, after 1969, when an article entitled Chronobiology was published in the Annual Review of Physiology and became a Current Content citation classic [64-67]. Recent research, by Vanin et al., 2012 suggests that much of scientists' current understanding about *Drosophila* circadian behavioral rhythms may actually be wrong [68]. The experiment emphasizes the importance of studying organisms in natural environments and challenges 40 years of assumptions about circadian clocks studied in the laboratory [68].

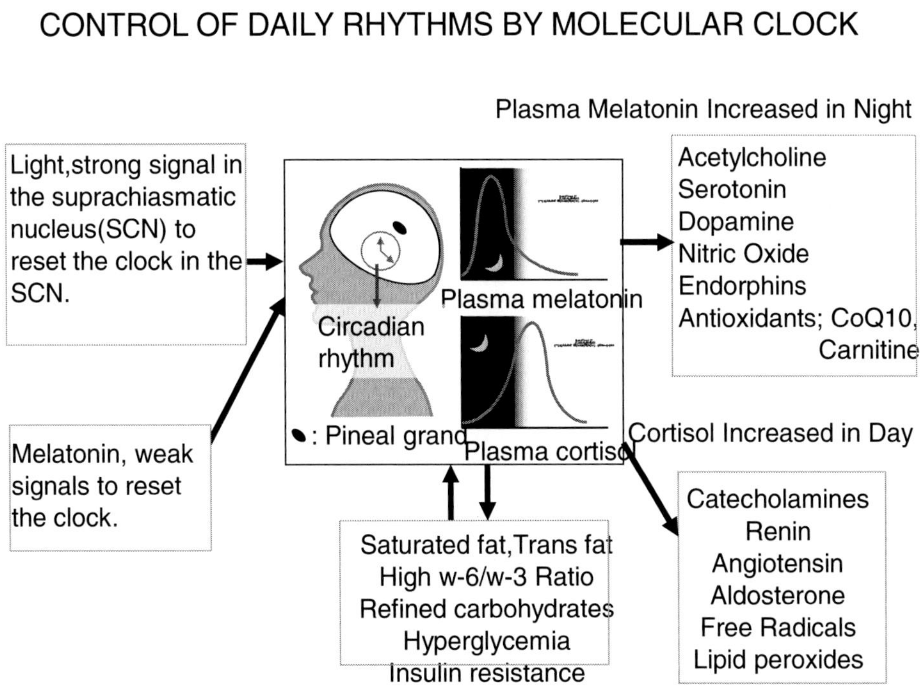

Figure 1. Effect of environmental factors on circadian rhythms of biological functions.

The mystery of flies behaving in the wild suggest that the body clock in a wild, natural environment, shows features, that are impossible to observe in simplified laboratory conditions. It seems that the behavior of the fly in the wild is quite different from what is seen in controlled conditions in the laboratory. Since insects and humans share many clock genes, scientists have used *Drosophila* as a model system for circadian studies for the last 40 years. In the laboratory, the flick of a switch turns the lights on suddenly in the morning and off suddenly at night, for example, without any dawn or dusk and without any change in temperature as observed in nature in the afternoon. The natural environment provides much richer cycling environmental stimuli than the laboratory, so it is of choice to examine fly locomotor rhythms in the wild. Vanin et al., observed an additional major locomotor component in addition to M and E, which they term 'A' (afternoon) [68]. It seems that a comprehensive re-examination of circadian behaviour and its molecular readouts under simulated natural conditions will provide a more authentic interpretation of the adaptive significance of this important rhythmic phenotype. The study may also indicate the underlying molecular and neuroanatomical substrates of the clock under natural protocols.

It is possible that this behavior of the flies may be due to variability in the underlying DNA or because of alteration in temperature or other environmental factors.

In the natural environment; flies in outdoors were most active during the day, rather than at dawn and dusk as seen in the laboratory, demonstrating the flies' diurnal (daytime) rather than crepuscular (from the Latin word for twilight) activity patterns. This difference stemmed in part from the fact that flies did not take an afternoon "siesta," a period of decreased activity in the middle of the day, as observed in the laboratory. The flies exhibited a burst of midday activity. Two other studies from the EUCLOCK project have made similar findings: both laboratory mice and hamsters are diurnal in outdoor settings, but nocturnal in the laboratory

[68]. It seems that outdoors, fly activity in the morning appears to be a physical response to changing temperature, or other environmental factor like geophysical forces, rather than a clock-controlled behavior. However, classical models suggest that there are morning and evening circadian oscillators controlling fly behavior. It is most surprising, that clock mutants did not demonstrate defective rhythms in a natural setting but they had normal behavioral cycles which was totally unexpected. In the laboratory, classical clock mutants have peculiar circadian behaviors that differ dramatically from wild type flies. But this was not so in the natural environment. It is possible that the complex environmental information that flies receive outdoors bypass mutations in single clock genes, which needs further research. The behavioral findings do not extrapolate to basic circadian molecular mechanisms underlying physiological changes, such as metabolism. Morning and evening oscillators are still doing their job modulating the biochemistry of the fly at the physiological level. Natural rhythm behaviors in the laboratory using the new simulations, have been produced among natural circadian phenotypes in a laboratory but outdoor experiments will yield the most illustrative results. If we want to understand exactly how the clock works, it is going to be very useful to have approaches, to observe organisms living in nature. It may be different from what we know in the laboratory because geophysical cycles of the earth as well as certain nutrients; flavonoids and w-3 fatty acids freely available in the wild, may influence clock activity better in the wild.

Circadian Rhythm, Nutrition and Cardiovascular Events

Circadian rhythm was known to ancient man from the time of *Homo erectus and Homo sapiens* who use to have intercourse in the early morning hours, before going for hunting to forests, causing increased secretion of testosterone in the morning to help the sexual activity, which evolved as a circadian rhythm. It has been also proposed that during human evolution, early man used to go for hunting near the river banks where animals come for drinking of water at sun rise [62, 63]. The early man had marked excitement during running, climbing and may be horse riding for hunting, causing marked increase in mental and physical stress, with increased release of catecholamines and cortisol and excessive sweating [62, 63]. These physiological and biochemical changes may have been responsible for the origin and the evolution of the circadian rhythm of increased cardiovascular biomarkers and cardiovascular events as well as greater basal metabolic rate in the second quarter of 24 hour cycle [62, 63]. In one study among 209 adults, increased heart rate and MESOR blood pressures were observed in the second quarter of the day [69]. Fruit and vegetable intake were inversely associated with MESOR hypertension [70]. In another study measuring heart rate and blood pressure, half hourly for 7 days, larger circadian amplitude of heart rate associated with Active Prayer were observed in Hindu Indians in Asia [71]. The circadian rhythm of antioxidant vitamins and oxidative stress have also been observed to protect against the oxidative stress caused by catecholamine in the second quarter of the 24 h cycle [72]. Figure 1 indicates that refined carbohydrates, w-6 fatty acids may have adverse effects on circadian rhythms whereas w-3 fatty acids and antioxidants may have beneficial effects.

Diet and lifestyle factors and geophysical cycles may trigger the circadian system neuroendocrines and may activate the pineal gland, pituitary functions and adrenal secretions, resulting in adverse effects on circadian variations, heart rate variability(HRV) and blood pressure variability (BPV), sleep disorders, obesity and cardiovascular disturbances [61-67]. Circadian rhythms (24-hour) and also possibly circaseptan rhythms of behavioral and physiological changes driven by oscillating biochemical mechanisms, are present in almost all organisms, including humans and even in some bacteria [62-65]. Many human health problems have a rhythmic component, including change in temperature, heart rate and blood pressures as well as other biomarkers as in cases of cancer and atherothrombosis [62, 63]. Exercise as a synchroniser of human circadian rhythms of coagulation and other biomarkers have been reported which may alter the natural course of circadian rhythms of cardiac dysfunctions [73-75]. One study shows that morning time exercise may be helpful in adaptation of our cardiovascular responses, causing less increase in MESOR hypertension and heart rate which may be protective against increased circadian rhythms of cardiovascular events [76]. In clinical studies circadian rhythms of cardiovascular events have been observed by several workers [62, 63, 77-81]. A Mediterranean style diet rich in fruits, vegetables, nuts and w-3 fatty acids and flavonoids may be protective against the adverse effects of circadian rhythms of cardiac events [77-82]. Extended consensus on need and means to detect vascular variability disorders (VVDs), and vascular variability syndromes (VVSs) and the role of brain-body interactions are poorly known to most experts but are wisely described by few workers [61-66]. Definition of ambulatory blood pressure targets for diagnosis and treatment of hypertension in relation to clinic blood pressures is based on 24-48 hour records which ignores the presence of circaseptan variability in blood pressures as well as VVS and VVDs indicating that 7-day record may be important preferably during working conditions to diagnose prehypertension or hypertension.

Finally, concerning (internal) evolution within organisms as well as about an external one with their environment, a distinction has been made [83]. Now Halberg make a third distinction, as we learn from Vernadsky's noosphere, which he regarded as a linear rather than cyclic entity, while we also further scrutinize the human mind's modification of the crust of the earth. Halberg also reported a transdisciplinary set of coperiodisms, a chrono-noosphere or rather, a chronousphere evolves. This is a feature of active rather than only passive bioevolution, a feature of cosmic evolution. There the problem of cyclic universes looms large.

Acknowledgments

Acknowledgements are due to International College of Nutrition for providing logistic support to write this article.

Conflict of interest has not been declared.

Contributions: The concept of the hypothesis was developed and the draft of the paper was made by Dr RB Singh and Dr Fabien De Meester. The hypothesis was extended and new concepts were added by Dr DW Wilson.

References

[1] Mckay, J. A., Mathers, J. C. Diet induced epigenetic changes and their implications for health. *Acta Physiologica*. 2011; 202(2):103–118.

[2] Diet, nutrition and the prevention of chronic diseases. *World Health Organization Technical Report Series*. 2003; 916(1–8):1–149.

[3] Davis, C. D., Ross, S. A. Dietary components impact histone modifications and cancer risk. *Nutr. Rev*. 2007 65(2): 88-94.

[4] Gabory, A., Attig, L., Junien, C. Sexual dimorphism in environmental epigenetic programming. *Molecular and Cellular Endocrinology*. 2009; 304(1-2):8–18.

[5] Meeran, S. M., Ahmed, A., Tollefsbol, T. O. Epigenetic targets of bioactive dietary components for cancer prevention and therapy. *Clin Epigenetics*. 2010; 1(3-4): 101–116.

[6] Bestor, T. The DNA methyltransferases of mammals. *Hum. Mol. Genet*. 2000; 9: 2395–2402. doi: 10.1093/hmg/9.16.2395.

[7] Tate, P., Bird, A. Effects of DNA methylation on DNA-binding proteins and gene expression. *Curr. Opin. Genet. Dev*. 1993; 3: 226–231. doi: 10.1016/0959-437X (93) 90027-M.

[8] Wade, P. Methyl CpG-binding proteins and transcriptional repression. *Bioessays*. 2001; 23: 1131–1137. doi: 10.1002/bies.10008.

[9] Ballestar, E., Wolffe, A. Methyl-CpG-binding proteins. Targeting specific gene repression. *Eur. J. Biochem*. 2001; 268:1–6. doi: 10.1046/j.1432-1327.2001.01869.x.

[10] Laird, P. Cancer epigenetics. *Hum. Mol. Genet*. 2005; 14 (Spec. No 1):R65–R76. doi: 10.1093/hmg/ddi113.

[11] Wolffe, A. P. and Matzke, M. A. (1999) Epigenetics: regulation through repression. *Science* (Wash., DC) 86:481-486.

[12] Jones, P. A. and Laird, P. W. (1999) Cancer epigenetics comes of age. *Nat. Genet.* 21:163-167.

[13] Wallwork, J. C. and Duerre, J. A. (1985) Effect of zinc deficiency on methionine metabolism, methylation reactions and protein synthesis in isolated perfused rat liver. *J. Nutr.*

[14] Dreosti, I. E. (2001) Zinc and the gene. *Mutat. Res.* 475:161-167.

[15] Davis, C. D., Uthus, E. O. and Finley, J. W. (2000) Dietary selenium and arsenic affect DNA methylation in vitro in Caco-2 cells and in vivo in rat liver and colon. *J. Nutr.*

[16] Simonetta Friso and Sang-Woon Choi, Gene-Nutrient Interactions and DNA Methylation. *J. Nutr.* August 1, 2002 -vol. 132 no. 8 2382S-2387S.

[17] El-Bayoumy, K. (2001) The protective role of selenium on genetic damage and on cancer. *Mutat. Res.*

[18] Piyathilake, C. J., Bell, W. C., Johanning, G. L., Cornwell, P. E., Heimburger, D. C. and Grizzle, W. E. (2000). The accumulation of ascorbic acid by squamous cell carcinomas of the lung and larynx is associated with global methylation of DNA. *Cancer* 89:171-176.

[19] Halliwell, B. (2001) Vitamin C and genomic stability. *Mutat. Res.* 475:29-35.

[20] Zardo, G. and Caiafa, P. (1998). The unmethylated state of CpG islands in mouse fibroblasts depends on the poly(ADP-ribosyl)ation process. *J. Biol. Chem.* 273:16517-16520.

[21] Hageman, G. J. and Stierum, R. H. (2001) Niacin, poly(ADP-ribose) polymerase-1 and genomic stability. *Mutat. Res.* 475:45-56.

[22] Steegers-Theunissen, R. P., Smith, S. C., Steegers, E. A., Guilbert, L. J., Baker, P. N. Folate affects apoptosis in human trophoblastic cells, *BJOG*. 2000; 107: 1513-5.

[23] Stefanska, B., Salamé, P., Bednarek, A., Fabianowska-Majewska, K. Comparative effects of retinoic acid, vitamin D and resveratrol alone and in combination with adenosine analogues on methylation and expression of phosphatase and tensin homologue tumour suppressor gene in breast cancer cells. *British Journal of Nutrition*/Volume 107/Issue 06/March 2012, pp. 781-790.

[24] Fang, M. Z., Chen, D., Yang, C. S. Dietary Polyphenols May Affect DNA Methylation, *J. Nutr.* 137: 223S–228S, 2007.

[25] Fang, M. Z., Wang, Y., Ai, N., Hou, Z., Sun, Y., Lu, H., Welsh, W., Yang, C. S. Tea Polyphenol (−)-Epigallocatechin-3-Gallate Inhibits DNA Methyltransferase and Reactivates Methylation-Silenced Genes in Cancer Cell Lines. *Cancer Res. November* 15, 2003 63:7563-7570.

[26] Lee, W., Shim, J., Zhu, B. Mechanisms for the inhibition of DNA methyltransferases by tea catechins and bioflavonoids. *Mol. Pharmacol.* 2005; 68:1018–1030. doi: 10.1124/mol.104.008367.

[27] Lee, W., Zhu, B. Inhibition of DNA methylation by caffeic acid and chlorogenic acid, two common catechol-containing coffee polyphenols. *Carcinogenesis.* 2006; 27:269–277. doi: 10.1093/carcin/bgi206.

[28] Majid, S., Kikuno, N., Nelles, J., Noonan, E., Tanaka, Y., Kawamoto, K., Hirata, H., Li, L., Zhao, H., Okino, S., Place, R., Pookot, D., Dahiya, R. Genistein induces the p21WAF1/CIP1 and p16INK4a tumor suppressor genes in prostate cancer cells by epigenetic mechanisms involving active chromatin modification. *Cancer Res.* 2008; 68: 2736–2744.

[29] Kikuno, N., Shiina, H., Urakami, S., Kawamoto, K., Hirata, H., Tanaka, Y., Majid, S., Igawa, M., Dahiya, R. Genistein mediated histone acetylation and demethylation activates tumor suppressor genes in prostate cancer cells. *Int. J. Cancer.* 2008; 123: 552–560.

[30] Liu, H., Chen, Y., Cui, G., Zhou, J. Curcumin, a potent anti-tumor reagent, is a novel histone deacetylase inhibitor regulating B-NHL cell line Raji proliferation. *Acta Pharmacol. Sin.* 2005; 26: 603–609.

[31] Bishayee, A. Cancer prevention and treatment with resveratrol: from rodent studies to clinical trials. *Cancer Prev. Res.* (Phila. PA) 2009; 2: 409–418.

[32] Rodenhiser, D., Mann, M. Epigenetics and human disease: translating basic biology into clinical applications. *CMAJ.* 2006; 174: 341–348.

[33] Kondo, Y. Epigenetic Cross-Talk between DNA Methylation and Histone Modifications in Human Cancers. *Yonsei Med. J.* 2009 August 31; 50(4): 455–463.

[34] Bannister, A. J., Kouzarides, T. Regulation of chromatin by histone modifications. *Cell Research.* 2011; 21(3):381–395.

[35] Link, A., Balaguer, F., Goel, A. Cancer chemoprevention by dietary polyphenols: promising role for epigenetics. *Biochemical Pharmacology.* 2010; 80(12):1771–1792.

[36] Mangelsdorf, D. J., Thummel, C., Beato, M., Herrlich, P., Schutz, G., Umesono, K., Blumberg, B., Kastner, P., Mark, M., Chambon, P., and Evans, R. M. (1995). The nuclear receptor superfamily: The second decade. *Cell* 83(6), 835–39.
[37] Wang, K., Chen, S., Xie, W., and Wan, Y. J. (2008 Jun. 1) Retinoids induce cytochrome P450 3A4 through RXR/VDR-mediated pathway. *Biochem. Pharmacol.* 75 (11), 2204–13.
[38] Müller, M., Kersten, S. Nutrigenomics: goals and strategies. *Nature Review/Genetics*, Vol. 4, 2003, 315-322.
[39] Lemon, B. D., Freedman, L. P. Nuclear receptor cofactors as chromatin remodelers. *Curr. Opin. Genet. Dev.* 1999 Oct.; 9(5): 499-504.
[40] Neil, J. McKenna, Rainer B. Lanz and Bert W. O'Malley. Nuclear Receptor Coregulators: Cellular and Molecular Biology. *Endocrine Reviews* 20(3): 321–344.
[41] Glass, C. K., Rosenfeld, M. G. The coregulator exchange in transcriptional functions of nuclear receptors. *Genes Dev.* 2000 Jan. 15; 14(2):121-41.
[42] Di Croce, L., Okret, S., Kersten, S., Gustafsson, J.-A., Parker, M., Wahli, W., Beato, M. 1999 Steroid and nuclear receptors. *EMBO J.* 18:6201–6210.
[43] Rachez, C., Freedman, L. P. 2000 Mechanisms of gene regulation by vitamin D3 receptor: a network of coactivator interactions. *Gene* 246:9–21.
[44] McKenna, N. J., Xu, J., Nawaz, Z., Tsai, S. Y., Tsai, M. J., O'Malley, B. W. 1999 Nuclear receptor coactivators: multiple enzymes, multiple complexes, multiple functions. *J. Steroid Biochem. Mol. Biol.* 69:3–12.
[45] Freedman, L. P. 1999 Increasing the complexity of coactivation in nuclear receptor signaling. *Cell* 97:5–8.
[46] Grant, D. Barish. Peroxisome Proliferator-Activated Receptors and Liver X Receptors in Atherosclerosis and Immunity. Symposium: Nutrients, Nuclear Receptors, Inflammation, and Immunity. *J. Nutr.* 136: 690–694, 2006.
[47] Bordoni, A., Di Nunzio, M., Danesi, F., Biag, P. L. Polyunsaturated fatty acids: from diet to binding to PPAR and other nuclear receptors. *Genes and Nutrition* Vol. 1, No. 2, pp. 95-106, 2006.
[48] Otaegui-Arrazola, A., Menéndez-Carreño, M., Ansorena, D., Astiasarán, I. Oxysterols: A world to explore. *Food Chem. Toxicol.* 2010 Dec.;48(12):3289-303. doi: 10.1016/j. fct.2010.09.023.
[49] Baranowski, M. Biological role of liver X receptors. *Journal of Physiology and Pharmacology.* 2008, 59, Suppl. 7, 31–55.
[50] Choi, K. C., Jung, M. G., Lee, Y. H., Yoon, J. C., Kwon, S. H., Kang, H. B., et al. Epigallocatechin-3-gallate, a histone acetyltransferase inhibitor, inhibits EBV-induced B lymphocyte transformation via suppression of RelA acetylation. *Cancer Res.* 2009; 69:583–592.
[51] Balasubramanyam, K., Varier, R. A., Altaf, M., Swaminathan, V., Siddappa, N. B., Ranga, U., et al. Curcumin, a novel p300/CREB-binding protein-specific inhibitor of acetyltransferase, represses the acetylation of histone/nonhistone proteins and histone acetyltransferase-dependent chromatin transcription. *J. Biol. Chem.* 2004; 279: 51163–51171.
[52] Chen, L., Fischle, W., Verdin, E., Greene, W. C. Duration of nuclear NF-kappaB action regulated by reversible acetylation. *Science.* 2001; 293: 1653–1657.

[53] Yun, J. M., Jialal, I., Devaraj, S. (2010) Epigenetic regulation of high glucose-induced proinflammatory cytokine production in monocytes by curcumin. *J. Nutr. Biochem.*

[54] Reuter, S., Gupta, S. C., Park, B., Goel, A., Aggarwal, B. B. Epigenetic changes induced by curcumin and other natural compounds. *Genes Nutr.* 2011 May; 6(2): 93–108.

[55] Parasramka, M. A., Ho, E., Williams, D. E., Dashwood, R. H. MicroRNAs, Diet, and Cancer: New mechanistic insights on the epigenetic actions of phytochemicals. *Molecular Carcinogenesis*, 2011.

[56] Sato, F., Tsuchiya, S., Meltzer, S. J., Shimizu, K. MicroRNAs and epigenetics. *FEBS J.* 2011 May;278(10):1598-609.

[57] Garzon, R., Pichiorri, F., Palumbo, T., et al. MicroRNA gene expression during retinoic acid-induced differentiation of human acute promyelocytic leukemia. *Oncogene* 2007; 26:4148–4157.

[58] Mohri, T., Nakajima, M., Takagi, S., Komagata, S., Yokoi, T. MicroRNA regulates human vitamin D receptor. *Int. J. Cancer* 2009;125:1328–1333.

[59] Gaedicke, S., Zhang, X., Schmelzer, C., et al. Vitamin E dependent microRNA regulation in rat liver. *FEBS Lett.* 2008;582:3542–3546.

[60] Humle, T., Matsuzawa, T. Ant-dipping among the chimpanzees of Bossou, Guinea, and some comparisons with other sites. *Am. J. Primatol.* 2002;58:133-48.

[61] Fox, S. W., Dose, K. *Molecular evolution and the Origin of Life.* WH Freeman Publication San Francisco, 1972.

[62] Singh, R. B., Singh, A. K., Sharma, J. P., Singh, R. K., Kumar, A., Rastogi, S. S., Singh, G., Kartikey, K., Singh, S., Fedacko, J., Pella, D., De Meester, F., Wilczynska, A., Wilson, D. W. Nutrition in Chronocardiology: We are Indebted Professor Franz Halberg. *The Open Nutraceuticals Journal,* 2012, 5, (Suppl. 1-M3) 45-65.

[63] Singh, R. B., Gupta, A., Pella, D., Circadian cardiology. *World Heart J.* 2008; 1: 101-26.

[64] Otsuka, K., Cornelissen, G., Halberg, F. Circadian rhythms and clinical chronobiology. *Biomed. Pharmacother.* 2001; 55: 7-18.

[65] Halberg, F., Stephens, A. N. Susceptibility to quabain and physiologic circadian periodicity. *Proc. Minn. Acad. Sci.* 1959; 27: 139-43.

[66] Halberg, F., Visscher, M. B. Regular diurnal physiological variation in eosinophil levels in five stocks of mice. *Proc. Soc. Exp. Biol.* (NY) 1950; 75: 846-7.

[67] Halberg, F., Cornelissen, G., Otsuka, K., Siegelova, J., Fiser, B., Dusek, J., Homolka, P., Sanchez de la Pena, S., Singh, R. B., BIOCOS project. Extended consensus on need and means to detect vascular variability disorders (VVDs) and vascular variability syndromes (VVSs). *World Heart J.* 2010; 2 (4): 279-305.

[68] Vanin, S., Bhutani, S., Montelli, S., Menegazzi, P., Green, E. W., Pegararo, M., Sandrelli, F., Costa, R., Kyriacou, C. P. Unexpected features of Drosophila circadian behavioural rhythms under natural conditions, *Nature,* doi:10.1038/nature10991, 2012.

[69] Singh, R. B., Cornelissen, G., Siegelova, J., Halberg, F. About half weekly pattern of blood pressure and heart rate in men and women of India. *Scripta Medica* (BRONO) 2002; 75: 125-128.

[70] Singh, R. B., Cornelissen, G., Otsuka, K., Juneja, L. R., Halberg, F. Coronary Risk Factors and ambulatory blood pressure and heart rate in Asian Indians. *The Open Nutra J.* 2012;5:79-80.

[71] Singh, R. B., Cornelissen, G., Kumar, A., Bathina, S., Halberg, F. Larger circadian amplitude of heart rate associated with active prayer in Hindu Indians in Asia. *World Heart J.* 2009;1:219-222.

[72] Singh, R. B., Niaz, M. A., Cornelissen, G., et al. Circadian rhythmicity of circulating vitamin concentrations. *Scripta Medica* (BRONO) 2001; 74: 93-6.

[73] van den Burg, P. J., Hospers, J. E., van Vliet, M., Mosterd, W. L., Bouma, B. N., et al. Changes in haemostatic factors and activation products after exercise in healthy subjects with different ages. *Thromb. Haemost.* 1995;74:1457–1464.

[74] Atkinson, G., Edwards, B., Reilly, T., Waterhouse, J. Exercise as a synchroniser of human circadian rhythms: an update and discussion of the methodological problems. *Euro. J. Applied Physiol.* 2007;99:331-341.

[75] Seifert, G., Kanitz, J. L., Pretzer, K., Henze, G., Witt, K., Reulecke, S., and Voss, A. Improvement of circadian rhythm of heart Rate variability by Eurythmy Therapy Training. *Evidence-Based Complementary and Alternative Medicine* 2013 (2013), Article ID 564340, 9 pages http://dx.doi.org/10.1155/2013/564340.

[76] Singh, R. B., Halberg, H., Cornelissen, G., Siegelova, J., Hristova, K., Toda, E., Toru, T., Fedacko, J., Otsuka, K. Effects of exercise according to time, on cardiovascular function in a subject doing exercise regularly in the morning and evening. *World Heart J.* 2013,4: (in press).

[77] Singh, R. B., Weydahl, K., Otsuka, K., et al. Can nutrition influence circadian rhythm and heart rate variability? *Biomed. Pharmacother.* 2001; 55: 115-24.

[78] Singh, R. B., Cornelissen, G., Otsuka, K., et al. Coronary risk factors and ambulatory blood pressures and heart rate in Asian Indians. *The Open Nutr. J.* 2012; 5:79-80.

[79] Singh, R. B., Pella, D., Rastogi, S., et al. Increased concentrations of lipoprotein(a),circadian rhythms, and metabolic reactants evoked by acute myocardial infarction in relation to large breakfast. *Biomed. Pharmacother.* 2004; 58: 116-22.

[80] Singh, R. B., Pella, D., Neki, N. S., et al. Mechanism of acute myocardial infarction (MAMI) study. *Biomed. Pharmacother.* 2004; 58: 111-5.

[81] Singh, R. B., Kartik, C., Otsuka, K., Pella, D., Pella, J. Brain-heart connection and the risk of heart attack (Editorial). *Biomed. Pharmacother.* 2002; 56: 257-65.

[82] Singh, R. B., Dubnov, G., Niaz, M. A., et al. Effect of an Indo-Mediterranean diet on progression of coronary disease in high risk patients: a randomized single blind trial. *Lancet* 2002; 360: 1455-61.

[83] Scheving, L. E., Halberg, F. *Principles and Applications to Shifts and Schedules.* Sijthoff and Noordoff, Alphen, Rockville, MD, US, 1980.

In: Circadian Cardiology with Focus on Both Prevention ... ISBN: 978-1-63463-956-9
Editors: K. Hristova, A. Shehab, G. Cornelissen et al. © 2015 Nova Science Publishers, Inc.

Chapter 12

Causes and Risk Factors of Chronic Heart Failure in India

***Krasimira Hristova*[1], *Ram B. Singh*[2,*], *Jan Fedacko*[3], *Eri Toda*[4], *Adarsh Kumar*[5], *Manoj Saxena*[2], *Anjum Baby*[6], *Takahashi Toru*[7], *Fabien De Meester*[8] and *Douglas W. Wilson*[9]**

[1]University National Heart Hospital, Department of Noninvasive Functional Diagnostic and Imaging, Sofia, Bulgaria
[2]Halberg Hospital and Research Institute, Moradabad, India
[3]Faculty of Medicine, PJ Safaric University, Kosice, Slovakia
[4]Department of Cardiology, Tokai University Hachioji Hospital, Tokyo, Japan
[5]Department of Medicine, Government Medical College, Amritsar, India
[6]Department of Physiology, KG Medical University, Lucknow, India
[7]Department of Nutrition, Graduate School of Human Environment Science, Fukuoka Women's University, Japan
[8]The TsimTsoum Institute, Krakow, Poland
[9]School of Medicine, Pharmacy, and Health, Durham University, Durham, UK

Abstract

Introduction. Recent evidence shows that chronic heart failure (CHF) is a major cause of morbidity and mortality. The prognosis in CHF may be dependent on cause, severity and presence of risk factors and drug therapy. In the present study, we examine the causes, risk factors, class and oxidative stress among CHF patients.

Subjects and methods. Of 127 patients with CHF, 2 were excluded and the remaining 125 patients (Men 61, women 64) with different aetiologies of CHF, and 250 age and sex matched control subjects, were evaluated in this case study. Severity of disease based on the New York Heart Association (NYHA) standards fell within functional classes II to

* Corresponding author: Ram B. Singh. University National Heart Hospital, Department of Noninvasive Functional Diagnostic and Imaging, Sofia, 1000, Bulgaria. E-mail: rbs@tsimtsoum.net.

IV. The diagnosis of HF was based on clinical manifestations as well as on echocardiographic heart enlargement.

Results. The causes of CHF were; CAD (n = 34, 27.2%), hypertensive heart disease (n = 10, 8.0%), valvular heart disease (n = 40, 32.0%) and idiopathic dilated cardiomyopathy (n = 38, 30.4%). Risk factors of HF were; CAD (n = 52, 41.6%), hypertension (>140/90mmHg) (n = 54, 43.2%), diabetes mellitus (n = 12, 9.6%), obesity (n = 43, 34.4%) and albuminurea (n = 12, 9.6%). Echocardiographic ejection fraction was 39.1 ± 8.2% (mean ± SD) in the study group, indicating class II-IV heart failure. There was a significant increase in biomarkers of oxidative stress, among HF patients compared to healthy subjects.

Conclusion. The findings indicate that HF has become a public health problem. The causes of HF appear to be CAD, cardiomyopathy and valvular heart disease. Severity of CHF, aetiology; CAD and cardiomyopathy appear to be important for increased oxidative stress among these patients.

Keywords: Aetiology, heart failure, risk factors, oxidative stress

Introduction

Heart failure (HF) is the third most common cardiovascular disease (CVD) in the US affecting about 2 per cent of the population, or almost 5 million people (1). The prevalence of HF increases with the age from less than 1 per cent in the 20-39 yr old age group to over 20 per cent in the people age 80 yr or older [1, 2].

The lifetime risk of developing HF at the age of 40 yr is 11.4 per cent for men and 15.4 per cent for women. The hospital discharges for HF remained essentially unchanged from 2000 to 2010 in US, with first-listed discharges of 1,008 000 and 1,023 000, respectively [1].

The deaths and disability due to HF are also quite high in the US (1). In 2009, 1 in 9 death certificates (274 601 deaths) in the US, mentioned HF and it was the underlying cause in 56410 of those deaths in 2009. The number of any-mentioned deaths attributable to HF was approximately as high in 1995 (287 000) as it was in 2009 (275 000). More than 7 million Europeans have congestive HF and it is a leading cause of hospitalization [1, 2]. The major risk factors for HF are hypertension, coronary artery disease (CAD), dyslipidaemia, diabetes, obesity, genetic predisposition, viruses, and toxins [1-4].

The major causes of HF in the Western World are CAD, cardiomyopathy, hypertension and diabetes mellitus [1, 2]. Although, no data are available from India, regarding the exact prevalence and incidence of CHF, it is possible that higher propensity for CVDs and ageing of population,may enhance the burden of CHF greater than that in the Western population. There is an urgent need to have HF registries in the secondary, tertiary care centers and at the national level. In earlier studies from India, rheumatic heart disease has been the major cause of HF [3, 4]. In later studies, CAD and cardiomyopathy have become major causes of HF in India [5, 6].

The prevalence and incidence of heart failure, based on disease-specific estimates may be conservatively attributedmainly due to CAD, hypertension, obesity, diabetes and rheumatic heart disease to range from 1.3 to 4.6 million, with an annual incidence of 491 600-1.8 million [5]. The heart failure registry may help and provide us the detailed information related to incidence, prevalence, and aetiology of HF in India. The mechanisms of HF may be related to oxidative stress, mitochondrial and microRNA dysfunction, inflammation and disturbed myocardial substrate metabolism as observed in various studies [6-12]. In the present study, we examine the risk factors and causes of CHF among Indian patients.

Subjects and Methods

Study Design

This case control study, including subjects of both sexes, affected by chronic heart failure (CHF) due to various aetiologies, was conducted at the Department of Cardiology, Government Medical College, Amritsar, and Halberg Hospital and Research Institute, Moradabad, India. After written informed consent and approval from the ethics committee, all patients were recruited to this study. We examined approximately 1110 adults with heart disease, during the recruitment period of two years to select 127 patients with diagnosis of CHF. Of total 127 patients, 2 were excluded due to various reasons, and the remaining 125 patients (Men 61, women 64) presenting with CHF and 250 (age and sex matched), control subjects were studied. Severity of disease was assessed according to the New York Heart Association (NYHA) standards and fell within functional classes II to IV. The diagnosis of CHF was based on clinical symptoms as well as on radiological and echocardiographic heart enlargement in presence of clinical signs of HF. Exclusion criteria: patients not willing to give consent (n=nil) were excluded. Patients presenting with shock (n=1), chronic renal failure (n=1), and cancer were also excluded before inclusion of the patients in this study.

All patients were administered standard drug therapy, including ACIs, betablockers, spironolactone, furosemide and receptor blockers wherever indicated. Patients were hospitalized for 3-7 days and then called for 1-4 weekly follow up depending upon clinical condition of the patients.

Laboratory Data

Baseline measurements were taken by a complete blood profile including; thiobarbituric acid reactive substances (TBARS) and malondialdehyde (MDA), which were obtained by colorimetric methods. MDA was quantified by a reaction with thiobarbituric acid [12, 13]. Resting electrocardiogram and prothrombin time (on anticoagulants) were done in all the patients at baseline and during follow-up as indicated. CoQ levels in the serum were measured by high-performance liquid chromatography based on a previously published method [14, 15]. The technicians measuring the data were blind to the groups.

Cytokine IL-10, was determined by using a solid phase, two-sided chemoluminescence enzyme immunometric assay (immulite automated analyzer) kit (DPC, Los Angeles, CA, US). The serum samples were kept at -24°C until the assay was performed. Results were expressed as pg/ml.

Follow Up. All patients were followed up for 3-7 days in the hospital. After discharge from the hospital, patients were called for follow up every 1-4 weekly depending up on clinical status and future complications. Two patients were lost in the follow up at the end of 2 years.

Statistical Analysis

Statistical analysis was performed by using analysis of variance for continuous variables. Categorical variables were described as frequencies and percentages, and differences in the two groups were assessed by using a chi square test. A two tailed t test for continuous normal data was considered significant at $p<0.05$.

Results

Of a total of 127 patients diagnosed with heart failure and recruited for this study, 2 were excluded, and the remaining 125 (men 61, women 64) were studied, as compared with 250 age and sex matched control group subjects. The aetiology of CHF were: CAD (n=34,27.2%), hypertensive heart disease (n=10,8.0%), valvular heart disease (n=40,32.0%) and idiopathic dilated cardiomyopathy (n=38,30.4%), hypertrophic cardiomyopathy (n=2,1.6%), amyloid cardiomyopathy (n=1,0.8%), as given in Table 1.

Two patients discontinued the medication for a few days, due to persistent vomiting whereupon other medicines were substituted.Echocardiographic ejection fraction was (mean ± SD) 39.1 ± 8.2% in the study group. However, echocardiographic data were not available in the healthy subjects group.

Table 1 shows the baseline clinical data in the study group and control subjects.

Mean age and height were comparable in the two groups. However mean body weight and waist girth were significantly lower in the heart failure group compared to healthy control subjects.

Valvular heart disease was the commonest cause of heart failure which was mainly due to rheumatic heart disease. Majority of the patients were given ACIs, beta blockers and furosemide for the treatment of HF. Digoxin was given only to those patients where heart failure was due to rheumatic heart disease and to a few more patients where heart rate was a problem.

The quality of life indicators; visual analogous scale revealed that dyspnoea, palpitation, and fatigue, were present at rest in all the patients during baseline measurements, which showed beneficial effects in the heart failure group whereas no such manifestations were observed among healthy subjects (Table 2). Heart rate showed a significant decline in the heart failure groups from the baseline during the next 3 days of follow up, compared with the healthy subject group. Distance walked in 6 minutes revealed that, at baseline; HF patients were unable to walk whereas healthy subjects walked significantly more distance than the HF patients.

Laboratory data (n=62 patients only in the HF group) showed that anti-inflammatory IL-10 was significantly lower among HF patients compared to healthy subjects (Table 3). Biomarkers of oxidative stress such as TBARS, MDA and diene conjugates were significantly higher among HF patients compared to healthy subjects whereas mean serum magnesium and coenzyme Q10 were slightly lower among CHF patients.

Table 1. Clinical data among subjects with heart failure and controls

Data	Heart failure (n=125)	Healthy subjects (n=250)
Men (Number %)	61 (48.8)	121 (48.4)
Women (Number %)	64 (51.2)	129 (51.6)
Mean age, years, mean +SD	51.8 ±7.7	52.5±10.5
Mean height, cm ditto	161.5 ±14.6	162.2±16.8
Mean Weight, Kg ditto	60.6 ± 8.2*	64.3±12.2
Waist girth, cm ditto	86.8±9.0*	88.7±11.6
Risk factors	(number %)	(Number %)
Coronary artery disease	52 (41.6)**	13 (10.0)
Hypertension (>140/90mmHg)	54 (43.2)**	25 (10.0)
Diabetes mellitus	12 (9.6)**	13 (5.2)
Obesity	43 (34.4)	80 (32.0)
Albuminurea	12 (9.6)*	6 (2.4)
Family history	4 (3.2)	2 (0.8)
Viral, toxins-carditis	6 (4.8%)	-
Causes of heart failure	(Number %)	-
Coronary artery disease	34 (27.2%)	-
Hypertensive heart disease	10 (8.0%)	
Valvular heart disease	40 (32.0%)	-
Dilated Cardiomyopathy	38 (30.4%)	-
Hypertrophic cardiomyopathy	2 (1.6)	-
Amyloid cardiomyopathy	1 (0.8)	-
Duration of heart failure, months Drug therapy given	19.5 (16.81%)	-
ACE-inhibiters (2.5-10mg/ day)	88 (75.53%)	-
Beta blockers (25-100mg/day,	40 (34.48%)	-
Furosemide (20-40mg/day)	116 (100%)	-
Digoxin (0.125 to 0.25mg/day)	60 (51.72%)	-
Angiotensin II receptor blockers (20-30mg/day)	11 (9.48%)	-

*p<0.05, **p<0.001 . p-value was obtained by analysis of variance or chi square test, by comparison of heart failure group and control subjects.

Table 2. Clinical manifestations of heart failure compared to healthy subjects

Clinical data	Heart failure (n=125)	Healthy subjects (n=250)
Heart rate (per minute)	107±18**	73.6±9.8
6- minute walk (Meters)	213 ±19.5**	452±26.6
Dyspnea (Number %)	125 (100%)	-
Fatigue (ditto)	125 (100%)	-
Palpitation (ditto)	125 (100%)	-
NYHA Class II (ditto)	24 (20.70%)	-
NYHA Class III and IV (ditto)	101 (80.8%)	-

p-values were obtained by analysis of variance. **p<0.01.

Table 3. Laboratory data among patients with heart failure

Data	Heart failure (n=62)	Healthy subjects (n=250)
Interleukin-10 (pg/ml)	3.3 ±1.9*	4.4±1.7
TBARS (μg/ml)	2.2±1.5*	1.2±0.67
(MDA (μg/ml),mean (SD)	3.4±1.3*	2.2±1.2
Diene conjugates (OD)	28.5±7.5*	22.7±5.5
Calcium (mg/dl)	9.0±2.2	9.1±2.5
Magnesium (mg/dl)	1.52±1.12*	1.64±1.23
Potassium (mEq/dl)	4.1±2.0	4.3±2.2
Coenzyme Q10 (g/ml)	0.21±0.11	0.28±0.18

Biochemical data were available among 62 randomly selected patients of CHF.
TBARS = thiobarbituric acid reactive substances, MDA = malondialdehyde.
TNF = tumour necrosis factor, CoQ = coenzyme Q10.
p-values were obtained by analysis of variance a by comparing heart failure and control subjects.
**p<0.01, *p<0.05.

Discussion

Reliable estimates of heart failure are lacking in India because of the absence of a surveillance programme to track incidence, prevalence, outcomes and key causes of heart failure [3-5].

It is possible that the incidence and prevalence rates of CHF and deaths due to CHF are rising due to population, epidemiological and health transitions. In a recent study, causes of deaths due to CVDs, CAD, stroke, heart failure, cardiomyopathy in relation to the social classes showed that deaths due to these CVDs were significantly more common among higher social classes 1-3, compared to social classes 4 and 5 and these were major causes of deaths [16].

This study shows that of 1110 patients of CVDs examined, 127 (11.4%) had CHF indicating that CHF is a common cause of morbidity and mortality and it has possibly become a major cause of deaths. The causes of CHF in majority of the patients were; CAD (n = 34,27.2%), hypertensive heart disease (n=10, 8.0%), valvular heart disease (n=40,32.0%) and idiopathic dilated cardiomyopathy (n=38,30.4%) (Table 1).

In 1949, Vakil reported a series of 1281 cases of heart failure out of 1860 patients of CVDs (3). The risk factors of CHF in this study were; 31.3% hypertensive coronary group, 29.3% rheumatic heart disease, 11.7% syphilitic group and 9.3% pulmonary group. The incidence of CHF in each group were; rheumatic heart disease (81.5%), hypertensive-coronary group (64.8%), pulmonary group (64%) and the syphilitic group 62.5%. The death-rate for all cases of CHF was 27.5%; highest (100%) in the bacterial group, 44.6% for the pulmonary group, 31.3% for the syphilitic group, 23.2% for the rheumatic group and only 18.3% for the hypertensive-coronary group. Singh et al. carried out the radioimmunoassay of serum digoxin in relation to digoxin intoxication among 98 patients, out of 125 patients of CHF [4]. Causes of CHF were; rheumatic heart disease (n=57, 58.2%), cor pulmonale (n=13, 13.3%), hypertensive heart disease (n=3,3.0%0, CAD (n= 4,4.0%) and Miscellaneous including cardiomyopathy (n=21, 21.4%). NYHA heart failure Class 1,2 were present among

69 (69.8 %), subjects and Class III+IV among 29(29.4%) patients. Among 22 patients, acute cardiac damage including myocardial infarction, carditis, as risk factors were also observed. Hypomagnesaemia was observed among 42 (42.8%), CHF patients, which may be due to diuretic therapy and poor dietary intake of magnesium among these patients which can predispose cardiac events. These are important risk factors of HF as well as mortality among these patients.

Risk factors of HF; CAD (n=52,41.6%), hypertension (>140/90mmHg) (n=54,43.2%), diabetes mellitus (n=12,9.6%), obesity (n=43,34.4%), and albuminurea (n=12,9.6%), were significantly more common among HF patients compared to control group (Table 1). These risk factors of CHF have also been observed in other studies [1-6]. All the patients had NYHA class II, III and IV heart failure in association with clinical manifestations of heart failure and in conjunction with low echocardiographic ejection fraction which confirms the diagnosis of CHF. Apart from other risk factors, a predictive value of depressive symptoms as risk factor and B-type natriuretic peptide for new-onset heart failure and mortality have also been observed [17].

There is a decline in overall HF hospitalization substantially from 1998 to 2008 but at a lower rate for black men [18]. However, decline in HF hospitalization and 1-year mortality rates were uneven across states which appear to be due to control of risk factors; hypertension, CAD, diabetes etc. Recent declines in the incidence of CAD in the developed countries, a common precursor for HF, may have led to declines in HF hospitalizations. Hypertension is also an important cause of HF, hence rates of blood pressure control can also help to facilitate thedecline in HF rate. Higher use of secondary prevention therapies, may have led to fewer recurrent HF hospitalizations. Secular trends in thresholds for hospital admission may also have a role, as clinicians substitute outpatient instead of in patient care when feasible [18]. Further studies indicate that presence of risk factors and lifestyle factors may also influence the quality of life, inflammation, incidence of heart failure as well as mortality which needs further study [19-21]. Male sex, less education, physical inactivity, cigarette smoking, overweight, diabetes, hypertension, valvular heart disease, and CAD are all independent risk factors for CHF. However, more than 60% of the CHF that occurs in the US general population might be attributable to CAD [19-21]. In a cohort Health ABC study among 2610 older persons with heart failure, during follow-up (median 9.4 years), HF developed in 311 (11.9%) participants [20, 21]. A variation in medication adherence in heart failure can also influence hospitalization rate and mortality [22].

In United States, the 30-day, 1-year, and 5-year case fatality rates after hospitalization for CHF were 10.4%, 22%, and 42.3%, respectively [23, 24]. The fatality rates after 5-years were greater in blacks than whites which are explained largely by blacks' greater levels of atherosclerotic risk factors. These rates are applicable for hospital deaths as well as for death rates in the community [23-26]. The percentage of deaths among Indians, may not be that high due to a lower proportion of patients having coronary risk factors, because a greater proportion of patients had HF due to rheumatic heart disease (Table 1).

The Atherosclerosis Risk in Communities (ARIC) cohort is a population-based study from 4 United States communities (1987 to 2002) included 750 patients with HF, from hospitalization (n = 1,206) or death certificate (n = 76) [24] records. There were 1,282 incident HF cases over 198,417 person-years. The age-adjusted incidence rate (per 1,000 person-years) for Caucasian women of 3.4, was significantly less compared with all other groups (Caucasian men, 6.0; African-American women, 8.1; African-American men, 9.1).

Age-adjusted HF incidence rates were greater for African-Americans than Caucasians, but adjustment for confounders attenuated the difference. HF incidence rates in African-American women were more similar to those of men than of Caucasian women. The greater HF incidence in African-Americans than in Caucasians is largely explained by African-Americans' greater levels of atherosclerotic risk factors [24].

In patients with heart failure (HF), mortality is lower in women versus men [25]. Of 5,888 CHS (Cardiovascular Health Study) subjects, 1,264 developed new HF and were followed up for 3 years. Mortality rates after HF were lower in women compared with men (for white women, African-American women, African-American men, and white men, total mortality was 35.5, 33.6, 44.4, and 40.5/100 person-years, and cardiovascular mortality was 18.4, 19.5, 20.2, and 22.7/100 person-years, respectively). After adjusting for covariates, women had a 15% to 20% lower risk of total and cardiovascular mortality compared with men, but there was no significant difference in outcome by race. The gender-race interaction for either outcome was not significant. It is possible that in older adults with HF, women had significantly better survival than men irrespective of race, suggesting that gender-based survival differences may be more important than race-based differences.

Prospective studies evaluating associations between food intake and risk of heart failure (HF) in diverse populations are needed [26]. In this large, population-based sample of African-American and white adults, whole-grain intake was associated with lower HF risk, whereas intake of eggs and high-fat dairy were associated with greater HF risk after adjustment for several confounders. In a further study [27], gender-specific long-term HF incidence and mortality in an urban community-based sample of middle-aged subjects were examined.Between 1974 and 1992, 33,342 HF-free subjects (10,900 [32.7%] women, mean age 45.7 ± 7.4 years) were included in the Malmö Preventive Project, on average 21.7 ± 4.3 years before study end. In a middle-aged, urban, community-based sample, women had a lower risk of HF, all-cause death and HF-related death over two decades of follow-up [27].

Female and male HF patients had similar mortality risk after the diagnosis of HF. In these comparatively young HF patients, few deaths were due to HF and more than 4 out of 10 deaths were non-cardiovascular.

Natriuretic peptides are potentially useful biomarkers for the evaluation and management of patients with HF, and the growing understanding of important and relevant comorbidities in HF [28]. Other candidate biomarkers from a number of classes: (a) myocyte stretch, (b) myocyte necrosis, (c) systemic inflammation, (d) oxidative stress, (e) extracellular matrix turnover, (f) neurohormones, and (g) biomarkers of extracardiac processes, such as renal function in CHF need further studies [28-30]. Further evidence indicate that myocardial injury (cardiac troponin), haemodynamic overload, genetic, neurohormonal, inflammatory, the components of the renin-angiotensin-aldosterone system and the catecholamines and biochemical factors are implicated in the development and progression of CHF [28-30]. Others sub-groups of biomarkers include extracellular matrix remodeling markers and myocardial injury markers (such as troponins I and T) which could provide a unique window into the pathophysiology of chronic HF, potentially improving our ability to predict adverse outcomes, provide novel drug targets, and even help gauge therapeutic efficacy [30].

Diet and lifestyle interventions, such as regulations to limit salt and tobacco consumption, are effective for primordial prevention and would have a wider impact on prevention of heart failure.

Clinical preventive interventions and clinical quality improvement interventions, such as treatment of hypertension, atherosclerotic disease, diabetes and acute decompensated heart failure may be effective for primary, secondary and even tertiary prevention.

In brief, this study showed that CAD, hypertension, cardiomyopathy and rheumatic heart disease are common causes of HF.

The risk factors of CHF were; CAD, hypertension, diabetes and obesity. ACE-inhibitors, betablockers, receptor blockers, spironolactone and furosemide are commonly used drugs for the treatment of CHF in India.

Acknowledgments

The International College of Nutrition is thanked for its financial support to conduct this study and write this article.

Conflict of interest has not been declared by the authors.

References

[1] Go, A. S., Mozaffarian, D., Roger, V. L., et al. Heart Disease and Stroke Statistics 2013 Update. A report from the American Heart Association. *Circulation* 2013;127:e6-e245.

[2] ESC Guidelines for the diagnosis and treatment of acute and chronic heart failure 2012. The Task Force for the Diagnosis and Treatment of Acute and Chronic Heart Failure 2012 of the European Society of Cardiology. Developed in collaboration with the Heart Failure Association (HFA) of the ESC. *Eur. J. Heart Fail.* 2012; 14: 803-869. doi: 10. 1093/eurjhf/hfs105.

[3] Vakil, R. J. A statistical study of 1281 cases of congestive cardiac failure or myocardial insufficiency in India. *Indian Physician.* 1949 Oct.;8(10):281-9.

[4] Singh, R. B., Rai, A. N., Dube, K. P., Srivastav, D. K., Somani, P. N., Katiyar, B. C. Radioimmunoassay of serum digoxin in relation to digoxin intoxication. *Brit. Heart J.* 1975; 37:619-623.

[5] Huffman, M. D., Prabhakaran, D. Heart failure: epidemiology and prevention in India. *Natl. Med. J. India.* 2010;23:283-8.

[6] Kumar, A., Singh, R. B., Saxena, M., Niaz, M. A., Josh, S. R., Chattopadhyay, P., Mechirova, V., Pella, D., Chopra, R. Effect of carni Q-gel (ubiquinol and carnitine) on cytokines in patients with heart failure in the Tishcon study. *Acta Cardiol.* 2007;62: 349-354.

[7] Shaw, S. M., Shah, M. K. H., Williams, S. G., Fildes, J. E. Immunological mechanisms of pentoxifylline in chronic heart failure. *Eur. J. Heart Fail.* 2009; 11(2): 113-118 doi: 10.1093/eurjhf/hfn040.

[8] Dickinson, B. A., Semus, H. M., Montgomery, R. L., Stack, C., Latimer, P. A., Lewton, S. M., Lynch, J. M., Hullinger, T. G., Seto, A. G., Rooij, E. V. Plasma microRNAs serve as biomarkers of therapeutic efficacy and disease progression in hypertension-

induced heart failure. *Eur. J. Heart Fail.* (2013) doi: 10.1093/eurjhf/hft018First published online: February 6, 2013.

[9] Ardehali, H., Sabbah, H. N., Burke, M. A., et al. Targeting myocardial substrate metabolism in heart failure: potential for new therapies, *Eur. J. Heart Fail.* 2012 14 () 120-129.

[10] Bayeva, M., Gheorghiade, M., Ardehali, H. Mitochondria as a therapeutic target in heart failure. *J. Am. Coll. Cardiol.* 2013;61(6):599-610. doi:10.1016/j.jacc.2012.08.102.

[11] Heymans, S. and Heart Failure Association of ESC. Inflammation as a therapeutic target in heart failure? A scientific statement from the Translational Research Committee of the Heart Failure Association of the European Society of Cardiology. *Eur. J. Heart Fail.* 2009;11:119-129,doi:10.1093/eurjhf/hfn043.

[12] Niato, C., Kawamura, M., Yamamoto, Y. Lipid peroxides as the initiating factor of atherosclerosis. *NY Acad. Sci.* 1993;676: 27-45.

[13] Esterbaur, H., Cheeseman, K. Determination of aldehyde peroxidation products malonaldehyde and 4-hydroxynonenal. *Methods Enzymol.* 1990; 186: 407-21.

[14] Chopra, R. K., Goldman, R., Sinatra, S. T., Bhagavan, H. N. Relative bioavailability of coenzyme Q formulations in human subjects. *Int. J. Vit. Nutr. Res.* 1998; 68: 109-13.

[15] Edlund, P. O. Determination of coenzyme Q, alpha-tocopherol and cholesterol in biological samples by coupled-column liquid chromatography with colorimetric and ultraviolet detection. *J. Chromatogr.* 1988; 425: 87-97.

[16] Fedacko, J., Vargova, V., Singh, R. B., Anjum, B., Takahashi, T., Tongnuka, M., Dharwadkar, S., Singh, S., Singh, V., Kulshresth, S. K., De Meester, F., Wilson, D. W. Association of high w-6/w-3 fatty acid ratio diet with causes of death due to noncommunicable diseases among urban decedents in North India. *The Open Nutra. Jour.* 2012;5:113-123.

[17] van den Broek, K. C., Defilippi, C. R., Christenson, R. H., Seliger, S. L., Gottdiener, J. S., Kop, W. J. Predictive value of depressive symptoms and B-type natriuretic peptide for new-onset heart failure and mortality. *Am. J. Cardiol.* 2011;107:723–729.

[18] Chen, J., Normand, S. L., Wang, Y., Krumholz, H. M. National and regional trends in heart failure hospitalization and mortality rates for Medicare beneficiaries, 1998-2008. *JAMA.* 2011;306:1669–1678.

[19] Djoussé, L., Driver, J. A., Gaziano, J. M. Relation between modifiable lifestyle factors and life time risk of heart failure. *JAMA.* 2009;302:394–400.

[20] Kalogeropoulos, A., Georgiopoulou, V., Kritchevsky, S. B., Psaty, B. M., Smith, N. L., Newman, A. B., Rodondi, N., Satterfield, S., Bauer, D. C., Bibbins-Domingo, K., Smith, A. L., Wilson, P. W., Vasan, R. S., Harris, T. B., Butler, J. Epidemiology of incident heart failure in a contemporary elderly cohort: the Health, Aging, and Body Composition study. *Arch. Intern. Med.* 2009;169:708–715.

[21] Kalogeropoulos, A., Georgiopoulou, V., Psaty, B. M., Rodondi, N., Smith, A. L., Harrison, D. G., Liu, Y., Hoffmann, U., Bauer, D. C., Newman, A. B., Kritchevsky, S. B., Harris, T. B., Butler, J.; Health ABC Study Investigators. Inflammatory markers and incident heart failure risk in older adults: the Health ABC (Health, Aging, and Body Composition) study. *J. Am. Coll. Cardiol.* 2010;55:2129–2137.

[22] Yuting Zhang, PhD; Shang-Hua Wu, M. S.; A. Mark Fendrick, MD; Katherine Baicker, PhD, Variation in medication adherence in heart failure. *JAMA Intern. Med.* 2013;():1-2. doi:10.1001/jamainternmed.2013.2509.

[23] Roger, V. L., Weston, S. A., Redfield, M. M., Hellermann-Homan, J. P., Killian, J., Yawn, B. P., Jacobsen, S. J. Trends in heart failure incidence and survival in a community-based population. *JAMA*. 2004;292:344–350.

[24] Loehr, L. R., Rosamond, W. D., Chang, P. P., Folsom, A. R., Chambless, L. E. Heart failure incidence and survival (from the Atherosclerosis Risk in Communities study). *Am. J. Cardiol.* 2008; 101:1016–1022.

[25] Parashar, S., Katz, R., Smith, N. L., Arnold, A. M., Vaccarino, V., Wenger, N. K., Gottdiener, J. S. Race, gender, and mortality in adults > or =65 years of age with incident heart failure (from the Cardiovascular Health Study). *Am. J. Cardiol.* 2009 Apr. 15; 103(8):1120-7.

[26] Nettleton, J. A., Steffen, L. M., Loehr, L. R., Rosamond, W. D., Folsom, A. R. Incident heart failure is associated with lower whole-grain intake and greater high-fat dairy and egg intake in the Atherosclerosis Risk in Communities (ARIC) study. *J. Am. Diet. Assoc.* 2008; 108(11):1881-7.

[27] Tasevska-Dinevska, G., Kennedy, L. M., Nilsson, P. M., Willenheimer, R. Gender aspects on heart failure incidence and mortality in a middle-aged, urban, community-based population sample: the Malmö preventive project. *Eur. J. Epidemiol.* 2009; 24(5): 249-57. Epub. 2009 Mar. 8.

[28] van Kimmenade, R. R., Januzzi, J. L. Jr. Emerging biomarkers in heart failure. *Clin. Chem.* 2012;58(1):127-38. doi: 10.1373/clinchem.2011.165720. Epub. 2011 Nov. 15.

[29] Tousoulis, D., Oikonomou, E., Siasos, G., Chrysohoou, C., Charakida, M., Trikas, A., Siasou, Z., Limperi, M., Papadimitriou, E. D., Papavassiliou, A. G., Stefanadis, C. Predictive value of biomarkers in patients with heart failure. *Curr. Med. Chem.* 2012;19 (16):2534-47.

[30] Ahmad, T., Fiuzat, M., Felker, G. M., O'Connor, C. Novel biomarkers in chronic heart failure. *Nat. Rev. Cardiol.* 2012 Mar. 27;9(6):347-59. doi: 10.1038/nrcardio.2012.37.

In: Circadian Cardiology with Focus on Both Prevention ... ISBN: 978-1-63463-956-9
Editors: K. Hristova, A. Shehab, G. Cornelissen et al.

Chapter 13

The Effects of Fireworks on Ambient Air and Possible Impact on Cardiac Health, during Deepawali Festival in North India

***Raina Pal*[1], *Mahima*[1], *A. Gupta*[1], *Chitra Singh*[1], *Anamika Tripathi*[1,*] *and R. B. Singh*[2]**
[1]Pollution Ecology Research Laboratory, Department of Botany, Hindu College, Moradabad, Uttar Pradesh, India
[2]Halberg Hospital and Research Institute, Moradabad, India

Abstract

Background: The previous studies reveal that the burning of crackers and sparklers on the occasion of Deepawali (Diwali) is a strong source of air pollution and may cause serious health hazards. Since fireworks displays are becoming more frequent and can influence the ambient air quality by Particulate Matter (PM), SO_2, NO_2 and trace elements, we examine the effect of firework during Diwali festival.

Subject and Methods: The effect of the burning of fireworks on ambient air quality was assessed in Moradabad city during Diwali festival in November 2012 i.e., one week before Diwali, on Diwali and one day after Diwali day. The monitoring was carried out using Respirable Dust Sampler (RDS) and Fine Dust Sampler (FDS). Particulate Matter i.e., TSP, PM_{10} and $PM_{2.5}$ were estimated by gravimetric method and gaseous pollutants namely SO_2 and NO_2 were measured by improved West - Geake and Jacob - Hochheiser modified method, respectively. Elemental content of PM_{10} has also been determined using ICP-OES in samples, taken during Diwali days.

Results: The 8 h concentration of TSP, PM_{10} and $PM_{2.5}$ on main festival night was found to be 856, 624 and 597 μgm^{-3}, respectively and these concentrations were found to

* Corresponding author: Dr Anamika Tripathi, PhD. Pollution Ecology Research Laboratory, Department Of Botany, Hindu College, Moradabad-244001, Uttar Pradesh, India. Tel.: 91-7599211137, e-mail: anamikambd @rediffmail.com.

be higher at 2.0, 2.56 and 5.52 times for TSP, PM_{10} and $PM_{2.5}$ when compared with the concentration of the respective night of the normal day and was significantly higher ($p<0.001$).

Gaseous concentration SO_2 and NO_2 was reported about two times higher than the Normal Day. On Diwali day, 24 h values for TSP, PM_{10}, $PM_{2.5}$, SO_2 and NO_2 were found to be higher than prescribed limit of National Ambient Air Quality Standard (NAAQS). The 24 h mean concentration of metals associated with PM_{10} on Diwali day was found to be in the order of Zn (17.74)>Al (13.36)> Fe (12.47)> Cu (10.47)> Pb (2.54)> Mn (0.42)> Cr (0.19)> Cd (0.18)> Ni (0.15) in μgm^{-3} and all these values were found to be higher (except Fe) than the normal day and post Diwali day. Correlation study shows a strong positive correlation between Cr with Cu and Cu with Zn. Lead and cadmium are known to have adverse effects on cardiovascular health.

Conclusion: It is possible that the burning of fireworks during Diwali festival may be the major source of emission and accumulation of fine particles in the atmosphere. The inhalation of PM may reach deep into the alveoli of lungs causing adverse effects on health; respiratory and cardiovascular diseases.

Keywords: Fireworks, TSP, PM_{10}, $PM_{2.5}$, Diwali festival, Heavy metals, Health impact

Introduction

Disability-adjusted life years (DALYs) assess disease burden, by combining the years of life lost and years lived with disability. The Global Burden of Disease Study group 2010, reported that indoor household and outdoor ambient particulate matter (PM) air pollution have a large effect on diminished quality of life [1]. Indoor and outdoor PM contributes 7.6 percent DALYs, compared to blood pressure at 7.0 percent, tobacco smoking including second-hand smoke at 6.3 percent, and alcohol use at 5.5 percent. In terms of DALYs, household air pollution was the second most important risk factor for women and girls, globally, and the largest environmental risk factor of those examined [2, 3]. Household air pollution from solid cooking fuels directly caused 3.5 million premature deaths in 2010, and another half a million through its contribution to outdoor air pollution [1, 2].

Deepawali (Diwali), the festival of lights, celebrated in Oct-Nov in India, with great enthusiasm and bursting of fire crackers, may cause air pollution and adverse effects on health. The important air pollutants generated during the fire crackers bursting are SO_2, NOx and smoke/dust containing respirable suspended particulates (RSPM or PM_{10}).

The Lantern festival in Beijing [4], in Yanshui, Taiwan [5], the Las Fallas in Spain [6] and the Diwali festival in India [7, 8] consume several hundreds of tons of fireworks and fire crackers every year. Fireworks contain chemicals, iron dust powder, and release various gaseous and particulate air pollutants as well as toxic metals of significant quantity which degrades the air quality as a whole. However, SO_2, NO_x, PM_{10} and $PM_{2.5}$ are of special interest, as they are known to be potentially injurious to the human health as well as plants.

Fireworks activities in Oahu in New Year celebration, lead to an increase in TSP and PM_{10} by an average of 300% and 700% at different locations [9]. The effect of the metal content of ambient air, on human health and the concentration of different metals in PM_{10} from fireworks appears to be interesting for a scientific study [6-12].

During Diwali festival, emission and accumulation of metals in the atmosphere was reported by a few workers from India [7, 13-16]. In Delhi, increase in ground level O_3 was reported, due to burning of colored sparkles from crackers, while Ba, K, Al and Sr went up to 1091, 25, 18 and 15 folds higher [14-16] and SO_2, NO_2, PM_{10} and TSP increased 2-10 folds higher [15]. These fine particles can reach into respiratory system by inhalation, causing bronchitis, pneumonia and lung cancer [17, 18].

Further studies indicate that air pollution can also predispose cardiovascular diseases (CVDs) [19-24]. The objective of the present study is to monitor the air pollutants (TSP, PM_{10}, $PM_{2.5}$, SO_2, NO_2) and heavy metals i.e., Zn, Cu, Pb, Cr, Ni, Al, Mn, Cd and Fe present in the air which may have adverse effects on cardiovascular health.

Subject and Methods

Study Area and Sample Collection

Moradabad, known as 'Brass city of India' is located in semi–arid zones towards the north–west of India with extreme summer and winter condition. It is located at an average height of 76.19 mts above sea level in the western gangetic plain of Indian subcontinent at latitude 28.15N and 74.49E. It covers an area of 3516.62 km^2 and the population of the city is 8,89,810 (as per 2011 census).

In order to elucidate the environmental issues concerning air pollution, during Diwali, air monitoring was conducted at four randomly selected residential areas, with possible number of subjects exposed, in four directions; east- Ram Ganga Vihar (RGV, 18000 subjects), west- Chandra Nagar (CN, 14000), north- Buddh Bazar (BB, 11000) and south- Police Training Centre (PTC, 6000). It is estimated that approximately 49,0000 subjects out of 8,89,810 subjects were exposed to firework pollution in the four areas.

We monitored the TSP, PM_{10}, $PM_{2.5}$, SO_2 and NO_2 simultaneously in the ambient air during the period of; before Diwali day (Normal day), Diwali day and post Diwali day in all the four areas. Trace metals namely Zn, Cu, Pb, Cr, Ni, Al, Mn, Cd, Fe associated with PM_{10} were also estimated. Samples were collected from different sites along with monitoring, analysis and data interpretation (Figure 1).

Monitoring and Analysis

Monitoring was carried out by using Respirable Dust Sampler (APM-460NL) with provision for gaseous sampling (APM-411, Envirotech, New Delhi) for measuring the concentration of TSP, PM_{10}, NO_2 and SO_2 at a flow rate of 1.0-1.2 m^{-3}min in the ambient air. $PM_{2.5}$ aerosols originating from the burning of fireworks was monitored over a 3-day period at urban area by using fine dust sampler (FDS), APM-550. The sampling instruments were fixed at breathing height of 1.6 m above the ground level at four residential sites. The samples of TSP and PM_{10} were taken on pre-weighed (W_1) glass microfiber filters (GF/A – Whatman) of 8×10 ins size and $PM_{2.5}$ samples were taken on teflon membrane paper (PTEF, 46.2mm). After sampling, we reweighed (W_2) the filters in order to determine the mass of the particles collected. The Particulate Matter concentration in ambient air was then computed by applying the following formula:

$\Delta W \times 10^6$/ volume of air, where $\Delta W = W_2 - W_1$

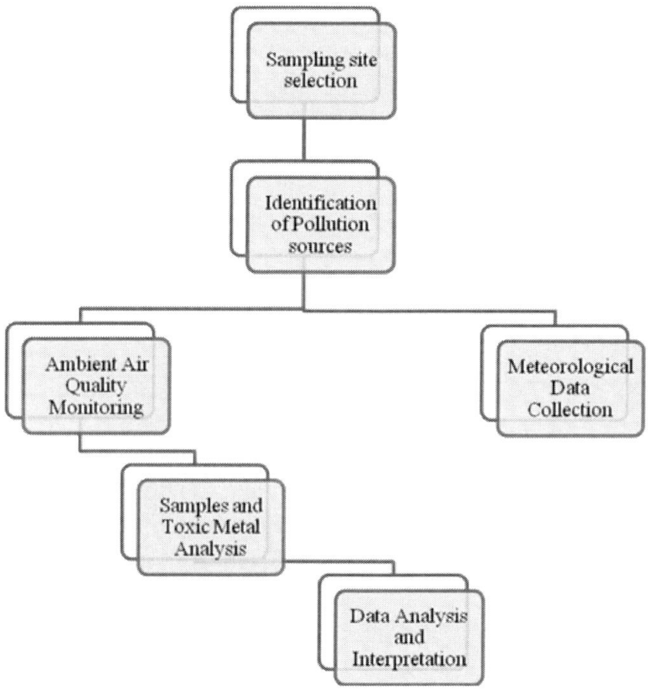

Figure 1. Flow chart of various steps of study protocol.

The samples of SO_2 and NO_2 were analyzed by employing the West-Geake method [25] and Jacob-Hochheiser modified method [26] respectively. A known quantity of air was passed through the impinger containing sodiumtetramercurate for SO_2 and sodium hydroxide for NO_2. Absorbing solution of SO_2 formed a dichlorosulphitomercurate complex which made to react with pararosaniline and methylsulphonic acid and absorbing reagent of NO_2 formed a stable solution of sodium nitrite by reacting the exposed absorbing solution with phosphoric acid, sulphanilamide and 1-napthyl ethylenediamine dihydrochloride (NEDA). The absorbing reagent of SO_2 and NO_2 were measured on Spectrophotometer (Systronics-166) at a wavelength of 560nm and 540 nm respectively.

The sampling duration for TSP, PM_{10} and $PM_{2.5}$ was 8 h, total three samples were taken in a day (24 h) and the gaseous samples were taken at the interval of every 4 h (six samples in a day). The dominant wind direction was westerly and north westerly. The temperature varied from 18°C – 30.2°C.

Metal Analysis

Total 40 square of 1×1 ins diameter (4 locations + 1 control/blank) of the fiber filter paper covered by particulates digested with nitric acid and perchloric acid in a ratio 1:3 on a 140°C hot plate till white fumes arose. Residues were then redissolved by 0.1M hydrochloric acid. The content was filtered through Whatman Filter no. 42 and made-up to 25 mL by double distilled water. The filtrate of each sample was examined for the concentrations of heavy metals by using Inductively Coupled Plasma-Optical Emission Spectrometer (ICP-OES; Spectro Analytical Instruments, West Midlands, UK) collected for each site.

To get the final concentration results of the blank samples are subtracted from the exposed samples for each of the metals the concentration of metals in the samples is then multiplied by the sample volume (i.e., 25 mL) to get the mass of each metal in the samples. These values are subsequently divided by corresponding total volume of sampled air to get the concentration of metals in the sampled air.

Statistical Analysis

Summary statistics were used to obtain the means and standard deviations for TSP, PM_{10} $PM_{2.5}$, SO_2 and NO_2 and trace metals. One–way analysis of variance (ANOVA) was applied to detect the significant difference between metals and within metals. Correlation matrix was used to determine the relative association among metals ($p \leq 0.001$).

Quality Assurance

Reliability in data collection is the most important criterion for such studies. The daily particulate samples was rejected if any disruption was noted the data was also discarded if the post-sampling flow rate was reduced by more than 10% of the pre-sampling flow rate, or if filters were found to be damaged during sampling. As many as 28 validated data samples were retained after application of the quality assurance procedures.

Results

Results of the monitoring work for all the three days for different particulate matter i.e., TSP, PM_{10} and $PM_{2.5}$ and gaseous pollutants i.e., SO_2 and NO_2 are presented in the study.

Particulate Pollutants

Comparison of 24hr means of $PM_{2.5}$ before Diwali day and post Diwali day, we observed more or less similar values but on Diwali Day the values showed 2-3 times higher than before Diwali day (Figure 2a). The higher concentration of $PM_{2.5}$ was found almost all the three days at the site RGV. The minimum concentration of $PM_{2.5}$ (99 ± 21.93 μgm^{-3}) was recorded at PTC site on before Diwali day and the highest concentration was recorded at RGV at Diwali Day i.e., 318 ± 284.8 μgm^{-3} (Figure 2a). All the values recorded during Diwali days were found more than the NAAQS (60 μgm^{-3}) [27]. Concentration of PM_{10} at all the four sites exceeded the maximum prescribed allowable limit of NAAQS i.e., 100 μgm^{-3} before and after Diwali [27]. During Diwali the maximum concentration of PM_{10} was found at RGV (612 ± 165.16

μgm^{-3}) followed by CN (414±132.71 μgm^{-3}), BB (372±55.32 μgm^{-3}) and PTC (307±80.54 μgm^{-3}) Figure 2b. The percent of PM_{10} have increased considerably on Diwali day.

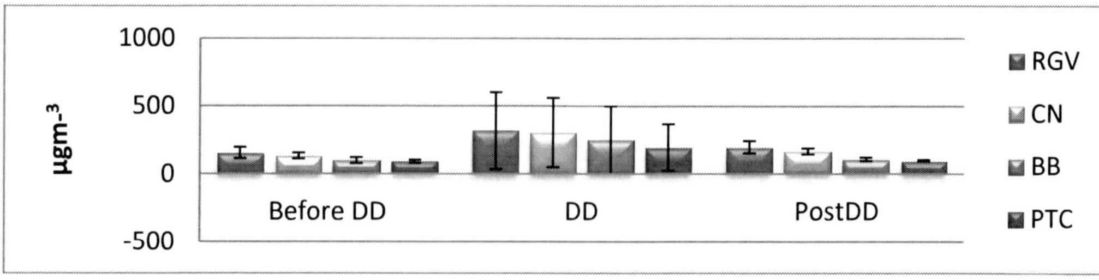

Figure 2(a). 24hrs mean concentration of fine particles ($PM_{2.5}$) with SD (±) at different sites during the period of Diwali festival.

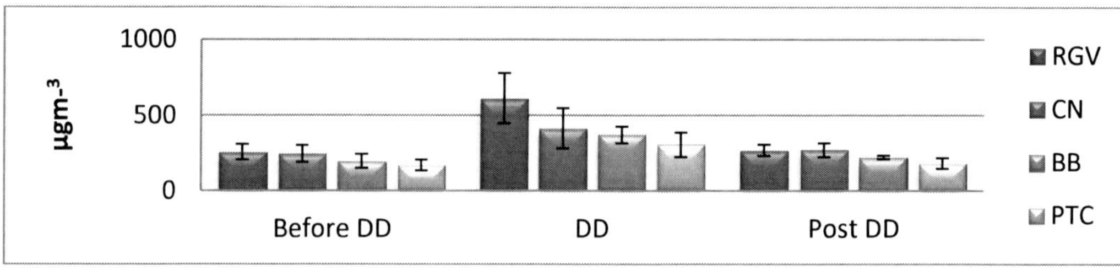

Figure 2(b). 24hrs mean concentration of RSPM (PM_{10}) with SD (±) at different sites during the period of Diwali festival.

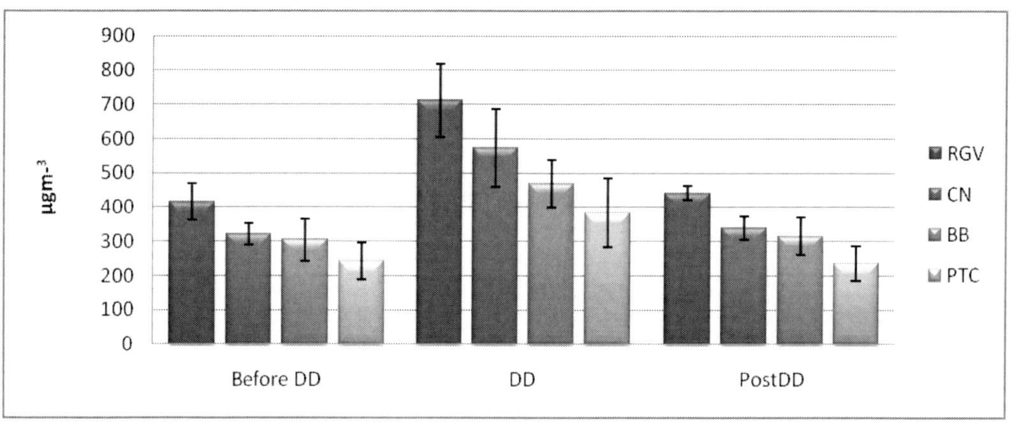

Figure 2(c). 24hrs mean concentration of TSP with SD (±) at different sites during the period of Diwali festival.

Comparison of PM_{10} on Diwali day showed a significant difference level ($p<0.05$) than the respective concentrations of other observed days and the level were greater during night time compared to day time with increase in fireworks.

TSP concentration exceeded the maximum prescribed limits (200 μgm^{-3}) at all the sites [27] shown in Figure 2c. Before Diwali, the daily average concentration was observed 629μgm^{-3}, while the maximum concentration 787μgm^{-3} was found on Diwali day during fireworks at site RGV. After Diwali the TSP concentration was slightly decreased may be due to local holiday [7]. Before Diwali, the daily average concentration was observed 629μgm^{-3}, while the maximum concentration 787μgm^{-3} was found on Diwali day during fireworks at site RGV. After Diwali the TSP concentration was slightly decreased may be due to local holiday [7].

Gaseous Pollutants

The concentration range of SO_2 before and after Diwali is 11μgm^{-3} to 71μgm^{-3} at all the study areas. On Diwali day the SO_2 concentration crossed the prescribed 24 hrs (80 μgm^{-3}) National Ambient Air Quality Standard [27] at all the four sites. The maximum concentration was found at site RGV whereas the minimum concentration was found at PTC. During Diwali day SO_2 concentration (42μgm^{-3}) was found highest at night time followed by noon (35μgm^{-3}) and morning time (22μgm^{-3}) which was more than two times in compared to the normal day before Diwali whereas on day time on Diwali day the highest SO_2 concentration (71μgm^{-3}) was found at site RGV. The 24 hrs average of the NO_2 concentration ranged from 17μgm^{-3} to 57μgm^{-3}. This concentration reported is about two to three times higher than the normal days (Table 1). The maximum concentration of NO_2 (54μgm^{-3}) was found in second shift (2 pm-10 pm) on pre Diwali day at CN. NO_2 concentration rapidly increase due to traffic density almost at both the sites in compared to the normal day and crossed the prescribed limit of NAAQS i. e., (80 μgm^{-3}) [27].

Heavy Metals in PM_{10}

The concentration of heavy metal such as Cu, Zn, Fe, Cd, Al, Mn, Pb, Cr and Ni for the three main festival days i.e., before one week of Diwali day, Diwali day and post Diwali day were also observed in PM_{10}. 24 hr. mean and standard deviation of elemental concentrations are reported in μgm^{-3} (Table 2). Among all the nine heavy metals, the highest concentration of Zn was reported followed by Al, Fe and Cu respectively.

An analysis of the result shows that the concentration of heavy metal ranges; for Zn between 5.65 -125.25μgm^{-3}. Al from 4.02 - 21.24μgm^{-3}, Fe from 7.45 - 19.0μgm^{-3}, Cu from 2.32 - 16.45μgm^{-3}. Pb from 0.92 - 2.50μgm^{-3}, Mn from 0.01-0.52, Cd from 0.001 - 0.48, Cr 0.003-0.32 and for Ni the range was found between 0.0003 - 0.32μgm^{-3} (Table 2).

The Pearson linear correlation matrix of the heavy metals in PM_{10} samples presented that Cr showed a very strong positive correlation with Cu (r=0.92) and Zn (r=0.92) and Cu with Zn (r=0.95) and Ni (r=0.9) and Ni with Al (r=0.92) was found, suggesting they might be originated from similar sources [10, 28] Table 3.

Similarly Cr showed a fairly strong correlation was observed with Ni (r=0.87), Mn (r=0.81). A positive correlation between Cu and Al (r=0.87) again suggests that these element could be derived from similar sources but probably different from that of Cr with Fe (r=0.17), Cr with Pb (r=0.2), Fe-Mn (r=0.02) and Ni with Cd (r=0.03) as the former elements showed week correlations with other elements (Table 3). Negative correlation was observed between Fe- Ni, Fe- Al, Ni- Pb, Al- Cd and between Cd- Pb.

Table 1. Mean concentration of 24hrs of gaseous pollutants with SD (±) at four sites during the period of Diwali festival

Gases/Days	SO_2				NO_2			
	RGV	CN	BB	PTC	RGV	CN	BB	PTC
Before DD	26±11.74	25±6.34	22±11.71	17±8.46	27±17.86	27±17.2	30±5.67	21±5.73
DD	60±23.56	55±14.78	51±24.3	52±32.54	42±36.87	57±6.47	42±6.88	31±5.98
Post DD	29±12.34	27±5.73	24±221.3	18±12.45	28±13.56	30±9.34	36±6.45	22±8.34

DD = Diwali Day, Values in μgm^{-3}.

Table 2. Concentration of heavy metals at different locations during Diwali Day

Sites/ Elements	RGV	CN	BB	PTC
Cr	0.12±0.17	0.085±0.11	0.075±0.12	0.004±0.004
Cu	10.42±5.29	7.423±3.9	7.06±2.82	2.55±0.58
Fe	11.22±1.32	8.326±0.88	17.84±1.38	8.96±1.34
Ni	0.14±0.15	0.103±0.12	0.01±0.02	0.002±0.001
Zn	16.13±7.36	10.72±6.81	14.14±6.44	5.39±1.05
Al	14.5±5.89	10.92±4.71	5.45±1.70	5.38±2.80
Cd	0.08±0.11	0.03±0.01	0.39±0.09	0.002±0.001
Mn	0.15±0.21	0.24±0.25	0.19±0.22	0.18±0.12
Pb	1.88±0.52	2.22±0.36	3.53±0.70	1.62±0.62

Values are mean ±Standard Deviation in μgm^{-3}.

Table 3. Spearman correlation coefficient (r)* matrix for PM_{10} and selected heavy metals in the airborne particulate matter during the period of Diwali festival

	PM_{10}	Cr	Cu	Fe	Ni	Zn	Al	Cd	Mn	Pb
PM_{10}	1									
Cr	**0.65**	1								
Cu	0.49	**0.92**	1							
Fe	0.28	0.17	0.23	1						
Ni	0.4	**0.87**	**0.9**	-0.01	1					
Zn	0.6	**0.92**	**0.95**	0.43	**0.79**	1				
Al	0.39	**0.77**	**0.87**	-0.17	**0.92**	**0.73**	1			
Cd	0.3	0.34	0.34	**0.92**	0.03	0.53	-0.1	1		
Mn	0.7	**0.81**	0.6	0.02	**0.63**	**0.62**	0.44	0.22	1	
Pb	0.33	0.2	0.2	**0.73**	-0.07	0.36	0.18	-0.83	0.27	1

*r-values shown in bold are significant at $p \leq 0.001$.

Equality of mean differences of trace metals concentration between metals and within metals was carried out by analysis of variance (ANOVA), Table 4. The one way analysis of variance shows that the calculated value of F is 14.04 which is larger than the table value of 2.31 at 5% level and 3.26 at 1% level with *d.f.* (degree of freedom) being $V_1 = 8$ and $V_2 = 9$. The study, therefore conclude that the difference in metal concentration due to fireworks is significant ($p<0.05$ and $p<0.01$, Table 4).

Table 4. One way analysis of variance Summary (ANOVA) of heavy metal concentration

Source of variation	SS	d.f.	MS	F-ratio	5%F-limit	1%F-limit
Between samples	833.81	(9-1) = 8	833.81/8=104.22	104.22/7.42 = 14.04	F (8,27) = 2.31*	F (8,27) = 3.26**
Within samples	200.41	(36-9) = 27	200.41/27=7.422			
Total	1034.22	(36-1) = 35				

SS = sum of squares, d.f. = degree of freedom, MS = mean square, *$p<0.05$, **$p<0.001$.

Discussion

Epidemiological studies indicate a consistent association of acute and chronic exposures to ambient air quality related to airborne particles with risk of morbidity and mortality due to cardiovascular and respiratory diseases [20-23]. In United States and other developed countries, the evidence on adverse effects of particulate air pollution on public health has led to more stringent standards for levels of PM in outdoor air. The US National Ambient Air Quality Standard for airborne PM was revised, maintaining the previous indicator of particulate matter of less than or equal to 10μm in aerodynamic diameter (PM_{10}) and creating a new indicator for fine PM of less than or equal to 2.5μm in aerodynamic diameter ($PM_{2.5}$). There is evidence of an association between recently measured $PM_{2.5}$ concentrations and daily hospitalizations on a national scale from various countries [20-24].

Our findings using PM_{10}, may be complementary to substantial evidence on PM and hospitalization for respiratory or cardiovascular causes using exposure measures other than $PM_{2.5}$ and the more limited evidence using $PM_{2.5}$ specifically. The PM_{10} which have greater impacts on health, have been found to increase by four fold during fireworks and about three fold after fireworks period. We found three fold increase in PM at the sites; RGV, CN and BB, during and after fireworks [29]. Compared to normal days, on the day before Diwali and post Diwali day showed an increase of 24 and 40% respectively. Sharp peaks, especially during night hours of all the observations suggest that there was less dispersion of particulates during the festival period due to low wind conditions [30]. In general, the particulate concentration seems to be very high during Diwali festival but the concentration of TSP, PM_{10} and $PM_{2.5}$ showed significant changes during day and night.

High levels of SO_2 are particularly dangerous in the presence of PM, because it slowly adsorbs on fine atmospheric particles and can be transported very deep into lungs causing damage to the lungs tissues. On post Diwali day, the concentration of SO_2 was lower than Diwali day (Table 1). The concentration showed a slight increase in night time as compared to day, indicating increased firework events during the night time. The NO_2 concentration also increased three fold (60 μgm^{-3}) at BB and (57μgm^{-3}) at PTC on Diwali day [31].

Pb, Cd, Hg and As are heavy metals and are known to have adverse effects on health [32-34]. We also observed high concentrations of metals in the PM in our study (Tables 2, 3, 4). Similar studies carried out at Hyderabad [14], Lucknow [7] and Howrah [13] during Diwali reported metal concentrations in coarser and fine fraction of particulates ($PM_{2.5}$ in Hyderabad and Howrah and PM_{10} in Lucknow). However, in the present study we studied, the coarser fraction of PM_{10} which are known for greater adverse effects on body systems. Since the particulate sizes are different the results are not truly comparable. Cd and Lead levels were higher during firework in our study which is consistent with the findings at Lucknow and Howrah [7, 13]. High concentration of Zn, may be attributed to its maximum use in fireworks for creating smoke effect, Al as white silver and Cu as fireworks tracer has been reported [12, 35, 36]. Among all the monitoring days, the highest concentration of the heavy metals was found on Diwali day followed by post Diwali day and before Diwali day as observed in other studies [14].

Arsenic, cadmium, lead, and mercury present potential health risks to children who are exposed through inhalation or ingestion [32, 34, 37]. Emerging Market countries; Brazil, China, India, Indonesia, Mexico, Poland, Russia, South Korea, Taiwan, and Turkey published studies on pediatric exposure to metals and reported concentrations of metals in blood and urine from these studies. The levels were generally higher than US reference values, and many studies identified adverse health effects associated with metals exposure.

Evidence of exposure to metals indicates a need for interventions to reduce exposure and efforts to establish country-specific reference values through surveillance or biomonitoring. The findings from review of these 10 countries also suggest the need for country-specific public health policies and clinician education in Emerging Markets. Lead enters in biological systems via food, water, air and soil. Children are particularly vulnerable to lead poisoning-more so than adults because they absorb more lead from their environment and their nervous system and blood get affected [32]. As first described by Friedrich Stromeyer (Gottingen, Germany) in 1817, cadmium intoxication can lead to kidney, bone, and pulmonary damages [37].

In a cross-sectional survey [33], an interquartile range increase in toenail arsenic was associated with higher systolic BP [0.93 mm Hg; 95% confidence interval (CI): 0.25, 1.62] and pulse pressure (0.76 mmHg; 95% CI: 0.22, 1.30). Positive associations between arsenic and BP and negative associations between manganese and BP were strengthened in models adjusted for other toenail metals.

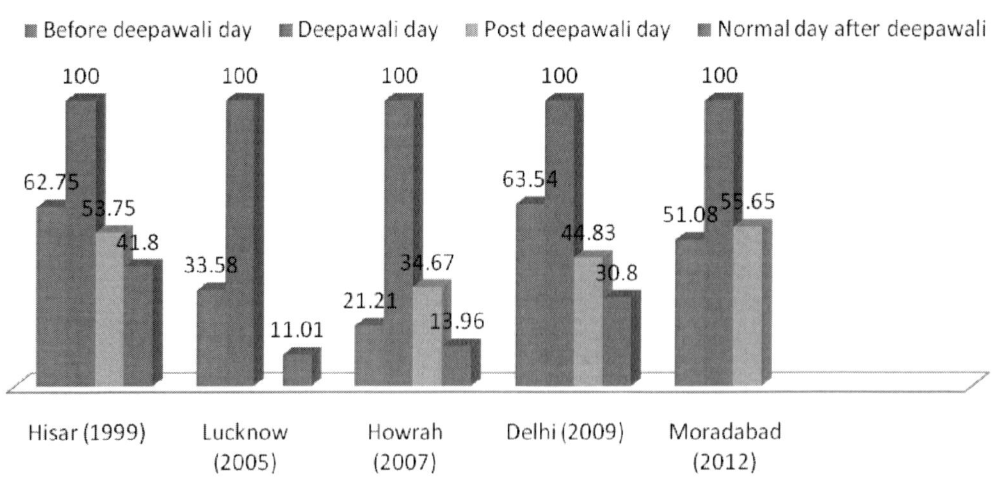

Figure 4. Comparison of PM_{10} among different cities of India during Diwali festival.

These findings indicate an associations between BP and arsenic and manganese. This may be of public health importance because of prevalence of both metal exposure and cardiovascular disease.

Figure 4 shows the comparison of PM_{10} among different cities of India during Diwali festival by various researchers [7, 13, 15]. Figure 4 clearly shows that percentage rise of PM is significantly greater at Moradabad than Howrah, Hisar, Lucknow and Delhi which indicates that pollution by firework is common in most urban areas of India [7, 13, 15].

The ambient concentration of pollutants may be under influence of meteorological factors. Since Diwali falls in pink winter season, when calm condition prevails throughout India, it may predispose to aggravate the air quality status. Figure 4 shows our findings, along with results in other studies, having similar meteorological conditions at all the sites, so the values are also comparable. The PM_{10} concentrations along with $PM_{2.5}$ during the festival are of episodic scale. According to USEPA the threshold episodic condition is $1000\mu gm^{-3}$ [38] which is reduced lately [39] and this is noticeable and alarming that the episode limit set by USEPA is routinely crossed during festival fireworks.

In US, fireworks usually are displayed during Independence Day celebration on 4th July [40] which contain black powder leaving a huge amount of smoke into atmosphere. Studies from different authors have been reported the short term effects of these fine particles on cardiovascular respiratory mortality [9, 18-19]. On Lantern festival in Beijing, primary components of Ba, K, Sr, Cl, Pb, Mg and secondary components of $C_5H_6O_4^{2-}$, $C_3H_2O_4^{2-}$, $C_2O_4^{2-}$, $C_4H_4O_4^{2-}$, SO_4^{2-}, NO^{3-} was reported about five times higher than the normal day [4].

The mechanisms underlying the adverse effects of PM on the respiratory and cardiac systems indicate inflammatory responses in the lung and release of cytokines with local and systemic consequences [19, 41-43]. PM may promote inflammation in the lung tissues and thereby exacerbate underlying lung disease and reduce the efficacy of lung-defense mechanisms. Adverse effects on cardiovascular system may reflect neurogenic and inflammatory cytokines [33]. Experimental studies of atherosclerosis using genetically susceptible mice also suggest that PM may accelerate atherogenesis leading to heart attack. The entry of PM in the body tissues may cause tachycardia leading to decline in heart rate variability which is a risk factor for sudden cardiac death and heart attack. Therefore, it is very important to monitor inhaled aerosol particles which may be smaller than 10μ and may reach in vital organs such as alveoli of the lung. Fine particles have a relatively large surface area compared to large inhaled particles as this has a more profound effect on the surface chemistry [36]. Epidemiological studies indicate the acute adverse effects on health, following exposure to particulates in several studies [19, 20].

Individuals, who were elderly or had preexisting lung or heart disease, appeared to be more susceptible than other to adverse effects of PM_{10} [19, 20]. $PM_{2.5}$ is more harmful particulate than the others as it reaches up to alveoli of lungs and gets solubilized leading to inflammation [20, 21]. Sulfur dioxide can damage lungs and cause coughing, wheezing and shortness of breath. Oxides of Nitrogen can cause skin problems, eye irritation, and respiratory problems in children [22]. Chemicals used in crackers like lead, magnesium, cadmium, nitrate, sodium, and others can have various harmful effects e.g., lung cancer, cardiovascular diseases, chronic respiratory diseases and allergies in adults [15, 19, 20].

The environmental authorities have been putting much effort to collect air quality data specially on the particulate matter of the size range of PM_{10} and $PM_{2.5}$. The adverse health effects of PM on cardiovascular disease and respiratory disease, and hospital admissions in the elderly were mainly mediated by fine particles (PM_{10}) which is mainly contributed by smoke [24].

Burning of fireworks might be a strong source of Suspended particles and large amount of soot and various metal oxides could be emitted into the air during burning. These high temperature derived particles are mostly in the fine mode and the large amount of smoke and metals in them could catalyze atmospheric reactions and thus might play a crucial role in the development of a more harmful PM which can easily damage body systems causing diseases. A short-term exposure to these pollutants above the permissible limit can increase the likelihood of acute health effects. Further, this study may be useful for public education, to develop public opinion against firework during Dewali festival to save our environment.

Short-term and long-term effects of pollutants have been observed on cardiovascular diseases and respiratory diseases; chronic obstructive pulmonary disease, asthma, and lung cancer. Emerging evidence suggests possible effects on premature births, lung-function development in children, and accelerated progression of atherosclerosis and cognitive impairment. The risk may exist at low levels of air pollution and that there is no safe threshold level, rather a linear concentration-response relationship. The current EU limit for PM2·5 at 25 μg/m3 annual average is already higher than the WHO Air Quality Guidelines (10 μg/m3), a discrepancy that needs urgent attention. The EU has declared 2013 "The Year of the Air", there is hope that air pollution—the most important environmental risk factor for the health of Europeans will get the attention it deserves [44].

In brief, a huge amount of burning of fireworks covers the city and surrounding areas over the next several days. Concentration of SO_2 and heavy metals at Diwali increased many fold, while the level of NO_2 approximately doubled the concentration reported during a normal day. Concentration of TSP, PM_{10} and $PM_{2.5}$ increased two to ten times respectively on main festival day. Increased levels of Pb and Cd in PM may have damaging effects on cardiovascular and respiratory system of our body. High concentration of pollutants was noted at Ramganga Vihar street which seems to be related to the economic status of the citizens of this area. Further studies should be conducted to find out the number of people exposed to pollution by the fireworks and what illnesses develop due to these pollutants.

Acknowledgments

The authors gratefully acknowledge for the financial assistance provided by University Grant Commission, New Delhi and Uttar Pradesh Pollution Control Board, Lucknow (India). Thanks are also due to Mr. V. K. Rajput, Atul Kumar and Mahesh Kumar for their kind support in sampling and analyses.

References

[1] Lim, S. S., Vos, T., Flaxman, A. D., Danaei, G., et al. for Global Burden of Disease Study Group. A comparative risk assessment of burden of disease and injury attributable to 67 risk factors and risk factor clusters in 21 regions, 1990-2010: a

systematic analysis for the Global Burden of Disease Study 2010. *Lancet* 2012; 380: 2224-2260. doi:10.1016/S0140-6736 (12) 61766-8.

[2] Environmental Factor - February 2013: Air pollution pinpointed as www.niehs.nih. gov/news/newsletter/2013/2/science-air/index.htm, http://ehs.sph.berkeley.edu/krsmith/ ?page_id=34).

[3] Murry, R. L., Britton, J. and Leonardi-Bee, Jo. Second hand smoke exposure and the risk of invasive meningococcal disease in children: systematic review and meta-analysis. *BMC Public Health*, 2012; 12: 1062.

[4] Wang, Y., Zhuang, G., Xu, C., and An, Z. The air pollution caused by the burning of fireworks during the lantern festival in Beijing. *Atmos. Environ.* 2007, 41: 417-431.

[5] Do, T.-M., Wang, C.-F., Hsich, Y.-K., Hsich, H.-F. Metal present in ambient air before and after a firework festival in Yanshui, Tainan, Taiwan. *Aerosol and Air Quality Research,* 2012; 12: 981-993.

[6] Moreno, T., Querol, X., Alastuey, A., Minguillon, M. C., Pey, J., Rodriguez, S., Miro, J. V., Felis, C., and Gibbons, W. Recreational Atmospheric Pollution Episodes: Inhable metalliferous particles from Firework displays. *Atmos. Environ.* 2007; 41: 913-922.

[7] Barman, S. C., Singh, R., Negi, M. P. S., Bhargava, S. K. Ambient air quality of Lucknow city (India) during use of fireworks on Diwali festival. *Environ. Monitor. Assess.* 2008; 137: 495-504.

[8] Padma, S., Gajghate, D. G., Gavane, A. G., Suryawanshi, P., Chauhan, C., Mishra, S., Gupta, N., Rao, C. V. C., Wate, S. R. Air Quality status during festival of India: A case study. *Bull. Environ. Toxicol.*, 2012, 89: 376-379.

[9] Bach, W., Daniels, A., Dickinson, L., Hertlein, F., Marrow, J., Margolis, S., Dinh, V. Fireworks pollution and health. *Intern. J. Studies* 1975; 7: 183-192.

[10] Tripathi, A. Airborne lead pollution in the city of Varanasi. *Atmos. Environ.*, 1994; 28: 2317-2323.

[11] Salve, P. R., Maurya, A. and Wale, S. R. Distribution of trace metal in atmosphere particulate matter of rural environment. *IJEP,* 2006; 26 (B): 673-677.

[12] Singh, R., Barman, S. C., Negi, M. P. S., Bhargava, S. K. Metals concentration associated with respirable particulate matter (PM10) in industrial area of eastern UP, India. *J. Environ. Biol.,* 2008; 29: 63-68.

[13] Thakur, B., Chakraborty, S., Debsarkar, A., Chakrabarty, S., Srivastava, R. C. Air pollution from fireworks during festival of lights (Deepawali) in Howrah, India- a case study. *Atmosfera* 2010; 23(4): 347-365.

[14] Kulshrestha, U. C., Rao, T. N., Azhaguvel, M. J. Emission and accumulation of metals in the atmosphere due to crackers and sparkles during Diwali festival in India. *Atmos. Environ.* 2004; 38: 4421-4425.

[15] Ravindra, K., Mor, S., Kaushik, C. P. Short-term variation in air quality associated with firework events- a case study. *J. Environ. Monit.*, 2003; 5: 260-264.

[16] Attri, A. K., Kumar, U., Jain, V. K. Microclimate formation of ozone by fireworks. *Nature,* 2001; 411(6841): 1015, 2001.

[17] Huang, S. L., Hsu, M. K. and Chan, C. C. Effects of submicrometer particle compositions on cytokine production and lipid peroxidation of human bronchial epithelial cells. *Environ. Health Prespect.* 2003; 111: 478-482.

[18] Vernath, J. M., Reilly, C. A., Vernath, M. M., Moss, T. A., Langelier, C. R., Lanza, D. L., and Yost, G. S. Inflammatory cytokines and cell death in BEAS-2B lung cells

treated with soil dust, lipopolysaccharide, and surface modified particles. *Toxicol. Sci.* 2004; 82: 88-96.

[19] Pope, C. A. III, Burnett, R. T., Thourton, G. D., Thun, M. J., Calle, E. E., Krewski, D., Godleski, J. J. Cardiovascular mortality and long-term exposure to particulate air pollution. *Circulation,* 2004; 109, 71-77.

[20] Pope, C. A. III, Dockery, D. W. Health effects of fine particulate air pollution: lines that connect. *J. Air Waste Manage. Assoc.* 2006; 56: 709-742.

[21] Dominici, F., Peng, R. D., Bell, M. L., Pham, L., McDermott, A., Zeger, S. L., Samet, J. M. Fine particulate air pollution and hospital admission for cardiovascular and respiratory diseases. *J. Am. Med. Assoc.* 2006; 295: 1127-1134.

[22] COMEAP. *A Report by the Committee on the Medical Effects of Air Pollutant's Cardiovascular Sub-Group.* Cardiovascular Disease and Air Pollution. United Kingdom Department of Health, London, United Kingdom, 2006; 208 pp.

[23] Analitis, A., Katsouyanni, K., Dimakopoulou, E., Samoli, A. K., Nikoloulopoulos, Y., Petasakis, G., Touloumi, J., Schwartz, H., Anderson, H. R., Cambra, K., Forastiere, F., Zmirou, D., Vonk, J. M., Clancy, L., Kriz, B., Bobvos, J., Pekkanen, J. Short-term effects of ambient particles on cardiovascular and respiratory mortality. *Epidemiol.* 2006; 17: 230-233.

[24] Anderson, H. R., Atkinson, R. W., Peacock, J. L., Sweeting, M. J., Marston, L. Ambient particulate matter and health effects: publication bias in studies of short-term associations. *Epidemiol.* 2005; 16: 155-163.

[25] West, P. W. and Gaeke, G. C. Fixation of sulphur dioxide as sulfitomercurate (II) and subsequent colorimetric determination. *J. Analy. Chem.* 1956; 28: 1816-1819.

[26] Jacob, M. B., Hochheiser, J. B. Continuous sampling and ultramicrodetermination of nitrogen dioxide in air. *J. Analy. Chem.* 1958; 30: 426-428.

[27] NAAQS. *The Gazette of India*, Ministry of Environment and Forest, GOI, New Delhi, 2009.

[28] Banarjee, A. D. K. Heavy metal levels and solid phase speciation in street dust of Delhi, India. *Environ. Pollut.* 2003; 123: 95-105.

[29] Clark, H. Air pollution from fireworks. *Atmos. Environ.* 1997; 31: 2893-2894.

[30] Barman, S. C., Singh, R., Negi, M. P. S., Bhargava, S. K. Fine particles (PM2.5) in ambient air of Lucknow city due to fireworks on Diwali festival. *J. Environ. Biol.* 2009; 30 (5): 625-632.

[31] Mandal, P., Prakash, M., Bassin, J. K. Impact of Diwali celebrations on urban air and noise quality in Delhi City, India. *Environ. Monit. Assess.* 2012; 184, 209-215.

[32] Wilhelm, M., Schulz, C., Schwenk, M. Revised and new reference values for arsenic, cadmium, lead, and mercury in blood or urine of children: basis for validation of human biomonitoring data in environmental medicine. *Int. J. Hyg. Environ. Health.* 2006 May; 209(3):301-5.

[33] Mordukhovich, I., Wright, R. O., Hu, H., Amarasiriwardena, C., Baccarelli, A., Litonjua, A., Sparrow, D., Vokonas, P., Schwartz, J. Associations of toenail arsenic, cadmium, mercury, manganese, and lead with blood pressure in the normative aging study. *Environ. Health Perspect.* 2012;120:98-104. doi: 10.1289/ehp.1002805.

[34] Horton, L. M., Mortensen, M. E., Iossifova, Y., Wald, M. M., Burgess, P. What do we know of childhood exposures to metals (arsenic, cadmium, lead, and mercury) in emerging market countries? *Int. J. Pediatr.* 2013; 2013:872596. Epub. 2013 Jan. 8.

[35] Vecchi, R., Bernardoni, V., Cricchio, D., Alessandro, A. D., Fermo, P., Lucarelli, F., Nava, S., Piazzalunga, A., Villia, G. The impact of fireworks on airborne particles. *Atmos. Environ.* 2008; 42: 1121-1132.

[36] Perrino, C. S., Tiwari, M., Catrambone, S. D., Torre, E., Rantica, S., Canepari. Chemical characterization of atmospheric PM in Delhi, India, during different periods of the year including Diwali festival. *Atmos. Pollu. Res.*, 2011; 2: 418-427.

[37] Johannes Godt, Franziska Scheidig, Christian Grosse-Siestrup, Vera Esche, Paul Brandenburg, Andrea Reich and David A Groneberg. The toxicity of cadmium and resulting hazards for human health. *Journal of Occupational Medicine and Toxicology* 2006, 1:22. doi:10.1186/1745-6673-1-22.Original T.

[38] USEPA. United States Environmental Protection Agency. *EPA Defines Air Pollution Danger level-EPA press release,* October 19, 1971. Available online: http://www.epa.gov/history/topics/caa70/09.htm.

[39] RIDEM. Rhode Island Department of Environmental Management - Division of Air Resources. *Air Pollution Control Regulation No. 10, 2003*. Available online: http://www.epa.gov/region 1/topics/air/sips/ri/2003_RI_reg_10.pdf.

[40] Liu, D. Y., Rutherford, D., Kinsey, M., and Prather, K. A. Real-time monitoring of pyrotechnically derived aerosol particles in the atmosphere. *Anal. Chem.* 1997; 69: 1808-1814.

[41] National Research Council; Committee on Research Priorities for Airborne Particulate Matter. *Research Priorities for Airborne Particulate Matter,* (2004) IV: Continuing Research Progress. Washington, DC: National Academies Press.

[42] Pope, C. A. III, Hansen, M. L., Long, R. W., et al. Ambient particulate air pollution, heart rate variability, and blood markers of inflammation in a panel of elderly subjects. *Environ. Health Perspect.* 2004; 112:339-345.

[43] Brook, R. D., Franklin, B., Cascio, W., et al. Air pollution and cardiovascular disease. *Circulation.* 2004; 109:2655- 2671.

[44] Editorial. Air pollution: Europe's avoidable health risk. *The Lancet*, Volume 381, Issue 9870, Page 876, 16 March 2013 <Previous Article|Next Article, doi:10.1016/S0140-6736(13)60656-X.

In: Circadian Cardiology with Focus on Both Prevention ... ISBN: 978-1-63463-956-9
Editors: K. Hristova, A. Shehab, G. Cornelissen et al.

Chapter 14

Prevalence and Risk Factors of Hypertension in Rangpur, Bangladesh

Ratindra Nath Mondal[1,*], *Md. Ashraful Haque*[2], *Shah Md. Sarwer Jahan*[2], *Abul Kalam Azad*[2], *Md. Mahfuzer Rahman*[2], *Moni Rani*[3], *B. D. Bidhu*[2], *A. K. M. Shaheduzzaman*[2], *Md. Zakir Hossain*[2], *Md. Kumruzzaman Sarker*[2], *Amaresh Chandra Shaha*[2] *and R. B. Singh*[4]

[1]RCMC, Rangpur, Bangladesh
[2]RpMC, Rangpur, Bangladesh
[3]RpMCH, Rangpur, Bangladesh
[4]Halberg Hospital and Research Institute, Moradabad, India

Abstract

Background. Hypertension is a major public health problem nowadays. It plays a major etiologic role in the pathogenesis of coronary artery disease, stroke, heart failure and renal failure.

Subjects and Methods. A cross-sectional survey was carried out among subjects aged18 years and above at Rangpur division, of Bangladesh from 1st July, 2012 to 30th April, 2013. After written informed consent and approval from ethic committee, subjects were selected (total n=481, 52.8% female, males 47.2%) randomly based on voters list. The sampling frame consisted of the total final population of the Rangpur city according to the 2011 census of Bangladesh. Evaluation was by a physician and a dietitian administered questionnaire, a physical examination and using a sphygmomanometer. The diagnosis of prehypertension and hypertension were based on modified JNC-VII criteria.

Results. The prevalence of hypertension and pre-hypertension in the study population were 33.3% and 29.9% respectively without any gender difference. Prevalence of

* Corresponding author: Dr. Ratindra Nath Mondal, Assistant professor of Medicine, Rangpur Community Medical College, Rangpur, Bangladesh. E-mail: dr.ratinmondal @gmail.com.

hypertension in 18-29 year age group was 9.35% and ≥70 years age group 53.84% (P value 0.00016134). Majority of the study subjects were illiterate (48.6%).

Risk factors of hypertension and prehypertension were; age, obesity, added salt, diabetes mellitus, physical inactivity, prolonged use of contraceptive. Only 25.2% of the study populations had knowledge (high blood pressure is hypertension) of hypertension. Among the previously diagnosed hypertensive patients, 46.67% (n=28) took the antihypertensive drugs regularly and blood pressure was at controlled level (<140/90 mm of Hg) in 36.06%. Only 61.66% patients were coming for regular follow up and ignorance and poverty were the main causes of irregular follow up.

Conclusion. Prevalence of hypertension and its risk factors in Rangpur division of Bangladesh is high. Co-ordinated approach from local and Government level may be needed to combat this situation.

Keywords: Prevalence, hypertension, blood pressure, Rangpur division

Introduction

Non-communicable diseases (NCDs) are already of major health importance in developed countries and are rapidly becoming a major public health threat in the developing world [1]. NCDs were estimated to have contributed to almost 60% of deaths in the world and among them about 80% occurred in the developing countries [2]. Like many other countries, Bangladesh has been experiencing an epidemiological transition from poverty to affluence and from communicable disease to NCDs [3] due to economic development and better control and treatment of infectious diseases [4]. Of total number of deaths, 51% of deaths in Bangladesh occur due to NCDs and other chronic health conditions [5].

Among the NCDs, hypertension is the major risk factor for developing ischemic heart disease, stroke, heart and renal failures and peripheral vascular diseases [4]. Hypertension has been reported to be responsible for 57 per cent of all stroke deaths and 24 per cent of all cardiovascular deaths in East Asians [6]. Although, blood pressure is easily measurable it has taken several decades to realize that hypertension is a frequent worldwide health disorder [7]. Hypertension affects nearly 26 per cent of the adult population worldwide.

Kearney and colleagues estimated that the prevalence of hypertension in 2000 was 26% of the adult population globally and that in 2025 the prevalence would increase by 24% in developed countries and 80% in developing countries [8].

The prevalence of cardiovascular diseases and hypertension is rapidly increasing in developing countries [9]. However, representative national level data on prevalence of hypertension and their risk factors are still inadequate in Bangladesh [10-13].

In addition, pre-hypertension is associated with increased risk of developing hypertension and major cardiovascular events, independent of other cardiovascular risk factors [14, 15]. This study was undertaken to determine prevalence, awareness and control status of hypertension at Rangpur division of Bangladesh.

Subject Methods

A community-based cross-sectional survey was carried out among the population of Rangpur division, from 1st July, 2012 to 20th April, 2013. Among the 8 divisions of Bangladesh, Rangpur, northern division, consists of 8 districts and 59 sub-districts, which is further divided into unions and ward/villages, and has a population of 15,665,000. The study population included individuals of 18 years and above of both sexes, residing in Rangpur division.

Sampling Methods and Sample Size

A multistage sampling procedure was used to select the study subjects. At the first stage, Rangpur division was divided into 8 sectors according to its administrative districts on a geographical map. In the second stage two sub-districts were taken by random sampling from each district and then 2 unions from each sub-district were selected further by random sampling. Finally, one village was chosen from each union by random sampling, from among the villages of the union.

In our study, village was the smallest sampling unit. From each village, study sample was taken by random sampling based on voter list. In places where the selected individual from voter list remained outside the village or not found during data collection, we have taken another individual by random sampling instead of that individual. Finally door to door survey was carried out to collect data from the selected people.

Assuming unknown prevalence of hypertension in Rangpur division, a sample size of 385 was calculated to give the true prevalence with a precision of 5% with 95% confidence level. Accounting non-response rate of 25% (96), finally the sample size was determined to be 481.

Exclusion criteria were: age below 18 years. (As 18 years age is the lower limit to enter into the voter list), those who refused to take part in the study, any acute illness of patient.

Data Collection

An informed written consent was obtained before measuring the blood pressure in all the subjects. Blood pressure was measured by auscultatory method (according to JNC-7), using the validated aneroid sphygmomanometer. All the participants were requested to take rest for ten minutes before measuring the blood pressure. The individual was seated in a chair with his back supported and his arms bared and supported at heart level and was refrained from the use of tobacco in any form or ingestion of caffeine 30 minutes preceding the measurement. Two separate readings were taken at an interval of minimum three minutes. If systolic blood pressure differ >10 mm of Hg and diastolic >5 mm of Hg, then more 2 readings were taken. The average of the two or 4 readings were used to classify the blood pressure. Systolic blood pressure measured at the appearance of the Korotkoff sounds (Phase I) and diastolic BP was taken at the point of disappearance of the sounds (Phase V).

Socioeconomic status, tobacco and alcohol consumption and salt intake were assessed by questionnaire. Clinical data; demographic characteris-tics, weight, height, (for body mass index (BMI), waist and hip circumference for waist-hip ratio, were measured in all the subjects. Data on knowledge about hypertension, treatment of hypertension, follow up and causes of irregular follow up (in diagnosed hypertensive) were collected in a pre-designed, pretested questionnaire. Random blood sugar (by glucometer), urine strip test for proteinuria, 12 lead ECG and fundoscopic examination (tropicamide eye drop was given 10 minutes before the procedure) was done in all the study population.

Criteria

Hypertension

Systolic or diastolic blood pressure or both ≥140/90 mm of Hg (according to NICE guideline). Or any individual diagnosed as hypertension and currently taking antihypertensive drugs.

Smoker: Those who currently smoke or have smoked tobacco in any form (cigarette, birri etc.) in last 6 months.

Ex-smoker; Ex-smoker who gave up smoking at least 6 months before.

Smokeless Tobacco (SLT): SLT are raw tobacco leafs taken other than inhalation route (usually taken with betel nut or use in inner aspect of lower lip).

Physical inactivity was diagnosed in presence sedentary occupations such as business and service, without any spare time activity.

Obesity: BMI level ≥30Kg/M^2 (according to WHO);

Central obesity: The absolute waist circumference >102 centimeters in men and >88 centi -meters in women and the waist-hip ratio >0.90 for men and >0.85 for women.

Statistical Analysis

The interested variables are processed, edited and analyzed by SPSS windows version 17.0. The socio-demographic data of the study population were expressed in frequency distribution and their observed difference was tested by one sample 't' test and 'chi square' test. P value <0.05 was considered as statistically significant with the 95% confidence interval. The results were presented in tables.

Results

In this study, we examined a total of 481 people of both sexes. Female were slightly more common than males (52.8% vs 47.2%). Mean age of the study population was 47.75 years (SD 15.52). Table 1 shows the socio-demographic characteristics of the study population.

In this study we have found 23.9% current smoker and 7.3% ex-smoker. Average duration of smoking was 15.63 pack year (range 0.30-75, SD 14.56). Smokeless tobacco (SLT) was taken by 36.6% (176) people for mean duration of 7.65 years (minimum 0.50-40, SD 6.08). (Table 2 shows frequency of use of different form of SLT). Among the SLT users majority 74.43% used to take jorda (Raw leaf of tobacco that people usually used to take with betel nut), and 23.68% used to take gul (Those are also raw tobacco leaf in powder form that people used to take inner side of the lower lip or undersurface of tongue). Among the current and ex-smoker majority were male 89.3%, female were 0.7%. But SLT intake was found to be equal in male and female.

Only 25.2% (121) of the study people had knowledge (knows high blood pressure is hypertension) about hypertension and among them 57.8% (70) had knowledge about factors related to hypertension. 28.6% reported stress and 24.3% reported excess salt intake is risk factor of hypertension.

Table 1. Socio-demographic characteristics of the study subjects (n=481)

Variables	Frequency	Percentage (%)
Mean age	47.75 years (SD 15.52) (age range 20-110 years)	
Sex Male	227	47.2%
Female	254	52.8%
Level of education		
Illiterate	234	48.6%
≤10 class	202	42%
>10	45	9.4%
Occupation		
Housewife	230	47.8%
Agriculture	156	32.4%
Business	29	6%
Service	21	4.4%
Others	45	9.35%
Monthly income		
<5000 taka*	416	86.5%
5001-10000 taka	49	10.2%
>10000 taka	16	3.3%

*1 dollar = 82 taka.

Table 2. Prevalence of hypertension in both sexes by age group

Age group	Male		Female		Total	
	Screened	Prevalence	Screened	Prevalence	Screened	Prevalence
18-29	26	3.84%	38	13.15%	64	9.35%
30-39	38	21.05%	49	10.20%	87	14.94%
40-49	49	28.57%	56	35.71%	105	32.38%
50-59	45	31.11%	46	36.95%	91	34.06%
60-69	45	55.56%	37	62.16%	82	58.53%
≥70	24	58.33%	28	50%	52	53.84%
Total 227		33.48%	Total 254	33.07%	Total 481	33.26%
P value 0.006			P value 0.000		P value 0.000	

Family history of hypertension was present in 22.6% (109). In this study we have found that 50.1% (241) people took added salt and 36% (173) people were physically inactive.

Among 481 people screened for hypertension, 160 were hypertensive, giving the prevalence rate of 33.3%. Prevalence of pre-hypertension was 29.9% (143). An upward trend in prevalence was observed with increase in age, especially in male. In females, prevalence of hypertension was increased after 40 years up to 70 years, thereafter prevalence was observed to be decreased but still higher than pre 40 years level (Table 2 and Figure 1).

Prevalence of hypertension was significantly higher among those having family history of hypertension (41.3% vs 30.9%, p value 0.012) as given in Table 3. Among the hypertensive only 11.9% taking regular 30 minutes exercise, 9.4% taking irregularly exercise and 78.8% did not taking any exercise.

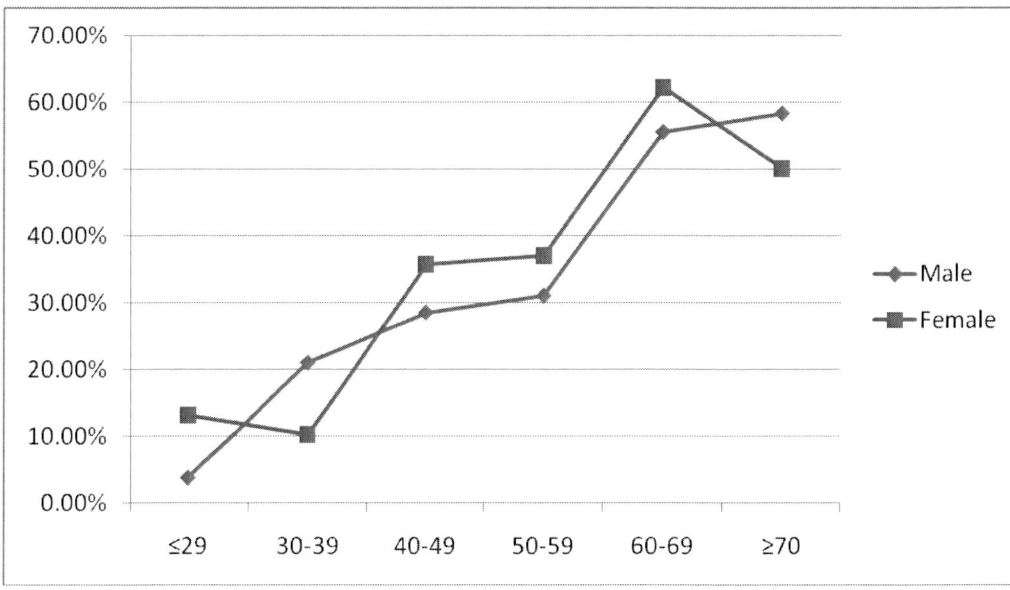

Figure 1. Linear chart showing prevalence of hypertension by age in male and female.

Table 3. Frequency of risk factors among hypertensive people

Variable	Frequency	Percentage
Age (years) Mean	55.21 (SD 13.74)	
Sex (M:F)	76: 84	47.5%: 52.5%
Current smoking	31	19.4%
SLT	56	35%
History of taking added salt	74	46.3%
Family history of hypertension	45	28.12%
Contraceptive use	15	17.85%
Sedentary lifestyle	48	30%
Obese (≥30 Kg/M2)	29	18.12%
Central obesity	100	62.5%
Mean Random Bood Sugar (mmol/L)	7.25(SD 3.58)	

Table 4. Showing different grade of hypertension

Grading of hypertension	Percentage	Male	Female
Grade 1	13% (18)	61.1% (11)	38.9% (7)
Grade 2	29% (40)	52.5%(21)	47.5% (19)
Systolic hypertension			
Grade 1	30.4% (42)	40.5% (17)	59.5% (25)
Grade 2	17.4% (24)	37.5% (9)	62.5% (15)
Diastolic hypertension	10.1% (14)	64.3% (9)	35.7% (5)

Among 254 female study population, 34.6% (87) were taking (oral and injectable) contraceptives and among them 83.90% (73) taking oral contraceptive pills and 16.09% (14) taking injectable contraceptives. Mean duration of use of contraceptive was 6.32 years. The prevalence of hypertension was significantly higher among those using contraceptive for more than 5 years (20.9% vs 14.3%, P<0.05)). The majority of the hypertensives (30.4%) had grade 1 systolic hypertension, followed by grade 2, both systolic and diastolic hypertension (29%) and 10.1% who had only diastolic hypertension (Table 4). Mean age of the population with isolated systolic hypertension, isolated diastolic hypertension and both systolic and diastolic hypertension were 58.69 (SD 14.35) years, 43.21 (SD 10.61) years and 54.70 (SD 13.29) years respectively. Isolated systolic hypertension was more prevalent in ≥50 years of age (80.30%). On the other hand diastolic hypertension more prevalent in ≤50 years of age (64.28%). Among the hypertensive patients previously diagnosed (those who were on antihypertensive drugs) case was 38.12% (61). Mean duration of hypertension was (6.85 years, SD 6.93). 46.67% (28) previously diagnosed hypertensive patients used to take the drugs regularly; regular adaptation of life style modification was found in 16.67% (10). Blood pressure was in controlled level (<140/90 mm of Hg) in 36.06% (22). Complications of hypertension are given in Table 5. Among the hypertensive heart disease LVH, ischemia, and MI was found in 10.62%, 10% and 1.3% respectively. Grade 2 hypertensive retinopathy was the maximum 11.9% followed by grade 3 (2.5%), grade 1 (1.3%) and grade 4 (0.6%).

Among the previously diagnosed hypertensive patients, 78.34% (47) patients did not come for regular follow up (Table 6 showing causes of irregular follow up).

In our study prevalence of diabetes mellitus was 4.78%. Newly diagnosed 3.32% (RBS ≥11.1 mmol/L; capillary blood testing by glucometer), and previously diagnosed (those were on treatment of diabetes mellitus) diabetes was 1.45%. Among the hypertensives, 7.5% were diabetic. Among the study population diabetes and hypertension co-exist in 2.49%.

Discussion

Hypertension is a major public health problem in developing countries like Bangladesh. According to WHO scientific group, 10-20 percent of adults in the world have hypertension based on the criteria suggested by their expert committee [16].

Table 5. Complications among hypertensive population

Variables	Newly diagnosed hypertensive (n=99)	Previously diagnosed hypertensive		Total hypertensive (n=160)
		Controlled hypertensive (n=22)	Uncontrolled hypertensive (n=39)	
Hypertensive heart disease (Ischaemia, LVH, MI)	20.20% (20)	18.18% (4)	28.20% (11)	21.87% (35)
Retinopathy	13.13% (13)	13.63% (3)	25.64% (10)	16.25% (26)
Proteinuria	5.05% (5)	4.5% (1)	7.69% (3)	5.625% (9)

LVH = left ventricular hypertrophy, MI = myocardial infarction.

Table 6. Causes of irregular follow up (n=47)

Variables	Frequency
Ignorance	27.65% (13)
Poor motivation	27.65% (13)
Economical problem	12.76% (6)
Lack of faith to the physician	8.51% (4)
Migration	2.12% (1)
Patient's busy schedule	2.12% (1)
Combination of above factors	9 (19.14%)

In the developing countries [17] the prevalence rate of hypertension varied from 15 to 33 percent [18]. Only a few studies have so far reported prevalence of hypertension in Bangladesh. A marked increase in prevalence of hypertension (from 11.3% to 17.9%) was observed in Bangladesh from 1999 to 2010 [10, 11].

According to one estimate in Bangladesh, 12 million people suffer from hypertension [12]. Persons with blood pressure above optimal levels, but not clinical hypertension (systolic blood pressure of 120-139 mm Hg or diastolic blood pressure of 80-89 mm Hg), are defined as having "prehypertension" [13]. Persons with pre-hypertension have a greater risk of developing hypertension than do those with lower blood pressure levels [14].

The prevalence of hypertension and pre-hypertension in this study population, were 33.3% and 29.9% respectively without any gender difference.

In most developing countries, hypertension has an increasing trend in both urban and rural areas. The prevalence of hypertension in this study was higher than other studies conducted in the country [10, 11]. However Rahim et al. [19] found similar high prevalence (30.64%) of hypertension in rural people of Joldhaka, Nilphamari of Rangpur division. In one study[19] they have taken only newly diagnosed cases of hypertension and the patients those were currently on hypertensive medication were excluded from the study. Singh et al. [20] examined the prevalence and determinants of hypertension among 6940 subjects aged 25 years and above. They also reported higher prevalence of hypertension in south (W 31.9%; M 35.5%) and west India (W 29.1%; M 35.6%) compared to north, east and central India (20). II Meshram reported similar high prevalence (40%) of hypertension in Kerala state of India

[21]. Recently a meta analysis done in Bangladesh revealed the prevalence of hypertension to be 13.5% (12.7%–14.3%) [22]. Therefore, from different studies, conducted across the country including ours, it is evident that prevalence of hypertension is increasing with time. Causes of increase in prevalence of hyper-tension may be multifactorial, rapid urbanization, ageing of population, sedentary behavior, increased consumption of fatty foods, less intake of vegetables, fruits and more common use of contraceptives, may be the contributory factors (20). Joshi et al. [23] in a study in Mumbai observed that there was an increasing trend of hypertension as age advances and even in post menopausal women, as the prevalence rate rose from 4% among young to 17.2% in 60 years of age group. Similar findings were also contributed by Chockalingam [24] among both Sexes. In our study, we have found similar results (prevalence of hypertension 9.35% in <29 years age group and 53.84% in >70 years age group). Thus, from our study, it is evident that increasing age has significant impact in development of hypertension. However, in females the prevalence increased after 40 years up to 70 years, after that the prevalence decreased; the cause of this variation needs to be examined by further studies. Singh et al. [25] observed that guava fruit intake can decrease blood pressures and blood lipids due to increased intake of potassium.

Tobacco intake may increase sodium absorption among smokers and tobacco chewers as compared to non-tobacco users. In our study, current smokers were lower (23.9%) than other [11] study (26.2%) conducted across the country. This may be due to different steps taken by the Government to reduce smoking; increased awareness regarding harmful effects of smoking etc. Guidelines [26] indicate significant role of smoking in development of hypertension, our study did not find any significant co-relation. In our study prevalence of hypertension was more in non-smoker (32.6% vs 26.95%).

Obesity is one of the important risk factor for hypertension. Friedman et al. [27] carried out a 6-year follow-up study on hypertension and obesity, and found that obesity and weight gain were clear precursors of hypertension. In our study, obesity was found to be more common than observed in another study [11] from Bangladesh (15.8% vs 3.6%). Prevalence of hypertension was significantly higher among obese (38.2%) and overweight (37.7%) population compared with those having normal BMI (30%) (P value 0.000). It seems that limit of higher BMI (>25 Kg/M 2) considered in our study for obese, is greater compared to other studies from South Asia [27, 28].

Populations in South Asia have increased susceptibility to central obesity and metabolic syndrome which are risk factors for hypertension and coronary artery disease and type 2 diabetes [28, 29]. Diet and lifestyle guidelines developed by Indian Consensus indicate that people of South Asian origin develop central obesity as soon as their BMI goes beyond 23Kg/M 2. Therefore, it has been suggested that subjects having BMI >23 to 25 Kg/ M2 should be considered obese and those having BMI 25-30 Kg/M 2 should be considered to have obesity [28, 29]. It is possible, that if the data are analyzed following these criteria, the prevalence of obesity among subjects above BMI 23 Kg/M 2 would be quite high and below this limit would be low. Singh et al. [20] has shown among 6940 subjects, that subjects with central obesity, sedentary lifestyle and family history of hypertension had increased risk of hypertension. Our study result is also consistent with this result.

In our study prevalence of hypertension was more in non-contraceptive user group than those using contraceptive. The impact of contraceptive as a risk factor of hypertension is not significant in our study which may be due to lesser duration of use of contraceptive in our study. Oral contraceptive usage was found to be a significant risk factor in causation of hypertension in other countries [30, 31]. Our study revealed that prevalence rate of hypertension was significantly higher with long term usage of oral contraceptive and the mean systolic and diastolic pressure in groups on oral contraceptive pills were higher by 6.5 mm of Hg and 3.24 mm of Hg respectively as compared to blood pressure readings in control group. The Five City Study in India among 6,940 subjects, aged 25 years and above, also found that oral contraceptive intake was a significant risk factor of hypertension [20].

Prevalence of diabetes mellitus in our study was higher than previous study [11] from our country. This may be due to the differences in criteria for labeling diabetes. The earlier study [11] taken only the self reported cases, but in our study we have taken the self reported cases (those who were on treatment of diabetes mellitus) as well as whose blood glucose was ≥11.1 mmol/L. Singh [31] et al. carried out a study among ethnic groups of rural and urban population of North India and showed that diabetes mellitus was a risk factor of hypertension. Our study results are also consistent with this study. Among the diabetes 52.17% were hypertensive in our study.

Studies from Southeast Asian and African Regions have shown that DM and HTN coexist in type 2 DM ranging from 20.6% in India to 78.4% in Thailand in the SEA region and ranging from 9.7% in Nigeria to 70.4% in Morocco in the AFR region [32]. Though in our study diabetes and hypertension co-exists only in 2.49% of the study population, study with large sample size will be needed to get a true picture of the twin diseases. The awareness (knowledge about hypertension), regular drug intake and control of hypertension (among treated hypertensives) in this study were low (25.2%, 46.8%, and 36.06%, respectively). In a study in Kerala, it was 39%, 29%, and 10%, respectively, in the urban elderly population [33]. In a study in US, control was only in 24% of all hypertensives and 45% of the treated hypertensives [34].

In Assam about 4% of all hypertensives and 18.1% of treated hypertensives had controlled blood pressure [35]. The low awareness and treatment of hypertension in our study might be due to competing health priorities with many communicable diseases and other important services. There is no free or subsidized treatment for non-communicable diseases (NCDs) through the public health system in Bangladesh but the incentivized programs exist for some communicable diseases and maternal and child health. Besides availability, accessibility, and lack of private services in rural areas and the silent nature of the problem plays an important role. Other factors like lower socioeconomic status, health perception and low level of education can also be cited. The low rates of awareness, treatment and control of hypertension are a matter of concern for the already outstretched public health facilities. Despite ignorance about awareness and treatment of hypertension and poverty, population may be educated about diet and lifestyle changes for prevention of NCDs via education, agriculture, food and health ministry of the governments and voluntary organizations [28, 29].

A randomized, controlled intervention trial, among 120 subjects with hypertension, on effects of guava intake on serum total and high density lipoprotein cholesterol levels and on systemic blood pressure in India reported significant benefits [25]. Fruits, vegetables, legumes and fish rich in potassium, magnesium, flavanols and w-3 fatty acids may advised in general along with increased physical activity and weight loss for prevention of hypertension in any country of the world [20, 25, 26, 28].

In brief, hypertension and prehypertension have become public health problems in Bangladesh. Obesity and central obesity, diabetes mellitus, added salt, physical inactivity, prolonged use of oral contraceptive may be the risk factors of hypertension, although multivariate regression analysis should have been done to demonstrate such association. Tobacco intake was not associated with hypertension. Most patients with a diagnosis of hypertension have poor adherence to treatment and follow up. Ignorance and poor motivation was the main cause of irregular follow up.

Acknowledgments

The authors of this study are grateful to Professor M. A. Jalil Chowdhury, Professor Dr. Md. Noor Islam, stuff of Hypertension and Research Centre, Rangpur; Department of Medicine, RpMCH, Department of Medicine of Rangpur Community Medical College particularly Dr. Md. Sayeed-uz-zaman for kind co-operation and Funding From Rangpur Hypertension Research Center.

Limitation. Failed to determine whether the high blood pressure was persistent or not. Multivariate analysis could not be conducted.

Future direction. Causes of increase prevalence of hypertension need to be studied.

There was no conflict of interest.

Contributions by the authors: Concept design by all authors except RBS who helped in the analysis of data and writing.

Dr. Ratindra Nath Mondal planned and supervised the study. Dr. Md. Ashraful Haque, Dr. Abul Kalam Azad analysed the data. Dr. Shah Md. Sarwer Jahan, Dr. Md. Mahfuzer Rahman, Dr. Md. Kumruzzaman Sarker, Dr. AKM Shaheduzzaman, Dr. Moni Rani helped to collect the data. Dr. B. D. Bidhu and Professor Dr. Md. Zakir Hossain, Professor Dr. Amaresh Chandra Shaha, Professor Ram B. Singh helped in writing of the manuscript and presentation of the data.

References

[1] World Health Organization. *WHO NCD Surveillance Strategy*. Available at <http://www.who.int/ncd_surveillance/strategy/en/print.html>. accessed on 22 Oct. 2010.

[2] World Health Organization. *Global status report on noncommunicable diseases 2010.* WHO Geneva 2011.

[3] World Health Organization. *Health profile of Bangladesh.* WHO Bangladesh 2003. Available at http://www.whoban.org/country_ health_ profile.html accessed on 22 Oct. 2010.

[4] World Health Organization. *World Health Report 2002: Reducing risks, promoting healthy life.* WHO Geneva 2003.

[5] Bangladesh Bureau of Statistics (2007) *Statistical Pocketbook of Bangladesh 2007.* Dhaka: Bangladesh Bureau of Statistics. Retrieved 13 May 2009 from http://www.bbs.gov.bd/dataindex/pb_wb_page.pdf.

[6] Rodgers, A., Lawes, C., MacMahon, S. Reducing the global burden of blood pressure related cardiovascular disease. *J. Hypertens.* 2000; 18 (Suppl. 1):S3-6.

[7] World Health Organization-Primary prevention of essential hypertension: *WHO tech. rep. ser. no:* 688, 1983.

[8] Kearney, P. M., Whelton, M., Reynolds, K., Muntner, P., Whelton, P. K., He, J. Global burden of hypertension: analysis of worldwide data. *Lancet* 2005; 365: 217-23.

[9] *The World Health Report 1999: The double burden: emerging epidemics and persistent problems.* Geneva: WHO; 1999. Available from: http://www.who.org/, accessed on June 2, 2007.

[10] M. M. Zaman and M. A. Rouf. *Journal of Human Hypertension* 1999; 13: 547-549.

[11] *NCD risk factor survey 2010 report published and disseminated published on 7th August* 2011. http://www.ban.searo.who.int/EN/Section31_295.htm.

[12] Sultana, M. H., *BMRC Bull.;* 2009; 35:76-78.

[13] Chobanian, A. V., Bakris, G. L., Black, H. R., Cushman, W. C., Green, L. A., Izzo, J. L. Jr, et al. The Seventh Report of the Joint National Committee on Prevention, Detection, Evaluation, and Treatment of High Blood Pressure: the JNC 7 Report. *JAMA* 2003; 289: 2560-72.

[14] Greenlund, K. J., Croft, J. B., Mensah, G. A. Prevalence of heart disease and stroke risk factors in persons with prehypertension in the United States, 1999-2000. *Arch. Intern. Med.* 2004; 164: 2113-8.

[15] Liszka, H. A., Mainous, A. G. 3rd, King, D. E., Everett, C. J., Egan, B. M. Pre-hypertension and cardiovascular morbidity. *Ann. Fam. Med.* 2005; 3: 294-9.

[16] JNC VI *Arch. Intern. Med.* 1997; 157: 2413-2418.

[17] Park, K. -*Text book of Preventive and Social medicine* 17th edition. P279-232.

[18] Burch, G. E. and De Pasquale, N. P.- *Primer of clinical measurement of blood pressure*; CV mc Soy company st lours(ed.) 1962, p. 15-35.

[19] Rahim, M. A., Rahman, M. M., Rahman, M., Ahmed, F., Chowdhury, J., Islam, F., *The Prevalence rate of Hypertension in Rural Population of Bangladesh*, available from: http://www.dnmc.edu.bd/journal/march/March-06.pdf.

[20] Singh, R. B., Fedacko, J., Pella, D., Macejova, Z., Ghosh, S., de Amit, K., Begom, R., et al., Prevalence and risk factors for prehypertension and hypertension in five Indian cities; *Acta Cardiol.* 2011; 66: 29-37.

[21] I. I. Meshram, N. Arlappa, N. Balkrishna, K. M. Rao, A. Laxmaiah, G. N. V. Brahmam; Prevalence of hypertension, its correlates and awareness among adult tribal population of Kerala state, India; *J. Postgraduate Med.* 2012; 58: 255-261.

[22] Moniruzzaman, Abu Taleb, Shahadur Rahman, Amitava Acharyya, Ferdous Ara Islam, MSA Mansur Ahmed and M Mostafa Zaman; Prevalence of hypertension among the Bangladeshi adult population: a meta - analysis; *J. Epidemiol. Community Health* 2011; 65:A405 doi:10.1136/jech.2011.142976n.89.

[23] Joshi, S. V., Patel, J. C., Dhar, H. C. Prevalence of Hypertension in Mumbai-Indian *Journal of Medical Sciences* 2000; 54: 380-83.

[24] Shakuntala chockalingam, Jayalakshmi, R., Arun Chockalingam, W. H. L. *News Letter* -69/2000-Scientific News-Prevalence of hypertension in a rural town in Tamilnadu. February 2000; 69.

[25] Singh, R. B., Rastogi, S. S., Ghosh, S., Singh, R., Niaz, M. A. Effects of guava intake on serum total and high density lipoprotein cholesterol levels and on systemic blood pressure. *Am. J. Cardiol.,* 1992, 70:1287-91.

[26] Mancia, G., Fagard, R., Narkiewicz, K., et al. 2013 ESH/ESC guidelines for the management of arterial hypertension. *Eur. Heart J.* 2013; DOI: 10.1093/eurheartj/eht. 151. Available here.

[27] Friedman, Gary D., Ghannem, H. Precursor of Essential Hypertension, body weight, alcohol and salt use and parental history. *Preventive Medicine* 1988; 17: 387-400.

[28] Singh, R. B., Rastogi, S. S., Rao, P. V., et al. Diet and lifestyle guidelines and desirable levels of risk factors for the prevention of diabetes and its vascular complications in Indians: a scientific statement of the International College of Nutrition. *J. Cardiovasc. Risk* 1997; 4: 201-8.

[29] Singh, R. B. and Indian Consensus Group. Indian consensus for prevention of hypertension and coronary artery disease. A scientific statement of the Indian Society of Hypertension and International College of Nutrition. *J. Nutr. Environ. Med.* 1996; 6: 309-18.

[30] Swedish International Development Cooperation Agency [SIDA]. *Reality Check Bangladesh 2008: Listening to the poor people's realities about primary healthcare and primary education.* Dhaka: Edita Communications. 2009.

[31] Singh, R. B., Bajaj, S., Niaz, M. A., Rastogi, S. S., Moshiri, M. - Prevalence of type II diabetes mellitus and risk of hypertension and coronary heart disease in rural and urban population with low rates of obesity. *Int. Jour. Cardiol.* 1998 Sep. 66; 65-72.

[32] Viswanathan Mohan,Yackoob K. Seedat, and Rajendra Pradeepa; The Rising Burden of Diabetes and Hypertension in Southeast Asian and African Regions: Need for Effective Strategies for Prevention and Control in Primary Health Care Settings; *International Journal of Hypertension* 2013 (2013), Article ID 409083, 14 pages, http://dx.doi.org/10.1155/2013/409083.

[33] Zachariah, M. G., Thankappan, K. R., Alex, S. C., Sarma, P. S., Vasan, R. S. Prevalence, correlates, awareness, treatment, and control of hypertension in a middle-aged urban population in Kerala. *Indian Heart J.* 2003; 55: 245-51.

[34] Hypertension Study Group. Prevalence, awareness, treatment and control of hypertension among the elderly in Bangladesh and India: A multicentre study. *Bull. World Health Organ.* 2001; 79:490-500.

[35] Hazarika, N. C., Narain, K., Biswas, D., Kalita, H. C., Mahanta, J. Hypertension in the native rural population of Assam. *Natl. Med. J. India* 2004; 17: 6: 300-4.

In: Circadian Cardiology with Focus on Both Prevention ... ISBN: 978-1-63463-956-9
Editors: K. Hristova, A. Shehab, G. Cornelissen et al.

Chapter 15

Hypertension in the Elderly

N. S. Neki [*1]***, Miki Tokunaga***[2]***,***
Toru Takahashi [3]***, Jan Fedacko***[4]
and Elena Gerasimova[5]

[1]Department of Medicine Government Medical College, Amritser, India
[2]Department of Food and Nutrition, Fukuoka Women's University, Japan
[3]Graduate School of Human Environment Science,
Fukuoka Women's University, Japan
[4]PJ Safaric University, Kosice, Slovakia
[5]State Medical Academy of NN Burdenko in Voronezh, Russia

Abstract

Background. Hypertension is a major problem in India and globally. Its prevalence is increasing amongst urban and rural population particularly among elderly subjects.

Methods. Internet search and discussion with friends.

Results. The prevalence of hypertension rises with age and the complications secondary to hypertension also accumu-late with time.

Amongst the various causes of hypertension in elderly population, isolated systolic hypertension (ISH) is the commonest cause which is associated with increased risk of outcomes as stroke, congestive heart failure and coronary artery disease etc. Also, there are other age related changes in cardiovascular system. A thorough clinical and investigative work is necessary in elderly. Besides pharmacological therapy, life style modification should be an integral part of the treatment regime.

Conclusions. Hypertension is a major cause of morbidity and mortality among elderly population. Treatment goals in the elderly are the same as for general population; however strict control is of great importance.

Keywords: Hypertension; elderly; isolated systolic hypertension

[*] Correspondence: Dr. N.S. NEKI, Professor of Medicine, Govt. Medical College and Guru Nanak Dev Hospital, Amritsar, India, 143001.Email: drneki123@gmail.com.

Introduction

Hypertension has become a major health problem in the world [1-4]. Reliable statistics related to the prevalence, incidence and mortality of hypertension and stroke, particularly among elderly subjects are not available from Asia. The mortality rate for stroke has been on the decline since the mid 1960s in the developed countries of Asia, such as Australia, New Zealand, and Japan, with some improvement in Singapore, Taiwan and Hong Kong, some areas of China and Malaysia about 15 years later. In India, China, Philippines, Thailand, Sri Lanka, Iran, Pakistan, Nepal, there has been a rapid increase in stroke mortality and prevalence of hypertension [5, 6]. Approximately one 4th of the population above 25 years of age appears to be hypertensive [1-4]. In elderly population, half of the population may have hypertension [2-5]. Hypertension is an important risk factor of stroke, heart attack, heart failure and kidney failure and is a major cause of mortality in the United States [2, 3]. Blood pressure trends, plasma insulin and risk factors in rural and urban elderly populations of north India have demonstrated that metabolic syndrome is a major risk factor of hypertension [7].

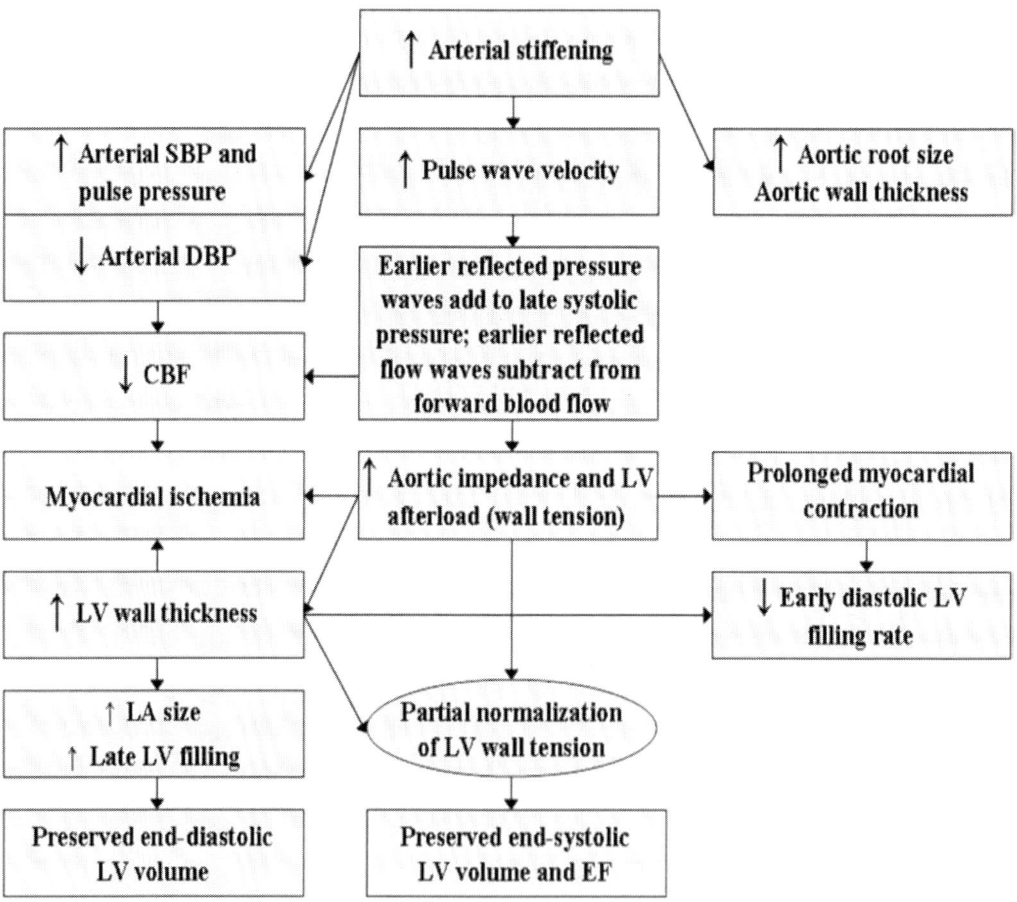

Figure 1. Conceptual Framework of Arterial Stiffening Occurring with Age.

In the developing countries, as the number of elderly are increasing due to improved health services and other modalities, so is the number of hypertensives along with other age related diseases [2-5]. According to the Framingham Heart Study, by the age 60 approximately 60% of the population develops hypertension, and by 70 years about 65% of men and about 75% of women have the disease. In the same study, 90% of those who were normotensive at age 55 were found to develop hypertension. Indian elderly were nearly 76.6 million in 2001 and it is estimated that their number will go up to 324 million by middle of this century [8]. According to JNC 7 classification, hypertension is defined as systolic BP of more than or equal to 140 mm of Hg and diastolic BP of more than or equal to 90 mm of Hg. Isolated systolic hypertension(ISH) is defined as systolic BP greater than or equal to 140 mm of Hg and diastolic BP in the normal range of less than 90 mm of Hg [9]. Grade 1 ISH is systolic BP 140 – 159 mm of Hg and grade 2 ISH is systolic BP >160 mm of Hg [9].

Pathophysiology of Hypertension in Elderly

Compared with younger patients with similar blood pressure, elderly hypertensive patients have lower cardiac output, higher peripheral resistance, wider pulse pressure, lower intravascular volume, and lower renal blood flow [10]. Age-associated increases in hypertension prevalence derive from changes in arterial structure and function accompanying aging. Large vessels become less distensible, which increases pulse wave velocity, causing late systolic blood pressure (SBP) augmentation and increasing myocardial oxygen demand. Reduction of forward flow also occurs, limiting organ perfusion. Autonomic dysfunction also occurs with age leading to orthostatic hypotension and hypertension leading to increased risk of cardiovascular and cerebrovascular diseases. Between 30 and 85 years of age, renal mass, particularly the cortex, declines by 20% to 25% [11]. Due to progressive development of glomerulo-sclerosis and interstitial fibrosis, there occurs a decline in GFR and reduction of other renal homeostatic mechanisms [11, 12]. Age-associated declines in membrane sodium/potassium–adenosine triphosphatase may also contribute to geriatric hypertension because this results in increased intracellular sodium that may reduce sodium– calcium exchange and thereby increase intracellular calcium and vascular resistance. After age 70 years, diastolic hypertension accounts for 10% of all patients with hypertension [13]. In addition, the relationship between DBP and cardiovascular risk is bimodal in older individuals, with DBP of 90 mm Hg associated with similar increased risk as that associated with DBP lower than about 70 mm Hg [14,15]. As a result, at any given level of SBP, CAD risk increases as DBP decreases [15, 16]. Hypertension prevalence is less in women than in men until 45 years of age, is similar in both sexes from 45 to 64 years of age, and is much higher in women than men 65 years of age [17].

Secondary causes of hypertension in elderly include renal artery stenosis, obstructive sleep apnea, primary aldosteronism, hypo/ hyperthyroidism, consumption of tobacco, alcohol, caffeine, NSAIDs, calcium, vitamin C & D, sex hormone & gluco-corticoids etc.

Clinical Assessment and End Organ Damage

Diagnosis of hypertension should be based on at least 3 different BP measurements, taken on 2 separate office visits. At least 2 measurements should be obtained once the patient is seated comfortably for at least 5 minutes with the back supported, feet on the floor, arm supported in the horizontal position, and the BP cuff at heart level. Hypertension in the elderly is a risk factor for both ischemic stroke and cerebral hemorrhage. ISH is as an important component of BP-related stroke risk [18]. The strength of the association between BP level and stroke decreases with increasing age [19]. CAD is highly prevalent among the elderly. Elderly patients with hypertension have a higher prevalence of MI than elderly patients without hypertension. Among autopsies in persons with average age 80 years, the age-related increase in atherosclerosis was evident even after age 80(20). Atherosclerosis was more severe in men than in women 60 to 70 years of age, but this gender difference diminished with increasing age and disappeared in the nineties. Aging and hypertension are both strongly associated with development of heart failure (HF) [21].

Abdominal aortic aneurysms, defined as dilation of the aorta with a minimum anteroposterior diameter of 30 mm, occur with increasing frequency with increasing age. The prevalence of abdominal aortic aneurysms is 12.5% among men 75 to 84 years of age [22]. It is estimated that 10 million persons have Peripheral arterial disease (PAD) in the United States, the prevalence among persons 75 years of age is approximately 20% and is 50% in persons 85 years of age [23, 24]. PAD is associated with a 4-fold increase in MI risk and a 3-fold increase in stroke or transient ischemic attack risk [25,26]. Hypertension and aging both impact renal function. Elderly patients are more likely to have chronic kidney disease (CKD), usually defined by a measured eGFR 60 mL/min/1.73 m^2. Prevalence of CKD ranges from 11% to 14% and 75% of the CKD population is 65 years of age [27].

The major cause of vision limitation in patients with hypertension of all ages is retinopathy, defined as arteriolar narrowing (generalized and focal), arteriovenous nicking, flame and blot hemorrhages, cotton-wool spots, and optic disk edema [28, 29].

Treatment

Drug treatment trials indicate that there is benefit in treating elderly and very elderly patients with hypertension to an SBP target lower than 160 mm Hg [30,31]. 2 treat-to-target trials provide evidence that lowering SBP to less than 140 mm Hg does not increase benefit compared with SBP targets of lower than 150 or 160 mm Hg [32,33].Hypertension in the Very Elderly Trial (HYVET) documented antihypertensive therapy benefits in persons 80 years of age [34]. Initially alpha blocker was preferred agent to treat hypertension in elderly with benign prostatic hypertrophy (BPH) but in ALLHAT study doxazosin is found to be associated with higher incidences of cardiovascular event rate in comparison to thiazides [35].

Principals of Hypertension Treatment in Elderly

In elderly, treatment of hypertension should begin with lifestyle modifications that may reduce the need of antihypertensive medication in this population [5, 36, 37]. The starting dose of medications should be half of that used in younger patients and in elderly without a compelling indication, thiazide/chlorthalidone diuretics is recommended as initial drug of choice. The recommended goal for BP control is < 140/90 mmHg while in those with diabetes and renal disease, the goal is < 130/80 mmHg and in patients with ISH, goal is to decrease systolic BP < 140mmHg but caution should be exercised if DBP falls below 55mHg while treating ISH.

The new ESC guidelines also make a host of lifestyle recommendations for lowering blood pressure [37]. They are recommending salt intake of approximately 5 to 6 g per day, in contrast with a typical intake of 9 to 12 g per day.

A reduction to 5 g per day can decrease systolic blood pressure about 1 to 2 mm Hg in normotensive individuals and 4 to 5 mm Hg in hypertensive patients, he said. The optimal body-mass index (BMI) is not known, the guidelines recommend getting BMIs down to 25 kg/m^2 and reducing waist circumferences to <102 cm in men and <88 cm in women. Losing about 5 kg can reduce systolic blood pressure by as much as 4 mm Hg, while aerobic endurance training in hypertensive patients can reduce systolic blood pressure 7 mm Hg [37]. Since fruit and vegetable intake were inversely associated with hypertension in India, eating 400g/day of these foods may be protective [1, 5]. Singh et al. [38] reported beneficial effects among 61 group A and 59 group B patients with essential hypertension who were administered guava fruit preferably before meals in a foods-to-eat approach rather than foods-to-restrict, in a randomized and single-blind fashion for 12 weeks. At entry into the study, mean age, male sex, mean body mass index, percentages of risk factors and mean levels of blood lipids were comparable between groups A and B. Adherence to guava consumption was assessed by questionnaires and weighing of guava intake by 24-hour recall after 12 weeks of follow-up. Nutrient intakes including saturated and total fat were significantly decreased; carbohydrates, total and soluble fiber and vitamins and mineral intakes were significantly higher in group A than in group B at 12 weeks. There was a significant net decrease in serum total cholesterol (9.9%), triglycerides (7.7%) and blood pressures (9.0/8.0 mm Hg) with a significant net increase in high-density lipoprotein cholesterol (8.0%) after 12 weeks of guava fruit substitution in group A than in group B. By adding moderate amounts of guava fruit in the usual diet, changes in dietary fatty acids and carbohydrates may occur, providing significant amounts of soluble dietary fiber and antioxidant vitamins and minerals without any adverse effects. There is a greater decrease in lipoprotein metabolism and blood pressures.

The Indian paradox in association of central obesity and insulin resistance with high prevalence of diabetes and cardiovascular disease in an elderly population with low fat intake and lower than normal prevalence of obesity has been observed [39].Therefore, it seems that guava intake in elderly subjects with hypertension can prevent cardiovascular diseases which is the major cause of morbidity and mortality in elderly [2,3,5,37]. Fruits, vegetables, legumes and fish rich in potassium, magnesium, flavanols and w-3 fatty acids may advised in general along with increased physical activity and weight loss for prevention of hypertension in any country of the world [5, 37, 38].

Lifestyle Modifications to Manage Hypertension

Modification	Recommendation	Approximate Systolic BP Reduction, Range
Weight reduction	Maintain normal body weight (BMI, 18.5–24.9 kg/m2)	5–20 mm Hg/10-kg weight loss
Adopt DASH eating plan As originally proposed in 1992 (5,38)	Consume a diet rich in fruits, vegetables, and low-fat dairy products with a reduced content of saturated and total fat	8–14 mm Hg
Dietary sodium reduction	Reduce dietary sodium intake to no more than 100 mEq/L (2.4 g sodium or 6 g sodium chloride)	2–8 mm Hg
Physical activity	Engage in regular aerobic physical activity such as brisk walking (at least 30 min/d, most days of the week)	4–9 mm Hg
Moderation of alcohol	consumption Limit consumption to no more than 2 drinks/d (1 oz or 30 mL ethanol [e.g., 24-oz beer, 10-oz wine, or 3-oz 80-proof whiskey]) in most men and no more than 1 drink/d in women and lighter-weight persons	2–4 mm Hg
Meditation, breathing exercises and yogasan	30-60 min daily	2-3 mm Hg

Pharmacological Treatment

Factors contributing to increased risk of pharmacological treatment in elderly include:

1. Diminished baroreceptor activity – orthostatic hypotension
2. Impaired cerebral autoregulation – cerebral ischaemia
3. Sensitivity to hypokalemia – arrhythmias and muscle weakness
4. Decreased renal and hepatic function – drug accumulation
5. Polypharmacy – drug interaction
6. CNS changes – depression and confusion

Major classes of drugs used for treatment of hypertension in elderly include Diuretics, ACE inhibitors and ARBs, Beta blockers, Calcium channel blockers, Aldosterone receptor antagonists etc.

Compelling Indications for Specific Classes of Antihypertensive Drugs As Initial Drug Therapy [36]

Heart Failure

Chlorthalidone, indapamide, beta-blockers, angiotensin- converting enzyme (ACE) inhibitors, angiotensin receptor blockers (ARBs), calcium channel blockers, aldosterone receptor antagonists.

After Myocardial Infarction

Beta-blockers, ACE inhibitors, ARBs, aldosterone receptor antagonists, nondihydropyridine calcium channel blockers

Coronary Artery Disease or High Cardiovascular Risk

Thiazides, ACE inhibitors, calcium channel blockers, beta-blockers

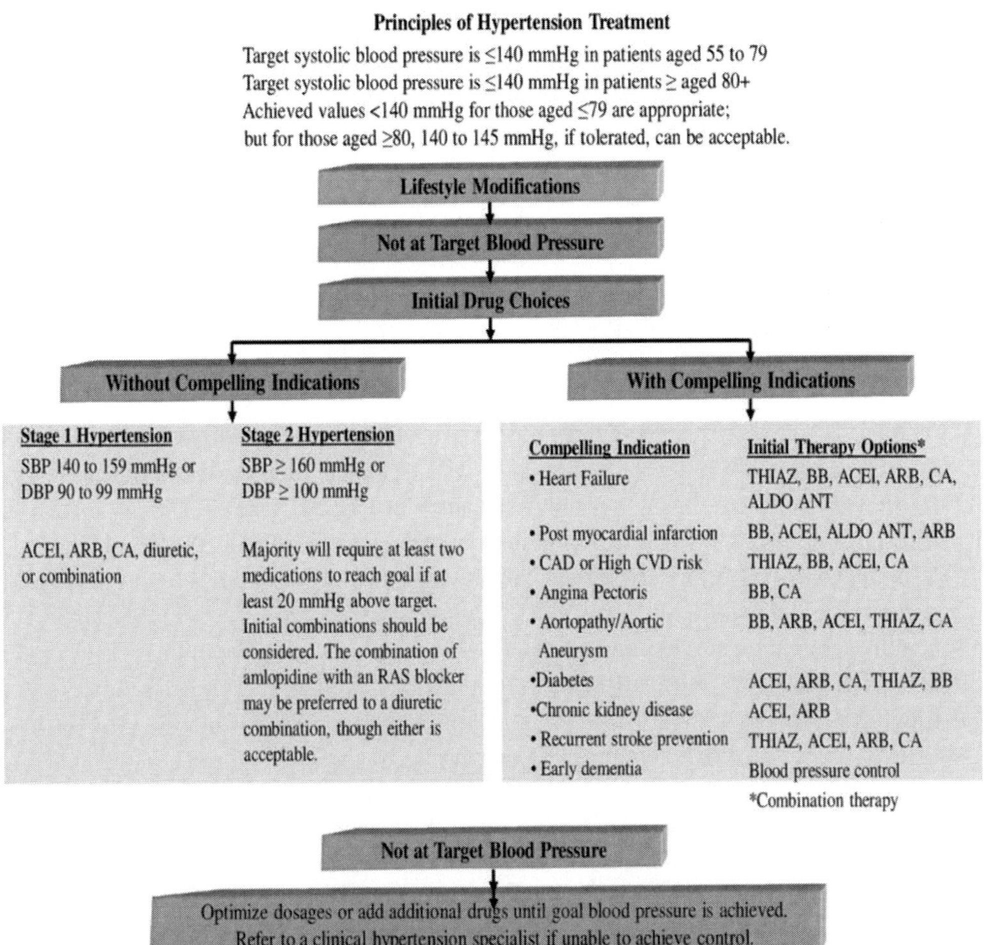

Angina Pectoris

Beta-blockers, calcium channel blockers

Aortopathy, Aortic Aneurysm

Beta-blockers, ARBs, ACE inhibitors, calcium channel blockers

Diabetes Mellitus

ACE inhibitors, ARBs, calcium channel blockers, nebivolol, carvedilol

Chronic Kidney Disease

ACE inhibitors, ARBs

Recurrent Stroke Prevention

Thiazides, ACE inhibitors, ARBs, calcium channel blockers

In conclusion, hypertension is a major problem in elderly population of India and isolated systolic BP is the commonest cause of raised BP in elderly. Increase in age is associated with higher incidence of hypertension and hypertension related complications in elderly. Initial treatment should be in the form of lifestyle modification and thiazide diuretics is usually the first drug to be used in elderly hypertensive with a goal of <140/90mmHg.

References

[1] Singh RB, Fedacko J, Pella D, Macejova Z, Ghosh S, Beegom R, Tumbi ZA, Haque M, Vajpeyee SK, De Meester F, Sergey C, Agarval R, Muthusamy VV, Gupta AA, and the Five City Study Group. Prevalence and risk factors of pre-hypertension and hypertension in five Indian cities. *Acta Cardiol* 2011; 66: 29-37.

[2] Heart Disease and Stroke Statistics 2013 Update. A report from the American Heart Association. *Circulation* 2013;127:e6-e245.

[3] The "Heart Disease and Stroke Statistics--2013 Update" and the Need for a National Cardiovascular Surveillance System Circulation. 2013;127:21-23,

[4] Danael G, Finucane MM, Lin JK, et al., on behalf of the Global Burden of Metabolic Risk Factors of Chronic Diseases Collaborating Group (Blood Pressure). National, regional, and global trends in systolic blood pressure since 1980: systematic analysis of health examination surveys and epidemiological studies with 786 country-years and 5·4 million participants. *Lancet* 2011;377: 568–77.

[5] Singh RB, Kumar A, Neki NS, Pella D, Rastogi SS,Basu TK, Otsuka K, De Meester F, Wilson DW. Acharya SN, Juneja L, Takahashi T, Diet and Lifestyle Guidelines and Desirable Levels of Risk Factors for Prevention of Cardiovascular Disease and Diabetes among Elderly Subjects. *A Revised Scientific Statement of the International College of Cardiology and International College of Nutrition-WHJ* 2011, 2011;3:305-320.

[6] Singh RB, Suh IL, Singh VP, Chaithiraphan S, Laothavorn P, Sy RG, Babilonia NA, Rahman ARA, Sheikh S, Tomlinson B. Hypertension and stroke in Asia: prevalence, control and strategies in developing countries for prevention. *Journal of Human Hypertension* 2000; 14: 749–763.

[7] Singh RB, Rastogi SS, Niaz MA, Ghosh S *et al.* Blood pressure trends, plasma insulin and risk factors in rural and urban elderly populations of north India. *Coro Art Dis* 1997; 8: 463–468.

[8] Puspanjali S, TP Sherin Raj. Demography of aging in India. *Indian Journal of Gerontology* 2005;19:327-42.

[9] Chobanian AV, Barkis GL, Black HR, et al. Seventh report of the Joint national committee on Prevention, Detection, Evaluation and Treatment of High Blood Pressure. *Hypertension.*2003;42:1206-52.

[10] Messerli FH, Sundgaard-Riise K, Ventura HO et al. Essential hypertension in the elderly: haemodynamics, intravascular volume, plasma renin activity, and circulating catecholamine levels. *Lancet* 1983; 2:983–986.

[11] Beck LH. The aging kidney: Defending a delicate balance of fluid and electrolytes. *Geriatrics.* 2000;55:26 –2.

[12] Fliser D, Ritz E. Relationship between hypertension and renal function and its therapeutic implications in the elderly. *Gerontology.* 1998;44:123–31.

[13] Franklin SS, Gustin W, Wong ND, et al. Hemodynamic patterns of age-related changes in blood pressure: the Framingham Heart Study. *Circulation.* 1997;96:308 –15.

[14] Staessen JA, Gasowski J, Wang JG, et al. Risks of untreated and treated isolated systolic hypertension in the elderly: Meta-analysis of outcome trials. *Lancet.* 2000;355:865–72.

[15] Franklin SS, Khan SA, Wong ND, et al. Is pulse pressure useful in predicting risk for coronary heart disease? The Framingham Heart Study. *Circulation.* 1999;100:354–60.

[16] Franklin SS. Hypertension in older people: part 1. *J Clin Hypertens* (Greenwich). 2006;8:444 –9.

[17] National Center for Health Statistics. Health, United States, 2007 With Chartbook on Trends in the Health of Americans. Hyattsville, Md: National Center for Health Statistics; 2007.

[18] Kannel WB, Dawber TR, Sorlie P, et al. Components of blood pressure and risk of atherothrombotic brain infarction: the Framingham study. *Stroke.* 1976;7:327–31.

[19] Cholesterol, diastolic blood pressure, and stroke: 13,000 strokes in 450,000 people in 45 prospective cohorts—prospective studies collaboration. *Lancet.* 1995; 346:1647–53.

[20] Sawabe M, Arai T, Kasahara I, et al. Sustained progression and loss of the gender-related difference in atherosclerosis in the very old: a pathological study of 1074 consecutive autopsy cases. *Atherosclerosis.* 2006;186:374 –9.

[21] Lakatta EG. Age-associated cardiovascular changes in health: impact on cardiovascular disease in older persons. *Heart Fail Rev.* 2002;7: 29–49.

[22] Hirsch AT, Haskal ZJ, Hertzer NR, et al. ACC/AHA 2005 practice guidelines for the management of patients with peripheral arterial disease (lower extremity, renal, mesenteric, and abdominal aortic): executive summary: a collaborative report from the American Association for Vascular Surgery/Society for Vascular Surgery, Society for Cardiovascular Angiography and Interventions, Society for Vascular Medicine and Biology, Society of Interventional Radiology, and the ACC/AHA Task Force on Practice Guidelines (Writing Committee to Develop Guidelines for the Management of Patients With Peripheral Arterial Disease). *J Am Coll Cardiol.* 2006;47:1239 –312.

[23] Meijer WT, Hoes AW, Rutgers D, et al. Peripheral arterial disease in the elderly: the Rotterdam Study. *Arterioscler Thromb Vasc Biol.* 1998;18:185–92.

[24] Criqui MH, Fronek A, Barrett-Connor E, et al. The prevalence of peripheral arterial disease in a defined population. *Circulation.* 1985;71:510 –5.

[25] Wilterdink JL, Easton JD. Vascular event rates in patients with atherosclerotic cerebrovascular disease. *Arch Neurol.* 1992;49:857–63.

[26] Criqui MH, Langer RD, Fronek A, et al. Mortality over a period of 10 years in patients with peripheral arterial disease. *N Engl J Med.* 1992;326:381– 6.

[27] Hallan SI, Coresh J, Astor BC, et al. International comparison of the relationship of chronic kidney disease prevalence and ESRD risk. *J Am Soc Nephrol.* 2006;17: 2275– 84.

[28] Gunn R. Ophthalmoscopic evidence of (1) arterial changes associated with chronic renal diseases and (2) of increased arterial tension. *Trans Ophthalmol Soc UK.* 1892;12:124 –5.

[29] Wong TY, Mitchell P. Hypertensive retinopathy. *N Engl J Med.* 2004;351:2310 –17.

[30] Musini VM, Tejani AM, Bassett K et al. Pharmacotherapy for hypertension in the elderly. *Cochrane Database Syst Rev* 2009;(4):CD000028.

[31] Mancia G, Laurent S, Agabiti-Rosei E et al. Reappraisal of European guidelines on hypertension management: a European Society of Hypertension Task Force document. *J Hypertens* 2009;27(11):2121-58.

[32] Ogihara T, Saruta T, Rakugi H et al. Target blood pressure for treatment of isolated systolic hypertension in the elderly: Valsartan in Elderly Isolated Systolic Hypertension Study. *Hypertension* 2010;56(2):196-202.

[33] JATOS Study Group. Principal results of the Japanese trial to assess optimal systolic blood pressure in elderly hypertensive patients. *Hypertens Res* 2008;31(12):2115-27.

[34] Beckett NS, Peters R, Fletcher AE, et al.; HYVET Study Group. Treatment of hypertension in patients 80 years of age or older. *N Engl J Med* 2008; 358:1887–1898.

[35] The ALLAHAT officers and coordinators for the ALLAHAT collaborative research group. Major cardiovascular events in hypertensive patients randomized to doxazocin vs chlorthalidone. The antihypertensive and lipid lowering treatment to prevent heart attack trial (ALLAHAT). *JAMA* 2000: 286; 1967 – 75.

[36] Chaudhry KN, Chavez P, Gasowski J et al. Hypertension in the elderly: some practical considera-tions. *Cleveland Clinic J Med* 2012: 74; 694 – 704.

[37] Mancia G, Fagard R, Narkiewicz K, et al. 2013 ESH/ESC guidelines for the management of arterial hypertension. *Eur Heart J* 2013; DOI: 10.1093/eurheartj /eht.151.

[38] Singh RB, Rastogi SS, Ghosh S, Singh R, Niaz MA. Effects of guava intake on serum total and high density lipoprotein cholesterol levels and on systemic blood pressure. *Am. J. Cardiol,* 1992, 70 :1287-91.

[39] Singh RB, Rastogi SS, Niaz MA, Postiglion A. Association of central obesity and insulin resistance with high prevalence of diabetes and cardiovascular disease in an elderly population with low fat intake and lower than normal prevalence of obesity. The Indian paradox. *Coro Art Dis* 1998; 9: 559-65.

In: Circadian Cardiology with Focus on Both Prevention ... ISBN: 978-1-63463-956-9
Editors: K. Hristova, A. Shehab, G. Cornelissen et al. © 2015 Nova Science Publishers, Inc.

Chapter 16

Effect of Environmental Factors on New Biomarkers in Relation to Cardiovascular Diseases and Type 2 Diabetes

***Miki Tokunaga*[*1], *Toru Takahashi*[2], *R. B. Singh*[3], *Elena Gerasimova*[4], *Lekh Juneja*[5], *Baby Anjum*[6] *and N. S. Verma*[6]**

[1]Department of Food and Nutrition, Fukuoka Women's University, Japan
[2]Graduate School of Human Environment Science, Fukuoka Women's University, Japan;
[3]Halberg Hospital and Research Institute, Moradabad, India
[4]State Medical Academy of NN Burdenko in Voronezh, Russia
[5]Taiyo Kagaku Co. Ltd, Yokkaichi, Japan
[6]Department of Physiology, KG Medical College, Lucknow, India

Introduction

Biological markers play an important role in understanding the relationships between exposure to environmental chemicals, and the development of health or cardiovascular diseases (CVDs) [1-3]. They are key molecular or cellular events that link a specific environmental exposure to a health outcome which may be clinical or biochemical. There is a continuous interaction between biological environment of our body with environmental factors; diet, tobacco, physical activity, alcohol, radiation, pesticides, and geo-magnetic forces [1-3]. Recently WHO and UNO-HLM have emphasized that non-communicable diseases (NCDs); CVDs and other chronic diseases have become a worldwide problem resulting into major causes of morbidity and mortality which may be due to rapid changes in diet and lifestyle factors [4,5]. A biomarker appears to be the earliest stage in the development of a

* Correspondence: Dr Toru Takahashi, Graduate School of Human Environment Science, Fukuoka Women's University, Japan; Email; takahashi@fwu.ac.jp.

disease from physiological limit to pathological and clinical state. Therefore diagnosis of a chronic disease at earliest stage would need to consider abnormalities in the biological state of the biomarker.

There has been a significant progress in identifying and validating new biomarkers that can be used in population-based studies of environmental disease. Blood glucose, angiotensin, insulin, cytokines, free fatty acids are important early markers of a disease which develop in to secondary biomarkers and risk factors; pre-diabetes, hypercholesterolemia, coronary disease and pre-hypertension. These are important indicators of diet, physical inactivity and tobacco consumption to which humans are most commonly exposed [6-10] (Figures 1, 2).

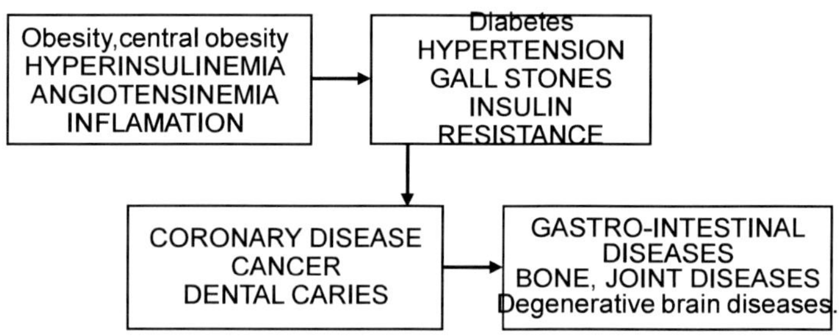

Figure 1. Emergence of chronic diseases due to interactions of gene and environment (Singh et al., 1999. modified from Burkitt and Trowell).

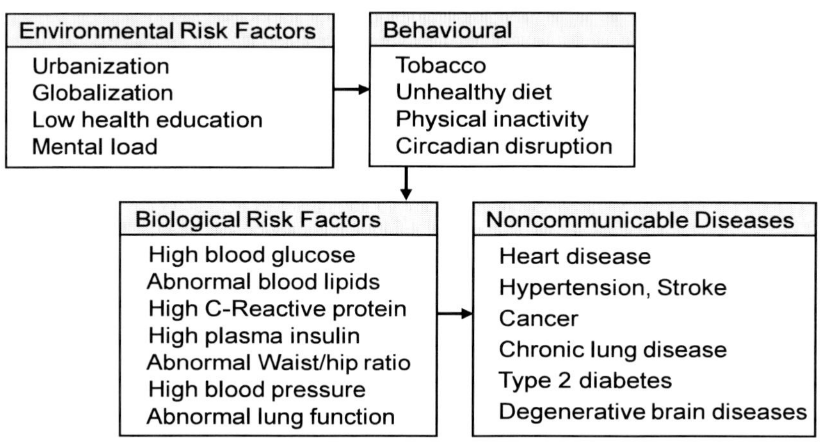

Figure 2. Pathway for development of noncommunicable disease.

Clinical biomarkers and other risk factors and the identification of subgroups that increase the risk for CVDs should be obtained in early stage in the individuals and population for prevention of CVDs.

As part of the new Genes and Environment Initiative, the NIEHS is supporting the development of new technologies that will enable scientists to study how the body responds to environmental agents. Such technologies include sensitive biomarkers that can measure subtle changes in inflammation, oxidative damage, and other pathways that can lead to disease.

There is a controversy regarding the physiological limits of blood glucose and serum cholesterol. In this review, we discuss the role of environmental factors; diet and physical inactivity on common biomarkers and new biomarkers and their limitations as well as prevention [1-3].

Effects of Diet on Biomarkers

Assessment of dietary fat, refined carbohydrates and energy intake are essential to examine the relationships between diet and disease risk but the process of estimating individual intakes of fat quality by dietary assessment is open to bias [6-10]. Tissue and blood fatty acids, are mainly derived from the diet, which have been used as biomarkers of dietary intake for a number of years. Nutritional biomarkers; biochemical, functional, or clinical indices of nutrient intake, status, or functional effects can reveal information about biological or physiological responses to dietary behavior or pathogenic processes, and can be used to monitor responses to therapeutic interventions and to provide information on inter individual differences in response to diet and nutrition [11]. Many nutritional biomarkers are available; yet there has been no formal mechanism to establish consensus regarding the optimal biomarkers for particular nutrients and applications.

Clinical studies in which dietary intake was strictly controlled, confirm that fatty acid biomarkers can complement dietary assessment methodologies and have the potential to be used more quantitatively. Factors affecting adipose tissue and blood lipid composition are discussed, such as the physical properties of triacylglycerol, total dietary fat in-take and endogenous fatty acid synthesis [8]. The relationship between plasma lipoprotein concentrations and total plasma fatty acid composition, and the use of fatty acid ratios as indices of enzyme activity are interesting.

In an earlier study [8], factor analysis was used to define major dietary patterns for a subsample of men (n = 466) from the Health Professionals Follow-up Study by using dietary information collected from food-frequency questionnaires (FFQs) in 1994. Two major dietary patterns were identified which were generally reproducible over time. The first pattern (prudent) was characterized by higher intakes of fruit, vegetables, whole grains, and poultry. The second pattern (Western) was characterized by higher intakes of red meats, high-fat dairy products, and refined grains. Using pattern scores from 1994 and adjusting for potential confounders, there was a significant positive correlation between the Western pattern and insulin, C-peptide, leptin, and homocysteine concentrations, and an inverse correlation with plasma folate concentrations. The prudent pattern was positively correlated with plasma folate and inversely correlated with insulin and homocysteine concentrations. It is clear that dietary patterns are predictors of plasma biomarkers of CVD and obesity risk, suggesting that the effect of overall diet on CVD risk may be mediated through these biomarkers. Apart from these biological biomarkers pro-inflammatory biomarkers; IL-6, TNF-alpha, C-reactive proteins are important biomarkers of CVDs, metabolic syndrome and type 2 diabetes mellitus which are considered pro-inflammatory conditions [9,10]. There is evidence that sedentary behavior can increase pro-inflammatory biomarkers and physical activity can modulate these factors. The ability to develop evidence-based clinical guidance and effective programs and policies to achieve global health promotion and disease prevention goals depends on the availability of valid and reliable data [12]. The role of food and nutrition in achieving those goals, relevant data are developed with the use of biomarkers that reflect nutrient exposure, status, and functional effect on biological functions. There is a need to promote the discovery, development, and use of biomarkers across a range of applications. A process is needed to harmonize the global health community's decision making about what biomarkers are best suited for a given use under specific conditions and settings.

A recent study revealed circulating levels of soluble receptor for advanced glycation end products (sRAGE) which are likely to comprise both a secreted isoform (esRAGE) and wild-type RAGE cleaved from the cell membrane [13]. Both sRAGE and esRAGE have been proposed as biomarkers of cardiovascular disease (CVD), but prospective data are limited. The relationship of sRAGE and esRAGE to incident CAD and stroke in type 2 diabetic patients followed for 3.9 years in a trial of atorvastatin: the Collaborative Atorvastatin Diabetes Study (CARDS). The sRAGE and esRAGE were strongly correlated ($\rho = 0.88$) and were both higher in those with lower BMI ($P < 0.001$), higher adiponectin ($P < 0.001$), lower estimated glomerular filtration rate ($P = 0.009$), and white ethnicity ($P < 0.001$). Both sRAGE and esRAGE were associated with incident CHD events, independently of treatment allocation and the above factors; hazard ratio (HR) = 1.74 (95% CI 1.25–2.41; $P = 0.002$) for a doubling of the sRAGE level; HR = 1.45 (1.11–1.89; $P = 0.006$) for a doubling of the esRAGE level. There was no significant association with stroke; HR for sRAGE = 0.66 (0.38–1.14). Atorvastatin, 10 mg daily, did not alter sRAGE. Higher levels of sRAGE and esRAGE are associated with incident CAD but not stroke in type 2 diabetes mellitus.

In comparison with earlier criteria, troponin testing in the diagnosis of myocardial infarction substantially increases the number of cases diagnosed as myocardial infarction among suspected cases in comparison with previous criteria (14). However, the impact of troponin testing on rates reported in national statistics that use routinely collected hospital morbidity data is uncertain. The authors developed Poisson regression models to estimate the effect of troponin testing on long-term trends in hospital admission rates in Perth, Western Australia, from 1980 to 2004. Troponin tests were used for 10.5% of patients with suspected myocardial infarction in 1996, rising rapidly to more than 90% of patients from 2001 onward. Fitted models that assumed a continuing linear decline estimated that 100% use of troponin testing in cases of suspected myocardial infarction would lead to an apparent increase in hospital admission rates of 42% (95% confidence interval (CI): 28, 56) in men and 21% (95% CI: 4, 41) in women as compared with rates that would be expected if previous linear trends had continued. Smaller effects of 30% (95% CI: 14, 48) in men and −2% (95% CI: −21, 20) in women were found in fitted models that assumed an underlying attenuating trend in the rates. Similarly constructed logistic regression trend models found no significant effect of troponin testing on trends in 28-day case-fatality.

Recently, a total antioxidant content of more than 3100 foods, beverages, spices, herbs and supplements used worldwide have been reported [15]. These foods rich in antioxidants can alter the oxidant biomarkers in our body tissues. In a meta analysis, fruit and vegetable intake were inversely associated with type 2 diabetes indicating that these foods can influence the oxidative damage of the beta cells of pancreas [16]. The relationship between intake of these phytochemicals; flavonols, flavones and isoflavones and various anthropometric and blood chemistry data were analyzed among 115 women, aged 29-78 years, from north Japan, in a cross-sectional study [17]. Each subject completed a 3-d dietary record and received a health check up, including urine and blood sampling for biochemical analysis. Total mean intakes of flavonoids (sum of flavonols and flavones) and isoflavones were 16.7 and 47.2 mg/d, respectively. The major source of flavonoids was onions (45.9%) and that of isoflavones was tofu (37.0%). Total intake of isoflavones exceeded that of other dietary antioxidants, such as flavonoids, carotenoids (3.5 mg/d) and vitamin E (8.2 mg/d), and was approximately one half of the vitamin C intake (109 mg/d). The total intake of flavonoids was inversely correlated with the plasma total cholesterol concentration (TC) (r = -0.236, P: < 0.05) and plasma LDL cholesterol concentration (LDL-C) (r = -0.220, P: < 0.05), after the adjustment for age, body mass index and total energy intake. As a single component, quercetin was inversely correlated with both TC (r = -0.261, P: < 0.01) and LDL-C (r = -0. 263, P: < 0.01). Among Japanese, flavonoid and isoflavone intake is the main component among non-nutrient phytochemicals with antioxidant potential in the diet. These results suggest that a high consumption of both flavonoids and isoflavones by Japanese women may inhibit biomarkers responsible for oxidation which is important in pathogenesis of CVDs and diabetes.

The researchers examined data from 73 308 Seventh-day Adventist men and women over age 25 years, mean age 57 years, 66% women, who were living in the US from 2002 to 2007 [18]. The participants were classed into the following dietary patterns:Vegan: Ate eggs, dairy products, fish, and meat less than once a month (n=5548; 7.6%). Lacto-ovo-vegetarian: ate eggs and dairy products once a month or more; ate fish and meat less often (n=21 177; 28.9%). Pescovegetarian: ate fish once a month or more; ate meat less often (n=7194; 9.8%). Semivegetarians: ate meat once a month or more; ate fish or meat no more than once a week (n=4031; 5.5%). Nonvegetarians: ate fish or meat more than once a week (35 359; 48.2%). Over a follow-up of a mean of 5.79 years, 2570 participants died. Compared with nonvegetarians, the hazard ratio (HR) for all-cause mortality for all vegetarians combined was significant: 0.88 (95% CI 0.80-0.97). The HR for all-cause mortality ranged from 0.81 to 0.92 in the different vegetarian types and was significant only for lacto-ovo-vegetarians and pescovegetarians. Some vegetarian diets were significantly associated with lower mortality from CVD, CAD, renal disease, and endocrine disease (diabetes), but not from cancer. The associations were larger and more likely to be significant in men than in women. In men, following any type of vegetarian diet was associated with a significant lower risk of dying from CVD or ischemic heart disease, but in women, this type of diet was linked with a non-significant lower risk of these outcomes. Men who were vegetarians had a non-significant lower risk dying from stroke, but among women, this type of diet appeared to increase the odds of dying from stroke. Effect of an Indo-Mediterranean diet on biomarkers and progression of coronary disease in high risk patients has been examined in various studies indicating beneficial role [19,20]. Association of inflammation, heavy meals, magnesium, nitrite, and coenzyme Q10 deficiency and circadian rhythms with risk of acute coronary syndromes have also been observed [21,22]. Table 1 and 2 show risk factors and biomarkers commonly observed in various countries.

Apart from diet, sedentary behavior and tobacco can also alter physiological limits of biomarkers [23-29]. Sedentary behavior is a risk factor for obesity, hypertension, type 2 diabetes and coronary artery disease (CAD). The biomarkers could be dyslipidemia, hypercholesterolemia, hyperglycemia and increase in C-reactive proteins as well as IL-6 and tumor necrosis factor-alpha. Tobacco consumption can cause a rise in proinflammatory biomarkers, oxidative stress and total cholesterol and a decrease in HDL cholesterol.

Table 1. Prevalence % of coronary artery disease and type 2 diabetes and biomarkers in south Asians and British

	North Indians		South Indians Urban	Sri Lanka Urban	UK South Asians	Native British
	Rural	Urban				
Coronary Artery Disease	3.0	9.0	13.9	11.5	15.0	6.0
Smoking (%) (Men)	30 2	5.6	44.6	57.4	22	30
Diabetes (%)	2.3	6.0	12.0	5.8	19.0	5.0
Hypertension (<140/90 mmHg) (%)	10.0	25.0	30.0	27.2	28.0	30.0
Total Fat (kcal daily)	14.8	24.7	28.8	28.0	38.8	42.2
Saturated Fat (% kcal daily)	4.9	9.2	14.4	13.6	13.7	18.5
Cholesterol in diet (mg daily)	53	166	240	210	200	405
Body Mass Index (kg/m2)	21.6	22.9	22.7	21.0	26	26
Waist-Hip Ratio (Male)	0.86	0.89	0.89	-	0.98	0.94
Fasting Plasma Insulin (Male) (mU/L)	4.8	5.1	16.6	-	9.8	7.2
2-Hour after Glucose (mU/L)	11.8	24.2	60.6	-	41.0	19.0
Serum Cholesterol (mg/dl)	168	190	210	196	230	236
HDL Cholesterol (mg/dl)	43.5	44.2	45.1	45.2	45.0	46.2
Angiotensin Converting Enzyme (IU)	37.5	63.4	72.0	-	80.0	-
TNF-alpha(pg/ml)	4.5	15.6	18.8			
IL-6(pg/ml)	2.8	11.5	14.4	-	-	-
Nitrite(umol/L)	1.0	0.76	0.67	-	-	-

Values are means, Modified from Singh et al., Lancet, 1995,J Nutr Environ Med 2001,Biomed Pharmaco 2004.HDL = High Density Lipoprotein Cholesterol.

Prevention of Cardiovascular Diseases and Diabetes

Increased consumption of salt, tobacco, fat, sugar, omega-6 rich oils and energy in the last four decades are important environmental factors for emergence of biological markers responsible for CVDs and type 2 diabetes [30,31]. There is increase in per capita income, gross domestic product, food production and automobile production in the last four decades. The period from 1970 to 2010 has witnessed marked changes in diet and lifestyle, particularly in the urban populations which may be the cause of the epidemic of CVD and diabetes among many developing countries [30,31]. Japan is a developed country and the only country in the world, which had transition from poverty to affluence but without significant increase in CVDs and diabetes [29,32,33].The goals for ideal clinical and biochemical markers; blood pressure, blood glucose, Hb 1C, and blood cholesterol have been altered by most US health agencies as well as by WHO and International College of cardiology and International College of Nutrition, depending upon new evidence [30,31]. Hence any guidelines for prevention of CVDs and diabetes do not appear to be final and are likely to be changed in near future. Effects of guava intake causing decrease in serum total and high density lipoprotein cholesterol levels and on systemic blood pressure has been observed [23]. Diet and lifestyle changes have been demonstrated to cause a decline in blood lipoproteins, blood glucose and blood pressures and an increase in HDL cholesterol and Apo A which are beneficial [24,25].

However, none of these guidelines emphasize on the role of various acute or chronic diseases such as acute myocardial infarction (AMI) and diabetes mellitus which can alter various markers as complications of these diseases [26]. It is easy to change health behaviours of the victims when they are admitted in the intensive cardiac care unit, to follow the same diet and lifestyle when they go back home. AMI is associated with hyperglycemia, hyper-insulinemia, hypertriglyceridemia, free radical stress, rise in free fatty acid and proinflammatory cytokines, leading to endothelial dysfunction [9,10]. There is an acute generation of proinflammatory milieu among AMI patients which is known to cause disruption of atheroma plaque, resulting into reinfarction and death. Most experts advise dietary patterns; including grains, vegetables, fruits, nuts, seeds and legumes, fat and oils based on clinical studies for prevention of CVDs and diabetes [21-26]. There is no recommendation for refined starches which is a most wise step in the prevention of endothelial dysfunction. However, there is no guideline about the type of oil and type of nuts depending upon the w-3 fat and monounsaturated fatty acid (MUFA) content of these foods. While foods and beverages with added sugars and refined starches as well as excess of w-6, total and saturated fat and trans fatty acids, may be proinflammatory, increased intake of w-3 fatty acid and MUFA may be protective against surge of TNF-alpha,IL-6,IL-18 and adhesion molecules like VCAM-1(vascular cell adhesion molecule-1) and IVAM-1 caused by high glycemic, rapidly absorbed proinflammatory foods. These foods are known to initiate a proinflammatory milieu in the body which is similar to that of AMI, diabetes mellitus causing further increase in complications, among these patients.

A low w-6/w-3 ratio and polyunsaturated/saturated fatty acid ratio in the diet of 1:1 has been proposed by the Columbus Paradigm Institute (www.columbus-concept.com) for prevention of dyslipidemia and CAD. It seems that this weakness in the guideline may be due to missing of related work. In fact one of the greatest weakness of the American guidelines is that there is no discussion on proinflammatory foods, so there is opportunity for the industry to influence the consumers without consideration of adverse effects of these foods on health. The increased servings of grains, vegetables, fruits, nuts and fish is good for prevention of weight gain, hyperlipidemia and hypertension but it is also good for prevention of dyslipidemia and metabolic syndrome and CAD by inhibiting the rise in FFA, hyperglycemia and maintaining good endothelial function .There has been overemphasis on dietary cholesterol, in the recommendations, although it is well known that dietary cholesterol has only little influence on atherothrombo-genesity of serum LDL which can be further decreased by substituting functional foods. Functional foods are natural foods, rich in phytochemicals which are slowly absorbed and have a beneficial effects on our biomarkers without causing any abnormal increase in blood glucose, insulin, proinflammatory cytokines and free fatty acids [21-26]. The Columbus® Concept stands for a return of alpha-linolenic acid (ALA, C18:3ω3) - herein referred to as *wild*- or *game*-type land-based fatty acid - into the feed ration of land-based bred animals to such an extent that their fat depots (white adipose tissue) exhibit a balanced ratio of *essential* fatty acids, i.e., ω6:ω3 = 1:1, characteristic of fat depots in wild animals or game [27]. This return to the wild standard translates into a substantial reduction in long chain omega-6 fatty acids and a moderate species-specific increase in long chain omega-3 fatty acids in organs and peripheral tissues of these domesticated animals or livestock. The ω6:ω3 = 1:1 ratio is also taken as reference for the design of composite plant-derived table oils and fats as these represent other primary sources of energy in the modern human diet. As observed from the overall distribution of fats in a natural untamed environment, nature recommends the consumption of a balanced ratio of saturated and polyunsaturated fatty acids as part of a dietary lipid pattern rich in monounsaturated fatty acids (P:M:S = 1:6:1).The value of such diet has also been proven in long term clinical trials [21-27].

Table 2. Clinical biomarkers and other risk factors common in various countries for unexplained risk of heart disease

Genetic or Environmental	Clinical biomarkers and other risk factors
1. Insulin resistance and hyperinsulinemia 2. Poor beta cell function 3. Increased prevalence of type II diabetes 4. Increased lipoprotein(a) 5. Increased angiotensin converting enzyme 6. High Apo-B	1. Increased plasminogen activator inhibitor-1 2. Decreased tissue plasminogen activator 3. Decreased antioxidant vitamins A, C, beta-carotene, E, Se, Zn, flavonoids, other polyphenolic substances 4. Low high density lipoprotein cholesterol 5. Elevated homocysteins 6. Low Apo A1 7. Increased heart rate and BP variability 8. Increased small dense LDL cholesterol 9.Increased w-6/w-3 ratio in the diet. 1:50 due to increased intake of sun flower, corn and soya bean oils. 10. Decreased tissue antioxidant potential.

Halberg et al. have suggested the need for consideration of blood pressure and blood glucose variability as important causes for target organ damage among patients of hypertension or normotension and type 2 diabetes mellitus [28]. In one experience (RBS), eating 50-60g/day of walnut caused significant reduction in blood pressures from 141/92 mm Hg to 130/ 88 mm Hg in about 8 weeks. Intake of Fish and n3 Fatty Acids and Risk of Coronary Heart Disease Among Japanese: The Japan Public Health Center-Based (JPHC) Study Cohort reported that high –omega-3 fatty acids diet was beneficial against CVDs [29].

Effect of a Mediterranean-style diet on endothelial dysfunction and markers of vascular inflammation in the metabolic syndrome indicated that w-3 fatty acid rich Mediterranean diet can protect against proinflammatory risk of CVDs [9,10,20-33]. Recently, circulating cardio-enriched microRNAs have been associated with long-term prognosis following myocardial infarction [34].

Dietary Guidelines in Japan (jda.com)

There has been substantial changes in the diets of Japanese population in the last three decades.

Which have been associated with poor nutritional balance, attributed to a higher average fat intake, and greater waste of food, a consequence of people preparing more food and throwing away more leftovers. There have been other changes in eating habits; skipping meals; eating alone; separate meals for individual family members, with each member eating what he or she likes; and inadequate instruction of children on recommended eating habits, owing largely to fewer opportunities for family members to eat together.

The Ministry of Agriculture, Forestry and Fisheries, the Ministry of Welfare and the Ministry of Education have prepared dietary guidelines for Japanese. On March 24, the Cabinet adopted a proposal for measures aimed at diffusing the dietary guidelines. Accordingly, various organizations within the government will work together to promote the reassessment of dietary patterns at the national level for promotion of better dietary patterns and the more effective use of agricultural resources.

Enjoy the meals means; have delicious and healthy meals that are good for your mind and body, to achieve a longer healthy life through daily meals. Enjoy communication at the table with your family and/or other people and participate in the preparation of meals. Establish a healthy rhythm by keeping regular hours for meals. Have breakfast containing functional foods to make a good start to the day and to protect against circadian rhythm of cardiovascular events in the second quarter of 24 hours. Avoid large snacks before bedtime and between meals and if you drink alcoholic beverages, do so in moderation preferably red wine rich in flavonoids to have beneficial effects on several biomarkers of our body due to its nitric oxide activating effects. Eat well-balanced meals with staple food, as well as main and side dishes, making a good combination of various foods. Combining wisely home-made meals with eating out, eating processed and prepared foods which may be consumed along with functional foods like olive oil and cocoa products and red wine. Combine vegetables, fruits, whole grains, milk products, beans and fish in the diet. Avoid too much of salt and fat by checking nutrition labels for choosing foods. A healthy body weight may be maintained by balance the calories we eat with physical activity. Chewing of food decreases the intake of calories. Take advantage of our dietary culture and local food products, while incorporating new and different dishes. Reduce leftovers and waste through proper cooking and storage methods. Assess daily eating by setting your own health goals and have a habit to assess your diet by thinking and talking about your diet with your family and friends. It is important to learn healthy eating habits at school as well as at home by promotion of eating habits at early age.

In brief, it is possible to advise 400g/day of fruits, vegetables and nuts (almonds and walnuts 50g/day), another 400g/day of whole grains including legumes in conjunction with 25-50g/day of mustered oil or Columbus oil (olive oil+linseed oil) for prevention of type 2 diabetes, hypertension and CAD. (Table 3) It is important to do regular physical activity; spot running and meditation (each 30min) and cessation of tobacco intake for prevention of adverse effects on biomarkers responsible for these problems. Moderate alcohol intake, particularly red wine rich in flovonoids appear to be more beneficial as proposed in the

French paradox by Searge Renaud because alcoholism has adverse effects on all the biomarkers causing CVDs and type 2 diabetes.

Table 3. Dietary guidelines and desirable level of biomarkers and other risk factors for South Asians

Factors	Desirable Values
Energy (k calories/day)	1900-2500
Total Carbohydrate (k calories/day)	65.0
Complex Carbohydrate ((k calories/day)	55.0
Total Fat (k calories/day) (Olive oil may be more)	21.0
Saturated Fatty Acids (k calories/day)	7.0
Polyunsaturated Fatty Acids (k calories/day)	7.0
Polyunsaturated/Saturated Fat Ratio	1.0
n 6/n-3 Fatty Acid Ratio	<5.0
Dietary Cholesterol (mg/day)	100
Whole Grains (wheat, rice, corn, legumes) (g/day)	400-500
Fruit, vegetables and nuts (g/day)	400-500
Salt (g/day)	<6.0
Brisk Walking (km/day)	9.0
Meditation/pranayam (minutes/day)	30.0
Body Mass Index (kg/m^2)	
Range	19.0-23.0
Average	21.0
Waist-Hip Girth Ratio	
Male	<0.90
Female	<0.85
Serum Total Cholesterol (mg/dl) (4.42 mmol/L)	<170
Mild Hypercholesterolemia (mg/dl) (4.42-5.20	170-200
mmol/L)	>200
Hypercholesterolemia (mg/dl) (>5.20 mmol/L)	<90
Low Density Lipoprotein Cholesterol (mg/dl)	90-110
(2.32 mmol/L)	>110
Borderline High (mg/dl) (2.32-2.84 mmol/L)	<150
High (mg/dl) (2.84 mmol/L)	>40 men, >50women
Triglycerides (mg/dl) (1.7 mmol/L)	<1.00mg/L
High Density Lipoprotein Cholesterol (mg/dl) (0.9	<125/85
mmol/L)	Amblodipine,ACE-I , receptor blockers
C-Reactive proteins	and new beta-blockers?
Blood Pressure (mmHg)	Fish oil, aspirin, statins, nicotinic acid,
Drug therapy in view of high risk of diabetes and	fibrates.
CAD.	

Table 4. Antioxidant potential of various foods

Food groups	Antioxidant value(mmol/100g)
Plant based foods	11.57
Berries and berry products	9.86
Beverages	8.30
Breakfast cereals	1.09
Chocolates and sweets with cocoa	4.93
Fruit and fruit juices	1.25
Nuts and seeds	4.57
Spices and herbs	29.02
Vegetable and vegetable products	0.8
Grains and grain products	0.34
Legumes	0.48
Condiments	0.77
Fats and oils	0.51
Fish and sea foods	0.11
Egg	0.04
Dairy products	0.14
Meat and meat products	0.31
Poultry and poultry products	0.23
Animal foods	0.18
Mixed foods	0.91

Carlsen et al., Nutrition Journal 2010.

Conflict of interest has not been declared by the authors.

Acknowledgment

Acknowledgment is due to International College of Nutrition for providing logistic support to write this article.

References

[1] Moodie R, Stuckler D, Monteiro C, Sheron N, Thamarangsi , Lincoln P, Casswel S on Behalf of the Lancet NCD Group. Profits and pandemics: prevention of harmful effects of tobacco, alcohol, and ultra-processed food and drink industries. *The Lancet*, Early Online Publication, 12 February 2013,doi:10.1016 /S0140-6736(12)62089.

[2] Singh RB, Takahashi T, Nakaoka T, Otsuka K,Toda E, Shin HH, Kyu Lee M, Beeharry V, Hristova K, Fedacko J, Pella D, De Meester F, Wilson DW, Juneja LR. Nutrition in transition from Homo sapiens to Homo economicus. *The Open Nutra J* 2013;6:6-17.

[3] Taylor AW; Price K; Fullerton S. A survey to assist in targeting the adults who undertake risky behaviours, know their health behaviours are not optimal and who acknowledge being worried about their health. *BMC Public Health* 2013; 13, 120

[4] 4.WHO. Mortality and burden of disease estimates for WHO Member States in 2008. Geneva: *World Health Organization,* 2010.

[5] Beaglehole R, Bonita R, Alleyne G, et al., for the Lancet NCD Group. UN High-Level Meeting on Non-Communicable Diseases: addressing four questions. *Lancet* 2011; 378: 449-55.

[6] Tokunaga M, Takahashi T, Singh RB, Rupini D, Toda E, Nakamura T, Mori H, Wilson DW. Diet, Nutrients and Noncommunicable Diseases. *The Open Nutraceuticals Journal,* 2012, 5, 146-159.

[7] Fedacko J, Vargova V, Singh RB, Anjum B, Takahashi T, Tongnuka M, Dharwadkar S, Singh S, Singh V, Kulshresth SK, De Meester F, Wilson DW. Association of high w-6/w-3 fatty acid ratio diet with causes of death due to noncommunicable diseases among urban decedents in North India. *The Open Nutra Jour* 2012;5:113-123.

[8] Fung TT, Rimm EB, Spiegelman DS, Rifai N, Tofler GH, Willet WC, Hu FB. Association between dietary patterns and plasma biomarkers of obesity and cardiovascular disease risk. *Am J Clin Nutr* 2001;73: 61-67.

[9] Esposito K,Glugliano D.Diet and inflammation:a link to metabolic and cardiovascular diseases. *Euro Heart J.*, 2006,27:15-20.

[10] Esposito K, Marfella R, Ciotola M, Di Palo C, Giugliano G, D'Armiento M, D'Andrea F, Giugliano D. Effect of a Mediterranean-style diet on endothelial dysfunction and markers of vascular inflammation in the metabolic syndrome: a randomized trial. *JAMA* 2004;292:1440-1446.

[11] Combs GF Jr, Trumbo PR, McKinley MC, Milner J, Studenski S, Kimura T, Watkins SM, Raiten DJ. Biomarkers in nutrition: new frontiers in research and application. *Ann N Y Acad Sci;* 2013 Mar;1278:1-10.

[12] Raiten DJ, Namasté S, Brabin B, Combs G Jr, L'Abbe MR, Wasantwisut E, Darnton-Hill I. Executive summary–Biomarkers of Nutrition for Development: Building a Consensus. *Am J Clin Nutr*; 2011;94:633S-50S.

[13] Colhoun HM, Betteridge DJ, Durrington P, Hitman G, Neil A, Livingstone S, Charlton-Menys V, Bao W, DeMicco DA, Preston GM, Deshmukh H, Tan K and Fuller JH. Total Soluble and Endogenous Secretory Receptor for Advanced Glycation End Products as Predictive Biomarkers of Coronary Heart Disease Risk in Patients With Type 2 Diabetes: An Analysis From the CARDS Trial, *Diabetes;*2011;60:2379-2385.

[14] Sanfilippo FM, Hobbs MST, Knuiman MW and Hung J. Impact of New Biomarkers of Myocardial Damage on Trends in Myocardial Infarction Hospital Admission Rates from Population-based Administrative Data. *Am J Epidemiol* 2008;168:225-233.

[15] Carlsen MH, Halvorsen LB, Holte K, Dragland S, Sampson L, Willey C, Senoo H, Umezono Y, Sanada C, Barikmo I, Berhe N, Willett WC, Phillips KM, Jacobs DR, and Blomhoff R. The total antioxidant content of more than 3100 foods, beverages, spices, herbs and supplements used worldwide. *Nutrition Jour* 2010;9:3-10, doi:10.1186/1475-2891-9-3.

[16] Carter P, Gray LJ, Troughton J, Khunti K, Davies MJ. Fruit and vegetable intake and incidence of type 2 diabetes mellitus: systematic review and meta-analysis. BMJ 2010;4229, *BMJ.* 2010;341:c4229. doi: 10.1136/bmj.c4229.

[17] Arai Y, Watanabe S, Kimira M, Shimoi K. Dietary intakes of flavonols, flavones and isoflavones by Japanese women and the inverse correlation between quercetin intake and plasma LDL cholesterol concentration. *J Nutr.* 2000;130:2243-50.

[18] Orlich MJ, Singh PN, Sabaté J, et al. Vegetarian dietary patterns and mortality in Adventist Health Study 2. *JAMA* Intern Med: DOI:10.1001/jamainternmed.2013.6473. Available at: http://archinte. jamanetwork. com/journal.aspx.

[19] Singh RB,Dubnov G,Niaz MA,Ghosh S,Singh R,Rastogi SS,Manor O,Pella D,Berry EM. Effect of an Indo-Mediterranean diet on progression of coronary disease in high risk patients: a randomized single blind trial. *Lancet* 2002,360:1455-1461.

[20] Singh RB, Rastogi SS, Niaz MA, Ghosh S, Singh R : Effects of fat-modified and fruits vegetable enriched diets on blood lipids in the Indian diet heart study. *Am. J. Cardiol,* 1992, 69: 869-874.

[21] Singh RB, Fedacko J, Sharma JP, Vargova V, Sharma, Moshiri M, De Meester F, Otsuka K. Association of inflammation, heavy meals, magnesium, nitrite, and coenzyme Q10 deficiency and circadian rhythms with risk of acute coronary syndromes. *World Heart J* 2010;2:219-228.

[22] Singh RB, Pella D, Sharma JP, Rastogi S, Kartikey K, Goel VK, Sharma R, Neki NS, Kumar A, Otsuka K. Increased concentrations of lipoprotein(a),circadian rhythms and metabolic reactions, evoked by acute myocardial infarctions, associated with acute reactions in relation to large breakfast. *Biomed Pharmacother J,* 2004,38 (suppl) : 116-123.

[23] Singh RB, Rastogi SS, Ghosh S, Singh R, Niaz MA : Effects of guava intake on serum total and high density lipoprotein cholesterol levels and on systemic blood pressure. *Am. J. Cardiol,* 1992, 70 :1287-91.

[24] Singh RB,Rastogi SS, Ghosh S, Niaz MA, Singh NK: The diet and moderate exercise trial (DAMET): Results after 24 weeks. *Acta Cardiol*, 1992, 48: 543-557.

[25] Singh RB,Rastogi SS, Sircar AR,Mani UV, Singh NK, Niaz MA :Effect of lifestyle changes on atherosclerotic risk factors in the Indian diet heart study. *Am. J. Cardiol,* 1993, 71: 1283-88.

[26] Singh RB, Rastogi SS, Verma R,Bolaki L, Singh R,Ghosh S. An Indian experiment with nutritional modulation in acute myocardial infarction. *Am. J. Cardiol.* 1992; 69:879-85.

[27] De Meester F, Wild-type land based foods in health promotion and disease prevention: the LDL-CC:HDL-CC model. In *Wild Type Foods in Health Promotion and Disease Prevention*, editors Fabien DeMeester and RR Watson, Humana Press, NJ 2008,p 3-20.

[28] Halberg F, Cornelissen G, Otsuka K, Siegelova J, Fiser B, Dusek J, Homolka P, Sanchez de la Pena S, Singh RB, BIOCOS project. Extended consensus on need and means to detect vascular variability disorders (VVDs) and vascular variability syndromes (VVSs). *World Heart J* 2010; 2 (4): 279-305.

[29] Hiroyasu Iso, Minatsu Kobayashi, Junko Ishihara, Satoshi Sasaki, Katsutoshi Okada, Yoshikuni Kita, Yoshihiro Kokubo, Shoichiro Tsugane for the JPHC Study Group. Intake of Fish and n3 Fatty Acids and Risk of Coronary Heart Disease Among Japanese: The Japan Public Health Center-Based (JPHC) Study Cohort I. *Circulation* 2006;113 195-202.

[30] Singh RB, Mori Hideki, Chen Junshi, Mendis S, Moshiri M, Shoumin Z, Kim SH, Faruqui AMA. Recommendations for the prevention of coronary artery disease in

Asians: a scientific statement of the International College of Nutrition. *J. Cardiovas. Risk* 1996, 3:489-494.

[31] Singh RB, Rastogi SS, Rao PV, Das S, Madhu SV, Das AK, Sahay BK, Fuss SM, Begum R, Sainani GS, Shah NA. Diet and lifestyle guidelines and desirable levels of risk factors for the prevention of diabetes and its risk factors in Indians: A scientific statement of the International College of Nutrition. *J Cardiovasc Risk* 1997; 4:201-08.

[32] Yokoyama M, Origasa H, Matsuzaki M, Matsuzawa Y, Saito Y, Ishikawa Y, Oikawa S, Sasaki J, Hishida H, Itakura H, Kita T, Kitabatake A, Nakaya N, Sakata T, Shimoda K, Shirata K, for the Japan EPA lipid intervention study (JELIS) Investigation. Effects of eicosapentaenoic acid on major coronary events in hypercholesterolaemic patients (JELIS): a randomised open-label, blinded endpoint analysis. *Lancet* 2007;369:1090–1098.

[33] Takahashi T, Toda E, Singh RB, De Meester F, Wilczynszka A, Wilson DW, Juneja LR. Essential and Non-Essential Amino Acids in Relation to Glutamate. *The Open Nutraceuticals Journal,* 2011, 4, 205-212 205

[34] Gidlöf O, Smith JG, Miyazu K, Gilje P, Spencer A, Blomquist S, Erlinge D. Circulating cardio-enriched microRNAs are associated with long-term prognosis following myocardial infarction. *BMC Cardiovasc Disord.* 2013 Feb 28;13(1):12.

In: Circadian Cardiology with Focus on Both Prevention ... ISBN: 978-1-63463-956-9
Editors: K. Hristova, A. Shehab, G. Cornelissen et al. © 2015 Nova Science Publishers, Inc.

Chapter 17

Nutrition in Transition from *Homo erectus* to *Homo modestis*

Fabien De Meester[8,1,*], Toru Takahashi[2], Ram B. Singh[3], Eri Toda[3], Hyun Ho Shin[4], Moon-Kyu Lee[4], Vicky Beeharry[5], Krisimira Hristova[6], Jan Fedacko[7], Daniel Pella[7], Douglas W. Wilson[9], Lekh R. Juneja[10] and Danik M. Martyrosyan[11,12]

[1]Halberg Hospital and Research Institute, India
[2]Graduate School of Human Environment Science, Fukuoka Women's University, Japan
[3]Department of Internal Medicine, Tokai University Hachioji Hospital, Tokyo, Japan
[4]Korean Society of Lipidology and Atherosclerosis, Korea
[5]Faculty of Medicine, Russian Peoples Friendship University, Moscow, Russia
[6]University National Heart Hospital, Sofia, Bulgaria
[7]Faculty of Medicine, PJ Safaric University, Kosice, Slovakia
[8]The Tsim Tsoum Institute, Krakow, Poland
[9]School of Medicine, Pharmacy and Health, Durham University, UK
[10]Taiyo Kagaku Co. Ltd, Yokkaichi, Japan
[11]Department of Internal Medicine, UT Southwestern Medical Center, Dallas, TX, US
[12]Functional Food Center, TX, US

Abstract

The food and nutrient intake of Paleolithic hunter-gatherers (*Homo erectus*) and of Western populations (*Homo economicus*) show marked variations. With increase in wealth and affluence there is a decrease in the intake of omega-3 fatty acids, vitamins,

* Corresponding author: Fabien De Meester, The TsimTsoum Institute, Krakow, Poland, fdm@tsimtsoum.net.

antioxidants and amino acids and a significant increase in the intake of refined carbohydrates, fats (saturated & trans fats, omega-6 fatty acids) and salt in comparison with those of the Paleolithic period. The protein or amino acid intake was 2.5 fold greater (33 vs. 13%) in the Paleolithic diet of *Homo erectus* compared to that of the modern Western diet. Prior to the Agricultural Revolution, man's diet was based on an enormous variety of wild plants, eggs, fish and seeds. In comparison, today about 17% of plant species provide 90% of the world's food supply which is mainly contributed by grains produced from fertilizer-based rapidly grown crops potentially lower in nutrient density and higher in energy. Grains are high in omega-6 fatty acids and carbohydrates and low in omega-3 fatty acids and antioxidants compared to leafy green vegetables. The appropriate diet for *Homo sapiens* is characterized by high levels of protective essential nutrients; amino acids, minerals, vitamins, flavonoids, omega-6/3 fatty acids. Whereas the average diet of *Homo erectus* did comply with this evolutionary pattern, the modern Western dietary pattern of *Homo economicus* has excess of energy-rich refined carbohydrates, omega-6, trans and saturated fats. The consumption of such foods in wealthy countries in conjunction with sedentary behavior are associated with increased risk of deaths due to cardiovascular (CVDs) and other non-communicable diseases (NCDs).

Keywords: Diet, social class, functional foods, ancient diet, Mediterranean diet

Introduction

Modern man, *Homo economicus*, has evolved from his primate precursor as an ever more skilled and organized hunter-gatherer [1-3]. Man started farming some 10,000 years ago, an adaptation which in turn led to unprecedented technical development, agriculture, industry and commerce. From that perspective, it seems that the biological functions of most social groups is to trigger the what/when hunter-gatherers' anticipation, socially and physiologically [1-5]. *Homo sapiens* and its predecessors, namely *Homo erectus* and *Homo habilis* were primarily vegetarians according to Paleontological records [3]. Consciousness of space and time led to perspective and organization, and hunting groups developed gradually as man moved away from other primates and became skilled in tool making [1-3]. The Paleolithic diet was about naturally available wild foods: fruits, vegetables, green leaves, seeds, honey, eggs, fish, and meat from running animals [3-10]. These foods were also available to pre-agricultural humans and obviously shaped modern man's genetic nutritional requirement. Early man like hunter-gatherers also had excellent health behaviors; enormous physical activity with limited or no mental stress, alcoholism and tobacco intake, which are now important behavioral risk factors of cardiovascular diseases (CVDs). After 1910, in the 20th century, industrialization and urbanization added stress, started refining and processing of foods, storing and distributing them, in a continuous search for a better economic model [10-14]. The Japanese and the Mediterranean traditional diets appear to share a common standard (omega-6/3 ~ 1/1) with the Paleolithic diet, although diet and lifestyle in Japan have been under transition from poverty in the 1950s to affluence by the year 1980s, without increase in most CVDs. This modern pattern of food availability, changes in lifestyle and in political system, led by profit, translates into new unprecedented challenges, opportunities and threats, for mankind in the whole of the Western world [14, 15]. The impact of these changes on our total health are clear from a recent study showing that Americans die sooner, are least likely to reach the age of 50 years, and have higher rates of disease or injury than people in 16 other economically wealthy countries [16]. There appears to be a globalization of wealth without much improvement in health behavior because the health budget in the United States is much higher than in any other country [17].

Primary Risk Factors and Genetic Variations

Excess intake of Western type foods, sedentary behavior, tobacco, alcoholism and psychosocial stress are behavioral risk factors which may be also called primary risk factors. It has been proposed that physical activity, sleep, sun exposure, and dietary needs of humans are genetically determined. The profound changes in diet and lifestyle, which occurred after the Neolithic Revolution (first agricultural revolution: *c*12000BC) and more particularly after the Industrial Revolution and the Modern Age are too recent on an evolutionary time scale for the human genome to have fully adapted [2, 3, 7, 18-20]. It is the genes selected during the Paleolithic period (from about 2.5 million years ago to 11,000 years ago) which make the human genome despite various alleles being targets of selection since the Agricultural Revolution [3, 21, 22]. In fact, all human beings living in Europe, Asia, Oceania, and the Americas share a common African *Homo sapiens* ancestor, although they may have genetic diversity [23-25]. These post agriculture alleles were induced by changes in pathogens, fatal diseases, and harsh environments without any impact of sleep, exercise, and diet [25-28]. There are very few primitive hunter-gatherers in the world today. India, with its approximately 1.22 billion population mainly has farmers, but the rapidly growing towns have an increasing number of poor urban slum dwellers and a big affluent society [1, 5]. Population studies reveal that the levels of physical inactivity, dietary patterns, salt intake, alcohol consumption, tobacco use and stress are important characteristics of various populations [17-21]. Social autopsy studies by verbal autopsy questionnaires indicate that the social determinants of health and behavioral risk factors may be dependent on various attributes of social classes [2, 17-29].

Nutritional Transition from *Homo erectus* to *Homo modestis*

Humans live in an environment that completely differs from that for which his genetic constitution was selected. The food and nutrient intake among humans during various periods of evolution are given in the Tables 1-5. There is marked reduction in consumption of omega-3 fatty acids, antioxidants, vitamins and minerals, proteins, and significant increase in intakes of carbohydrates (mainly refined), fats (saturated, trans, omega-6) and salt compared to the Paleolithic period [1-10]. However, major dietary changes have occurred during the last 100-150 years causing increased intake of saturated fatty acids (SFA), trans fatty acids and linoleic acid, and meat from grain fed cattle tamed at farm houses rather than meat from running animals. In general, there is increase in the intake of refined carbohydrates and decrease in the intake of complex carbohydrates, essential amino acids, minerals, omega-3 fatty acids, vitamins and antioxidants. These dietary changes in conjunction with sedentary behavior, mental strain, pollution, tobacco consumption and alcoholism, particularly after 1910, may have caused damage to our genes, leading to emergence of phenotypes of CVDs among *Homo economicus* populations. Industrialization and urbanization have been associated with a greater consumption of pro-atherogenic foods during transition from lower social classes 4-5 (poverty) to higher social classes 1-3 (30-38). These differences can cause differences in the food and nutrient intake among hunter-gatherers and among Western and Asian populations during transition from poverty to affluence and from Homo erectus to Homo economicus populations (Table 6). Western type diet consumed by lower social classes in the developed countries and higher social classes in developing countries may be associated with marked reductions in the consumption of omega-3 fatty acids, vitamins, antioxidants, and amino acids and significant increases in the intakes of carbohydrates, (mainly refined), fats (saturated, trans fat, and linoleic acid), and salt compared to the Paleolithic period (Tables 1-5). The protein or amino acid intake was 2.5 fold greater (33 vs. 13 %) in the Paleolithic diet compared to modern diet (Table 3, Figure 1). However, today about 17% of plant species provide 90% of the world's food supply which is mainly contributed by grains. Recent additions to dietary patterns are cereal grains (refined), and vegetable oils that are rich in omega-6 fatty acids and trans fats and low in amino acids which represent dramatic departure from those foods and nutrients to which man is adapted [1-3].Wheat, corn and rice account for three quarters of the world's grain production on which humans are dependent for their food supply. Several experts have estimated higher intakes for protein, calcium, potassium and ascorbic acid and a lower intake of sodium in the diet of late Paleolithic period than the current diets of the developed and developing countries [1-5]. Green leafy vegetables are also rich sources of antioxidants, magnesium, omega-3 fatty acids and carotenoids which appear to be high in the Mediterranean region [7-10]. The protective effects of Paleolithic diets in the form of Mediterranean diet, Indo-Mediterranean diet, Japanese diet, and DASH diet have been documented in several previous studies [1-12].

Table 1. Food and nutrient intake among various social groups

Food and nutrient	Hunter-gatherer society	Western society	Asian societies
Energy density	Low	High	Low
Protein	High	Low-moderate	Low
Animal	High	Low–moderate	Low
Vegetable	Very low	Low–moderate	Low
Carbohydrate	Low-moderate(slowly absorbed)	Moderate-rapidly absorbed	High-Slow
Fiber	High (>30g)	Low (<15g)	High
Fat	Low	High	Low
Animal	Low	High	Low
Vegetable	Very low	Low	High
Total omega-3	High(2.3g/day)	(0.2g/day)	0.5-0.85g
Ratio omega-6/3	Low 2.4	High 15-20	25-50
Vitamins and minerals	High	Low	moderate

Modified from Eaton et al. [2, 3] and Singh et al. [6] references [1, 10].

Table 2. Estimated fatty acid consumption in the late Paleolithic period social group

Sources	Fatty acids(g/day) en 35.65/day
Plants	
Linoleic acid	4.28
Alpha-linoleic acid	11.40
Animal	
Linoleic acids	4.56
Alpha-linolenic acid	1.21
Total	
Linoleic acid	8.84
Alpha linolenic acid	12.60
Animal	
Arachidonic acid(w-6) (AA)	1.81
Long chain w-3 fatty acids	
Eicosapentaenoic acid(w-3)(EPA)	0.39
Docosatetraenoic acid(w-6) (DTA)	0.12
Docosapentaenoic acid(w-3)(DPA)	0.42
Docosahexaenoic acid(w-3)(DHA)	0.27
Total long chain w-3 fatty acids	1.20
Ratios of w-6/w-3	0.70
Linoleic acid/alpha linolenic acid+ AA+DTA/EPA+DPA+DHA	1.79
Total w-6/w-3	0.77

Modified from Eaton et al. [2, 3] and Singh et al. [1, 6].

Table 3. Nutrient composition in the late Paleolithic society and current recommendations

Nutrient per day	Late Paleolithic society	Current recommendation
Total dietary energy%		
Protein	33	12
Carbohydrate	46	58
Fat	21	30
Alcohol	-0	moderate alcohol
P/S ratio	1.41	1.00
Cholesterol, mg	520	300
Fiber, g	100-150	30-60
Sodium, mg	690	1100-3300
Calcium, mg	1500-2000	800-1600
Ascorbic acid, mg	440	60

Modified from Eaton et al. [2, 3] their references 1, 2, and Singh et al. [1, 6].

Table 4. Ethnic differences in fatty acid levels in thrombocytes phospholipids and percentage of all deaths from cardiovascular disease

	Europe and USA %	Japanese society %	Greenland Eskimos %
Arachidonic acid(20:4w6)	26	21	8.3
Eicosapentaenoic acid(20:5w-3)	0.5	1.6	8.0
Ratio of w-6/w-3	50	12	1
Mortality from cardiovascular disease	45	12	7

Modified from Singh et al. [1, 6].

Table 5. Fatty acids ratio in the diets of various societies

Subjects	w-6/w-3	
Paleolithic	0.79	Estimated
Greece prior to 1960	1.00-2.00	Current 7.10
Japan	4.00	Early 1-2
India, rural	5-6.1	Prior to 1960, 3-4
India urban	38-50	Prior to 1960, 5-10
UK	15.00	Prior to 1960,10.00
Northern Europe	15.00	Prior to 1960,10.00
USA	16.74	Prior to 1950 7-8
Eastern Europe	20-25	Estimated
Indian hunter-gatherers	1.00-2.00	Estimated

Modified from Singh et al. [1, 6, 10].

Table 6. Nutrition in transition and emergence of non-communicable diseases

Homo sapiens diet given in tables 1-5	Pattern 1: Hunter-gatherers	Pattern 2: Food Scarcity-Poverty	Pattern 3: Receding Food Scarcity & Poverty	Pattern 4: More food, less exercise- *Homo economicus*	Pattern 5: Healthy Behavior-*Homo modestis*
Nutrition profile Diet	Plants, low-fat wild animals, diet diversity by collecting foods.	Cereals predominant, diet less varied	Fewer starchy staples; more fruit, vegetables, animal protein; low variety continuesContinued MCH[1] nutrition problems, many deficiencies disappear, weaning diseases emerge, stature grows	More fat (animal products, trans fat, w-6 fat), sugar, processed foods; less fiber, less w-3 fat and flavonoids	Higher-quality fats, reduced refined carbohydrates, more whole grains, fruit, vegetables rich in w-3 and flavonoids
Nutritional status	Robust, lean population; few nutritional deficiencies	Children and women suffer most from low fat intake, nutritional-deficiency disease emerge, stature declines		Obesity, problems for elderly (Osteoporosis, fractures etc), type 2 diabetes, hypertension, stroke, heart attack, brain degeneration, Psychological disorders, and cancer	Reduction in body fat and obesity, and NCDs, improvement in bone health (Epigenetic modulation and trans-generational epigenetic inheritance-natural selection.)
Economy	Hunter-gatherers	Agriculture, animal husbandry, homemaking begin; shift to monocumono cultures	Second agricultural revolution (crop rotation, fertilizer), Industrial Revolution, women join labor force	Fewer jobs with heavy physical activity, service sector and mechanization, household technology revolution	Service sector mechanization and industrial robotization dominate, increase in leisure exercise offsets sedentary jobs
Household	Primitive, onset of fire	Labor-intensive, primitive technology begins (clay cooking vessels)	Primitive water systems, clay stoves, cooking technology advances	Household technology mechanizes and proliferates	Significant reduction in food preparation costs as a result of technologic change

Homo sapiens diet given in tables 1-5	Pattern 1: Hunter-gatherers	Pattern 2: Food Scarcity-Poverty	Pattern 3: Receding Food Scarcity & Poverty	Pattern 4: More food, less exercise- *Homo economicus*	Pattern 5: Healthy Behavior-*Homo modestis*
Income and assets	Subsistence, primitive stone tools	Subsistence, few tools	Increases in income disparity and agricultural tools industrialization	Rapid growth in income and income disparities, technology proliferation	Decrease in income growth, increase in home and leisure technologies
Professional skill/Education	Hunting	Stock breeding , cultivation	Industry, intensive agriculture	Processed unhealthy foods increased	Functional foods availability increases
Demographic profile					
Mortality	Low fertility, high mortality, low life expectancy	Age of Malthus; high natural fertility, short life expectancy, high infant and maternal mortality	Mortality declines slowly, then rapidly; fertility static, then declines; small, cumulative population growth, which later explodes	Life expectancy hits unique levels (ages 60–70), huge decline and fluctuations in fertility (eg, postwar baby boom)	Life expectancy extends to ages 70 and 90 y, disability-free period increases
Age structure	Young population	Young, very few elderly	Chiefly young, shift to older population begins	Rapid decline in fertility, rapid increase in proportion of elderly person	Increases in the proportion of elderly >75 y of age
Housing	Rural, low density	Rural, a few small, crowded cities Food storage begins	Chiefly rural, move to cities increases, international migration begins, megacities develop	Dispersal of urban population decrease in rural green space	Lower-density cities rejuvenate, increase in urbanization of rural areas encircling cities
Food processing	Nonexistent	None	Storage processes (drying, salting) begin, canning and processing technologies emerge, increases in food refining and milling	Numerous food-transforming technologies	Technologies create functional foods and food constituent substitutes (i.e., macronutrient substitutes)

[1] MCH, maternal and child health, modified from Popkin 2006, References [20, 38, 42].

These diets and lifestyles adopted by *Homo economicus* will help evolve *Homo modestis*, either directly or by epigenetic inheritance or by natural selection (Fabien De Meester, 2011, personal communication).

Nutrition in Transition and Diet-Linked Vascular Diseases

In wealthy countries, there is marked reduction in deaths due to communicable diseases and emergence of morbidity and mortality due to CVDs (15-21). With increase in income, there is change in food quality towards energy rich Western type diet and lifestyle characterized with use of automobiles, lack of physical activity and increase in occupational stress, alcoholism and tobacco consumption [15-21, 30-37]. The main attributes of social classes are: general education and health education, occupation, household income, housing and availability of automobiles, television, car and other luxury items [30-37]. In developed countries, consumer durables and housing are important determinant of lifestyle: physical activity, occupational stress and social health [30-37]. These are also important determinants of food consumption patterns as well as social behaviors, which may be societal determinants of health and diseases (misuse of any substance independently can cause morbidity and mortality) [30-37]. Therefore, the pathways for development of CVDs among individuals (*Homo economicus*) in wealthy populations are sedentary behavior, excess eating of Western type foods, tobacco use, stress, and alcoholism, with underlying lack of general and health education. Nutrition and health education appears to be a most important attribute; knowledge and attitude about healthy foods and physical activity can substantially reduce CVDs and improve physical, social, mental and spiritual health leading to emergence of *Homo modestis* populations (Table 6).

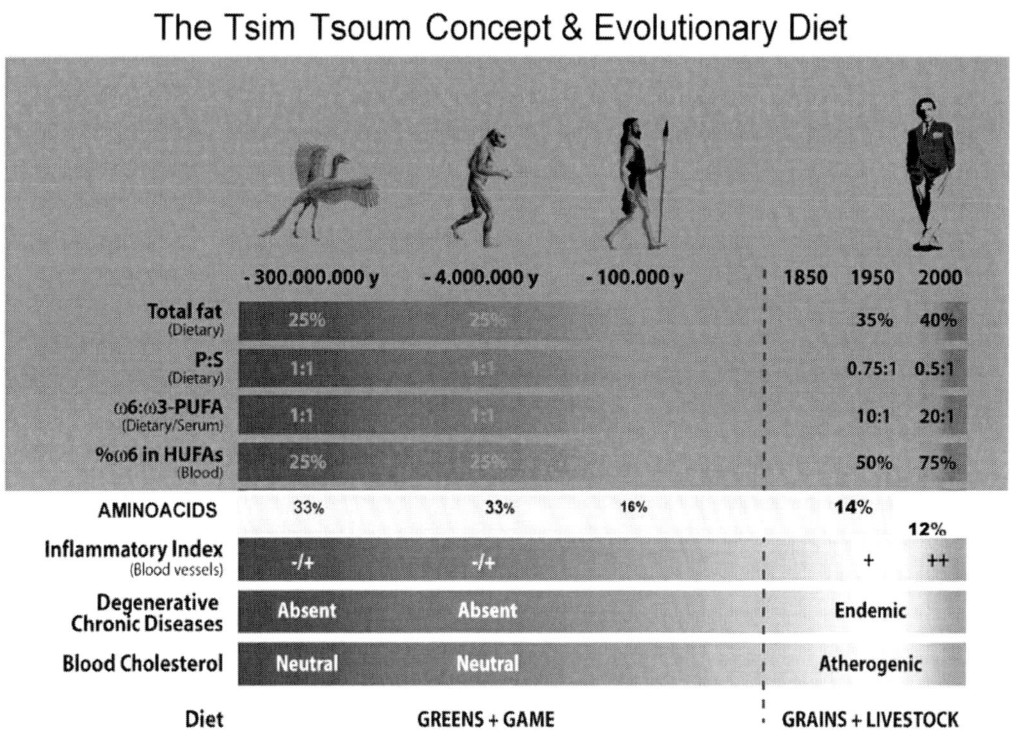

Figure 1. Nutrient intake among societies: *Homo sapiens* and *Homo economicus.*

In developing countries like India, China, Indonesia and Brazil, higher social classes, particularly with an increase in income and greater availability of ready prepared foods, have been observed to have higher risk of NCDs, CVDs including coronary artery disease (CAD), stroke, hypertension and type 2 diabetes mellitus [30-36]. However, some experts have proposed that both poverty as well as wealth may be the primary causes of disability and deaths due to NCDs among *Homo economicus* populations [16, 17, 39-41].

Globalization of Wealth but Not Health

US health spending was US$2.7 trillion in 2011, which is $8700 for every person in the country, and represents 17.9% of the economy. This is far greater than any other economically advanced country. Unfortunately, spending on health care bears little relation to good health. "Wealth but not health in US" and globalization of wealth reported that despite spending more on health care per person than other high-income affluent democratic countries (Australia, Canada, France, Italy, most of the Nordic countries, Spain, and the UK), life expectancy is shorter at birth for American men than for men in any of the other 16 countries and American women fare little better [16, 17]. Americans score lower in certain areas of health: infant mortality and low birth weight, injuries and homicides, teenage pregnancies and sexually transmitted infections, HIV/AIDS prevalence, drug-related deaths, obesity and diabetes, heart disease, chronic lung disease and disability. This disadvantage in health applies to those with health insurance, college education, higher incomes and healthy behaviors, as well as to those without. However, those Americans who reach 75 years of age live longer than their peers in other countries and Americans have low death rates from stroke and cancer. The current smoking rates are low and household income is relatively high, which should lead to future health benefits. It is not absolutely clear why Americans are at a health disadvantage compared with those in other countries. Lack of insurance or inadequate insurance restricts access to health care for many Americans and, in particular, poor access to health care and to primary care is partly to blame. Apart from the system, unhealthy behaviors abound in the USA, particularly overeating, drug abuse, and other risk-taking activities such as not wearing motorcycle helmets, drinking and driving, and using firearms. There is a great disparity in social and economic conditions and standard of education in the USA which contribute to high incomes for some, but to high poverty and income inequality for others [16, 17].

A recent study among randomly selected death records of 2,222 (1,385 men and 837 women) decedents aged 25-64 years at time of death were studied. Poverty was considered if the total family income was <US$300per month. Lack of knowledge on health education about the role of exercise, prudent diet, adverse effects of tobacco use and alcoholism was studied by validated questionnaires [33]. The findings showed that sedentary behavior, excess salt intake and other typical Western dietary habits were significantly more common among decedents belonging to higher social classes 1-3 compared to those within lower social classes 4 and 5. Lack of knowledge regarding health education was significantly more common among decedents in lower social classes who died more often due to communicable diseases. The study also revealed that deaths associated with diabetes mellitus and due to circulatory diseases were significantly more common among higher social classes 1-3 compared to lower social classes 4 and 5. However, deaths due to malignant diseases and chronic lung diseases were not associated with social class (except the social class of women with breast cancer), but total proportion of deaths due to NCDs including these causes were significantly greater among higher social classes 1-3 compared to lower social classes 4 and 5. The findings indicate that sedentary behavior, typical Western diet and excessive salt intake, in conjunction with underlying lack of health education, could well be the predisposing factors for deaths among decedents of higher social classes 1-3. Among lower social classes 4 and 5, general lack of health education may have caused more deaths due to

communicable diseases, as well as injury and accidents. Malignant diseases and chronic lung diseases were common among all social classes. It is clear that knowledge about health education on diet and lifestyle appears to be most important attribute which should be assessed to determine social class of the subjects and among all *Homo economicus* populations.

The INTERHEART study [41] among 11,119 patients with a first myocardial infarction and 13,648 age matched (up to 5 years older or younger) and sex-matched controls from 262 centers in Asia, Europe, the Middle East, Africa, Australia, and North and South America was undertaken. Subjects with myocardial infarction (cases) reported higher prevalence of all four stress factors ($p<0{\cdot}0001$). Of those subjects still working, 23.0% (n=1249) experienced several periods of work stress compared with 17.9% (1324) of controls, and 10.0% (540) experienced permanent work stress during the previous year versus 5.0% (372) of controls. After adjustment for age, sex, geographic region, and smoking, odds ratios were 1.38 (99% CI 1.19-1.61) for several periods of work stress and 2.14 (1.73-2.64) for permanent stress at work, 11.6% (1288) of cases had several periods of stress at home compared with 8.6% (1179) of controls (odds ratio 1.52 [99% CI 1.34-1.72]), and 3.5% (384) of cases reported permanent stress at home versus 1.9% (253) of controls (2.12 [1.68-2.65]). General stress (work, home, or both) was associated with an odds ratio of 1.45 (99% CI 1.30-1.61) for several periods and 2.17 (1.84-2.55) for permanent stress. Severe financial stress was more typical in cases than controls (14.6% [1622] vs. 12.2% [1659]; odds ratio 1.33 [99% CI 1.19-1.48]). Stressful life events in the past year were also more frequent in cases than controls (16.1% [1790] vs. 13.0% [1771]; 1.48 [1.33-1.64]), as was depression (24.0% [2673] vs. 17.6% [2404]; odds ratio 1.55 [1.42-1.69]). It is clear that the presence of psychosocial stressors were associated with increased risk of acute myocardial infarction, suggesting that new approaches aimed at modifying social and mental health as well as health behavior should be developed for prevention of CVDs.

In high income countries, lower social classes 3 and 4 have poor health behavior and greater risk factors of CVD and cancer mortality, as well as all-cause mortality than higher social classes [16, 17, 37-41]. The higher social classes (1 and 2) appear to have greater access to health education, spare time to increase physical activity and additional resources to maintain prudent diets than lower social classes in the developed countries [39]. This situation is in contrast with lower social class 3-5 in developing countries who are living with scarcity of foods and irregular employment. Physically demanding occupations are common in developing countries, but do not exist in developed countries, where they have only social classes 1-4 [18-21, 30-41]. It seems that in developing societies, urban populations have a double burden of diseases related to overeating as well as malnutrition because occupational physical activity decreases along with change up the ladder in social classes [30-35].

Popkin has also agreed with the above global nutrition dynamics in which the world is shifting rapidly toward a diet linked with NCDs [42] (Table 6).The role of fatty acids, essential and non-essential amino acids, antioxidants and vitamins in the prevention and pathogenesis of CVDs, type 2 diabetes and insulin resistance are well known [43-45]. Humans appear to live in a nutritional environment which completely differs from that for which our genetic constitution was selected [19-29]. The role of a low omega-6/3 (w-6/w-3) ratio Paleolithic-type of diet (by increasing w-3 and by decreasing w-6 fatty acid) can cause significant decline in cardiovascular and all causes mortality [45] (Figure 2). In this randomized controlled trial, the experimental group received significantly greater amount of fruits, vegetables and whole grains, nuts and mustard oil, and lower amount of refined bread, biscuits and sugar, butter and clarified butter compared to the control diet group at one year of follow up [45]. Total adherence score to Paleolithic-style diet and prudent diet were significant in both groups. The dietary omega-6/3 fatty acid ratio which was much higher before entry to the study (32.5±3.3) was brought down significantly in the Paleolithic-style diet group A (n = 204, compared to control group diet B (n = 202) at entry to the study (3.5±0.76 vs. 24.0±2.4 KJ/day, $p<0.001$). The fatty acid ratio remained significantly much lower in the experimental group compared to the control group after one year of follow up (4.4±0.56 vs. 22.3±2.1,KJ/day, $p<0.001$). Total mortality was 14.7% in the Paleolithic-style diet group and 25.2% in the control group after a follow up of two years. The association w-6/w-3 ratio of fatty acids with mortality showed a gradient in both groups independently, as well as among total number of deaths. A lower w-6/w-3 ratio of fatty acids from 1-10 was associated with a significantly lower mortality whereas increase in w-6/w-3 fatty acid ratio to more than 10 was associated with greater mortality (Figure 2, Tables 7, 8, 9)

Halberg's group demonstrated in 1950s that eating moderate breakfast in the morning may lead to less increase in body weight compared to same amount for dinner [46]. This observation is quite important for developing societies who had under-nutrition in 1950s after World War II, but eating a larger breakfast comprising of 1000 calories may result into circadian increase in cardiovascular events in between 6.00-1200 hours among societies with over nutrition such as in affluent societies now [47]. It would be pertinent to select functional foods for the breakfast comprising of about 500 KCal to have the beneficial effects of functional nutrients as well as of increased energy metabolism in the morning as advised by Halberg et al, causing no increase in body weight compared to same amount of energy in the evening [45-48]. Diet can also influence brain-liver and heart connection leading to NCDs [51, 52]. However, an Indo-Mediterranean diet and other nitric oxide activating foods may be protective against CVDs as well as other NCDs (Table 10) [53, 54].

Adverse effects of Tamasic foods characteristics of the Western diet were also proposed by Indian ancient physicians Charak and Sushruta in 600 BC as well as by Confucius in China (500 BC) and by the Greek physician Hippocrates (500 BC) who proposed lets "FOOD BE OUR MEDICINE" [49, 50]. Around 5000 years ago, Indians were aware of the harmful effects of dietary ingredients which are evident from ancient scripture of the Bhagwata Gita (3100 BC). Atherosclerosis has been identified in human remains from an elite Chinese burial (around 700 BC) and among Canadian Eskimos (1520-400 BC) whose diet was almost entirely meat [55]. Arterial lesions including dense calcium deposits in dozens of Egyptian mummies have been described in 1911, and Shattock noted atheromatous deposits in the aorta of King Menephtheh in 1909 [55].

We (FDM) have proposed the "Mind-Body Index or BMI" to address total health because physical, social, mental and spiritual health may depend on body composition. It is remarkable that animals including man in the wild do not suffer overweight. Even modern husbandry animals don't. In contrast, companion pets may and societal man does. The human part – the mind – appears responsible for the disease (Figure 3).

Table 7. Effect of w-3 fatty acid rich Paleolithic style diet in patients with acute myocardial infarction

	Paleolithic style diet group (n=204)		Standard diet group (n=202)	
Foods and nutrients	4-7 days	After 1 years	4-7 days	After 1 years
Fruits and vegetables (g. day-1)	508.4(28.66) *	575(91.4) **	254.4(17.2)	220.5(19.6)
Potato, radish,	60.5(6.8)	115(12.7) **	72.0(12.5)	155.6(32.5)
Legumes and pulses (g. Day-1)	80.5(6.6) **	95.0(8.9) **	52.5(4.6)	45.6(5.6)
Almonds and walnuts (g. Day-1)	82.4(5.7) **	75.5(5.2) **	-	-
Fish (g. Day-1)	52.5(6.5) **	22.4(4.1) **	20.2(3.1)	10.5(3.5)
Chicken (g. Day-1)	-	10.2(3.2)*	76.2(6.5)	66.5(10.5)
Mustard or soybean oil	18.4(3.9)*	31.5(5.5) **	10.5(2.3)	6.8(2.8)
Butter or clarified butter (g. Day-1)	2.5(0.6) **	3.3(0.71)*	10.5(2.6)	12.6(3.5)
Skim milk (ml day-1)	161.2(12.0)	152(14.5)*	150.2(8.0)	165.5(16.1)
Wheat chapatti	5.5(1.6) **	30.6(5.5)	50.6(6.6)	55.6(7.8)
Bread, biscuits (g. Day-1)	10.6(2.2)*	25.5(6.2) **	230.6(20.1)	212.2(18.1)
Rice and wheat cereals (g. Day-1)	25.6(2.4)	30.6(5.5)	30.2(3.1)	35.6(4.8)
Honey or raisins (g. Day-1)	2.6(0.8)	5.5(1.2)	-	-
Sugar (g. Day-1)	16.4(3.7)*	12.6(3.4)*	25.5(5.4)	30.5(7.6)
Total Adherence score (%)	**65.2(17.2)**	**63.9(14.8)**	**123.0(30.0)**	**71.0(30.0)**
Total foods	**1027(232)**	**1184.6(254)**	**983.4(213)**	**862(204)**

P values for mean (standard deviation) were obtained by comparison of intervention and control groups after 1 week, and after 1 year.*=P<0.05, **P<0.01, Singh et al. (45) their reference 40.

Table 8. Fatty acid consumption in the Paleolithic style diet group and standard diet group

	Before entry	At entry		After one year	
Fatty acid KJ/day	All patients (n=406)	Paleolithic (n=204)	Standard (n=202)	Paleolithic (n=204)	Standard (n-202)
Saturated	10.0 (0.39)	7.0(0.22)	10.0(0.38)	7.2(0.24)	10.8(0.36)
Monounsaturated	9.3(0.38)	9.5(0.37)	7.6(0.26)	8.0(0.35)	10.2(0.32)
Polyunsaturated	6.7 (0.30)	8.1(0.44)	6.5(0.39)	8.6(0.39)	7.0(0.26)
W-6	6.5(0.29)	6.3(0.28)	6.3(0.29)	7.0(0.36)	6.2(0.24)
W-3	0.2(0.07)	1.8(0.13)**	0.2(0.082)	1.6(0.12)**	0. 3(0.083)
W-6/W-3 ratio	32.5(3.3)**	3.5(0.76)**	31.5(2.4)	4.4(0.56)**	20.6(2.1)
Main dietary oil	Peanut	Mustard	Sunflower	Mustard	Sunflower

Values are mean± Standard deviation **=p<0.01, Singh et al. 2012 [45], their reference 40.

Table 9. Numbers and rate ratios for end points in the Paleolithic style diet group and standard diet group after 2 years of follow up

Events	Paleolithic style diet(n=204)	Control diet (n=202)	Adjusted Rate Ratio (95% confidence interval)
Total Cardiac mortality	27(13.2)**	45(22.3)	0.59(0.52-0.67)
Fatal myocardial infarction	18(8.8)**	27(13.3)	0.66(0.61—0.73)
Sudden cardiac death	9(4.4)*	18(8.9)	0.50(o.38-0.73)
Total cardiovascular mortality	30(14.7)**	50(24.7)	0.50(0.42-0.59)
Total mortality	30(14.7)**	51(25.2)	0.59(0.51-0.67)

Values are number (%), **=p<0.001, *=p<0.01, Singh et al, 2012, (45) Total deaths; adjustment made for base line age, gender, body mass index, cholesterol and blood pressure.

Table 10. Total cardiovascular events in the Paleolithic style diet and control group

Data	Indo-Mediterranean diet (n=499)	Standard Diet (n=501)
Non fatal myocardial infarction	21 (4.2%)*	43 (8.6%)
Fatal myocardial infarction	12 (2.4%)	17 (3.4%)
Sudden cardiac death	6 (1.2%)	16 (3.2%)
Total cardiac events	39 (7.8%)**	76 (15.2%)
Stroke	7 (1.4%)	13 (2.6%)
Stroke death	2 (0.4%)	3 (0.6%)
Total cardiovascular events	48 (9.6%)**	92 (18.3%)
Total deaths	24 (4.8%)	38 (8%)

Values are number (%), *=P <0.01, **= P<0.001, Singh et al, 2002 [53] their references [35].

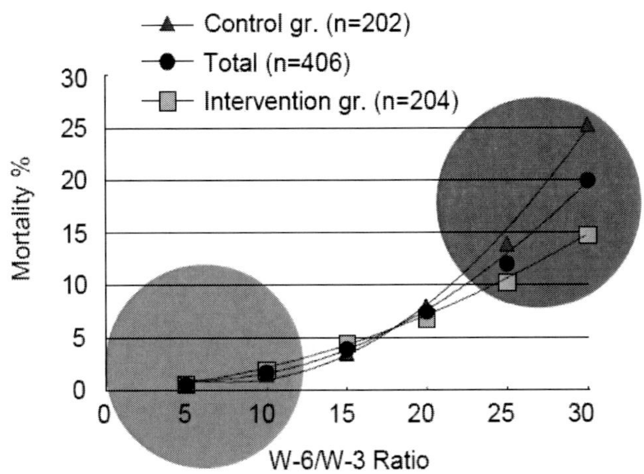

Figure 2. Effect of low w-6/w-3 fatty acid ratio diet on mortality (Reference [35], Singh et al.).

Figure 3. Effects of diet and lifestyle on body composition from *Homo erectus* to *Homo economicus,* indicating body-mind index (Fabien De Meester, 2013).

It is important to analyze facts as primary and secondary risk factors. Food is here secondary. It contributes, yet does not cause the problem. Just as cholesterol contributes, but does not cause heart disease (www.columbus-concept.com). Once understood and accepted, such basic principle allows one to take the right decision. This is an important explanation of why there is wealth, but only limited health in the United States [16].

In brief, food consumption patterns and health behavior have changed significantly in various populations during the mind transition from *Homo erectus* to *Homo economicus* and we need to have better health behaviors to develop into *Homo modestis* population. The nutritional transition from wild to modern food has been quite rapid during the last 100–150 years, causing increased intake of saturated fatty acids (SFA), trans fat, refined carbohydrates and linoleic acid, and decreased omega-3 fatty acids and flavonoids; this has resulted in marked increase in morbidity and mortality due to CVDs. Behavioral characteristics such as excess dietary intakes in conjunction with sedentary behavior appear to be the main causes of poor social, mental and spiritual health, as well as of hyperlipidemia, hyperglycemia, oxidative stress and inflammation, which are important mechanisms in the pathogenesis or prevention of diet-related CVDs [40-44]. What is needed now through education is a cultural change of families to take heed of advice on dietary intervention/prevention of CVDs and gives each generation healthier outcomes, a reduced health burden to individual and state, and even a world unification on this issue – a Human World Army as proposed by The TsimTsoum Institute at the 6th World Congress of Clinical Nutrition in Delhi in 2011, which would decide on its fate, ie what the world should do to help *Homo modestis*, at Sofia, Bulgaria on October 24-26, 2013.

Competing Interests: Conflict of interest has not been declared by the authors.

Acknowledgment

The International College of Nutrition and International College of Cardiology is thanked for supporting this study.

References

[1] Singh RB, Takahashi T, Nakaoka T, Otsuka K, Toda E, Shin HH, Kyu Lee M, Beeharry V, Hristova K, Fedacko J, Pella D, De Meester F, Wilson DW, Juneja LR. Nutrition in transition from Homo sapiens to Homo economicus. *The Open Nutra J* 2013;6: (in press).
[2] Marmot M, Allen J, Bell R, Bloomer E, Goldblatt P; Consortium for the European Review of Social determinants of Health and the Health Divide. WHO European review of social determinants of health and the health divide. *The Lancet* 2012;380:1011-1029. doi:10.1016/S0140-6736(12)61228-8.
[3] Eaton SB, Konner M. Paleolithic nutrition. A consideration of its nature and current implications. *N Engl J Med*. 1985;312(5):283–289.
[4] Eaton SB, Konner M, Shostak M. Stone agers in the fast lane: chronic degenerative diseases in evolutionary perspective. *Amer J Med* 1988. 84:739-749.
[5] Lindeberg S. Food and Western Disease: Health and Nutrition from an Evolutionary Perspective. Chichester, UK: Wiley-Blackwell; 2010.

[6] Singh RB, De Meester F, Wilczynska A. The Tsim Tsoum approaches for prevention of cardiovascular diseases. *Cardiology Research and Practice Volume*, 2010, Article ID, 824938, 18 pages,doi:10.4061/2010/824938.

[7] Eaton SB, Eaton SB III, Sinclair AJ, Cordain I, Mann NJ. Dietary intake of long chain polyunsaturated fatty acids during the Paleolithic period. In: Simopoulos AP Ed. The return of w-3 fatty acids in the food supply. Land based Animal Food Products and their Health Effects. *World Rev Nutr Dietetics* 1998; 83: 12-23.

[8] Carrera-Bastos P, Fintess-Villaba M, O'Keefe JH, Lindeberg S, Cordain L. The Western diet and lifestyle and diseases of civilization. *Res Rep Clin Cardiol* 2011;2:15-35.

[9] Singh RB, Moshiri M, De Meester F, Juneja L, Muthusamy V, Manoharan S. The evolution of low w-6/w-3 ratio dietary pattern and risk of cardiovascular diseases and diabetes. *J Altern Med Res* 2011; 3:45-57.

[10] De Meester F. Wild-type land based foods in health promotion and disease prevention: the LDL-CC:HDL-CC model. In: *Wild Type Foods in Health Promotion and Disease Prevention*, Ed. Fabien De Meester, RR Watson, Humana Press, NJ 2008, 3-20.

[11] Esposito K, Glugliano D. Diet and inflammation: a link to metabolic and cardiovascular diseases. *Eur. Heart J* 2006; 27: 15-20.

[12] Katcher HI, Legro RS, Kunselman AR et al. The effects of whole grain- enriched hypocaloric diet on cardiovascular disease risk factors in men and women with metabolic syndrome. *Am. J. Clin. Nutr.* 2008;87:79-90.

[13] Fung TT, Malik V, Rexroad KM, Manson JE, Willett WC, Hu FB. Sweetened beverage consumption and risk of coronary heart in women. *Amer J Clin Nutr* 2009;89: 1037-42.

[14] De Meester F. Progress in lipid nutrition: the Columbus concept addressing chronic diseases. *World Rev Nutr Diet* 2009; 100: 110-21.

[15] Marmot M, Allen J, Bell R, Bloomer E, Goldblatt P; Consortium for the European Review of Social determinants of Health and the Health Divide. WHO European review of social determinants of health and the health divide. *The Lancet* 2012; 380:1011-1029.

[16] Editorial. Wealth but not health in USA, Lancet 2013; 381:177, doi: 10.1016/S0140-6736(13)60069-0

[17] Singh RB, De Meester F, Wilson DW. Globalization of wealth but not health in the world. (Under consideration)

[18] Pettee KK, Ainsworth BE. The Building Healthy Lifestyles Conference: modifying lifestyles to enhance physical activity, diet, and reduce cardiovascular disease. *Am J Lifestyle Med.* 2009;3 (1 Suppl):6s–10s.

[19] Frassetto LA, Schloetter M, Mietus-Synder M, et al. Metabolic and physiologic improvements from consuming a paleolithic, hunter gatherer type diet. *Eur J Clin Nutr.* 2009;63(8):947–955.

[20] Singh RB, Reddy KK, Fedacko J, De Meester F, Wilczynska A and D.W. Wilson, Ancient concepts in nutrition and diets in hunter-gatherers. *The Open Nutra* J 2011;4:130-135.

[21] Carrera-Bastos P, Fintess-Villaba M, O'Keefe JH, Lindeberg S, Cordain L. The Western diet and lifestyle and diseases of civilization. *Res Rep Clin Cardiol* 2011;2:15-35.

[22] Pritchard JK. How we are evolving. *Sci Am.* 2010;303(4):40–47.

[23] White TD, Asfaw B, DeGusta D, et al. Pleistocene Homo sapiens from Middle Awash, Ethiopia. *Nature.* 2003;423(6941):742–747.
[24] McDougall I, Brown FH, Fleagle JG. Stratigraphic placement and age of modern humans from Kibish, Ethiopia. *Nature*. 2005;433:733–736.
[25] Manica A, Amos W, Balloux F, Hanihara T. The effect of ancient population bottlenecks on human phenotypic variation. *Nature.* 2007;448(7151):346–348.
[26] Wang ET, Kodama G, Baldi P, Moyzis RK. Global landscape of recent inferred Darwinian selection for Homo sapiens. *Proc Natl Acad Sci USA*. 2006;103(1):135–40.
[27] Voight BF, Kudaravalli S, Wen X, Pritchard JK. A map of recent positive selection in the human genome. *PLoS Biol.* 2006;4(3):e72.
[28] Pollard KS, Salama SR, Lambert N, et al. An RNA gene expressed during cortical development evolved rapidly in humans. *Nature* 2006;443(7108):167–172.
[29] Luca F, Perry GH, Di Rienzo A. Evolutionary adaptations to dietary changes. *Annu Rev Nutr.* 2010;30:291–314.
[30] Singh RB, Singh V, Kulshrestha SK, Singh S, Gupta P, Kumar R, Krishna A, Srivastav SSL, Gupta SB, Pella D, Cornelissen G. Social class and all cause mortality in an urban population of north India. *Acta Cardiol* 2005: 60;611-617.
[31] Singh RB, Fedacko J, Vargova V, Kumar A, Mohan A, Pella D, De Meester F, Wilson DW. Singh's verbal autopsy questionnaire for assessment of causes of death, social autopsy, tobacco autopsy, and dietary autopsy based on medical records and interview. *Acta Cardiol* 2011:66; 471-481.
[32] Singh RB, Ghosh S, Niaz MA, Rastogi V. Validation of physical activity and socioeconomic questionnaire in relation to food intakes for the five city study and a proposed classification for Indians. *J Assoc Phys India* 1997:45;603-607.
[33] Singh RB, Anjum B, Takahashi T, Martirosyan D, Pella D, De Meester F, Wilson DW. Poverty is the absolute cause of deaths due to noncommunicable diseases. *World Heart J* 2012: V4 (2-3) (in press).
[34] Singh RB, Beegom R, Mehta AS, Niaz MA, De AK, Mitra RK, Haque M, Verma SP, Dube GP, Siddiqui HM, Wander GS, Janus ED, Postiglion A, Haque MS. Social class, coronary risk factors and undernutrition, a double burden of diseases, in women during transition, in five Indian cities. *Int J Cardiol*, 1999:69; 139-147.
[35] Singh RB, Sharma JP, Rasogi V, Niaz MA, Ghosh S, Beegom R, Janus ED. Social class and coronary disease in a rural population of north India. *Eur Heart J* 1997:18; 588-95.
[36] Pednekar MS, Gupta R, Gupta PC. Illiteracy, low educational status and cardiovascular mortality in India. *BMC Public Health*. 2011:11; 567.
[37] Teo K, Chow CK, Vaz M, Rangarajan S, Yusuf S. The Prospective Urban Rural Epidemiology (PURE) study: examining the impact of societal influences on chronic noncommunicable diseases in low, middle, and high-income countries. *Am Heart J.* 2009:158;1–7.
[38] Gillum RF. The epidemiological evolution in pattern of cardiovascular diseases in blacks. *N Engl J Med* 1996:335:1597-99.
[39] Kesteloot H. Social class, all cause and cardiovascular mortality. *Acta Cardiol* 2004:59;117.
[40] WHO. Mortality and burden of disease estimates for WHO Member States in 2008. Geneva: *World Health Organization,* 2010.

[41] Rosengren A, Hawken S, Ounpuu S, Sliwa K, Zubaid M, Almahmeed WA, Blackett KN, Amom CS, Sato H, Ysuf S. Association of psychosocial risk factors with risk of acute myocardial infarction in 11119 cases, 13648 controls from 52 countries (the INTERHEART Study): case control study. *Lancet* 2004; 364: 937-52.

[42] Popkin BM. Global nutrition dynamics; the world is shifting rapidly toward a diet linked with noncommunicable diseases. *Am J Clin Nutr* 2006:83:289-298.

[43] Simopoulos AP. Evolutionary aspects of the dietary omega-6/omega-3 fatty acid ratio: medical implications. In Simopoulos AP, De Meester F (eds), A Balanced Omega-6/Omega-3 Fatty acid Ratio. Cholesterol and Coronary Heart Disease. *World Rev Nutr Diet,* Basel, Karger 2009;100: 1-21.

[44] Singh RB, DeMeester F, Mechirova V, Pella D, Otsuka K. Fatty acids in the causation and therapy of metabolic syndrome. In Wild type foods in health promotion and disease prevention, editors Fabien De Meester and RR Watson, *Humana Press,* NJ 2008. pp 263-284.

[45] Singh RB, Fedacko J, Vargova V, Pella D, Niaz MA, Ghosh S. Effect of Low W-6/W-3 Fatty Acid Ratio Paleolithic Style Diet in Patients with Acute Coronary Syndromes: A Randomized, Single Blind, Controlled Trial, *World Heart J* 2012;4:71-84.

[46] Cornelissen G. When you eat matters: 60 years of Franz Halberg's nutrition chronomics. *The Open Nutra J* 2012;5(Suppl) 16-44.

[47] Singh RB, Singh AK, Sharma JP, Singh RK, Kumar A, Singh G, Kartikey K, Rastogi SS, Singh S, Fedacko J, Pella D, De Meester F, Wilczynska A, Wilson DW. Nutrition in Chronocardiology: We are Indebted Professor Franz Halberg. *The Open Nutr. J* 2012; 5 (supple) 45-65.

[48] Toda E, Toru T, Singh RB, Alam SE et al. Can low w-6/w-3 ratio Paleolithic style diet stop cardiovascular diseases? Tissue is the Issue. *Amer. Med. J.* 2012; 3:183-193.

[49] Charaka Samhita (Charka, 600 BC). Varanasi, Delhi: Chaukhambha Orientalia, 1981, translated by Sharma PV. *Harvard Oriental Series*. Motilal Banarsidass, Delhi, 1984.

[50] Sushruta Samhita (Sushruta, 600 BC), vol. 1–3, Varanasi: Sanskrit Series Office, 1999, translated by Bhisagratna Chaukhambha, Varanasi 1984.

[51] Wang PY, Caspi L, Lam CK, Chari M, Li X, Light PE, Gutierrez-Juarez R, Ang M, Schwartz GJ, Lam TK. Upper intestinal lipids trigger a gut-brain-liver axis to regulate glucose production. *Nature* 2008;452: 1012-16.

[52] Singh RB, De Meester F, Wilczynska A, Wilson DW, Hungin APS. The liver-pancreas and brain connection in the pathogenesis of obesity and diabetes mellitus. *World Heart J* 2010;2:319-326.

[53] Singh RB, Dubnov G, Niaz MA,Ghosh S,Singh R,Rastogi SS,Manor O,Pella D,Berry EM. Effect of an Indo-Mediterranean diet on progression of coronary disease in high risk patients: a randomized single blind trial. *Lancet* 2002,360:1455-1461.

[54] Hristova K, Nakaoka T, Otsuka K, Fedacko J, Singh R, Singh RB, De Meester F, Wilczynska A, Wilson DW. Perspectives on Chocolate Consumption and Risk of Cardiovascular Diseases and Cognitive Function. *The Open Nutraceuticals Journal,* 2012, 5, 207-212.

[55] Heagerty AM. Scanning ancient history for evidence of modern diseases. *The Lancet, Early Online Publication,* 11 March 2013,doi:10.1016/S0140-6736(13)60639-XCite or Link Using DOI.

In: Circadian Cardiology with Focus on Both Prevention ... ISBN: 978-1-63463-956-9
Editors: K. Hristova, A. Shehab, G. Cornelissen et al. © 2015 Nova Science Publishers, Inc.

Chapter 18

Association of Cocoa Consumption and Risk of Cardiovascular Diseases and Other Chronic Diseases

Shabnam Omidvar[1], *Sanghamitra Pati*[2], *R. B. Singh*[*3], *Toru Takahashi*[4], *Hyun Ho Shin*[5], *Moon-Kyu Lee*[5], *Sung-Rae Kim*[5], *Jan Fedacko*[6], *Reema Singh*[7], *Narcis Tribulova*[8], *Krasimira Hristova*[9], *Fabien De Meester*[3], *Agnieszka Wilczynska*[3], *D. W. Wilson*[10], *Danik Martyrosyan*[11], *R. B. Singh*[12], *Rakesh Sharma*[3] *and Lekh R. Juneja*[13]

[1]Department of Food Science and Nutrition and Faculty of Nursing and Midwifery Department in Babol University of Medical Sciences, Iran
[2]Public Health Foundation of India, Bhubaneshwar, India
[3]The Tsim Tsoum Institute, Krakow, Poland
[4]Graduate School of Human Environment Science, Fukuoka Women's University, Japan
[5]Korean Society of Lipidology and Atherosclerosis, Seoul, Korea
[6]Faculty of Medicine, PJ Safaric University, Kosice, Slovakia
[7]Reema Singh's 'The Reim's Chocolates (Chocolatier), Mumbai, India
[8]Institute for Heart Research, Slovak Academy of Sciences, Slovakia
[9]University National Heart Hospital, Department of Noninvasive Functional Diagnostic and Imaging, Sofia, Bulgaria
[10]School of Medicine, Pharmacy, and Health, Durham University, Stockton-on-Tees, UK
[11]Functional Food Center Inc. Dallas, TX, US
[12]Center of Agriculture and Functional Food Security Policy, New Delhi, India
[13]Taiyo Kagaku Co. Ltd, Yokkaichi, Japan

[*] Corresponding author: Prof. Dr. R.B. Singh, MD Halberg Hospital and Research Institute, Civil Lines, Moradabad-10(UP) 244001, India. Email: rbs@tsimtsoum.net, drkk@dataone.in, Tel/Fax 00915912417437.

Abstract

Background. Antioxidant flavonoid deficiency may be risk factor for non-communicable diseases (NCDs). Further studies indicate that increased consumption of cocoa products; cocoa, and chocolates, etc. may be associated with decreased risk of NCDs; cardiovascular diseases (CVDs), coronary artery disease (CAD), stroke, hypertension, insulin resistance, memory dysfunction and cancers. In the present review, on supporting evidence for such claims.

Methods. Internet search and discussion with experts. Parts of meta analysis have been briefly and partly reproduced in the Results Section of the Abstract to maintain impact and impartiality throughout the paper.

Results. Cocoa flavonol appears to have potential beneficial effects against the risk of, metabolic syndrome, hypertension, blood lipids, stroke, coronary artery disease (CAD), cancer, cognitive function and dementia due to its antioxidant, anti-inflammatory and activation of nitric oxide effects. In a meta-analysis, epidemiological studies, comprising of 114 009 participants, The highest levels of chocolate consumption, were associated with a 37% reduction in CVD (relative risk 0.63 (95% confidence interval 0.44 to 0.90)) and a 29% reduction in stroke compared with the lowest levels. Another meta-analysis showed a significant blood pressure-reducing effect of cocoa-chocolate compared with control but it was significant only for the hypertensive or prehypertensive subgroups (SBP: -5.0 ± 3.0 mmHg; P = 0.0009; DBP: -2.7 ± 2.2 mm Hg, P = 0.01). BP was not significantly reduced in the normotensive subgroups. Clinical studies among subjects showed that cocoa intake can improve endothelial function by activation of nitric oxide. In a double-blind, parallel arm study in 90 elderly individuals with mild cognitive impairment, randomized to consume once daily for 8 weeks a drink containing ≈990 mg (high flavanols), ≈520 mg (intermediate flavanols), or ≈45 mg (low flavanols) of cocoa flavanols per day. The time required to complete Trail Making Test A and Trail Making Test B was significantly (P<0.05) lower in subjects assigned to high flavanols (38.10±10.94 and 104.10±28.73 seconds, respectively) and intermediate flavanols (40.20±11.35 and 115.97±28.35 seconds, respectively) in comparison with those assigned to low flavanols (52.60±17.97 and 139.23±43.02 seconds, respectively). Similarly, verbal fluency test score was significantly (P<0.05) better in subjects assigned to high flavanols in comparison with those assigned to low flavanols (27.50±6.75 versus 22.30±8.09 words per 60 seconds). Insulin resistance, blood pressure, and lipid peroxidation also decreased among subjects in the high-flavanol and intermediate-flavanol groups.

Conclusion. Dietary supplementation with cocoa (300-1000mg/day) in the form of dark chocolates can protect against NCDs; hypertension, stroke, atherosclerosis, insulin resistance, memory dysfunction and cancer. Further research is necessary to prove this finding.

Keywords: Flavonoids, flavonols, antioxidants, diet, nutrition, hypertension, stroke, dementia, cancer, cognitive function

Introduction

The beneficial effects of hot chocolate on health and non-communicable diseases (NCDs) are considered secondary, although these are recently reported in the world class journals. However, cocoa chocolate is consumed mainly for taste and flavor in most of the countries of the world. Recent studies indicate that it has beneficial biochemical and clinical effects as well as being an enjoyable drink as recorded by 16th century explorers and in early civilizations [1]. The scientific name of the cocoa tree is derived from Theobroma cacao, from the Greek words theo (god) and broma (drink) and therefore Incas considered it, the drink of gods [2]. In Honduras, archeologists uncovered elaborately designed bowls of this period that are believed to have been used by the Aztecs to drink liquid cocoa thousands of years ago [2-4]. The Aztec Emperor Montezuma (16th century) was a keen admirer of cocoa, calling it a divine drink, which builds up resistance and fights fatigue. A cup of this precious drink permits a man to walk for a whole day without food. In the language of the Aztecs (Nahuatl) this drink was called chocolatl ("xocolatl" comprising "xococ" (sour or bitter) and "atl" meaning water or drink from the cocoa tree. However, theobroma cacao was known much earlier, perhaps not originally consumed by the Paleo-tribal Indians gatherers of Mesoamerica as an infusate, who became hunter-gatherers. The evidence regarding cocoa consumption by Incas, Aztecs and Mayans is abundant and early consumption can be traced to the Mokaya and other pre-Olmec people, the cacao beverages dating back even earlier (1900 BC) [2-4].

Chocolate became a luxury item in the 19th century; and its consumption was considered to have adverse effects on health because it was a part of Western diet and lifestyle [1]. Western diet in conjunction with ordinary chocolate intake may be associated with dental caries, obesity, high blood pressure, and diabetes; hence many physicians currently tend to warn patients about the potential health hazards of consuming large amount of chocolate-based energy rich foods [1, 5]. The polyphenolic flavonoid constituents of cocoa was consumed as an unsweetened drink of raw, dried cocoa powder, traditionally by native Indians [1-4]. It was sweetened, after it was brought to Europe, and other processing steps became common in order to reduce bitterness and to provide European taste and flavour. The flavanol content of cocoa, the likely polyphenolic having potential antioxidant effects was reduced due to this modification. It seems that, all chocolates are not equal and the protective effect of high cocoa consumption by native Indians may not be achievable with Western chocolate products unless the products are duly fortified and more research is needed on its (plant) epidemiology and manufacturing processes and interactions with co-ingredients and subsequent consumption.

Cocoa Flavanoids and Risk Factors

The recent discovery of biologically active phenolic compounds in cocoa has changed the above perception and stimulated research on its effects on oxidative stress, blood lipids, insulin resistance, blood pressure, atherosclerosis and ageing [6-14]. Early evidence of a protective effect of cocoa was obtained in Kuna Indians, a native population living on islands off the coast of Panama. The Kuna belong to one of the few cultures that are protected against the age-dependent increase in blood pressure and the development of arterial hypertension. Cocoa flavonolappears to have potential beneficial effects against the risk of, metabolic syndrome, hypertension, blood lipids, stroke, coronary artery disease (CAD), cancer, cognitive function and dementia [6-25]. Polyphenolics including flavonoids share a common chemical structure: C6-C3-C6 (Figures 1, 2).

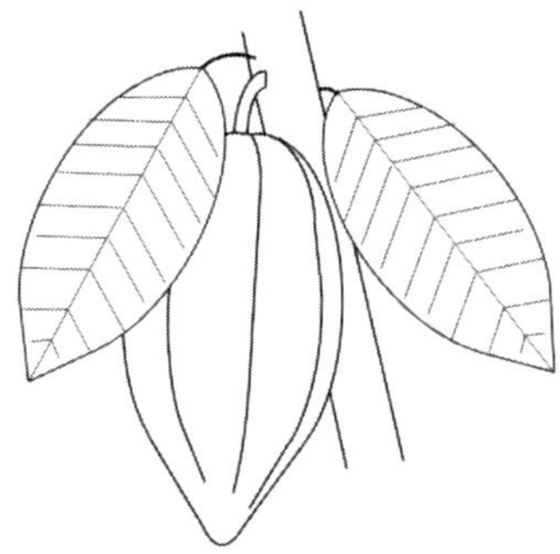

Figure 1. Cocoa fruit and leaves.

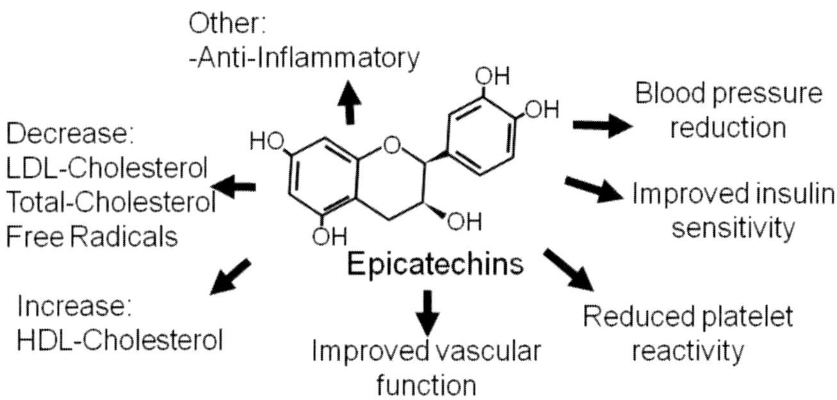

Figure 2. Effects of cocoa epicatechin on biomarkers of health and disease.

Flavonoids can be classified into: flavonols, flavones, isoflavones, flavanones, anthocyanidins, and flavanols [26-29]. Fruits and vegetables are rich sources of flavanols (also called flavan-3-ols) which are well known to provide protective effects against deaths due to NCDs (5). Major sources of flavonols are certain teas, grape juice, wine, various berries, and especially cocoa (Table 1).

Table 1. Flavonoid contents in various foods

Foods.	Flavanol content,(mg/kg or mg/L)
Beans	350-550
Appricots	100-250
Cherries	50-220
Peaches	50-140
Blackberries	130
Apples	20-120
Green tea	100-800
Black tea	60-500
Red wine	80-300
Cider	40
Chocolate	460-610

Flavanols occur as the monomers epicatechin and catechin, which are the main flavanols in fruits [26-28]. These monomers can form links between C4 and C8, allowing them to assemble as dimers, oligomers, and polymers of catechins, the so-called procyanidins [18].

Procyanidins are also known as condensed tannins, which, through the formation of complexes with salivary proteins, are responsible for the bitterness of cocoa [26].

Pharmakokinetics

The flavanol content and the total antioxidant capacity in plasma increase after oral consumption of flavonol [26, 27]. However, if cocoa is consumed with milk or if cocoa is ingested as milk chocolate, these effects appear to be markedly reduced; but the finding is controversial [29-31]. The highest plasma peak concentrations of flavanols are obtained 2 to 3 hours after ingestion in a dose-dependent manner and are still measurable after 8 hours. [29, 32, 33]. The smaller the polyphenol is, the higher the concentration in blood, so it is the molecular size of the flavonol that also matters. A single measurement of plasma levels at 2 hours may be a check for compliance [34]. There is limited usefulness of this measure in finding out the bioavailability and bioactivity of flavonols. The metabolic conversion of flavonols in the intestinal cells, liver, and other tissues; their binding to proteins; their accumulation in cells; and the urinary elimination rate are important to find out bioactivity [34, 35]. The concentration of flavanols markedly decreases, during the conventional chocolate manufacturing process from fresh cocoa seeds to the final product [36]. Fermentation and roasting may also have a detrimental impact on the final flavanol content of foods and the flavanol concentrations may depend on the agricultural origin of the raw cocoa [37]. It should be noted that milk chocolate has the lowest flavanol content compared with cocoa powder and dark chocolate [38].

Mechanisms of Action

Several potential mechanisms of actions which flavanols and cocoa might exert for their benefits on cardiovascular health have been proposed (Figure 3). Activation of nitric oxide (NO) and antioxidant, anti-inflammatory and antiplatelet effects, which in turn might improve endothelial function, lipid levels, blood pressure, insulin resistance, and eventually clinical outcome, have been established [39]. Among these beneficial effects of cocoa, improvement in endothelial function, characterized with a continuous, smooth, nonthrombogenic surface of all blood vessels that exhibit a highly selective permeability in its healthy state appears to be most important. The endothelium synthesizes and releases abroad range of vasoactive substances. The structural atherosclerotic changes may develop, long after functional impairment of the vascular endothelium in response to injury by the oxidants and toxicants which impair the release of NO, synthesized by endothelial NO synthase (eNOS) from L-arginine [40], in the presence of the cofactor the trahydrobiopterin. The circulating blood or receptor ligand substances such as acetylcholine, bradykinin, or serotonin induce shear stress responsible for the release of NO from the endothelial cells [41]. NO causes vasodilation as

Table 2. Effects of cocoa on endothelial function

Authors	Year	n	Animals/patients	Duration	Intervention	Outcome
Karim et al. [51]	2000	5	Arotic rings from rats	Immediately	Procyanidins derived from cocoa	Endothelium-derived relaxation mediated by activation of NOS
Fisher et al. [52]	2003	27	Healthy people	5 d	Flavanol-rich cocoa (821 mg/d)	Peripheral vasodilatation , improvement in vasodi- lator response to ischemia assessed by pulse-wave amplitude on the finger
Engeler et al. [53]	2004	21	Healthy subjects	2 wk	High-flavonoid chocolates (213 mg procyani dins,46 mg epicatechin) vs low-flavonoid chocolate	Improvement in flow –mediated vasodilatation of the brachial artery, increase in epicatechin concentrations
Schroeter et al. [54]	2006	16	Healthy subjects, isolated rabbit rings		Drink with high flavonoid content	Improvement in flow-mediated vasodilatation paralleled the appearance of flavanols in plasma;concentrations in plasma enough to mediate ex vivo vasodilatation;pure epicatecins mimic vascular effects of cocoa;high-flavanol diet is associated with high urinary excretion of NO metabolites
Heiss et al. [55]	2003	26	Patients with atleast 1 cardiovascular risk factor	2 h (crossover)	Flavanol-rich cocoa drink (100 ml)	Improvement in flow-mediated vasodilatation, increase in levels of nitrosated and nitrosylated species
Hermann et al. [56]	2006	20	Healthy smokers	2 h	40 g commercially available dark chocolate vs white chocolate	Increase in flow-mediated vasodilatation of the brachial artery, improvement in antioxidant status and platelet function
Grassi et al. [57]	2005	20	Patients with untreated essential hypertension	15d (crossover)	100 g dark chocolate (21.91 mg catechin,65.97 mg epicatechins) vs flavanol free white chocolate	Increase in flow-mediated vasodilatation of the brachial artery, decrese in blood pressure and LDL cholesterol, increase in insulin sensivity
Heiss et al. [58]	2005	11	Smokers	2 h (crossover)	100 ml cocoa drink with high (176-18 mg) or low (<11 mg) flavanol content	Increase in flow-mediated vasodilatation and circulating NO pool, increase in flavanol metabolities
Balzer et al. [59]	2008	41	Diabetes	4 wk	Drink with high flavanol content	Improvement in flow-mediated vasodilatation (acute and chronic)
Flammer et al. [82]	2007	22	Heart transplant recipients	2 h	40 gm commercially available dark chocolate vs flavonoid-free placebo chocolate	Induction of coronary vasodilation, improvement in coronary endothelial function and platelet function
Grassi et al. [92]	2008	19	Hypertensives, impaired glucose tolerance patients	2 wk	Flavanol-rich dark chocolate	Improvement in flow-mediated vasodilatation, insulin sensitivity, β-cell function, and blood pressure
Shina et al. [84]	2009	39	Healthy subjects	2 wk	45 gm commercially available dark chocolate vs white chocolate	Improvement in coronary circulation as measured by coronary velocity flow reserve
Grassi et al. [13]	2012	12	healthy volunteers	3 days	100 g/d dark chocolate	improved flow-mediated dilation, wave reflections, and endothelin-1

well as prevents leukocyte adhesion and migration, smooth muscle cell proliferation, and platelet adhesion and aggregation. Decrease in eNOS expression and/or NO bioavailability is associated with endothelial dysfunction and eventually atherothrombosis [42]. After diffusion from endothelial to vascular smooth muscle cells, within a few seconds it increases intracellular cGMP concentrations and, in turn, induces a relaxation of vascular smooth muscle cells [43]. Epidemiological studies indicate that forearm endothelial dysfunction is predictive of coronary vascular dysfunction and predispose future coronary events [44-46].

There is evidence that increased consumption of flavanol rich red wine, black tea and green tea can enhance endothelial function among patients with CAD [47-50]. An experimental study in rats indicated that cocoa can induce NO-dependent vasodilation [51]. Cocoa can also cause vasodilation in the finger or forearm circulation of healthy subjects [52-54] and among patients with coronary risk factors including diabetes mellitus [55-59] (Table 2). High flavanol diet on long-term consumption may cause an increased release of NO metabolites consistent with an augmented NO production or diminish its degradation [54].

Epicatechins are known to closely mimic the vascular effects of flavanol-rich cocoa in humans, indicating that these antioxidants represent the primary mediator of the beneficial effect of cocoa flavanols on vascular function [54]. In vitro experiments in cultured endothelial cells and rat aorta indicate that plant extracts rich in flavanoids enhance eNOS activity [60-62]. Further study showed that incubation of endothelial cells with flavonoid-rich red wine upregulates eNOS mRNA and protein expression, most likely via stabilization of eNOS mRNA [63]. It is known that endothelial cells can produce up to three fold more bioactive NO compared to control cells [63].

Similarly, cocoa flavanols can induce NOS in vitro and can lower vascular arginase activity in human endothelial cells in vitro, thus augmenting the local levels of L-arginine [64, 65]. Further studies indicate that epicatechin elevates nitric oxide in endothelial cells via inhibition of NADPH oxidase [66] and pure dietary flavonoids quercetin and epicatechin augment nitric oxide products and reduce endothelin-1 acutely in healthy men [67].

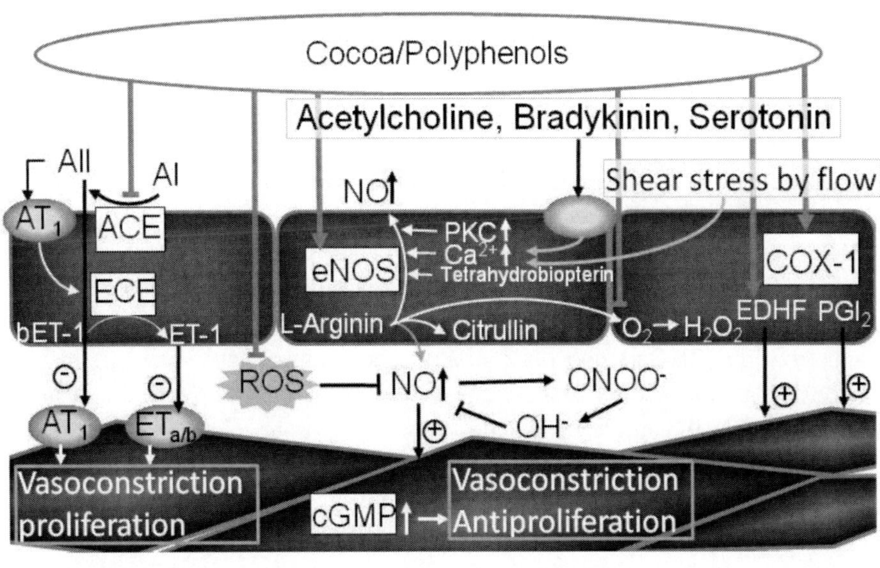

Figure 3. Mechanism of action of cocoa-epicatechin showing release of nitric oxide.

A recent study (23) indicates that low-dose epicatechin, which does not have significant antioxidant activity, may be also protective. The mechanism by which low-dose epicatechin induces this effect is unknown. It is possible that low-dose epicatechin mediates cardiac protection via opioid receptor activation. In this experimental study, C57BL/6 mice were randomly assigned to 1 of 10 groups: control, epicatechin, naloxone (nonselective opioid receptor antagonist), epicatechin + naloxone, naltrindole (δ-specific opioid receptor antagonist), epicatechin + naltrindole, norbinaltorphimine (nor-BNI, κ-specific opioid receptor antagonist), epicatechin + nor-BNI, 5-hydroxydecanoic acid [5-HD, ATP-sensitive potassium channel antagonist], and epicatechin + 5-HD. Epicatechin (1 mg/kg) or other inhibitors (5 mg/kg) were administered by oral gavage or intraperitoneal injection, respectively, daily for 10 days. In the experiment, mice were subjected to 30 min coronary artery occlusion followed by 2 h of reperfusion, and infarct size was determined via planimetry. The homogenates from whole heart were assayed for downstream opioid receptor signaling targets. There was a significant reduction in infarct size in epicatechin- and epicatechin + nor-BNI-treated mice compared with control mice. This protection was blocked by naloxone, naltrindole, and 5-HD. Epicatechin and epicatechin + nor-BNI increased the phosphorylation of Src, Akt, and IκBα, while simultaneously decreasing the expression of c-Jun NH_2-terminal kinase and caspase-activated DNase. All signaling effects are consistent with opioid receptor stimulation and subsequent cardiac protection. Naloxone, naltrindole, and 5-HD attenuated these effects. It is clear that epicatechin acts via opioid receptors and more specifically through the δ-opioid receptor to produce cardiac protection from ischemia-reperfusion injury. These beneficial effects of cocoa and dark chocolates have been observed also in the epidemiological studies [68-73].

Platelet Dysfunction

Platelet dysfunction may predispose atherothrombotic vascular disease [74]. Apart from providing antioxidant vitamins, certain fruits and vegetables and legumes may also protect against thrombosis because of their high flavanol content [5, 75]. The antiplatelet effects of cocoa have been demonstrated in many studies [76-78]. Cocoa reduces ADP/collagen activated, platelet-related primary hemostasis within hours of ingestion. The antiplatelet effects may be explained, at least in part, by a reduction in the ADP-induced expression of the activated conformation of glycoprotein IIb/IIIa surface proteins. In vitro experiment indicate that catechin and epicatechin reduce glycoprotein IIb/IIIa expression, thereby exerting antiplatelet effects [77]. In a clinical study among healthy volunteers, 100 g dark chocolate intake reduced platelet aggregation [79]. This effect was not observed among subjects eating white chocolate or milk chocolate [79]. Cocoa decreases platelet aggregation as well as platelet adhesion [79-82]. Dark chocolate intake among young healthy smokers may decrease platelet adhesion as assessed by a shear stress–dependent platelet test [82]. Stearic acid, a saturated fat found in Indian clarified butter as well as in chocolate, reduces mean platelet volume, an index of platelet activation, in humans [80, 81].

Epidemiological Studies

Epidemiological studies of human consumption, in particular observations in native Kuna Indians living on Central American islands, suggest that cocoa-rich products reduce cardiovascular mortality [5, 68-70]. The Kunas consume enormous amounts of cocoa daily which may be rich in salt similar to India rural populations with enormous physically demanding occupations. Clinical studies revealed that the Kunas indeed have lower blood pressure values and no age-dependent decline in kidney function [69, 70]. Randomized controlled trials and cohort studies further confirm that polyphenolic antioxidant flavanoid consumption can have beneficial effects on morbidity and mortality due to NCDs; CAD, stroke, hypertension, metabolic syndrome and degenerative diseases of the brain. It is believed that the high content of polyphenolic antioxidants in some fruit, vegetables, whole grains, nuts, dark chocolate and tea may contribute to their cardioprotective effects. In a recent study [5], randomly selected records of death of 2222 (1385 men and 837 women) decedents, aged 25-64years, were examined. Clinical data, dietary intakes and causes of death were assessed using data from questionnaires. The score for intake of prudent foods; fruits, vegetables, whole grains, nuts, chocolates was significantly greater and the ratio of w-6/w-3 fatty acids of the diet significantly lower for deaths due to 'injury' and accidental causes compared to deaths due to NCD. Multivariate logistic regression analysis revealed that after adjustment for age, total prudent foods(OR, CI: 1.11;1.06-1.18 men; 1.09;1.04-1.16 women) as well as fruits, vegetables, legumes and nuts (1.07; 1.02-1.12men; 1.05; 1.99-1.11 women) were independently, inversely associated whereas Western type foods (1.02; 0.95-1.09men; 1.00; 0.94-1.06 women); meat and eggs(1.00-0.94-1.06 men; 0.98; 0.93-1.04 women) and refined carbohydrates(0.98; 0.91-1.05 men, 0.95; 0.89-1.02 women) and high w-6/w-3 ratio of fatty acids were positively associated with deaths due to NCDs. The findings indicated that polyphenolic flavanoids and w-3 fatty acids may be protective against deaths due to NCDs. The Zutphen Elderly Study [15] provided important further data. in a cross-sectional analysis, that cocoa intake was inversely related to blood pressure. In a prospective analysis, the study showed that intake of cocoa was associated with a reduction of cardiovascular and all-cause mortality [15]. This study, involving 470 elderly men free of chronic disease, highlighted the protective effects of cocoa intake. After adjustment for age, body mass index, lifestyle factors, drug use, food, and caloric intake, the risk of cardiovascular mortality for men in the highest tertile of cocoa intake was reduced by 50% compared with the lowest tertile (P=0.004). The adjusted relative risk for all-cause mortality was 0.53(95% CI, 0.39 to 0.72; P=0.001). In the native population, mortality resulting from cardiovascular events is significantly lower compared with other Pan-American citizens (9.2-3.1 versus 83.4-0.7 age-adjusted deaths per 100,000) [71]. It is possible that the factors involved are environmental rather than genetic because this protection is lost on migration to urban Panama City, where other cocoa products food with a lower flavanol content are used [69]. In the Iowa Women's Health Study [72], among 34,489 postmenopausal women who were free of CVDs were examined. After 16 years of follow up, foods rich in flavonoids were associated with a decreased risk of death caused by CAD. There was a significant inverse association between chocolate consumption and CVDs mortality after multivariate adjustment.

Cocoa is highly rich in flavonoids; catechin, epicatechin, and procyanidins which predominate in antioxidant activity [73]. The tricyclic structure of the flavonoids determines antioxidant effects that scavenge reactive oxygen species, chelate Fe2+ and Cu+, inhibit enzymes, and upregulate antioxidant defenses. The epicatechin content of cocoa upregulates nitric oxide production which is primarily responsible for its favorable impact on vascular endothelium. Other beneficial effects are mediated through anti-inflammatory effects of cocoa polyphenols, and modulated through the activity of NF-κB.

Cocoa can stimulate changes in redox-sensitive signaling pathways involved in gene expression and the immune response. It may also protect neurons from injury and inflammation as well as protect the skin from oxidative damage from UV radiation in topical preparations. Cocoa consumption have beneficial effects on satiety, cognitive function, and mood and regular intake has been found toimprove cognitive function in elderly subjects with mild cognitive impairment, a condition thought to be associated with (micro-)vascular dysfunction [1, 24].

Insulin Resistance, Diabetes Mellitus and Vascular Disease

The antioxidant activity of cocoa may directly influence insulin resistance by protecting beta cells of pancreas against free radicals resulting in increased insulin release. Decrease in insulin resistance in turn, reduces the risk for type 2 diabetes mellitus. However, as cocoa is predominantly consumed as energy-dense chocolate, potential detrimental effects of overconsumption exist, including increased risk of weight gain. Overall, research to date suggests that the benefits of moderate dark chocolate consumption or cocoa intake is likely to outweigh the risks of increased energy intake and sugar induced hyperglycemia, oxidative stress, hypertriglyceridemia and free fatty acids. It seems that chocolate's potential benefits should be explored in outcome studies looking at these biomarkers and NCDs; diabetes mellitus, dementia, heart attack, stroke, cancer and death [2-9].

The overall prognosis among patients with diabetes mellitus is unfavorable due to deteriorated cardiovascular function, despite optimal control of hyperglycemia by optimal drug therapy [59]. Epidemiological data indicate that diets rich in flavanols are associated with a reduced cardiovascular risk. This study examined the feasibility and efficacy of a dietary intervention based on daily intake of flavanol-containing cocoa for improving vascular function of patients with type 2 diabetes patients on optimal drug therapy [59]. The study included 10 diabetic patients, in which vascular function were assessed as flow-mediated dilation (FMD) of the brachial artery, plasma levels of flavanol metabolites, and tolerability after an acute, single-dose ingestion of cocoa, containing increasing concentrations of flavanols (75, 371, and 963 mg). In a subsequent efficacy study, changes in vascular function in 41 medicated diabetic patients were assessed after a 30-day, thrice-daily dietary intervention with either flavanol-rich cocoa (321 mg flavanols per dose) or a nutrient-matched control (25 mg flavanols per dose). Both studies were undertaken in a randomized, double-blind fashion. Primary and secondary outcome measures included changes in FMD and plasma flavanol metabolites, respectively. A single ingestion of flavanol-containing cocoa was dose-dependently associated with significant acute increases in circulating flavanols and FMD (at 2 h: from 3.7 - 0.2% to 5.5 -0.4%, $p = 0.001$). A 30-day, thrice-daily consumption of flavanol-containing cocoa increased baseline FMD by 30% ($p = 0.0001$), while acute increases of FMD upon ingestion of flavanol-containing cocoa continued to be manifest throughout the study. Treatment was well tolerated without evidence of tachyphylaxia. Endothelium-independent responses, blood pressure, heart rate, and glycemic control were unaffected.

The effect of flavanol-rich dark chocolate on endothelial function, insulin sensitivity, b-cell function, and blood pressure (BP) were examined in hypertensive patients with impaired glucose tolerance (IGT) [56, 57]. After a run-in phase, 19 hypertensives with IGT (11 males, 8 females; 44.8 +/-8.0 y) were randomized to receive isocalorically either dark or flavanol-free white chocolate at 100 g/d for 15 d. After a washout period, patients were switched to the other treatment. Clinical and 24-h ambulatory BP was determined by sphygmometry and oscillometry, respectively, flow-mediated dilation (FMD), oral glucose tolerance test, serum cholesterol and C-reactive protein, and plasma homocysteine were evaluated after each treatment phase. Dark chocolate but not white chocolate ingestion decreased insulin resistance (homeostasis model assessment of insulin resistance; P=0.0001) and increased insulin sensitivity (quantitative insulin sensitivity check index, insulin sensitivity index (ISI), ISI(0); P <0.05) and beta cell function (corrected insulin response CIR(120); P =0.035). Systolic (S) and diastolic (D) BP decreased (P =0.0001) after dark chocolate (SBP, -3.82 +/-2.40 mm Hg; DBP, -3.92 +/-1.98 mm Hg; 24-h SBP, -4.52 +/-3.94 mm Hg; 24-h DBP, -4.17 +/-3.29mm Hg) but not after white chocolate. Dark chocolate also increased FMD (P =0.0001) and decreased total cholesterol (-6.5%; P <0.0001), and LDL cholesterol (-7.5%; P <0.0001). Changes in insulin sensitivity (Delta ISI -Delta FMD: r =0.510, P =0.001;Delta QUICK1 -Delta FMD: r =0.502, P =0.001) and beta-cell function (D CIR(120) -Delta FMD: r =0.400, P =0.012) were directlycorrelated with increases in FMD and inversely correlated with decreases in BP (Delta ISI -Delta 24-h SBP: r =-0.368, P =0.022; Delta ISI -Delta 24-h DBP: r = -0.384, P = 0.017). Thus, dark chocolate ameliorated insulin sensitivity and beta-cell function, decreased blood pressures, and increased FMD in IGT hypertensive patients. These findings indicate that flavanol-rich, low-energy cocoa food products may have a positive impact on CVD risk factors. Another study aimed to test the hypothesis that consumption of cocoa may simultaneously lower blood pressure, improve endothelial dysfunction, and ameliorate insulin resistance in subjects with essential hypertension [58]. This randomized, placebo-controlled, double blind, crossover trial examined the effects of a flavanol-rich cocoa drink (150 mL twice a day, 900 mg flavanols/d) in 20 individuals with essential hypertension. Antihypertensive medications were discontinued before study enrollment. After a 7-d cocoa-free run-in period, cocoa orflavanol-poor placebo (28 mg flavanols/d)treatment for 2 wks. was followed by a 1-wk washout and then crossover to the other treatmentarm. Blood pressure was measured thrice weekly. At baseline and after each treatment period, those authors [58] assessed insulin sensitivity (hyperinsulinemic-isoglycemic glucose clamp) and insulin stimulated changes in brachial artery diameter and forearm skeletal muscle capillary recruitment (Doppler ultrasound with or without microbubble contrast). Cocoa treatment for 2 wks increased insulin-stimulated changes in brachial artery diameter when compared with placebo[median percentage increase from baseline (25th–75th percentile):8.3 (4.2–11.3) compared with 5.9 (-0.3 to 9.6); *P* =0.04]. Nevertheless, cocoa treatment did not significantly reduce blood pressure or improve insulin resistance and had no significant effects on skeletal muscle capillary recruitment, circulating plasma concentrations of adipocytokines, or endothelial adhesion molecules. Daily consumption of flavanol-rich cocoa for 2 wkswas not sufficient to reduce blood pressure or improve insulin resistancein human subjects with essential hypertension. It is clear that there is a controversy on the role of cocoa flavanol in the treatment of hypertension and insulin resistance. Therefore, meta-analysis of studies is conducted to find out the results of epidemiological studies and controlled trials.

In a recent meta-analysis published in 2011 [12], in which the association between chocolate consumption and the risk of outcomes related to cardiometabolic disorders were reported, the primary outcomes were CAD, stroke, type 2 diabetes, and metabolic syndrome. This meta-analysis assessed the risk of developing these disorders by comparing the highest and lowest level of chocolate consumption. Only seven studies met the inclusion criteria comprising of 114,009 participants. None of the studies was a randomised trial, six were cohort studies, and one a cross sectional study. Large variation was observed between these seven studies for measurement of chocolate consumption, methods, and outcomes evaluated. Five of the seven studies reported a beneficial association between higher levels of chocolate consumption and the risk of cardiometabolic disorders. The highest levels of chocolate consumption were associated with a 37% reduction in cardiovascular disease (relative risk 0.63 (95% confidence interval 0.44 to 0.90)) and a 29% reduction in stroke compared with the lowest levels (Figure 4).

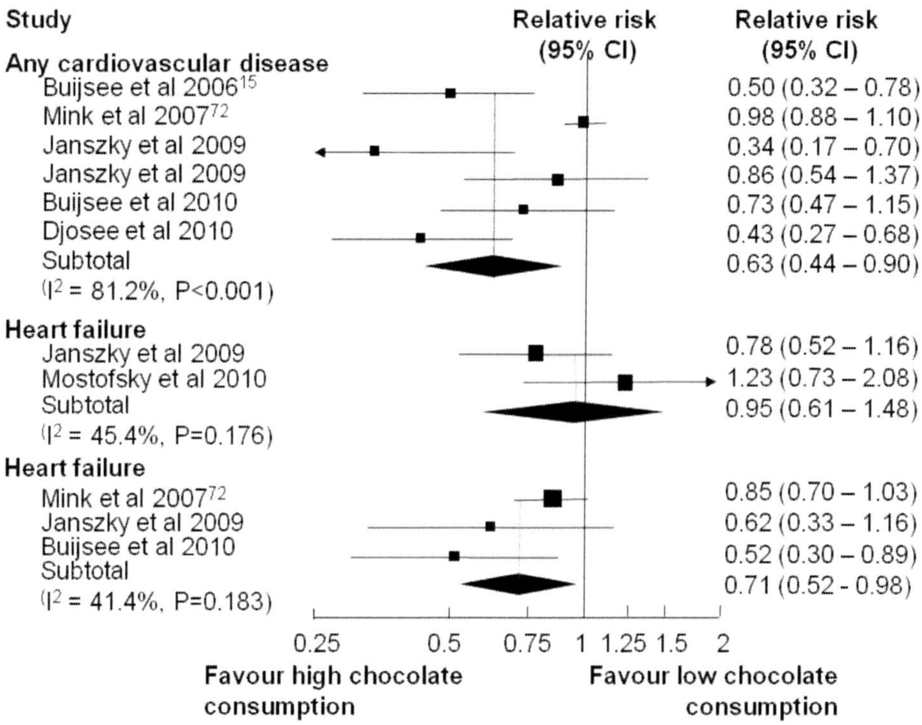

Figure 4. Relative risks for cardiovascular disease, heart failure, and stroke in adults with higher levels of chocolate consumption compared with lower levels. Buitrago-Lpez et al., BMJ 2011. Ref [12].

It is possible that chocolate consumption may be associated with a substantial reduction in the risk of cardiometabolic disorders. A more recentmeta-analysis published in 2012, included 42 acute or short-term chronic (≤18 wks.) randomized, controlled trials that comprised 1297 participants [7]. Insulin resistance (HOMA-IR: -0.67; 95% CI: -0.98, -0.36) was improved by chocolate or cocoa due to significant reductions in serum insulin. Flow-mediated dilatation (FMD) improved after chronic (1.34%; 95% CI: 1.00%, 1.68%) and acute (3.19%; 95% CI: 2.04%, 4.33%) intakes. Effects on HOMA-IR and FMD remained stable to sensitivity analyses. The investigators observed reductions in diastolic blood pressure (BP; -1.60 mm Hg; 95% CI: -2.77, -0.43 mm Hg) and mean arterial pressure (-1.64 mm Hg; 95% CI: -3.27, -0.01 mm Hg) and marginally significant effects on LDL (-0.07 mmol/L; 95% CI: -0.13, 0.00 mmol/L) and HDL (0.03 mmol/L; 95% CI: 0.00, 0.06 mmol/L) cholesterol. Chocolate or cocoa improved FMD regardless of the dose consumed, whereas doses >50 mg epicatechin/d resulted in greater effects on systolic and diastolic BP. GRADE (Grading of Recommendations, Assessment, Development and Evaluation, a tool to assess quality of evidence and strength of recommendations) suggested low- to moderate-quality evidence of beneficial effects, with no suggestion of negative effects. The strength of evidence was lowered due to unclear reporting for allocation concealment, dropouts, missing data on outcomes, and heterogeneity in biomarker results in some studies. There was a consistent acute and chronic benefit of chocolate or cocoa on FMD and previously unreported promising effects on insulin and HOMA-IR.

Hypertension

There is a controversy regarding the role of dark chocolate in reducing blood pressures. As mentioned earlier, one study [57] noted a decrease in daytime and night time blood pressures, as assessed by ambulatory 24-hour measurements, after intake of 100 g flavonoid-rich dark chocolate daily for 2 weeks. In this study, systolic blood pressure decreased after consumption of dark chocolate by 12 mm Hg, whereas white chocolate had no effect. Another study reported reductions in systolic and diastolic blood pressures in hypertensive elderly subjects [16]. Earlier meta-analyses concluded that cocoa-rich foods may reduce blood pressure. Recently, several additional trials have been conducted with conflicting results. Therefore, a random effects meta-analysis of all studies was necessary, fitting the inclusion criteria, as well as subgroup analysis by baseline blood pressure among patients with hypertension and normotension [17]. In this meta-analysis, current evidence on the effect of flavanol-rich cocoa products on blood pressure in hypertensive and normotensive individuals was reported [17]. Only randomized controlled trials investigating the effect of cocoa as food or drink compared with placebo on systolic and diastolic blood pressure (SBP/DBP) for a minimum duration of 2 wks were included in this study. Only fifteen trial arms of 13 assessed studies met the inclusion criteria. Nine trials used chocolate containing 50% to 70% cocoa compared with white chocolate or other cocoa-free controls, while six trials compared high- with low-flavanol cocoa products. Daily flavanol dosages ranged from 30 mg to 1000 mg in the active treatment groups, and interventions ran for 2 to 18 wks. Pooled meta-analysis of all trials revealed a significant blood pressure-reducing effect of cocoa-chocolate compared with control (mean BP change ±SE: Systolic BP: -3.2±1.9 mmHg, P = 0.001; Diastolic BP: -2.0 ± 1.3 mmHg, P = 0.003). However, subgroup meta-analysis was significant only for the hypertensive or pre-hypertensive subgroups (SBP: -5.0 ±3.0 mmHg; P = 0.0009; DBP: -2.7±2.2 mm Hg, P = 0.01), while BP was not significantly reduced in the normotensive subgroups (SBP: -1.6± 2.3 mmHg, P = 0.17; DBP: -1.3± 1.6 mmHg, P = 0.12). It is possible that dark chocolate is superior to placebo in reducing systolic hypertension or diastolic prehypertension. The flavanol-rich chocolate did not significantly reduce mean blood pressure below 140 mmHg systolic or 80 mmHg diastolic which indicated that it could be a supplement to other treatments of hypertension. Another meta-analysis of randomized controlled trials assessing the antihypertensive effects of flavanol-rich cocoa products was published in the same year in 2010 [18]. The primary outcome measure was the change in systolic and diastolic BP between intervention and control groups. In total, 10 randomized controlled trials comprising 297 individuals were included in the analysis. The populations studied were either healthy normotensive adults or patients with prehypertension/stage 1 hypertension. There may be significant statistical heterogeneity across studies, and questions such as the most appropriate dose and the long-term side effect profile warrant further investigation.

Obesity

The commercially available chocolates in the market have high fat content which may enhance dietary energy intake leading to obesity and dyslipidemia [21, 22]. The high caloric load of such chocolate (about 500 kcal/100 g) may induce weight gain, a risk factor for hypertension, stroke, dyslipidemia, CAD and diabetes. However, a study in 49 healthy women showed no weight gain after daily consumption of 41 g chocolate, 60 g almonds, or almonds and chocolate together for 6 weeks [21]. It is possible that weight gain may occur only with higher amounts of daily chocolate intake with very high fat content. In this study, the primary objective was to identify potentially synergistic or additive effects of combining consumption of dark chocolate with almonds as part of a low-fat diet on serum lipids and pro-inflammatory biomarkers of obesity and atherosclerosis. The biomarkers were; intercellular adhesion molecule (ICAM), vascular adhesion molecule, and high-sensitivity C-reactive protein. The duration of the study was 6 wks among 49 women with normal serum cholesterol, and a 4-armed parallel design was used. Women were randomized to 1 of 3 treatments: chocolate (41 g/d), almonds (60 g/d), chocolate and almonds, or control (no chocolate or almonds). All women improved dietary intakes in accordance with guidelines, and no subjects gained or lost weight. Serum cholesterol concentrations showed no changes after 6 weeks; however, triacylglycerol levels were reduced by approximately 21%, 13%, 19%, and 11% ($P < .05$), in the chocolate, almond, chocolate and almond, and control groups, respectively. Circulating ICAM levels decreased significantly by 10% in the treatment group consuming chocolate only ($P = .027$). Vascular adhesion molecule and high-sensitivity C-reactive protein levels showed no abnormality in any treatment group. It seems, that the consumption of chocolate and almonds as part of the Therapeutic Lifestyle Changes diet for 6 weeks showed no harmful effects in healthy women; all dietary modifications improved serum triacylglycerol levels, and consumption of chocolate, reduced the levels of circulating ICAM.

Blood Lipids

All chocolates contain saturated fat and are a major source of dietary calories. Cocoa butter is derived from cocoa plants and found predominantly in dark chocolate which contains an average of 33%monounsaturated oleic acid and 33% stearic acid. In general, plant stearic acid neither lowers high-density lipoprotein (HDL) nor increases LDL or total cholesterol but can increase platelet volume and function. As mentioned in the above studies [21, 22], it is important to determine whether consumption of dark chocolate adversely affects the blood lipid profile. Tokede et al. (6) reported that cocoa products, which are rich sources of flavonoids, can reduce blood lipids and the risk of CVDs. The primary effect measure was the difference in means of the final measurements between the intervention and control groups. In all, 10 clinical trials consisting of 320 participants with treatment duration of 2-12 weeks were included in the analysis. Intervention with dark chocolate/cocoa products significantly reduced serum low-density lipoprotein (LDL) and total cholesterol (TC) levels (differences in means (95% CI) were -5.90 mg/dl (-10.47, -1.32 mg/dl) and -6.23 mg/dl (-11.60, -0.85 mg/dl), respectively). No statistically significant effects were observed for high-density lipoprotein (HDL) (difference in means (95% CI): -0.76 mg/dl (-3.02 to 1.51mg/dl)) and triglyceride (TG) (-5.06 mg/dl (-13.45 to 3.32 mg/dl)). The findings suggest that cocoa and its products may have beneficial effects on total and LDL cholesterol and no major effects on HDL and TG in short-term intervention trials. However one earlier study found that dark chocolate consumption increases HDL cholesterol concentration and chocolate fatty acids may inhibit lipid peroxidation in healthy humans [19].

Stroke

Recent studies indicate that cerebral ischemia is a common, morbid condition accompanied by cognitive decline leading to stroke [14-18, 20]. Studies on the vascular health benefits of flavanol-containing foods signify a promising approach to the treatment of cerebral ischemia by improving endothelial function [13, 14]. This study was designed to investigate the effects offlavanol-rich cocoa consumption on cerebral blood flow in older healthy volunteers [14]. In this study, Transcranial Doppler (TCD) ultrasound was used to measure mean blood flow velocity (MFV) in the middle cerebral artery (MCA) in thirty-four healthy elderly volunteers (72 to 76 years) in response to the regular intake of cocoa or flavanol-poor cocoa. In response to two weeks of cocoa intake, MFV increased by 8% and 4% at one week (p =0.01) and 10% and 4% (p =0.04) at two weeks. In response to one week of cocoa, significantly more subjects in the high cocoa group, as compared with the low cocoa group had an increase in their MFV (p =0.05). It seems that dietary intake of high cocoa is associated with a significant increase in cerebral blood flow velocity in the MCA as measured by TCD. The data of this study and other studies suggest a promising role for regular cocoa flavanol consumption in the treatment of cerebrovascular ischemic syndromes and stroke including dementias and cognitive function [13-18].

Larsson et al. [8] investigated the association between chocolate consumption and risk of stroke in men and conduct a meta-analysis to summarize. They 'prospectively followed a cohort of 37,103 Swedish Men. A food frequency questionnaire was used to assess baseline chocolate consumption. The authors reported 1,995 incident stroke cases, including 1,511 cerebral infarctions, 321 hemorrhagic strokes, and 163 unspecified strokes during 10.2 years of follow up. A lower risk of stroke was reported in association with greater intake of chocolate. The association did not differ by stroke subtypes. In a meta-analysis of 5 studies, with a total of 4,260 stroke cases, the overall relative risk of stroke for the highest vs. lowest category of chocolate consumption was 0.81 (95% CI 0.73-0.90), without heterogeneity among studies (p = 0.47). The multivariable relative risk of stroke comparing the highest quartile of chocolate consumption (median 62.9 g/week) with the lowest quartile (median 0 g/week) was 0.83 (95 % CI 0.70-0.99).

It seems that moderate chocolate consumption may lower the risk of stroke, but more studies are needed to reduce fat and sugar content of chocolates to improve the efficacy and long term safety in the prevention of obesity and metabolic syndrome [8-11].

Endothelial Function

Endothelial function is an important determinant of hypertension, stroke and atherothrombosis. There is evidence that endothelial function is impaired during hyperglycemia which may be due to increase in reactive oxygen species, free fatty acids and decreased nitric oxide release [7, 13, 56-59]. The effects of pretreatment with dark chocolate on endothelial function and other vascular responses to hyperglycemia need further studies (Table 2). It is known that cocoa rich chocolate increases flow-mediated dilation in healthy and hypertensive subjects with and without glucose intolerance. Grassi et al. 2008 [92] sought to test the effect of flavanol-rich dark chocolate (FRDC) on endothelial function, insulin sensitivity, beta-cell function, and blood pressure (BP) in hypertensive patients with impaired glucose tolerance (IGT). After a run-in phase, 19 hypertensives with IGT (11 males, 8 females; 44.8 +/- 8.0 y) were randomized to receive isocalorically either FRDC or flavanol-free white chocolate (FFWC) at 100 g/d for 15 d. After a wash-out period, patients were switched to the other treatment. FRDC but not FFWC ingestion decreased insulin resistance (homeostasis model assessment of insulin resistance; P < 0.0001) and increased insulin sensitivity (quantitative insulin sensitivity check index, insulin sensitivity index (ISI), ISI(0); P < 0.05) and beta-cell function (corrected insulin response CIR(120); P = 0.035). Systolic (S) and diastolic (D) BP decreased (P < 0.0001) after FRDC (SBP, -3.82 +/- 2.40 mm Hg; DBP, -3.92 +/- 1.98 mm Hg; 24-h SBP, -4.52 +/- 3.94 mm Hg; 24-h DBP, -4.17 +/- 3.29 mm Hg) but not after FFWC. Further, FRDC increased FMD (P < 0.0001) and decreased total cholesterol (-6.5%; P < 0.0001), and LDL cholesterol (-7.5%; P < 0.0001). Changes in insulin sensitivity (Delta ISI - Delta FMD: r = 0.510, P = 0.001; Delta QUICKI - Delta FMD: r = 0.502, P = 0.001) and beta-cell function (Delta CIR(120) - Delta FMD: r = 0.400, P = 0.012) were directly correlated with increases in FMD and inversely correlated with decreases in BP (Delta ISI - Delta 24-h SBP: r = -0.368, P = 0.022; Delta ISI - Delta 24-h DBP r = -0.384, P = 0.017). Thus, FRDC ameliorated insulin sensitivity and beta-cell function, decreased BP, and increased FMD in IGT hypertensive patients. These findings suggest flavanol-rich, low-energy cocoa food products may have a positive impact on endothelial function and other CVD risk factors. Shiina et al. 2009 [84] assessed the effects of the oral intake of flavonoid-rich dark chocolate on coronary circulation, by measuring coronary flow velocity reserve

(CFVR) by noninvasive transthoracic Doppler echocardiography (TTDE) in healthy adult subjects in a randomized, single-blind design. After a follow up for 2 weeks in 39 healthy men (mean age 29.7+/-3.9 years, range 23-40 years). Subjects were randomly assigned a daily intake of either flavonoid-rich dark chocolate 45 g, including cacao polyphenol 550 mg/day, 200 kcal) or non-flavonoid white chocolate 35 g, including cacao polyphenol 0 mg/day, 140 kcal) as a control. After a follow up of 2 weeks, flavonoid-rich dark chocolate consumption significantly improved CFVR (3.38+/-0.49 before intake, 4.28+/-0.85 after intake; p<0.01), whereas non-flavonoid white chocolate consumption did not (3.28+/-0.49 before intake, 3.16+/-0.49 after intake; p=0.44). All predictor variables were used as dependent variables in a multiple regression model of the incremental change in CFVR after 2 weeks of chocolate intake. Intake of dark (but not white) chocolate, MDA-LDL, triglyceride (TG) and heart rate (HR) significantly influenced the change of CFVR after 2 weeks of intake (p<0.01) according to the multiple regression formula: Y=1.01X(1)-0.005X(2)-0.003X(3)-0.017X4 (Y=change in CFVR after 2 weeks of chocolate intake, X1=intake of dark (but not white) chocolate, X2=MDA-LDL, X3=TG, X4=HR). In a more recent study, the effects of flavanol-rich dark chocolate administration on flow-mediated dilation and wave reflections as well as on blood pressure, endothelin-1 and oxidative stress, before and after oral glucose tolerance test (OGTT) were examined [13]. The volunteers (n=12) (5 males, aged 28.2-2.7 years) randomly received either 100 g/d dark chocolate or flavanol-free white chocolate for 3 days. After 7 days washout period, volunteers were switched to the other treatment. After each treatment phase and OGTT, flow-mediated dilation, stiffness index, reflection index, peak-to–peak time, blood pressure, endothelin-1 and 8-iso-PGF2were evaluated. Compared with white chocolate, dark chocolate ingestion improved flow-mediated dilation (P=0.03), wave reflections, endothelin-1 and 8-iso-PGF2 (P=0.05). After white chocolate ingestion, flow-mediated dilation was reduced after OGTT from 7.88-0.68 to 6.07-0.76(P=0.027), 6.74-0.51 (P=0.046) at 1 and 2 h after the glucose load, respectively. Similarly, after white chocolate but not after dark chocolate, wave reflections, blood pressure, and endothelin-1 and 8-iso-PGF2- increased after OGTT. There is acute, transient impairment of endothelial function and oxidative stress, due to OGTT, which may be suppressed by flavanol-rich dark chocolate. The findings indicate that cocoa flavanols may improve vascular health by reducing the hyperglycemia induced decrease in arterial function associated with the mechanism of atherothrombosis' and cerebral blood flow response among healthy elderly humans [13]. Hooper et al. 2012, [7] review the effects of chocolate, cocoa, and flavan-3-ols on major CVD risk factors by studying the databases for randomized controlled trials (RCTs) of chocolate, cocoa, or flavan-3-ols. The meta analysis included 42 acute or short-term chronic (≤18 wk) trials that comprised 1297 participants. Insulin resistance (HOMA-IR: -0.67; 95% CI: -0.98, -0.36) was improved by chocolate or cocoa due to significant reductions in serum insulin. Flow-mediated dilatation (FMD) improved after chronic (1.34%; 95% CI: 1.00%, 1.68%) and acute (3.19%; 95% CI: 2.04%, 4.33%) intakes. Effects on HOMA-IR and FMD remained stable to sensitivity analyses. Reductions in diastolic blood pressure (BP; -1.60 mm Hg; 95% CI: -2.77, -0.43 mm Hg) and mean arterial pressure (-1.64 mm Hg; 95% CI: -3.27, -0.01 mm Hg) and marginally significant effects on LDL (-0.07 mmol/L; 95% CI: -0.13, 0.00 mmol/L) and HDL (0.03 mmol/L; 95% CI: 0.00, 0.06 mmol/L) cholesterol were also observed. Chocolate or cocoa improved FMD regardless of the dose consumed, whereas doses >50 mg epicatechin/d resulted in greater effects on systolic and diastolic BP. GRADE (Grading of Recommendations, Assessment, Development and Evaluation, a tool to assess quality of evidence and strength of recommendations) suggested low- to moderate-quality evidence of beneficial effects. The finding showed consistent acute and chronic benefits of chocolate or cocoa on FMD and previously unreported promising effects on insulin and HOMA-IR. However, larger, longer-duration, and independently funded trials are required to confirm the potential cardiovascular benefits of cocoa flavan-3-ols. Table 2 shows the studies indicating the role of cocoa in endothelial function [51-59, 82, 84, 92].

Memory Dysfunction and Dementia

Recent studies indicate that dietary flavonoids that are abundant in plant-based foods, can improve cognitive function, particularly a reduction in the risk of dementia, enhanced performance on some cognitive tests, and improved cognitive function in elderly patients. [14, 85-87]. Improved cognitive performance with the administration of a cocoa polyphenolic extract has also even been reported in aged Wistar–Unilever rats [24]. There is evidence that dietary flavanols might improve cognitive function in subjects with mild cognitive impairment. The memory functions are under influence of microcirculation of the neurons as well as NO release which may be impaired due to hyperglycemia, oxidative stress, atherosclerosis and inflammation. Since regular consumption of cocoa flavanols might be effective in improving cognitive function in subjects with mild cognitive impairment, it is possible that flavonol can protect the neurons from dementia and neurodegeneration. This effect appears to be mediated in part by an improvement in insulin sensitivity and NO release [14] A cross-sectional study in Norway, among 2031 elderly subjects, aged 70-74 years (55% women) examined the relation between intake of 3 common foodstuffs that contain flavonoids (chocolate, wine, and tea) [85]. Cognitive performance was examined via a cognitive test battery including the Kendrick Object Learning Test, Trail Making Test, part A (TMT-A), modified versions of the Digit Symbol Test, Block Design, Mini-Mental State Examination, and Controlled Oral Word Association Test. Poor cognitive performance was defined as a score in the highest decile for the TMT-A and in the lowest decile for all other tests. A self-reported FFQ was used to assess habitual food intake. Subjects who consumed chocolate, wine, or tea had significantly better mean test scores and lower prevalence of poor cognitive performance than those who were not consuming flavonoid rich foods. Those subjects who consumed all 3 studied items, had the best test scores and the lowest risks for poor test performance and the associations between intake of these foodstuffs and cognition were dose dependent. The effect was approximately linear for tea with maximum effect at intakes of approximately 10 g/d for chocolate and approximately 75-100 mL/d for wine and all cognitive functions tested were influenced by intake of these 3 foodstuffs. The effect was most pronounced for wine and modestly weaker for chocolate intake. It seems that in the elderly subjects, a diet high in some flavonoid-rich foods was associated with better performance in several cognitive abilities in a dose-dependent manner.

Desideri et al. conducted a double-blind, parallel arm study in 90 elderly individuals with mild cognitive impairment randomized to consume once daily for 8 weeks a drink containing ≈990 mg (high flavanols), ≈520 mg (intermediate flavanols), or ≈45 mg (low flavanols) of cocoa flavanols per day [14]. Mini Mental State Examination, Trail Making Test A and B, and verbal fluency test were the instruments for the study of Cognitive function. Afterthe follow-up period, Mini Mental State Examination was similar in the 3 treatment groups (P=0.13). There was a significant decrease in insulin resistance, blood pressure, and lipid peroxidation among subjects in the high-flavanol and intermediate-flavanol groups compared to low flavanol group. Alteration in insulin resistance explained ≈40% of composite z score variability through the study period (partial r(2)=0.4013; P<0.0001). The time required to complete Trail Making Test A and Trail Making Test B was significantly (P<0.05) lower in subjects assigned to high flavanols (38.10±10.94 and 104.10±28.73 seconds, respectively) and intermediate flavanols (40.20±11.35 and 115.97±28.35 seconds, respectively) in comparison with those assigned to low flavanols (52.60±17.97 and 139.23±43.02 seconds, respectively.). Similarly, verbal fluency test score was significantly (P<0.05) better in subjects assigned to high flavanols in comparison with those assigned to low flavanols (27.50±6.75 versus 22.30±8.09 words per 60 seconds).

Of all flavonoids, a subclass called flavanols, which are widely present in cocoa, green tea, red wine, and some fruits, seems to be effective in slowing down or even reversing the reductions in cognitive performance that occur with aging among elderly subjects. However, it is possible that flavonols can improve memory function at any age including children. There appear to be only nine published studies in the scientific literature that have examined the effects of cocoa- and chocolate-related products on neurocognitive functioning [87]. Seven of nine of the published studies that have examined the effects of cocoa- and chocolate-related products on neurocognitive processes have found that the compounds are associated with significantly improved and/or preserved aspects of cognitive functioning. Cocoa and chocolate consumption could hypothetically improve cognitive function in individuals as well as in whole populations [85, 87]. It is possible that there would be a correlation between a country's level of chocolate consumption and its population's cognitive function which may be examined, not withstanding other decision-making criteria, by studying the total number of Nobel Laureates per capita as a surrogate end point reflecting the proportion with superior cognitive function. It might give some measure of the overall cognitive function of a given country showing association of chocolate consumption and number of Nobel Laureates in a country (86). Recently, Messerli reported a significant linear correlation (r=0.791, P<0.0001) between chocolate consumption per capita and the number of Nobel Laureates per 10 million persons in a total of 23 countries ([86], Figure 2). He analyzed and observed a correlation between Countries' Annual Per Capita Chocolate consumption and the number of Nobel Laureates per 10 Million population. Switzerland was the top performer in terms of both the number of Nobel Laureates and chocolate consumption. The slope of the regression line allows us to estimate that it would take about 0.4 kg of chocolate per capita per year to increase the number of Nobel Laureates in a given country by 1. For the United States, that would amount to 125 million kg per year. The minimally effective chocolate dose seems to hover around 2 kg per year, and the dose–response curve reveals no apparent ceiling on the number of Nobel Laureates at the highest chocolate-dose level of 11 kg per year. Calculation with the exclusion of Sweden revealed that the correlation coefficient increased to 0.862 which may be due to general behavior. However, a much more rigorous association between chocolate consumption per capita and cognitive function needs to be undertaken, possibly with migrant or community based studies. American Indians and Kuna Indians who traditionally consumed chocolates, to the best of our knowledge, did not receive any Nobel prize so far, indicating, of course, that many confounders may be existing apart from memory function. Further evidence suggests that dietary flavonoids and methylxanthines, such as those commonly found in cocoa and chocolate products, may possess neuroprotective, neuro-enhancing, and neuro-stimulating and cardioprotective effects [26-29]. Addition of w-3 fatty acids to dark chocolates, which is rich in the cell membranes of neurons, can enhance its capability in enhancing memory function and in getting Nobel Prize as well as in the prevention of CVDs and other chronic diseases [5], (Figure 5).

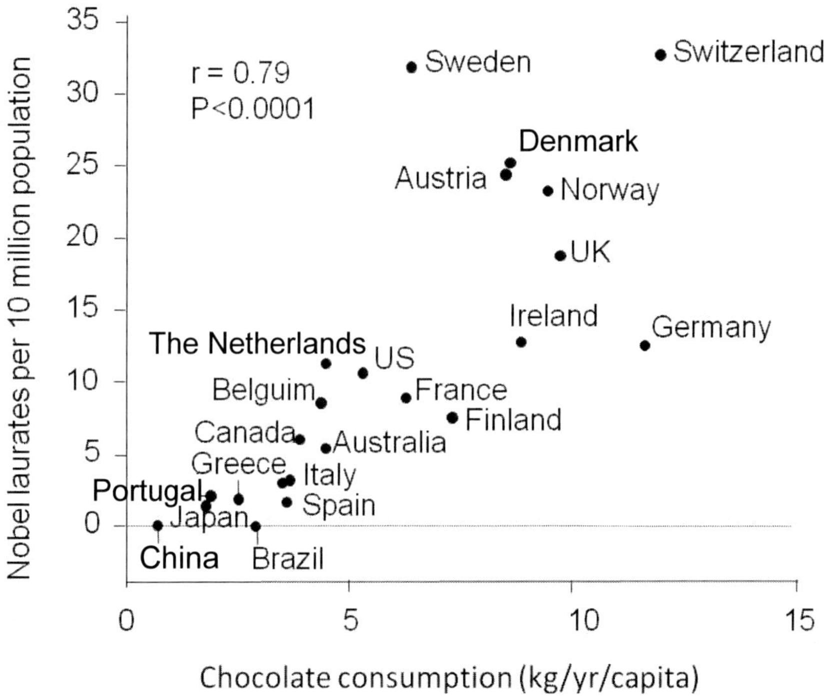

Figure 5. Correlation between countries annual per capita chocolate consumption and the number of Nobel laureates per 10 million population. Ref [86].

Other Chronic Diseases

Cocoa-rich chocolate has been known not only for its good taste but also for its proposed health effects on all other body systems related to NCD [1-20, 88-91]. Several supposed health effects of cocoa have been considered, including improved cardiac function and relief of angina pectoris, stimulation of the nervous system, facilitated digestion, and improved kidney and bowel function. In addition, cocoa has been used to treat anemia, mental fatigue, tuberculosis, fever, gout, kidney stones, and even poor sexual appetite. Substantial data suggest that flavonoid-rich food could help prevent CVDs, CAD, stroke, hypertension, hyperlipidemia, insulin resistance, diabetes mellitus and cancer [88-91]. Cocoa is one of the richest sources of flavonoids, but current processing reduces the content substantially. The Kuna living in the San Blas, drink a flavanol-rich cocoa as their main beverage, contributing more than 900 mg/day and thus probably have the most flavonoid-rich diet than any other population. However, the Nobel Prize would continue to be a dream for most of them even in the next one century.

In a recent study, diagnosis on death certificates for comparison of cause-specific death rates from year 2000 to 2004, in mainland and the San Blas islands where only Kuna live, were studied [88]. It is possible that high flavanoid intake and consequent nitric oxide system activation would be considered important, if the result would indicate, a reduction in the frequency of CAD, stroke, diabetes mellitus, and cancer that are all nitric oxide sensitive processes. The study included 77,375 deaths in mainland Panama and 558 deaths in the San Blas. In mainland Panama, as anticipated, CVD was the leading cause of death (83.4 ± 0.70 age adjusted deaths/100,000) and cancer was second (68.4 ± 1.6). In contrast, the rate of CVD and cancer among island-dwelling Kuna was much lower (9.2 ± 3.1 vs.4.4 ± 4.4) respectively. Similarly deaths due to diabetes mellitus were much more common in the mainland, than in the San Blas (24.1 ± 0.74 vs.6.6 ± 1.94). This comparatively lower risk of NCDs among Kuna in the San Blas from the most common causes of morbidity and mortality in most of the world, possibly reflects a very high flavanol intake and sustained nitric oxide synthesis activation. However, there are many risk factors and an observational study cannot provide definitive evidence; hence controlled trials among high risk populations would be necessary to provide causal proof regarding the role of cocoa flavanol in the prevention of NCDs.

Drug Development

The activation of nitric oxide (NO), antioxidant, anti-inflammatory, and antiplatelet effects, of cocoa appear to be important which in turn might improve endothelial function, lipid levels, blood pressure, insulin resistance, hyperglycemia and atherothrombosis and eventually clinical outcome in CVDs, diabetes, dementia and early carcinogenesis [2-7, 11-15]. The blood-pressure lowering effect is small (2-3mmHg) but significant. There is a need to develop concentrated forms of flavanol rich cocoa which may be used as softsules for conducting double blind randomized, controlled trials to prove the beneficial effects of cocoa on CVDs and metabolic syndrome, memory dysfunction and dementia. Such studies may also clarify both dose-response relationships and drug development, as well as address the optimal processing method of raw cocoa beans to optimize anti-hypertensive actions and other actions causing beneficial effects on obesity, diabetes, insulin resistance, blood lipids, hypertension and atherothrombosis, brain degeneration and carcinogenesis. Cocoa itself, unlike chocolate, can be recommended without hesitation because it is low in sugar and fat. Current evidence suggests that the beneficial effects of cocoa are attributed mainly to its flavanol content, especially epicatechin. Therefore, direct dietary supplementation with flavanols instead of chocolate consumption deserves further study. Protocols using epicatechin or other flavanoids specifically are now feasible and should clarify this question. Recommending dietary supplementation with flavanols, similar to vitamins, appears problematic because potential pro-oxidative effects of large quantities cannot be excluded (89).

Functional Food Development

It is important to assess the side effects of long-term daily chocolate consumption such as obesity, dental caries and type 2 diabetes. The fat and sugar content and thus energy density of chocolate may be reduced by incorporation of fruit juice or water acidified with ascorbic acid without impacting mouth-feel [9]. Cocoa flavanols may also up-regulate endogenous antioxidant enzymes and thereby lower reactive oxygen species (ROS) levels, apart from increased endothelial NO release. Since ROS interfere with NO's bioavailability [10], this may indirectly increase NO levels and its actions. Another natural product, nitrate-rich vegetables such as beetroot, also reduce blood pressure in hypertensives in a NO-dependent manner which may be incorporated in the cocoa powder to dilute its bitterness [11]. It may even be beneficial to combine beetroot or dietary nitrate (generating NO) with cocoa (increasing NO's bioavailability), and thus move from adverse effect feeling of fat and sugar containing chocolate to nitric oxide containing chocolate. Addition of flax seed or flax seed oil may also decrease the need for saturate fat in cocoa which would enhance w-3 fatty acid content, known to be protective against NCDs (Figure 6) [5].

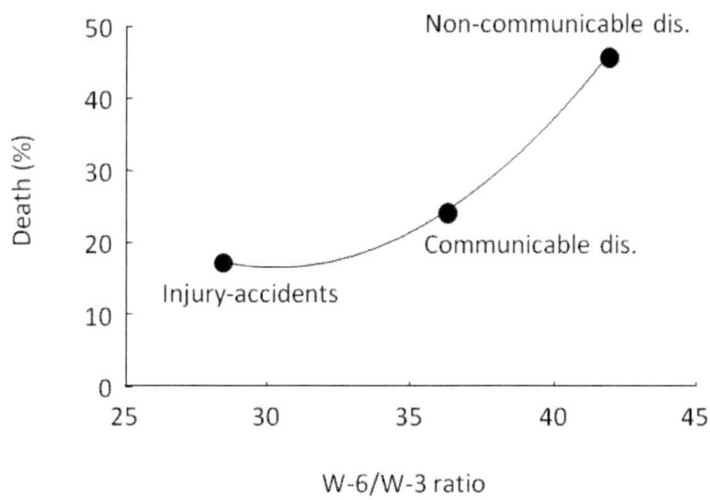

Figure 5. Association of W-6/W-3ratio of fatty acids with mortality in injury- accidents, communicable disease and non-communicable diseases. Ref [5].

Agricultural Policy

Cocoa farming should be encouraged in all the developing and developed countries. Since, protective effects of cocoa are due to the flavanol contents; catechin and epicatechin, these antioxidants need to be enhanced in the cocoa seeds by plant breeding and genetic modification. Efforts should be made to decrease bitterness of cocoa by genetic manipulations or by other methods, so that use of milk, saturated fat and sugar can be avoided from cocoa based dishes.

In view of the limited data, future studies should provide detailed information about the chocolate product used. The exact content of polyphenols, especially flavanols; and most importantly, the flavanol plasma concentrations achieved would be important. It has to be taken into account that cocoa contains many other potentially active substances, e.g., theobromine or magnesium, which may be responsible for a significant part of the benefits. In brief, the available evidence indicates that increased consumption of dark chocolates or cocoa

(300-1000mg/day) can protect against CVDs and other chronic diseases including metabolic syndrome and memory dysfunction.

Conflict of Interests: None declared by the authors.

Acknowledgments

The International College of Nutrition, International College of Cardiology and the Tsim Tsoum Institute gave ogistic support to write this article.

References

[1] Hristova K, Nakaoka T, Otsuka K, Fedacko J, Singh R, Singh RB, Meester F, Agnieszka Wilczynska A, Wilson DW. Editorial: Perspectives on Chocolate Consumption and Risk of Cardiovascular Diseases and Cognitive Function. *The Open Nutr J* 2012;5:207-212.

[2] Dillinger TL, Barriga P, Escarcega S, Jimenez M, Salazar Lowe D, Grivetti LE. Food of the gods: cure for humanity? A cultural history of the medicinal and ritual use of chocolate. *J Nutr*. 2000; 130: 2057S–2072S.

[3] Grivetti, Louis E.; Howard-Yana Shapiro (2009). *Chocolate: history, culture, and heritage*. John Wiley and Sons. ISBN 9780470121658.2009.

[4] Henderson JS, Joyce RA, Hall GR, Hurst WJ, McGovern PE. Chemical and archaeological evidence for the earliest cacao beverages. *Proc Natl Acad Sci USA*. 2007; 104: 18937–18940.

[5] Fedacko J, Vargova V, Singh RB, Anjum B, Takahashi T, Tongnuka M, Dharwadkar S, Singh S, Singh V, Kulshresth SK, De Meester F, Wilson DW. Association of high w-6/w-3 fatty acid ratio diet with causes of death due to noncommunicable diseases among urban decedents in North India. *The Open Nutra Jour* 2012;5:113-123.

[6] Tokede OA, Gaziano JM, Djoussé L. Effects of cocoa products/dark chocolate on serum lipids: a meta-analysis. *Eur J Clin Nutr*. 2011 Aug;65(8):879-86. doi: 10.1038/ejcn.2011.64. Epub 2011 May 11.

[7] Hooper L, Kay C, Abdelhamid A, Kroon PA, Cohn JS, Rimm EB, Cassidy A. Effects of chocolate, cocoa, and flavan-3-ols on cardiovascular health: a systematic review and meta-analysis of randomized trials. *Am J Clin Nutr. 2012 Mar;95(3):740-51. Epub 2012 Feb*

[8] *Larsson SC*, Virtamo J and Wolk A. Chocolate consumption and risk of stroke A prospective cohort of men and meta-analysis. *Neurology* 2012;79:1223-29 Published online before print August 29, 2012, doi: 10.1212/WNL.0b013e31826aacfa

[9] Skelhon T, Grossiord N, Morgan A, Bon S. Quiescent water-in-oil Pickering emulsions as a route toward healthier fruit juice infused chocolate confectionary. *J Mater Chem* 2012; 22:19289-95 DOI: 10.1039/c2jm34233B.

[10] Armitage ME, Wingler K, Schmidt HH, La M. Translating the oxidative stress hypothesis into the clinic: NOX versus NOS. *J Mol Med* (Berl) 2009 Nov; 87(11):1071-6 PMID: 19834654 DOI: 10.1007/s00109-009-0544-2.

[11] Webb AJ, Patel N, Loukogeorgakis S, Okorie M, Benjamin N, MacAllister R, Hobbs AJ, Ahluwalia A. Acute blood pressure lowering, vasoprotective, and antiplatelet properties of dietary nitrate via bioconversion to nitrite. *Hypertension* 2008 Mar; 51(3):784-90 PMID: 18250365 DOI: 10.1161/HYPERTENSIONAHA.107.103523.12.

[12] Buitrago-Lopez A, Sanderson J, Johnson L, Warnakula S, Wood A, Di Angelantonio E, Franco OH. Chocolate consumption and cardiometabolic disorders: systematic review and meta-analysis. *BMJ.* 2011;343:d4488. doi: 10.1136/bmj.d4488. 2012.

[13] Grassi D, Desideri G, Necozione S, Ruggieri F, Blumberg JB, Stornello M, Ferri C. Protective Effects of Flavanol-Rich Dark chocolate on endothelial function and wave reflection during acute hyperglycemia. *Hypertension.* 2012;60:827-832.

[14] Desideri G, Kwik-Uribe C, Grassi D, et al. Benefits in cognitive function, blood pressure, and insulin resistance through cocoa flavanol consumption in elderly subjects with mild cognitive impairment: the Cocoa, Cognition, and Aging (CoCoA) Study. *Hypertension* 2012;60:794-801.

[15] Buijsse B, Feskens EJ, Kok FJ, Kromhout D. Cocoa intake, blood pressure, and cardiovascular mortality: the Zutphen Elderly Study. *Arch Intern Med.* 2006;166:411–417.

[16] Taubert D, Roesen R, Schömig E. Effect of cocoa and tea intake on blood pressure: a meta-analysis. *Arch Intern Med.* 2007 Apr 9;167(7):626-34.

[17] Ried K, Sullivan T, Fakler P, Frank OR, Stocks NP. Does chocolate reduce blood pressure? *A meta-analysis. BMC Med* 2010; 8:39, PMID: 20584271 DOI: 10.1186/1741-7015-8-39.

[18] Desch S, Schmidt J, Kobler D, Sonnabend M, Sareban M, Rahimi K, Schuler G, Thiele H. Effect of cocoa products on blood pressure: systematic review and meta-analysis. *Am J Hypertens* 2010 Jan; 23(1):97-103, PMID: 19910929 DOI: 10.1038/ajh.2009.213.

[19] Mursu J, Voutilainen S, Nurmi T, Rissanen TH, Virtanen JK, Kaikkonen J, Nyyssonen K, Salonen JT. Dark chocolate consumption increases HDL cholesterol concentration and chocolate fatty acids may inhibit lipid peroxidation in healthy humans. *Free Radic Biol Med.* 2004;37: 1351–1359.

[20] Sorond FA, Lipsitz LA, Hollenberg NK, Fisher ND. Cerebral blood flow response to flavanol-rich cocoa in healthy elderly humans. *Neuropsychiatric Disease and Treatment* 2008:4(2) 433–440.

[21] Kurlandsky SB, Stote KS. Cardioprotective effects of chocolate and almond consumption in healthy women. *Nutr Res.* 2006;26:509–516.

[22] Baba S, Osakabe N, Kato Y, Natsume M, Yasuda A, Kido T, Fukuda K, Muto Y, Kondo K. Continuous intake of polyphenolic compounds containing cocoa powder reduces LDL oxidative susceptibility and has beneficial effects on plasma HDL-cholesterol concentrations in humans. *Am J Clin Nutr.* 2007;85:709 –717.

[23] Panneerselvam M, Tsutsumi YM, Bonds JA, Horikawa YT, Saldana M, Dalton ND, Head BP, Patel PM, David M. Roth DM, and Hemal H. Patel HM. Dark chocolate receptors: epicatechin-induced cardiac protection is dependent on δ-opioid receptor stimulation *Am J Physiol Heart Circ Physiol.* 2010;299(5): H1604–H1609. PMCID: PMC2993198.

[24] Bisson JF, Nejdi A, Rozan P, Hidalgo S, Lalonde R, Messaoudi M. Effects of long-term administration of a cocoa polyphenolic extract (Acticoa powder) on cognitive performances in aged rats. *Br J Nutr* 2008;100:94-101.

[25] Corti R, Flammer AJ, Hollenberg NK, Luscher TF. Cocoa and cardiovascular health. *Circulation* 2009;119:1433-1441.

[26] Manach C, Scalbert A, Morand C, Remesy C, Jimenez L. Polyphenols:food sources and bioavailability. *Am J Clin Nutr.* 2004;79:727–747.

[27] Arts IC, van de Putte B, Hollman PC. Catechin contents of foods commonly consumed in The Netherlands, 1: fruits, vegetables, staple foods, and processed foods. *J Agric Food Chem.* 2000;48:1746 –1751.

[28] Adamson GE, Lazarus SA, Mitchell AE, Prior RL, Cao G, Jacobs PH, Kremers BG, Hammerstone JF, Rucker RB, Ritter KA, Schmitz HH. HPLC method for the quantification of procyanidins in cocoa and chocolate samples and correlation to total antioxidant capacity. *J Agric Food Chem.* 1999;47:4184–4188.

[29] Serafini M, Bugianesi R, Maiani G, Valtuena S, De Santis S, Crozier A. Plasma antioxidants from chocolate. *Nature.* 2003;424:1013.

[30] Roura E, Andres-Lacueva C, Estruch R, Mata-Bilbao ML, Izquierdo-Pulido M, Waterhouse AL, Lamuela-Raventos RM. Milk does not affect the bioavailability of cocoa powder flavonoid in healthy human. *Ann Nutr Metab.* 2007;51:493– 498.

[31] Schroeter H, Holt RR, Orozco TJ, Schmitz HH, Keen CL. Nutrition: milk and absorption of dietary flavanols. *Nature.* 2003;426:787–788.

[32] Rein D, Lotito S, Holt RR, Keen CL, Schmitz HH, Fraga CG. Epicatechinin human plasma: in vivo determination and effect of chocolate consumption on plasma oxidation status. *J Nutr.* 2000;130: 2109S–2114S.

[33] Richelle M, Tavazzi I, Enslen M, Offord EA. Plasma kinetics in man of epicatechin from black chocolate. *Eur J Clin Nutr.* 1999;53:22–26.

[34] Cooper KA, Donovan JL, Waterhouse AL, Williamson G. Cocoa and health: a decade of research. *Br J Nutr.* 2008;99:1–11.

[35] Schewe T, Steffen Y, Sies H. How do dietary flavanols improve vascular function? A position paper. *Arch Biochem Biophys.* 2008;476:102–106.

[36] Andres-Lacueva C, Monagas M, Khan N, Izquierdo-Pulido M, Urpi-Sarda M, Permanyer J, Lamuela-Raventos RM. Flavanol and flavonol contents of cocoa powder products: influence of the manufacturing process. *J Agric Food Chem.* 2008;56:3111–3117.

[37] Counet C, Ouwerx C, Rosoux D, Collin S. Relationship between procyanidin and flavor contents of cocoa liquors from different origins. *J Agric Food Chem.* 2004;52:6243–6249.

[38] Miller KB, Stuart DA, Smith NL, Lee CY, McHale NL, Flanagan JA, Ou B, Hurst WJ. Antioxidant activity and polyphenol and procyanidin contents of selected commercially available cocoa-containing and chocolate products in the United States. *J Agric Food Chem.* 2006;54: 4062–4068.

[39] Ding EL, Hutfless SM, Ding X, Girotra S. Chocolate and prevention of cardiovascular disease: a systematic review. *Nutr Metab (Lond).* 2006;3:2-28.

[40] Palmer RM, Ashton DS, Moncada S. Vascular endothelial cells synthesize nitric oxide from L-arginine. *Nature.* 1988;333:664–666.

[41] Boulanger C, Luscher TF. Release of endothelin from the porcine aorta: inhibition by endothelium-derived nitric oxide. *J Clin Invest.* 1990;85:587–590.

[42] Joannides R, Richard V, Haefeli WE, Linder L, Luscher TF, Thuillez C. Role of basal and stimulated release of nitric oxide in the regulation of radial artery caliber in humans. *Hypertension.* 1995;26:327–331.

[43] Oemar BS, Tschudi MR, Godoy N, Brovkovich V, Malinski T, Luscher TF. Reduced endothelial nitric oxide synthase expression and production in human atherosclerosis. *Circulation.* 1998;97:2494 –2498.

[44] Celermajer DS, Sorensen KE, Bull C, Robinson J, Deanfield JE. Endothelium-dependent dilation in the systemic arteries of asymptomatic subjects relates to coronary risk factors and their interaction. *J Am Coll Cardiol.* 1994;24:1468 –1474.

[45] Schachinger V, Britten MB, Zeiher AM. Prognostic impact of coronary vasodilator dysfunction on adverse long-term outcome of coronary heart disease. *Circulation.* 2000;101:1899 –1906.

[46] Suwaidi JA, Hamasaki S, Higano ST, Nishimura RA, Holmes DR Jr, Lerman A. Long-term follow-up of patients with mild coronary artery disease and endothelial dysfunction. *Circulation.* 2000;101:948 –954.

[47] Duffy SJ, Keaney JF Jr, Holbrook M, Gokce N, Swerdloff PL, Frei B, Vita JA. Short- and long-term black tea consumption reverses endothelial dysfunction in patients with coronary artery disease. *Circulation.* 2001;104:151–156.

[48] Karatzi K, Papamichael C, Aznaouridis K, Karatzis E, Lekakis J, Matsouka C, Boskou G, Chiou A, Sitara M, Feliou G, Kontoyiannis D, Zampelas A, Mavrikakis M. Constituents of red wine other than alcohol improve endothelial function in patients with coronary artery disease. *Coron Artery Dis.* 2004;15:485– 490.

[49] Whelan AP, Sutherland WH, McCormick MP, Yeoman DJ, de Jong SA, Williams MJ. Effects of white and red wine on endothelial function in subjects with coronary artery disease. *Intern Med J.* 2004;34:224 –228.

[50] Nagaya N, Yamamoto H, Uematsu M, Itoh T, Nakagawa K, MiyazawaT, Kangawa K, Miyatake K. Green tea reverses endothelial dysfunction in healthy smokers. *Heart.* 2004;90: 1485–1486.

[51] Karim M, McCormick K, Kappagoda CT. Effects of cocoa extracts on endothelium-dependent relaxation. *J Nutr.* 2000;130:2105S–2108S.

[52] Fisher ND, Hughes M, Gerhard-Herman M, Hollenberg NK. Flavanol-rich cocoa induces nitric-oxide-dependent vasodilation in healthy humans. *J Hypertens.* 2003;21:2281–2286.

[53] Engler MB, Engler MM, Chen CY, Malloy MJ, Browne A, Chiu EY, Kwak HK, Milbury P, Paul SM, Blumberg J, Mietus-Snyder ML. Flavonoid-rich dark chocolate improves endothelial function and increases plasma epicatechin concentrations in healthy adults. *J Am Coll Nutr.* 2004;23:197–204.

[54] Schroeter H, Heiss C, Balzer J, Kleinbongard P, Keen CL, Hollenberg NK, Sies H, Kwik-Uribe C, Schmitz HH, Kelm M. (-)-Epicatechin mediates beneficial effects of flavanol-rich cocoa on vascular function in humans. *Proc Natl Acad Sci USA.* 2006;103:1024 –1029.

[55] Heiss C, Dejam A, Kleinbongard P, Schewe T, Sies H, Kelm M. Vascular effects of cocoa rich in flavan-3-ols. *JAMA.* 2003;290:1030–1031.

[56] Hermann F, Spieker LE, Ruschitzka F, Sudano I, Hermann M, Binggeli C, Luscher TF, Riesen W, Noll G, Corti R. Dark chocolate improves endothelial and platelet function. *Heart.* 2006;92:119 –120.

[57] Grassi D, Desideri G, Necozione S, Lippi C, Casale R, Properzi G, Blumberg JB, Ferri C. Blood pressure is reduced and insulin sensitivity increased in glucose-intolerant, hypertensive subjects after 15 days of consuming high-polyphenol dark chocolate in hypertensives. *Hypertension. J Nutr.* 2008 Sep;138(9):1671-6.47.

[58] Heiss C, Kleinbongard P, Dejam A, Perre S, Schroeter H, Sies H, Kelm M. Acute consumption of flavanol-rich cocoa and the reversal of endothelial dysfunction in smokers. *J Am Coll Cardiol.* 2005;46:1276 –1283.

[59] Balzer J, Rassaf T, Heiss C, Kleinbongard P, Lauer T, Merx M, Heussen N, Gross HB, Keen CL, Schroeter H, Kelm M. Sustained benefits in vascular function through flavanol-containing cocoa in medicated diabetic patients a double-masked, randomized, controlled trial. *J Am Coll Cardiol.* 2008;51:2141–2149.

[60] Choi YJ, Kang JS, Park JH, Lee YJ, Choi JS, Kang YH. Polyphenolic flavonoids differ in their antiapoptotic efficacy in hydrogen peroxide treated human vascular endothelial cells. *J Nutr.* 2003;133:985–991.

[61] Leikert JF, Rathel TR, Wohlfart P, Cheynier V, Vollmar AM, Dirsch VM. Red wine polyphenols enhance endothelial nitric oxide synthase expression and subsequent nitric oxide release from endothelial cells. *Circulation.* 2002;106:1614 –1617.

[62] Benito S, Lopez D, Saiz MP, Buxaderas S, Sanchez J, Puig-Parellada P, Mitjavila MT. A flavonoid-rich diet increases nitric oxide production in rat aorta. *Br J Pharmacol.* 2002;135:910 –916.

[63] Wallerath T, Poleo D, Li H, Forstermann U. Red wine increases the expression of human endothelial nitric oxide synthase: a mechanism that may contribute to its beneficial cardiovascular effects. *J Am Coll Cardiol.* 2003;41:471– 478.

[64] Schnorr O, Brossette T, Momma TY, Kleinbongard P, Keen CL, Schroeter H, Sies H. Cocoa flavanols lower vascular arginase activity in human endothelial cells in vitro and in erythrocytes in vivo. *Arch Biochem Biophys.* 2008;476:211–215.

[65] Vinson JA, Proch J, Zubik L. Phenol antioxidant quantity and quality in foods: cocoa, dark chocolate, and milk chocolate. *J Agric Food Chem.* 1999;47:4821– 4824.

[66] Steffen Y, Schewe T, Sies H. (-)-Epicatechin elevates nitric oxide in endothelial cells via inhibition of NADPH oxidase. *Biochem Biophys Res Commun.* 2007;359:828–833.

[67] Loke WM, Hodgson JM, Proudfoot JM, McKinley AJ, Puddey IB, Croft KD. Pure dietary flavonoids quercetin and (-)-epicatechin augment nitric oxide products and reduce endothelin-1 acutely in healthy men. *Am J Clin Nutr.* 2008;88:1018 –1025.

[68] Keli SO, Hertog MG, Feskens EJ, Kromhout D. Dietary flavonoids, antioxidant vitamins, and incidence of stroke: the Zutphen study. *Arch Intern Med.* 1996;156:637–642.

[69] Hollenberg NK, Martinez G, McCullough M, Meinking T, Passan D, Preston M, Rivera A, Taplin D, Vicaria-Clement M. Aging, acculturation, salt intake, and hypertension in the Kuna of Panama. *Hypertension.* 1997; 29:171–176.

[70] Hollenberg NK, Rivera A, Meinking T, Martinez G, McCullough M, Passan D, Preston M, Taplin D, Vicaria-Clement M. Age, renal perfusion and function in island-dwelling indigenous Kuna Amerinds of Panama. *Nephron.* 1999;82:131–138.

[71] Bayard V, Chamorro F, Motta J, Hollenberg NK. Does flavanol intake influence mortality from nitric oxide-dependent processes? Ischemic heart disease, stroke, diabetes mellitus, and cancer in Panama. *Int J Med Sci.* 2007;4:53–58.
[72] Mink PJ, Scrafford CG, Barraj LM, Harnack L, Hong CP, Nettleton JA, Jacobs DR Jr. Flavonoid intake and cardiovascular disease mortality: a prospective study in postmenopausal women. *Am J Clin Nutr*. 2007;85:895–909.
[73] Katz DL, Doughty K, Ali A. Cocoa and chocolate in human health and disease. *Antioxid Redox Signal* 2011;15:2779-811.
[74] Fuster V, Moreno PR, Fayad ZA, Corti R, Badimon JJ. Atherothrombosis and high-risk plaque, part I: evolving concepts. *J Am Coll Cardiol.* 2005;46:937–954.
[75] Keen CL, Holt RR, Oteiza PI, Fraga CG, Schmitz HH. Cocoa antioxidants and cardiovascular health. *Am J Clin Nutr.* 2005;81:298S–303S.
[76] Holt RR, Schramm DD, Keen CL, Lazarus SA, Schmitz HH. Chocolate consumption and platelet function. *JAMA*. 2002;287:2212–2213.
[77] Pearson DA, Paglieroni TG, Rein D, Wun T, Schramm DD, Wang JF, Holt RR, Gosselin R, Schmitz HH, Keen CL. The effects of flavanol-rich cocoa and aspirin on ex vivo platelet function. *Thromb Res*. 2002;106:191–197.
[78] Rein D, Paglieroni TG, Wun T, Pearson DA, Schmitz HH, Gosselin R, Keen CL. Cocoa inhibits platelet activation and function. *Am J Clin Nutr*. 2000;72:30 –35.
[79] Innes AJ, Kennedy G, McLaren M, Bancroft AJ, Belch JJ. Dark chocolate inhibits platelet aggregation in healthy volunteers. *Platelets.* 2003;14:325–327.
[80] Kelly FD, Sinclair AJ, Mann NJ, Turner AH, Abedin L, Li D. A stearic acid-rich diet improves thrombogenic and atherogenic risk factor profiles in healthy males. *Eur J Clin Nu*tr. 2001;55:88 –96.
[81] Kelly FD, Sinclair AJ, Mann NJ, Turner AH, Raffin FL, Blandford MV, Pike MJ. Short-term diets enriched in stearic or palmitic acids do not alter plasma lipids, platelet aggregation or platelet activation status. *Eur J Clin Nutr.* 2002;56:490–499.
[82] Flammer AJ, Hermann F, Sudano I, Spieker L, Hermann M, Cooper KA, Serafini M, Luscher TF, et al. Dark chocolate improves coronary vasomotion and reduces platelet reactivity. *Circulation* 2007;116:2376-82.
[83] Grassi D, Desideri G, Necozione S, Lippi C, Casale R, Properzi G, Blumberg JB, Ferri C. Blood pressure is reduced and insulin sensitivity increased in glucose intolerant hypertensive subjects after 15 days of consuming high polyphenol dark chocolate. *J Nutr* 2009;138:1671-76.
[84] Shiina Y, Funabashi N, Lee K, Murayama T, Nakamura K, Wakatsuki Y, Daimon M, Komuro I. Acute effect of oral flavonoid-rich dark chocolate intake on coronary circulation, as compared with nonflavonoid white chocolate, by transthoracic Doppler echocardiography in healthy adults. *Int J Cardiol*. 2009;131:424–429.
[85] Nurk E, Refsum H, Drevon CA, et al. Intake of flavonoid-rich wine, tea, and chocolate by elderly men and women is associated with better cognitive test performance. *J Nutr* 2009;139:120-127.
[86] Messerli FH. Chocolates consumption, cognitive function and Nobel laureate Laureates. *N Engl J Med* 2012; October 10, 2012DOI: 10.1056/NEJMon1211064.
[87] W. David Crews Jr. Ph.D., David W. Harrison Ph.D., Kim P. Gregory M.A., Bon Kim B.S., Allison B. Darling B.S. The effects of cocoa- and chocolate-related products on neurocognitive functioning. *Nutrition and Health*. 2013, 7: pp 369-379.

[88] Bayard V, Chamorro F, Morro J, and Hollenberg NK. Does flavanol intake influence mortality from nitric oxide-dependent processes? Ischemic Heart Disease, Stroke, Diabetes Mellitus, and Cancer in Panama. *Int J Med Sci.* 2007; 4(1): 53–58. PMCID: PMC1796954

[89] Halliwell B. Dietary polyphenols: good, bad, or indifferent for your health? *Cardiovasc Res.* 2007;73:341–347.

[90] Bonanome A, Grundy SM. Effect of dietary stearic acid on plasma cholesterol and lipoprotein levels. *N Engl J Med.* 1988;318:1244 –1248.

[91] Mensink RP, Zock PL, Kester AD, Katan MB. Effects of dietary fatty acids and carbohydrates on the ratio of serum total to HDL cholesterol and on serum lipids and apolipoproteins: a meta-analysis of 60 controlled trials. *Am J Clin Nutr.* 2003;77:1146 – 1155.

[92] Grassi D, Desideri G, Necozione S, Lippi C, Casale R, Properzi G, Blumberg JB, Ferri C. Blood pressure is reduced and insulin sensitivity increased in glucose-intolerant, hypertensive subjects after 15 days of consuming high-polyphenol dark chocolate. *J Nutr.* 2008;138: 1671–1676.

In: Circadian Cardiology with Focus on Both Prevention ... ISBN: 978-1-63463-956-9
Editors: K. Hristova, A. Shehab, G. Cornelissen et al.

Chapter 19

You Are What You Eat, Which Depends on Available Food and Agriculture?

Rukam S. Tomar[1], Ran S. Tomar[2], Raina Pal[4], Anamika Tripathi[4] and R. B. Singh[3,5]*

[1]Department of Biotechnology, Junagarh Agricultural University
[2]Directorate of Groundnut Research, ICAR, Junagadh, (Gujrat)
[3]National Academy of Agriculture Science, New Delhi
[4]Department of Botany, Hindu College
[5]Halberg Hospital and Research Institute, Moradabad, India

Introduction

You are what you eat was proposed by Hippocrates in the 5th century BC. Hippocrates wrote "Positive health requires a knowledge of man's primary constitution (which today we call genetics) and of the powers of various foods, both those natural to them and those resulting from human skill (today's processed food), but apart from eating, exercise is also important for health.

It seems that the above statement is true even today, because our total health, particularly heart and brain health, is dependent on physical, social, mental and spiritual wellbeing. Diet and nutrient can influence physical and mental performance as well as behavior which are characteristics of total health including heart health. The food and nutrient intake among Paleolithic *Homo sapiens*, hunter-gatherers and among early peasants, Asian and *Homo economicus* Western populations show marked variations [1-3] (Table 1,2). While foods available to Hunter-gatherers were heart friendly, foods consumed by modern men have adverse effects on risk factors of cardiovascular diseases.

* Correspondence: Prof. Dr. R.B. Singh, MD, FICC Halberg Hospital and Research Institute, Civil Lines, Moradabad-10(UP) 244001, India Email: rbs@tsimtsoum.net, drkk@dataone.in. Tel/fax 00919512417437.

The evolution of men and their diet from a nutritional and Agricultural view point can be divided into four categories; Homo sapiens, hunter-gatherers, peasant farmers and modern farmers which have different attributes of social classes, food intakes and behavior. There are very few primitive hunter-gatherers in the world today, but India with its approximately 1.22 billion populations mainly has farmers and the rapidly growing towns having increasing number of poor urban slum dwellers and a big affluent society [1,4]. Cardiovascular diseases (CVDs) and other chronic diseases have become major causes of deaths among affluent society, whereas urban slum dwellers and farmers continue to die more often due to communicable diseases [5-7]. The causes of death among hunter-gatherers are mainly related to injury; falls, bites and poisonings without any evidence of CVDs and type 2 diabetes [1-4].

Man started farming in Africa for the first time, some 10,000 years ago, an adaptation which in turn led to unprecedented technical development, agriculture, industry and commerce. From that perspective, it seems that the biological functions of most social groups are to trigger the what/when hunter-gatherers' anticipation, socially and physiologically [1-4]. Prior to the Agricultural Revolution and notwithstanding the Neolithic Revolution, sometimes called the Agricultural Revolution concerned with the initial transition from hunter-gather to settled agriculture (BP 10,100), our diet was based on an enormous variety of whole grains and wild plants [1-4]. Hunter-gatherers and farmers also had excellent health characteristics; enormous physical activity with limited or no mental stress, alcoholism, and tobacco intake that are now important behavioural risk factors of cardiovascular diseases (CVDs) and other non-communicable diseases (NCDs). The foods available to early peasants, were not as healthy as during the Paleolithic period but also not too unhealthy like today's modern foods which are high energy but have poor nutrient density and are rapidly absorbed [1-4] (Tables 1, 2). In this editorial, we discuss the functions of various foods and methods of agriculture, in relation to nutrients, available during early period, compared to modern foods and agriculture.

Table 1. Food and nutrient intake among various social groups

Food and nutrient	Hunter-gatherer society	Western society	Asian societies
Energy density	Low	High	Low
Protein	High	Low-moderate	Low
Animal	High	Low –moderate	Low
Vegetable	Very low	Low –moderate	Low
Carbohydrate	Low-moderate (slowly absorbed)	Moderate-rapidly absorbed	High-Slow
			High
Fiber	High (>30g)		Low
Fat	Low	Low (<15g)	Low
Animal	Low	High	High
Vegetable	Very low	High	0.5-0.85g
Total w-3	High(2.3g/day)	Low (0.2g/day)	25-50
Ratio w-6:w-3	Low 2.4	High 15-20	Moderate
Vitamins and minerals	High	Low	

Modified from Eaton et al. [2,3] and Singh et al. [6] references [1,10].

Table 2. Nutrition in transition and emergence of non-communicable diseases

Homo sapiens diet given in tables 1-5.	Pattern 1: Huntergatherers	Pattern 2: Food Scarcity-Poverty	Pattern 3: Receding Food Scarcity & Poverty	Pattern 4: More food, less execise- *Homo economicus*	Pattern 5: Healthy Behavior-*Homo modestis*
Nutrition profile					
Diet	Plants, low-fat wild animals, diet diversity by collecting foods.	Cereals predominant, diet less varied	*Fewer starchy staples; more fruit, vegetables, animal protein; low variety continues*	More fat (animal products, trans fat, w-6 fat), sugar, processed foods; less fiber, less w-3 fat and flavonoids	Higher-quality fats, reduced refined carbohydrates, more whole grains, fruit, vegetables rich in w-3 and flavonoids
Nutritional status	Robust, lean population; few nutritional deficiencies	Children and women suffer most from low fat intake, nutritional-deficiency disease emerge, stature declines	Continued MCH[1] nutrition problems, many deficiencies disappear, weaning diseases emerge, stature grows	Obesity, problems for elderly (Osteoporosis, fractures etc), type 2 diabetes, hypertension, stroke, heart attack, brain degeneration, Psychological disorders,cancer	Reduction in body fat and obesity, and NCDs,improvement in bone health Epigenetic modulation and transgenerational epigenetic inheritance-natural selection.
Economy	Hunter-gatherers	Agriculture, animal husbandry, homemaking begin; shift to monocumono cultures	Second agricultural revolution (crop rotation, fertilizer), Industrial Revolution, women join labor force	Fewer jobs with heavy physical activity, service sector and mechanization, household technology revolution	Service sector mechanization and industrial robotization dominate, increase in leisure exercise offsets sedentary jobs
Household	Primitive, onset of fire	Labor-intensive, primitive technology begins (clay cooking vessels)	Primitive water systems, clay stoves, cooking technology advances	Household technology mechanizes and proliferates	Significant reduction in food preparation costs as a result of technologic change
Income and assets	Subsistence, primitive stone tools	Subsistence, few tools	Increases in income disparity and agricultural tools industrialization	Rapid growth in income and income disparities, technology proliferation	Decrease in income growth, increase in home and leisure technologies

Table 2. (Continued)

Homo sapiens diet given in tables 1-5.	Pattern 1: Huntergatherers	Pattern 2: Food Scarcity-Poverty	Pattern 3: Receding Food Scarcity & Poverty	Pattern 4: More food, less execise- *Homo economicus*	Pattern 5: Healthy Behavior-*Homo modestis*
Professional skill/Education	Hunting	Stock breeding , cultivation	Industry, intensive agriculture	Processed unhealthy foods increased	Functional foods availability increases
Demographic profile Mortality	Low fertility, high mortality, low life expectancy	Age of Malthus; high natural fertility, short life expectancy, high infant and maternal mortality	Mortality declines slowly, then rapidly; fertility static, then declines; small, cumulative population growth, which later explodes	Life expectancy hits unique levels (ages 60–70), huge decline and fluctuations in fertility (eg, postwar baby boom)	Life expectancy extends to ages 70 and 90 y, disability-free period increases
Age structure	Young population	Young, very few elderly	Chiefly young, shift to older population begins	Rapid decline in fertility, rapid increase in proportion of elderly person	Increases in the proportion of elderly >75 y of age
Housing	Rural, low density	Rural, a few small, crowded cities	Chiefly rural, move to cities increases, international migration begins, megacities develop	Dispersal of urban population decrease in rural green space	Lower-density cities rejuvenate, increase in urbanization of rural areas encircling cities
Food processing	Nonexistent	Food storage begins	Storage processes (drying, salting) begin, canning and processing technologies emerge, increases in food refining and milling	Numerous food-transforming technologies	Technologies create functional foods and food constituent substitutes (eg, macronutrient substitutes)

[1] MCH, maternal and child health, Modified from Popkin 2006, Singh et al. 2013, References [7, 8].

Food and Agricultural Transition

The modern affluent society of the 20th century in the Western World started large scale use of fertilizers and biotechnology for rapid growth of crops for greater yield of foods. Refining and processing of foods, storing and distributing them became widespread in the continuous search of a better economic model [5-8]. Industrialization and urbanization leading to economic development and affluence were associated with greater availability of foods to populations in middle and high income countries. However these foods are high in energy and fat but poor in nutrient density, resulting in to a decrease in the consumption of omega-3 fatty acids, vitamins, antioxidants and amino acids and significant increase in the intakes of carbohydrates, (mainly refined), fat (saturated, trans fat and linoleic acid) and salt compared to the early peasants and Paleolithic period (Table 1,2). The protein or amino acid intake was 2.5 fold greater (33 vs. 13%) in the Paleolithic diet of *Homo sapiens* compared to modern Western diet consumed by *Homo economicus* populations.

It has been estimated that diets of *Homo sapiens* was characterized by higher intakes by essential and non-essential amino acids, calcium, potassium, magnesium, flavonoids and w-3 fatty acids whereas modern Western diet of *Homo* economicus has excess of energy-rich refined carbohydrates, w-6, trans fat and saturated fat and low in protective nutrients (Table 1,2). Food and nutrient intakes have changed significantly, during the last 100-160 years causing increased intake of saturated fatty acids (SFA), trans fatty acids and linoleic acid and meat from grain fed cattle, tamed at farm houses, rather than meat from running animals [6].

There is a greater consumption of pro-atherogenic foods during transition from lower social classes 4-5 (poverty) to higher social classes 1-3 [5-8]. These differences may be associated with differences in the food and nutrient intake among peasants and among Western and Asian populations during transition from poverty to affluence (Tables 1, 2). Western type foods consumed by lower social classes in the developed countries and higher social classes in developing countries may cause marked reductions in the consumption of anti-atherogenic protective foods and significant increases in the intakes of pro-atherogenic foods. These dietary changes in conjunction with sedentary behavior, mental strain, pollution, tobacco consumption and alcoholism, particularly after 1910, during the last 100 years may have caused damage to our genes, leading to emergence of phenotypes of CVDs and other chronic diseases. It is possible that the lower social classes 4 and 5 need food security as well as functional food security, so that they do not suffer from CVDs like higher social classes 1-3, who became victims of food security during transition from poverty to affluence.

Food Security and Functional Food Security

FAO's latest estimates indicate that the proportion of the world's population suffering from undernourishment is around 12.5 percent, down from almost half of the world's population in 1947 [9].This is a remarkable achievement, yet 868 million people remain hungry, an estimated 2 billion people suffer from one or more micronutrient deficiencies and an estimated 1.4 billion people are overweight, of whom 500 million people are obese. Today, 90% of the world's food supply is from 17% of plant species, which is mainly contributed by grains produced by fertilizer based on rapidly grown crops.

Wheat, corn and rice account for three fourths of the world's grain production on which humans are dependent for their food supply [6-9]. Grains are high in omega-6 fatty acids and carbohydrates and low in omega-3 fatty acids and antioxidants compared to leafy green vegetables. Green leafy vegetables are also rich sources of antioxidants, magnesium, w-3 fatty acids and carotenoids which appear to be high in the Mediterranean region [1-5]. There are fundamental contributions of Agriculture to human nutrition through production, prices and incomes.

However, agriculture and the broader food system including post-harvest processing, distribution and retailing, can contribute much more. Food systems as a whole, from production through consumption, can be made more nutrition-enhancing and more environmentally sustainable through a number of specific actions that are identified in the report [9]. FAO of the UNO emphasizes the persistence of undernourishment, under-nutrition and micronutrient deficiencies and the emergence of overweight and obesity in many parts of the world. Multiple forms of malnutrition can coexist within the same country, household and individual. Functional food security recognizes the potential of food systems to become both more sustainable and more supportive of good nutritional outcomes and the need for a multi-sectoral approach that includes agriculture and food systems, health, sanitation, social protection, employment and education. There is a need to collaborate all relevant stakeholders; health care professionals, food and Agricultural scientists to make a concerted effort to close the gaps that remain regarding basic data on diets and on nutritional status and development health or diseases. Further studies to conduct impact evaluation of agricultural food-based interventions on nutritional outcomes; encourage management practices and technologies to improve sustainability and nutrition; evaluate the impacts of markets, trade and market structure on environmental sustainability and human nutrition; and recognize the roles of consumer choice in achieving nutritional and sustainability objectives [9].

Effects of Functional Foods on Cardiovascular Diseases

The protective effects of Paleolithic diets in the form of Mediterranean diet, Indo-Mediterranean diet, Japanese diet, and DASH diet have been documented in several previous studies [10-19].

Therefore, food security of functional diets rather than modern foods, appear to be important considerations for evolving human health and for prevention of NCDs. It is expected that if and only if, these diets are adopted by *Homo economicus*, and as such these standards will help evolve *Homo modestis,* either directly or by epigenetic inheritance or by natural selection [5,8].

Therefore, it is food security of functional foods, rather than modern foods, which appear to be important consideration for evolving human health and for prevention of NCDs.

Utility and Necessity of Biotechnology for Food Production

Biotechnological discoveries and inventions have been observed to improve food production qualitatively as well as quantitatively [9,20-22]. Plant breeding and genetic modifications are latest technologies for developing diversity in foods by altering nutrient content of foods, to solve the problem of functional food security [8]. Green revolution had contributed greatly to staple plant food productivity but without consideration on micronutrient. In certain specific cases, the improvement in the quality and nutrition of foods by altering their composition were also monitored. Agriculture played a vital role during the era of green revolution initiated by Norman Borlaug, the "Father of the Green Revolution" who saved over a billion people from starvation and developed high-yielding varieties of cereal grains. However, these food varieties were deficient in micronutrients which are known to have beneficial effects on health. Therefore, the practice of biotechnology has also upraised concerns about its potential risks to the environment as well as human being resulting in to emergence of NCDs. The stable crops can be enriched with micronutrients using plant breeding strategies. Now at this point of time, agriculture can again play a vital role by developing food varieties high in micronutrients, macronutrients, organic acids, trace mineral, antioxidants, flavonoids and vitamins leading to functional food security. The traits responsible for micronutrient enrichment in crops exist in their wild types or can be said that these traits are present within their genomes which can be used for substantially increasing micronutrient levels in crops without any negative impact on crop productivity [20-22]. The fruits and vegetables which are consumed raw can be subjected for the improvement of micronutrients through imparting new technologies of plant breeding and genetic engineering.

Genetic Engineering provides resources to host genes into plants via mechanisms, different in some respects from classical breeding. A number of genetically engineered variety of foods have been developed, which have become important nutraceuticals; most notably canola, cotton, maize and soybean, were developed employing this modern technology, and at present the traits introduced are herbicide and/or pest tolerance. Gene transfer technology is employed to alter the physical and chemical composition with nutraceutical worth. This technology leads to increase the production in the species, as well as the elevation of resistance to pests, viruses, frost, etc.

In brief, biotechnology can be used to increase protective nutrients; omega-3 fatty acids, monounsaturated fatty acids, flavonoids, amino acids, vitamins and antioxidants and soluble fiber and decrease harmful nutrient contents; saturated fat, linoleic acid, sugar and erucic acid in a food by these two methods. In future, attempt should be made urgently, to develop omega-3 fatty acid and flavonoid rich slowly absorbed foods to provide functional food security, for control of under-nutrition so that there is no increase in obesity which is important for prevention of CVDs and other chronic diseases.

Conflict of interest has not been declared by the authors.

Acknowledgment

Acknowledgment is due to the International College of Nutrition and International College of Cardiology for providing logistic support to write this article.

References

[1] Singh RB, Reddy KK, FedackoJ, De MeesterF, WilczynskaA and D.W. Wilson, Ancient concepts in nutrition and diets in hunter-gatherers. *The Open Nutra* J 2011;4:130-135.

[2] Eaton SB, Konner M, Shostak M. Stone agers in the fast lane: chronic degenerative diseases in evolutionary perspective. *Amer J Med* 1988. 84:739-749.

[3] Eaton SB, Konner M. Paleolithic nutrition. A consideration of its nature and current implications. *N Engl J Med.* 1985;312(5):283–289.

[4] Lindeberg S. Food and Western Disease: Health and Nutrition from an Evolutionary Perspective. Chichester, UK: Wiley-Blackwell; 2010.

[5] Singh RB, Hristova K, Muthusamy, Rastogi SS, Basu TK, Toda E, Takahashi T, Fedacko J, Pella D, De Meester F, Wilson DW, Mondal RN, Ishaq M, Mohideen MR. The adverse effects of wealth on cardiovascular health: a scientific statement of the international college of cardiology. *Cardiology and Angiology: an international journal.*2013;1(1):9-22, 2013, article no.ca.2013.002.

[6] De Meester F. Wild-type land based foods in health promotion and disease prevention: the LDL-CC:HDL-CC model. In: Wild TypeFoods in *Health Promotion and Disease Prevention,* Ed. Fabien DeMeester, RR Watson, Humana Press, NJ 2008, 3-20.

[7] Singh RB, Takahashi T, Nakaoka T, Otsuka K,Toda E, Shin HH, Kyu Lee M, Beeharry V, Hristova K, Fedacko J, Pella D, De Meester F, Wilson DW, Juneja LR. Nutrition in transition from Homo sapiens to *Homo economicus. The Open Nutra J* 2013;6:6-17.

[8] Popkin BM. Global nutrition dynamics; the world is shifting rapidly toward a diet linked with noncommunicable diseases. *Am J Clin Nutr* 2006:83:289-298.

[9] FAO. UNO. The State of Food and Agriculture: Sustainable Food Systems for Food Security and Nutrition http://www.fao.org/docrep/meeting/028/mg413e01.pdf.

[10] Singh RB, Rastogi SS, Niaz MA, Ghosh S, Singh R: Effects of fat modified and fruits vegetable enriched diets on blood lipids in the Indian diet heart study. *Am. J. Cardiol,* 1992, 69: 869-874.

[11] Singh RB, Rastogi SS, Verma R,Bolaki L, Singh R,Ghosh S. An Indian experiment with nutritional modulation in acute myocardial infarction. *Am. J. Cardiol.* 1992; 69 : 879-85.

[12] Singh RB, Rastogi SS, Verma R, Laxmi B, Singh R,Ghosh S, Niaz MA. Randomized, controlled trial of cardioprotective diet in patients with acute myocardial infarction : results of one year follow up. *BMJ* 1992;304:1015-9.

[13] De Lorgeril M, Renaud S, Mamelle N et al. Mediterranean alpha-linolenic acid-rich diet in secondary prevention of coronary heart disease. *The Lancet,* vol. 343, no. 8911, pp. 1454–1459, 1994.

[14] Singh RB, Dubnov G, Niaz MA,Ghosh S,Singh R,Rastogi SS,Manor O,Pella D,Berry EM. Effect of an Indo-Mediterranean diet on progression of coronary disease in high risk patients: a randomized single blind trial. *Lancet* 2002,360:1455-1461.

[15] Trichopoulou A, Bamia C, Trichopoulos D. Anatomy of health effects of Mediterranean diets. Greek epic prospective heart study. *BMJ* 2009;338: Cite this as: BMJ 2009;338:b2337,doi:10.1136/bmj.b2337.

[16] Singh RB, Fedacko J, Vargova V, Pella D, Niaz MA, Ghosh S. Effect of Low W-6/W-3 Fatty Acid Ratio Paleolithic Style Diet in Patients with Acute Coronary Syndromes: A Randomized, Single Blind, Controlled Trial, *World Heart J* 2012;4:71-84.

[17] Pella D, Dubnow G, Singh RB, Sharma R. Effects of an Indo-Mediterranean diet on the w-6/w-3 ratio in patients at high risk of coronary artery disease : The Indian paradox. *World Rev Nutr Diet*, Basel, Karger 2003; 92:74-80.

[18] De Meester F. Progress in lipid nutrition.In Simopoulos AP, De Meester F (eds), A Balanced Omega-6/Omega-3 Fatty acid Ratio.Cholesterol and Coronary Heart Disease. World Rev Nutr Diet, Basel, Karger 2009; 100:110-121.

[19] Simopoulos AP. Evolutionary aspects of the dietary omega-6/omega-3 fatty acid ratio: medical implications. In Simopoulos AP, De Meester F (eds), A Balanced Omega-6/Omega-3 Fatty acid Ratio.Cholesterol and Coronary Heart Disease. *World Rev Nutr Diet*, Basel, Karger 2009; 100:1-21.

[20] Ross MW, Graham RD. Breeding for micronutrients in staple food crops from a human nutrition perspective. *J Expl Botany*. 2003;55(396):353-364.

[21] Bouis HE. Enrichment of food staples through plant breeding: a new strategy for fighting micronutrient malnutrition. *Nutrition* 2000. 16, 701-704.

[22] Mishra S, Singh RB. Physiological and biochemical significance of genetically modified foods. *The Open Nutraceuticals Journal*, 2013;6, 18-26.

In: Circadian Cardiology with Focus on Both Prevention ... ISBN: 978-1-63463-956-9
Editors: K. Hristova, A. Shehab, G. Cornelissen et al.

Chapter 20

Plasma C Reactive Protein and Homocysteine Concentration in Relation to Increased Intake of Fruits and Vegetables

Eagappan Kannan*[1], Philip G. Mamatha[2] and Sangeetha C. Deena[3]

[1]Associate Professor, Department of Clinical Nutrition & Dietetics,
PSG College of Arts and Science, Coimbatore, India
[2]Ph.D., Scholar, Department of Foods and Nutrition
(R& D), Bharathiar University, Coimbatore, India
[3]Assistant Professor, Department of Pharmacology,
PSG Institute of Medical Sciences & Research, Coimbatore, Tamil Nadu, India

Abstract

Elevated C-reactive protein (CRP) and plasma total homocysteine have been identified as risk factors for coronary heart disease (CHD) by recent meta-analyses. Improper diet and lifestyle and certain diseases have been found to increase homocysteine levels. Fruits and vegetables, by acting as a source of Vitamin B_6, Vitamin B_{12} and folate lower plasma homocysteine levels. Metabolic syndrome and atherosclerosis in particular, cause a mild elevation in CRP, not to the extent of acute inflammatory conditions. This creates a positive feedback loop, mediated by pro-inflammatory cytokines, to worsen the disease process. Refined sugars, excess salt, and ω-6 fatty acids contribute to the development of atherosclerosis, while complex carbohydrates, ω-3 fatty acids and fruits and vegetables, as sources of vitamins, antioxidants and dietary fiber protect against these diseases.

Keywords: C-reactive protein, Homocysteine, Folic Acid,Vitamin B12, CVD, Fruits and Vegetables

---

* Corresponding Author: Dr Kannan Eagappan, Associate Professor, Department of Clinical Nutrition & Dietetics, PSG College of Arts and Science, Coimbatore, India, email ID: dtkannan@gmail.com.

Introduction

Elevated C-reactive protein (CRP) and plasma total homocysteine (Hcy) are identified as risk factors for cardiovascular disease (CVD). [1] A meta-analysis showed that individuals with CRP plasma concentration (2.4mg/L) were twice as likely to have coronary heart disease (CHD) relative to those with CRP concentration (1.0 mg/L). [2]

Two recent meta-analyses based on prospective studies demonstrated that every 5 mol/L increase in Hcy increased the risk of coronary heart disease (CHD) by 20% [3,4]. In an overview of 17 studies (both retrospective and prospective) examining the relation between total homocysteine and coronary heart disease, Boushey et al. found that 14 studies showed a positive association, with an increased odds of 1.6 for men and 1.8 for women each 5 μmol/l increase in total homocysteine [5].

A study known as the Reykjavik study showed that of over 700 nurses, those in the highest quartile of trans fat consumption had blood levels of CRP that were 73% higher than those in the lowest quartile [6,7,8,9].

Factors That Influence Homocysteine Levels in Blood

The metabolism of homocysteine is shown in Figure 1. It is important to identify modifiable factors that influence plasma total homocysteine levels, because elevated concentrations are associated with an increased risk of cardiovascular disease. [12, 13].

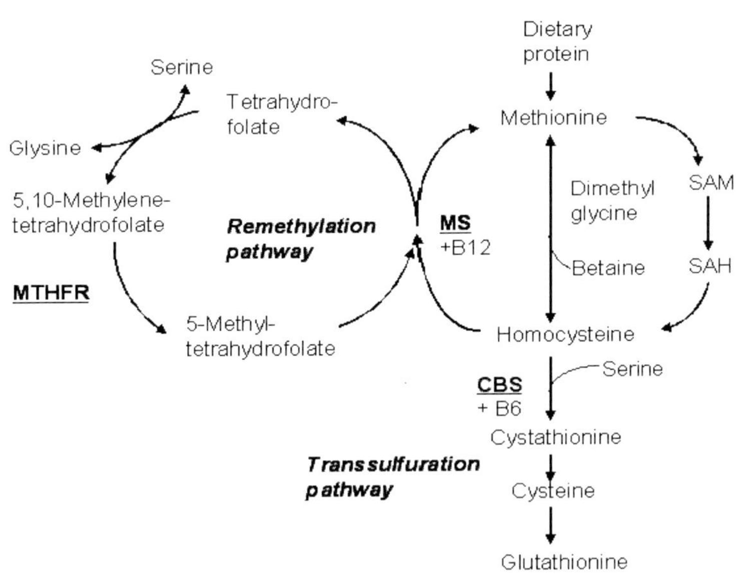

Figure 1. Metabolism of homocysteine (CBS, cystathionine beta-synthase; MS, methionine synthase; MTHFR, methylenetetrahydrofolate reductase; SAH, S-adenosylhomocysteine; SAM, S-adenosylmethionine. Adapted from Fowler B (1997).

Folate intake is identified as the most important dietary determinant of plasma total homocysteine concentration. [14, 15] Besides diet, other lifestyle factors such as smoking, [16] physical activity [17-19] and consumption of alcohol, coffee and tea may have an effect on plasma total homocysteine levels in the general population. [20-23] Studies on the effect of alcohol and coffee consumption on homocysteine levels lack representative populations with a broad age range and create the need to investigate further. In addition, only two studies had corrected for the confounding effect of folate intake. [24] High homocysteine levels are also seen in the following conditions - kidney disease, hypothyroidism, medications (such as antiepileptic drugs and methotrexate) Methylene tetrahydrofolate reductase (MTHFR) genetic mutations and psoriasis (Figure 2). [25]

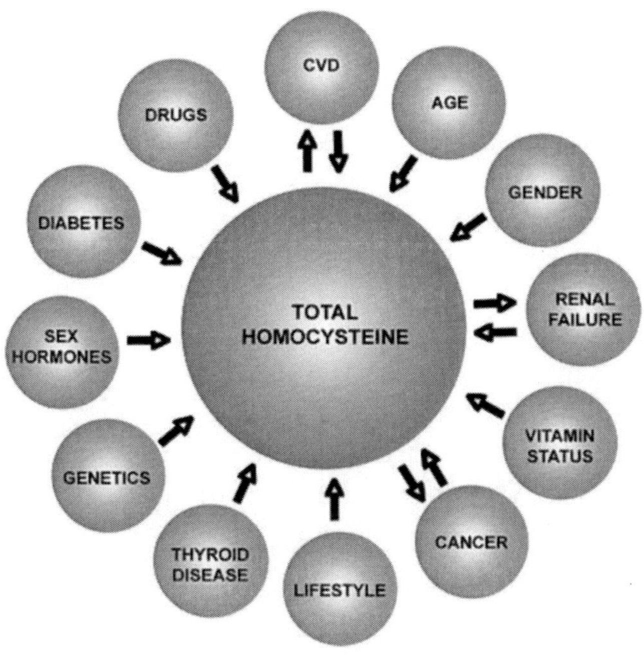

Figure 2. Determinants of total plasma homocysteine levels in blood.

Normal Homocysteine Levels in Blood

Normal and abnormal levels are set by individual laboratories. Typically a low level less than 13 μmol/L is considered normal. A level between 13 and 60 μmol/L is considered moderately elevated and a value greater than 60-100 μmol/L is severely elevated. [26]

Factors That Influence C-Reactive Protein in Blood

The acute phase response develops in a wide range of acute and chronic inflammatory conditions like bacterial, viral, or fungal infections, rheumatic and other inflammatory diseases, malignancy and tissue injury or necrosis. These conditions cause release of interleukin-6 and other cytokines that trigger the synthesis of CRP and fibrinogen by the liver. During the acute phase response, levels of CRP rapidly increase within 2 hours of acute insult, reaching a peak at 48 hours. With resolution of the acute phase response, CRP declines with a relatively short half-life of 18 hours. Measuring CRP level is a screen for infectious and inflammatory diseases. A recent meta-analysis comprising >7000 patients with coronary events showed that subjects with elevated CRP have increased atherosclerosis. It has been well established that several metabolic disturbances that characterize insulin resistance, Type 2 diabetes and hypertriglyceridemia are associated with increased CRP levels. CRP is only a marker of metabolic disturbances increasing the risk for development of cardiovascular disease but has in itself no association with actual cardiovascular disease process. [25, 27].

Normal Levels of Plasma C Reactive Protein in Blood

Levels of CRP less than 1 mg/L are desirable and reflect a low overall cardiovascular risk. Levels of CRP between 1 and 3 mg/L are indicative of moderate risk, while levels of CRP in excess of 3 mg/L suggest quite elevated vascular risk (Figure 3).

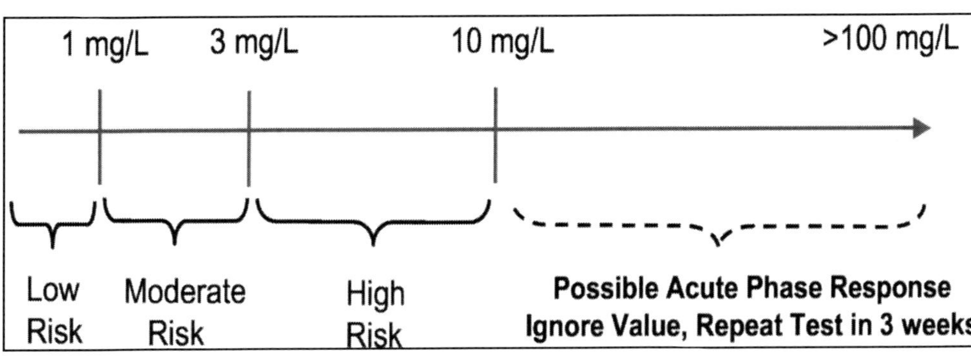

Figure 3. Clinical interpretation of CRP for cardiovascular risk prediction. Adapted from Yeh and Willerson 2003.

Cardio Vascular Disease due to Increased CRP

1. Metabolic disturbances cause atherosclerotic plaque development in arteries & release of CRP from the liver
2. Cytokines are released from atherosclerotic arteries and induce CRP formation in the liver
3. Increased release of CRP from liver, subsequently promotes plaque development in the arteries [28] (Figure 4)

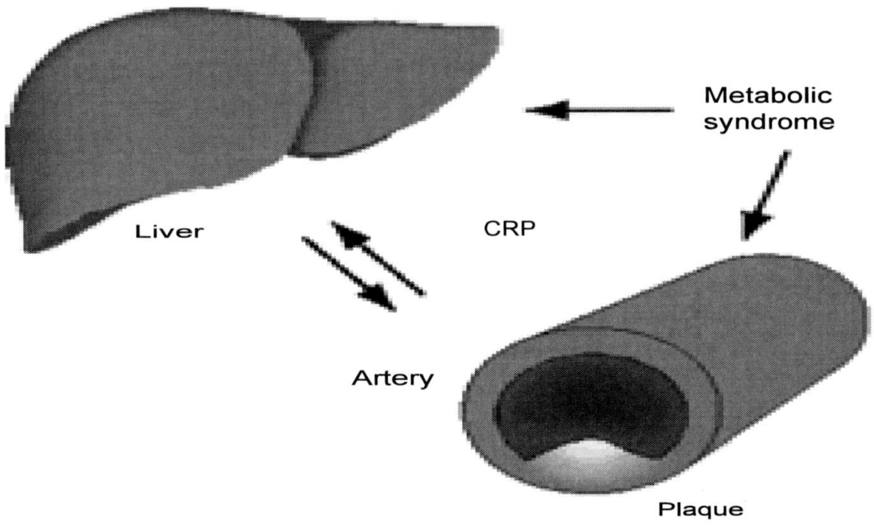

Figure 4. Mechanism of athero-thrombosis due to increased CRP.

Effect of Fruits and Vegetables on Homocysteine Levels in Blood

The factors known to influence homocysteine metabolism are several dietary nutrients like folate, vitamin B_6 and vitamin $B_{12.}$ Homocysteine is re-methylated to methionine by methionine synthase in a reaction that requires methyltetrahydrofolate as a methyl donor and vitamin B_{12} as an enzyme cofactor. [29] Homocysteine is trans-sulfurated to cysteine in reactions that require vitamin B_6. In cross-sectional analyses from the Framingham Heart Study, elevated homocysteine was associated with low serum levels and low dietary intake of folate and vitamin B_6. [30]

Homocysteine Lowering Trialists' Collaboration assessed the effects of folic acid, vitamin B_6, and vitamin B_{12} supplements on homocysteine. In these trials, which typically enrolled persons with elevated homocysteine, high-dose folic acid supplements (often providing ≥ 1 mg/d) have reduced fasting levels of homocysteine. [31] Vitamin B_6 supplements have had little impact on fasting levels of homocysteine but reduced the level of homocysteine after methionine loading. [32] Vitamin B_{12} supplements have reduced homocysteine in individuals with vitamin B_{12} deficiency but have minimal impact in healthy populations. [33,34] However, because of differences in dose and bioavailability, the effects of vitamins derived from food would be different from that of high-dose vitamin supplements.

Current dietary guidelines recommend increased consumption of fruits, vegetables, and low-fat dairy products. [35, 36] One controlled feeding study conducted in a metabolic ward suggested that folate-deficient diets may raise homocysteine, but the diets were unusual because of the artificially low intake of Folate, just 25 and 90 μg/d. [37]

Clinical trials have using B-complex vitamins to reduce blood levels of homocysteine failed to prevent cardiovascular events in people with advanced atherosclerosis. [38] The Life Extension Foundation believes these studies were seriously flawed, most notably because they used doses of B vitamins that were too low to reduce homocysteine to Life Extension's recommended optimal range of <7-8 μmol/L. At present, medical testing laboratories consider a homocysteine number between 11-15 μmol/L as the upper limit of "normal" despite robust clinical data to the contrary. [39] Consequently, many doctors remain misinformed as to the optimal target range for homocysteine and the doses of homocysteine-lowering nutrients required to achieve this optimal range.

In another double-blind randomized controlled clinical trial in 2009, twenty hypercholesterolemic adults taking the drug Lovastatin were given a daily folate supplement (5 mg) for 8 weeks while 20 patients received a placebo; only the folate-supplemented group experienced decreased blood levels of homocysteine. [40]

A study on the Chinese population showed that reducing blood levels of homocysteine through B-vitamin therapy improved endothelial function in renal transplant recipients with hyperhomocysteinemia. Investigators assigned 36 stable renal transplant recipients with hyperhomocysteinemia to either a B-vitamin treatment group (5 mg folic acid, 50 mg vitamin B_6 and 1000 mg vitamin B_{12} per day) or to a control group (placebo only) for 6 months. Investigators found that homocysteine significantly decreased in the B-vitamin treatment group compared with baseline (12.6 vs. 20.1 μmol/l) and no significant changes in homocysteine levels were observed in the control group. Vasodilatation responses were significantly improved in the treatment group compared to controls. [41]

Fourteen interventional studies [42] showed substantial decreases in average total homocysteine levels after the administration of 650 to 10 000 μg of supplemental folic acid. Homocysteine concentrations do not appear to reach a plateau until folate intakes approach 400 μg/d and serum folate reaches ≥15 μmol/L. [43] Other studies have shown significantly lower homocysteine levels in persons taking supplements containing folic acid than in those relying on diet. Thus, a supplement containing 400 μg of folic acid is expected to produce an average reduction of 5 μmol/L of homocysteine than nonusers of supplements. [42, 44]

It has been suggested that plasma homocysteine concentration increases after tissue damage, and that elevated levels of homocysteine further promote endothelial damage. A high plasma homocysteine level would thus be an indicator of tissue damage and a promoter or enhancer of inflammatory thickening and vascular damage. [45] Findings from a Finnish study support the possibility that homocysteine is a consequence rather than a cause of vascular damage and disease [46] In addition, some epidemiological studies have observed that low serum folate and vitamin B_6 concentrations increase the risk for vascular diseases, and the elevated plasma homocysteine would thus be a marker of the low vitamin concentrations [47]. These findings are strengthened by a recent study showing that folic acid supplementation improves endothelial function in patients with coronary artery disease independently of plasma tHcy reduction. [48, 49]

Vitamin B supplementation is not the only option to increase plasma folate levels and to reduce plasma homocysteine concentration. Dietary vegetables and fruits are important contributors to dietary folate intake, which may influence the risk of CVD as well. It is, therefore, worthwhile to study in detail the effects of dietary factors and wholesome diets rich in fruits and vegetables on the plasma levels of folate and homocysteine. In addition, the influence of genetic factors, such as the polymorphisms of key enzymes in homocysteine metabolism, and the possible interaction between diet and genes on the dietary response of plasma homocysteine should be clarified.

A study investigates the influence of modifications in the intake of vegetables, berries, fruits and dietary fat on the plasma concentrations of antioxidants, lipids, lipoprotein (a), oxidised LDL, folate, and homocysteine. The effects of a daily consumption of common vegetables, berries, and citrus fruit (altogether about 700 g per day) along with a diet that was low in total and saturated fat and relatively high in polyunsaturated fat, on the plasma levels of antioxidants, folate, and homocysteine, on serum paraoxonase-1 activity, and on plasma oxidised LDL of healthy women was determined. The consumption of increased amounts of vegetables, berries, and citrus fruit resulted in enhanced plasma levels of antioxidants, such as alpha-carotene, beta-carotene, beta-cryptoxanthin, lutein-zeaxanthin, and vitamin C. The high vegetable diet decreased the plasma total and LDL cholesterol concentrations, but did not result in lower plasma levels of Lp (a) and circulating oxidised LDL, when compared to the low vegetable diet. In conclusion, the observed changes in response to the high vegetable diet were mainly positive: reduced plasma homocysteine and total and LDL cholesterol concentrations and enhanced plasma levels of several antioxidants and folate. [50]

Two studies in the Indian subcontinent showed that a diet rich in fruits and vegetables, low on calories, rich in complex carbohydrates polyunsaturated fatty acids and dietary fiber had a beneficial lowering effect on total serum cholesterol, LDL and triglycerides [51, 52]

Effect of Fruits and Vegetables on Plasma C-Reactive Protein Levels in Blood

Increasing fruit & vegetable consumption was shown to protect against CVD. The Framingham Heart study [53-56] showed that the age adjusted risk ratio was 0.78 for every additional 3 servings of fruit and vegetables. In the Nurses' Health Study and the Health Professionals follow up study, a protective effect of fruit and vegetables against risk for coronary heart disease was also seen. One of the mechanism through which fruits & vegetables prevent CVD is by reduction of plasma homocysteine [57-58].

Intake of fruits and vegetables also reduces the risk of metabolic syndrome through the beneficial combination of antioxidants, fiber, potassium, magnesium and phytochemicals. [59-61] However, only a few studies have related fruit and vegetable consumption to inflammatory status. [62-64]

Inflammation is regarded as a risk predictor for metabolic syndrome and atherogenesis. A meta-analysis of 18 randomised controlled trials showed no significant differences for CRP reduction between the subjects with vitamin-mineral supplementation and placebo control. Sun CH et al. recommended large and well-designed studies to confirm this conclusion. [65] Esmailzadeh et al. evaluated the relation between fruit and vegetable intakes and CRP concentrations and the prevalence of the metabolic syndrome in 486 women aged 40-60 years. After statistical control for age, body mass index, and waist circumference, mean plasma CRP concentrations were inversely proportional to fruit intake, and thus a lower risk of metabolic syndrome. [66] Intake of fruits, vegetables, vitamin C, E and fibre were negatively associated with CRP in men, but not in women in a cross-sectional study of 1060 persons. [67]

Tokunaga et al. surmise that all non-communicable disease is mediated by inflammation due to interaction of biological systems with dietary factors, including deficiency of nutrient-rich functional foods and an excess of rapidly absorbed energy-rich foods. They also identify from epidemiological studies, that the consumption of highly refined foods, trans and ω-6 fatty acids, salt and sugar in the form of processed and readily prepared refined foods increases as affluence increases, in preference to adhering to a whole-grain and vegetable based diet. [68]

A diet low in cholesterol, saturated and trans fat was found to significantly decrease CRP concentrations. Conversely a diet rich in polyunsaturated fatty acids, fish and fish oils was found to lower CRP levels. [69-73] High consumption of nuts that are rich in monounsaturated fatty acids, PUFA, and arginine has been inversely associated with risk of CVD. The beneficial effect is partially attributable to an anti-inflammatory effect. [74] High intake of refined carbohydrates, with the resulting increase of postprandial hyperglycemia, leads to the increased circulating concentrations of CRP. [75-76] Dietary fiber intake was inversely associated with serum CRP concentration among 3920 adult participants in the Nurses' Health Study. [77] The antiinflammatory effect of dietary fiber may be mediated through reduction of lipid oxidation, normalization of bowel flora, and inhibition of hyperglycemia. [78]

Conclusion

Scientific, clinical and epidemiological evidence indicates a positive association between the plasma level of the amino acid homocysteine and vascular disease. Homocysteine levels are inversely related to both intake and plasma levels of folate; higher intake of fruit & vegetables are associated with a lower risk of the metabolic syndrome. Lower risk of metabolic syndrome may result in low CRP concentration. In India, surveys carried out by the National Sample Survey Organization (NSSO), provide data on time trends & interstate differences in food consumption patterns. According to the NSSO, there is no change in consumption of fruits in the last 10 years in both urban and rural areas. The consumption is much higher in urban areas than in rural. Vegetable consumption in India is better than fruit, urban India consumes more vegetable than rural. Potato consumption is higher in rural areas. Poverty and lack or purchasing power have been identified as the two major factors responsible for the poor dietary intake in India. According to the NSSO, there has been a substantial increase in fat intake in both rural & urban areas, over the last three decades. In view of known inter-state differences in dietary intake & nutritional status, it is important to analyse the state-wise data and modify the intervention programmes at the state level to cope with problems. [79,80] These findings support the current dietary recommendation to increase the intake of fruits & vegetables as a primary preventive measure against cardiovascular disease. [66].

References

[1] Ridker PM, Buring J E, Cook NR & Rifai N; C-reactive protein, the metabolic syndrome, and risk of incident cardiovascular events: an 8-year follow-up of 14719 initially healthy American women; *Circulation;* 2000; 107:391-397.

[2] Danesh J, Whincup P, Walker M, Lennon L, Thomson A, Appleby P, Gallimore JR & Pepys MB; Low grade inflammation and coronary heart disease: prospective study and updated meta-analyses; *Br. Med. J.;* 2000; 321: 199–204.

[3] Ueland PM, Refsum H, Beresford SA & Vollset SE; The controversy over homocysteine and cardiovascular risk; *Am. J. Clin. Nutr.;* 2000; 72:324–332.

[4] Ford ES, Smith SJ, Stroup DF, Steinberg KK, Mueller PW &Thacker SB; Homocysteine and cardiovascular disease: a systematic review of the evidence with special emphasis on case-control studies and nested case-control studies; *Int. J. Epidemiol.;* 2002; 31: 59–70.

[5] Boushey CJ, Beresford SA, Omenn GS, Motulsky AG; A quantitative assessment of plasma homocysteine as a risk factor for vascular disease. Probable benefits of increasing folic acid intakes; *JAMA*; 1995 Oct 4; 274 (13):1049–1057.

[6] Lopez-Garcia E, Schulze MB, Meigs JB, Manson JE, Rifai N, Stampfer MJ, Willett WC, Hu FB; Consumption of trans fatty acids is related to plasma biomarkers of inflammation and endothelial dysfunction; *J. Nutr.*; 2005 March; 135 (3): 562–6.

[7] Danesh J, Wheeler JG, Hirschfield GM, Eda S, Eiriksdottir G, Rumley A, Lowe GD, Pepys MB et al; C-reactive protein and other circulating markers of inflammation in the prediction of coronary heart disease; *N. Engl. J. Med.*; 2005; 350 (14): 1387–97.

[8] Koenig, Wolfgang (2006). C-reactive protein - a critical cardiovascular risk marker; CRPhealth.com.

[9] Pepys MB, Hirschfield GM, Tennent GA, Gallimore JR, Kahan MC, Bellotti V et al; Targeting C-reactive protein for the treatment of cardiovascular disease; *Nature;* 2006; 440 (7088): 1217–21.

[10] Ridker PM, Danielson E, Fonseca FA, Genest J, Gotto AM, Kastelein JJ et al; Rosuvastatin to prevent vascular events in men and women with elevated C-reactive protein; *N. Engl. J. Med.*; 2008 Nov; 359 (21) 207.

[11] Jonathan Emberson; C-reactive protein concentration and the vascular benefits of statin therapy: an analysis of 20,536 patients in the Heart Protection Study; *Lancet;* 2011; 377 (9764): 469–76.

[12] Boushey CJ, Beresford SA, Omenn GS et al; A quantitative assessment of plasma homocysteine as a risk factor for vascular disease: probable benefits of increasing folic acid Christen WG, Ajani UA, Glynn RJ et al; Blood levels of homocysteine and increased risks of cardiovascular disease: causal or casual?; *Arch. Intern. Med.*; 2000; 160:422–34.

[13] Selhub J, Jacques PF, Wilson PW et al; Vitamin status and intake as primary determinants of homocysteinemia in an elderly population; *JAMA;* 1993; 270:2693–8.

[14] De Bree A, Verschuren WM, Blom HJ, et al; The association between B vitamin intake and plasma homocysteine concentration in the general Dutch population aged 20–65 years; *Am. J. Clin. Nutr.*; 2001; 73:1027–33.

[15] Nygard O, Refsum H, Ueland PM, et al. Major lifestyle determinants of plasma total homocysteine distribution: The Hordaland Homocysteine Study. *Am. J. Clin. Nutr.*; 1998; 67:263-70.

[16] Nygard O, Vollset SE, Refsum H, et al. Total plasma homocysteine and cardiovascular risk profile: The Hordaland Homocysteine Study; *JAMA;* 1995; 274:1526–33.

[17] Ubbink JB, Fehily AM, Pickering J, et al. Homocysteine and ischaemic heart disease in the Caerphilly cohort. *Atherosclerosis*; 1998; 140:349–56

[18] Giles WH, Kittner SJ, Croft JB, et al. Distribution and correlates of elevated total homocyst(e)ine: The Stroke Prevention in Young Women Study. *Ann. Epidemiol.* 1999; 9:307–13.

[19] Nygard O, Refsum H, Ueland PM, et al. Coffee consumption and plasma total homocysteine: The Hordaland Homocysteine Study. *Am. J. Clin. Nutr.* 1997;65:136–43.

[20] Stolzenberg, Solomon RZ, Miller ER, Maguire MG, et al. Association of dietary protein intake and coffee consumption with serum homocysteine concentrations in an older population. *Am. J. Clin. Nutr.* 1999;69:467–75.

[21] Oshaug A, Bugge KH, Refsum H. Diet, an independent determinant for plasma total homocysteine: a cross sectional study of Norwegian workers on platforms in the North Sea. *Eur. J. Clin. Nutr.* 1998;52:7–11.

[22] Fung TT, Rimm EB, Spiegelman DS, Rifai N, Tofler GH, Willet WC, Hu FB; Association between dietary patterns and plasma biomarkers of obesity and cardiovascular disease risk; *Am. J. Clin. Nutr.*; 2001; 73: 61-67.

[23] *Nieto FJ, Comstock GW, Chambless LE, et al. Coffee consumption and plasma homocyst(e)ine: results from the Atherosclerosis Risk in Communities Study. (Letter). Am. J. Clin. Nutr.* 1997*;66:*1475*–7.*

[24] *Malinow MR, Bostom AG, Krauss RM. Homocysteine, diet and cardiovascular diseases: a statement for healthcare professionals from the Nutrition Committee, American Heart Association. Circulation.* 1999*;* 99*:*178*–182*

[25] *Ross R. Atherosclerosis: an inflammatory disease. N. Engl. J. Med.* 1999*;*340*:* 115*–126.*

[26] *Danesh J, Wheeler JG, Hirschfield GM, Eda S, Eiriksdottir G, Rumley A, Lowe GD, Pepys MB, Gudnason V. C-reactive protein and other circulating markers of inflammation in the prediction of coronary heart disease. N. Engl. J. Med.*.2004*;* 350*:* 1387*–1397.*

[27] From the Department of Clinical Sciences, Malmö University Hospital, Lund University, Sweden. Department of Clinical Sciences, Malmoe University Hospital, E-mailjan.nilsson@medforsk.mas.lu.se

[28] Mason JB, Miller JW; The effects of vitamins B_{12}, B_6, and folate on blood homocysteine levels; *Ann. N Y Acad. Sci.*; 1992; 669:197–204.

[29] Selhub J, Jacques PF, Wilson PWF, et al; Vitamin status and intake as primary determinants of homocysteinemia in an elderly population; *JAMA*; 1993; 270: 2693–2698.

[30] Homocysteine Lowering Trialists' Collaboration. Lowering blood homocysteine with folic acid based supplements: meta-analysis of randomised trials; *BMJ;* 1998; 316: 894–898.

[31] Franken DG, Boers GHJ, Blom HJ et al; Effect of various regimens of vitamin B_6 and folic acid on mild hyper-homocysteinemia in vascular patients; *J. Inherit Metab. Dis.;* 1994; 17: 159–162.

[32] Brattstrom LE, Israelsson B, Jeppsson JO et al; Folic acid: an innocuous means to reduce plasma homocysteine; *Scand. J. Clin. Lab. Invest;* 1988; 48: 215–221.

[33] Sumner AE, Chin MM, Abrahm JL et al; Elevated methyl-malonic acid and total homocysteine levels show high prevalence of vitamin B_{12} deficiency after gastric surgery; *Ann. Intern. Med.;* 1996; 124: 469–476.

[34] US Department of Health and Human Services. Nutrition and Your Health: Dietary Guidelines for Americans; 4th ed. Washington DC: US Government Printing Office; 1995.

[35] Tucker KL, Selhub J, Wilson PWF et al; Dietary intake pattern relates to plasma folate and homocysteine concentrations; *J. Nutr.*; 1996; 126: 3025–3031.

[36] Jacob RA, Wu M, Henning SM et al; Homocysteine increases as folate decreases in plasma of healthy men during short-term dietary folate and methyl group restriction; *J. Nutr.*; 1994; 124: 1072–1080.

[37] Albert CM, Cook NR et al; Effect of folic acid and B vitamins on risk of cardiovascular events and total mortality among women at high risk for cardiovascular disease: a randomized trial; *JAMA;* 2008; 299: 2027-2036.

[38] Guo H, Chi J et al; Influence of folic acid on plasma homocysteine levels & arterial endothelial function in patients with unstable angina; *Indian J. Med. Res.;* 2009; 129: 279-284.

[39] Shidfar F, Homayounfar R et al; Effect of folate supplementation on serum homocysteine and plasma total antioxidant capacity in hyper-cholesterolemic adults under lovastatin treatment: a double-blind randomized controlled clinical trial; *Arch. Med. Res.;* 2009; 40: 380-386.

[40] Xu T, Zhang XW et al; Treatment of hyper-homocysteinemia and endothelial dysfunction in renal transplant recipients with B vitamins in the Chinese population; *J. Urol.*; 2008; 179: 1190-1194.

[41] Selhub J, Jacques PF, Bostom AG et al; Association between plasma homocysteine concentrations and extra-cranial carotid-artery stenosis; *N. Engl. J. Med.;* 1995; 332: 286–291.

[42] Franken DG, Boers GHJ, Blom HJ, et al. Effect of various regimens of vitamin B_6 and folic acid on mild hyperhomocysteinemia in vascular patients. *J. Inherit Metab. Dis.* 1994;17:159–162

[43] Boushey CJ, Beresford SAA, Omenn GS, Motulsky AG. A quantitative assessment of plasma homocysteine as a factor for vascular disease: probable benefits of increasing folic acid intakes. *JAMA.* 1995;472:1049–1057.

[44] Dudman NP. An alternative view of homocysteine. *Lancet* 1999;354: 2072–2074.

[45] Knekt P, Reunanen A, Alfthan G, Heliovaara M, Rissanen H, Marniemi J & Aromaa A. Hyperhomocystinemia: a risk factor or a consequence of coronary heart disease? *Arch. Intern. Med.* 2001;161: 1589–1594.

[46] Folsom AR, Nieto FJ, McGovern PG, Tsai MY, Malinow MR, Eckfeldt JH, Hess DL & Davis CE. Prospective study of coronary heart disease incidence in relation to fasting total homocysteine, related genetic polymorphisms, and B vitamins: the Atherosclerosis Risk in Communities (ARIC) study. *Circulation* 1998;98: 204–210.

[47] Doshi SN, McDowell IF, Moat SJ, Payne N, Durrant HJ, Lewis MJ & Goodfellow J. Folic acid improves endothelial function in coronary artery disease via mechanisms largely independent of homocysteine lowering. *Circulation* 2002a;105: 22–26.

[48] Doshi SN, Moat SJ, McDowell IF, Lewis MJ & Goodfellow J. Lowering plasma homocysteine with folic acid in cardiovascular disease: what will the trials tell us? *Atherosclerosis* 2002b;165: 1–3.

[49] Dietary effects on antioxidants, oxidised LDL and homocysteineMarja-Leena Silaste Biocenter Oulu, University of Oulu Department of Internal Medicine, University of Oulu.

[50] Singh RB, Rastogi SS, Niaz MA, Ghosh S, Singh R; Effects of fat-modified and fruits vegetable enriched diets on blood lipids in the Indian diet heart study; *Am. J. Cardiol.;* 1992; 69: 869-874.

[51] Singh RB, Rastogi SS, Verma R,Bolaki L, Singh R,Ghosh S. An Indian experiment with nutritional modulation in acute myocardial infarction. *Am. J. Cardiol.* 1992; 69:879-85.

[52] *Rissanen, T. H., Voutilainen, S., Virtanen, J. K., Venho, B., Vanharanta, M., Mursu, J. & Salonen, J. T. Low intake of fruits, berries and vegetables is associated with excess mortality in men: the Kuopio Ischaemic Heart Disease Risk Factor (KIHD) Study. J. Nutr.* 2003;133:199-*204.*

[53] *Liu, S., Manson, J. E., Lee, I. M., Cole, S. R., Hennekens, C. H., Willett, W. C. & Buring, J. E. Fruit and vegetable intake and risk of cardiovascular disease: the Women's Health Study. Am. J. Clin. Nutr.* 2003;72*:*922-*928.*

[54] *Joshipura, K. J., Ascherio, A., Manson, J. E., Stampfer, M. J., Rimm, E. B., Speizer, F. E., Hennekens, C. H., Spiegelman, D. & Willett, W. C. Fruit and vegetable intake in relation to risk of ischemic stroke. J. Am. Med. Assoc.*1999;282*:*1233-*1239.*

[55] *Bazzano, L. A., He, J., Ogden, L. G., Loria, C. M., Vupputuri, S., Myers, L. & Whelton, P. K. Fruit and vegetable intake and risk of cardiovascular disease in US adults: the first National Health and Nutrition Examination Survey Epidemiologic Follow-up Study. Am. J. Clin. Nutr.* 2002;76*:*93-*99.*

[56] *Gillman, M. W., Cupples, L. A., Gagnon, D., Posner, B. M., Ellison, R. C., Castelli, W. P. & Wolf, P. A. Protective effect of fruits and vegetables on development of stroke in men. J. Am. Med. Assoc.* 1995;273*:*1113-*1117.*

[57] *Joshipura, K. J., Hu, F. B., Manson, J. E., Stampfer, M. J., Rimm, E. B., Speizer, F. E., Colditz, G., Ascherio, A., Rosner, B., Spiegelman, D. & Willett, W. C. The effect of fruit and vegetable intake on risk for coronary heart disease. Ann. Intern. Med.* 2001;134*:*1106-*1114.*

[58] Liu S, Manson JE, Lee IM, et al. Fruit and vegetable intake and risk of cardiovascular disease: the Women's Health Study. *Am. J. Clin. Nutr.*. 2000;72:922– 8.

[59] Genkinger JM, Platz EA, Hoffman SC, Comstock GW, Helzlsouer KJ. Fruit, vegetable, and antioxidant intake and all-cause, cancer, and cardiovascular disease mortality in a community-dwelling population in Washington County, Maryland. *Am. J. Epidemiol.* 2004;160:1223–33.

[60] Bazzano LA, He J, Ogden LG, et al. Fruit and vegetable intake and risk of cardiovascular disease in US adults: the first National Health and Nutrition Examination Survey Epidemiologic Follow-up Study. *Am. J. Clin. Nutr.* 2002;76:93–9.

[61] Clifton PM. Diet and C-reactive protein. *Curr. Atheroscler Rep.* 2003;5: 431–6.

[62] Hilpert KF, Kris-Etherton PM, West SG. Lipid response to a low-fat diet with or without soy is modified by C-reactive protein status in moderately hypercholesterolemic adults. *J. Nutr.* 2005;135:1075–9.

[63] Esposito K, Glugliano D. Diet and inflammation: a link to metabolic and cardiovascular diseases. *Euro Heart J.,* 2006, 27:15-20.

[64] Sun CH, Li Y, Zhang YB, Wang F, Zhou XL, Wang F. The effect of vitamin-mineral supplementation on CRP and IL-6: a systemic review and meta-analysis of randomised controlled trials. *Nutr. Metab. Cardiovasc. Dis.* 2011 Aug; 21(8):576-83.

[65] Esmaillzadeh A, Kimiagar M, Mehrabi Y, Azadbakht L, Hu FB, Willett WC. Fruit and vegetable intakes, C-reactive protein, and the metabolic syndrome. *Am. J. Clin. Nutr.* 2006;84:1489–97.

[66] Oliveira A, Rodríguez-Artalejo F, Lopes C. The association of fruits, vegetables, antioxidant vitamins and fibre intake with high-sensitivity C-reactive protein: sex and body mass index interactions. *Eur. J. Clin. Nutr.*. 2009;63(11):1345-52.

[67] Tokunaga M, Takahashi T, Singh RB, Rupini D, Toda E, Nakamura T et al; Diet, Nutrients and Noncommunicable Diseases; *The Open Nutraceuticals Journal*, 2012, 5, 146-159.

[68] Shen J, Ordovas JM. Impact of Genetic and Environmenta Factors on hsCRP Concentrations and Response to Therapeutic Agents. *Clinical Chemistry*; 2009; 55:2: 256–264.

[69] Pirro M, Schillaci G, Savarese G, Gemelli F, Mannarino MR, Siepi D, et al. Attenuation of inflammation with short-term dietary intervention is associated with a reduction of arterial stiffness in subjects with hypercholesterolaemia. *Eur. J. Cardiovasc. Prev. Rehabil.* 2004;11:497–502.

[70] Baer DJ, Judd JT, Clevidence BA, Tracy RP. Dietary fatty acids affect plasma markers of inflammation in healthy men fed controlled diets: a randomized crossover study. *Am. J. Clin. Nutr.* 2004;79:969 –73.

[71] Lopez-Garcia E, Schulze MB, Manson JE, Meigs JB, Albert CM, Rifai N, et al. Consumption of (n-3) fatty acids is related to plasma biomarkers of inflammation and endothelial activation in women. *J. Nutr.* 2004;134:1806 –11.

[72] Zampelas A, Panagiotakos DB, Pitsavos C, Das UN, Chrysohoou C, Skoumas Y, Stefanadis C. Fish consumption among healthy adults is associated with decreased levels of inflammatory markers related to cardiovascular disease: the ATTICA study. *J. Am. Coll. Cardiol.* 2005;46:120–4.

[73] Giugliano D, Ceriello A, Esposito K. The effects of diet on inflammation: emphasis on the metabolic syndrome. *J. Am. Coll. Cardiol.* 2006;48:677– 85.

[74] Esposito K, Nappo F, Marfella R, Giugliano G, Giugliano F, Ciotola M, et al. Inflammatory cytokine concentrations are acutely increased by hyperglycemia in humans: role of oxidative stress. *Circulation* 2002;106:2067–72.

[75] Levitan EB, Cook NR, Stampfer MJ, Ridker PM, Rexrode KM, Buring JE, et al. Dietary glycemic index, dietary glycemic load, blood lipids, and C-reactive protein. *Metabolism* 2008;57:437– 43.

[76] Ajani UA, Ford ES, Mokdad AH. Dietary fiber and C-reactive protein: findings from National Health and Nutrition Examination Survey data. *J. Nutr* 2004;134:1181–5.

[77] King DE. Dietary fiber, inflammation, and cardiovascular disease. *Mol Nutr Food Res* 2005;49:594–600.

[78] National Accounts Statistics, CSO: http://mospi.nic.in/mospi_cso_rept_pubn.htm

[79] *National Family Health Survey (NFHS-3):* http://mohfw.nic.in/nfhsfactsheet.htm

[80] *National Sample Survey Organisation:* http://mospi.nic.in/mospi_nsso_rept_pubn.htm

[81] *FAO State of World Food Security, FAO Rome -2007*

In: Circadian Cardiology with Focus on Both Prevention ... ISBN: 978-1-63463-956-9
Editors: K. Hristova, A. Shehab, G. Cornelissen et al.

Chapter 21

Can a Changed Food Industry Prevent Cardiovascular and Other Chronic Diseases?

***R. B. Singh*[1], *Fabien De Meester**[1], *Agnieszka Wilczynska*[1], *Toru Takahashi*[2], *Lekh Juneja*[3] and *R. R. Watson*[4]**

[1]The Tsim Tsoum Institute, Krakow, Poland
[2]Graduate School of Human Environment Science, Fukuoka Women's University, Japan
[3]Taiyo Kagaku Co. Ltd, Yokkaichi, Japan
[4]Department of Health Promotion Sciences, Health Sciences Center, College of Public Health, School of Medicine, Tucson, AZ, US

We have published three multi-author reference books; how diet causes and prevents diseases [1-3]. Many of these book authors concluded that diet changes can improve physical, mental, social and spiritual health, by altering the ratio of omega-6/omega-3 fatty acids in the tissues, resulting in to a healthy human being [1-3]. Apart from these seminal volumes, six original research studies have been published in the last one year by the renouned investigators from all over the world, giving the same message, that a Mediterranean style diet particularly with low w-omega-6/omega-3 ratio of fatty acids near 1:1, is protective [4-9]. It seems logical to advise such diets, for prevention of morbidity and mortality due to noncommunicable disease (NCDs), which in the view of WHO and International College of Cardiology, is taking away resources, funds and brains that could have been allocated for human development [10]. The prevalence rates of major NCDs; cardiovascular disease (CVD), cancer, type 2 diabetes and chronic obstructive pulmonary diseases (COPD) are rapidly increasing in almost all countries and are now among the world's biggest killers [9,10]. NCDs are polygenic and multifactorial and pose a major challenge to the health, well-being and prosperity of populations across the world, and cause unnecessary suffering and

* Correspondence: Dr Fabien De Meester, Executive Director, The Tsim Tsoum Institute, Krakow, Poland, fdm@tsimtsoum.net.

premature death [10-15]. The contributing factors are multifaceted, complex. They include health behavior related to population ageing, urbanization, the globalization of trade and marketing, and the resulting progressive increase in unhealthy patterns of living [11-16].

Some of the most beloved foodstuffs; bacon, sausage, and ham are once again found to be key foods driving the association between meat consumption and the world's most common diseases [4] In this large study, involving 10 countries and almost half a million men and women, high consumption of processed meat by middle-aged adults was associated with a nearly two fold greater risk of all-cause mortality, compared with low consumption, over a mean of 12 years. Risk of cardiovascular death, after rigorous modeling, was increased by more than 70% among people eating more than 160 g/day, as compared with those eating 10 to 19.9 g/day. Risk of cancer deaths was also 43% higher among the highest consumers of processed meats. The interesting point was that a signal of increased mortality was seen among the highest consumers of red meat in general, the risk for red meat was much lower that of processed meats and lost statistical significance after correction for measurement error. With the same adjustments and corrections, high processed-meat consumption was associated with an 18% greater risk of all-cause mortality. It seems that processed meats tend to contain more peroxidized fatty acids, including saturated fat and salt, than unprocessed meat (where the fat is often trimmed off), as well as more oxidized cholesterol and additives, as part of the smoking or curing process which may be carcinogenic or precursors to carcinogenic processes and atherogenic. Heme iron is another mechanism, which may interact with oxidized products linking meat consumption to cardiovascular disease (CVD) risk. Eating unhealthy diets go hand in hand with other unhealthy behaviors, including smoking, low physical-activity levels, and low consumption of fruits and vegetables leading to NCDs [1-15].

The REGARD study enrolled black and white Americans aged 45 years or older between 2003-2007 [5]. The factor analysis identified five dietary patterns: Convenience "(Chinese and Mexican foods, pasta, pizza), Plant-based" (fruits, vegetables, legumes), and Southern "(fried foods, organ meats, sweetened beverages), Sweets/Fats" (desserts, added sugars, sweet snacks) and Alcohol/Salads" (alcohol, fats, vegetables). Subjects with a higher adherence to the Southern dietary pattern, experienced a 41% increased risk of stroke (comparing Q4 to Q1: HR=1.41; 95% CI = 1.07, 1.85). However, higher adherence to the plant-based pattern was associated with a 29% reduction in stroke risk (comparing Q4 to Q1: HR=0.71; 95% CI = 0.55, 0.91). The trend across quartiles was <0.001 indicating a dose response for adherence to each pattern. Adding socio-economic status, smoking, physical activity and total energy (calories) intake to the models attenuated the association but the direction remained the same and persisted in sub-group analysis examining only ischemic strokes. The Convenience, Sweets, and Alcohol patterns were not associated with stroke risk. The present study suggests that foods common to US Southern cuisine such as fried foods and sweetened beverages may increase the risk of stroke, while diets rich in legumes, fruits, vegetables, and fish may reduce stroke risk. Interventions focusing on increasing plant-based foods and fish while reducing fried foods and sweetened beverages are needed [5].

Long-term data from two large studies might have more people considering a switch to vegetarianism, as processed and unprocessed meat consumption is associated with a significantly increased risk of all-cause CVD mortality, death from cancer [4, 5]. After adjustment for multiple risk factors, eating one additional serving of meat daily was associated with a 16% increase in the risk of cardiovascular mortality and a 10% increased risk of death from cancer. The PREDIMED Study comprised of a total of 7447 persons, aged 55 to 80 years; including 57% women [6]. A primary end-point event occurred in 288 participants. The multivariable-adjusted hazard ratios were 0.70 (95% confidence interval [CI], 0.54 to 0.92) and 0.72 (95% CI, 0.54 to 0.96) for the group assigned to a Mediterranean diet with extra-virgin olive oil (96 events) and the group assigned to a Mediterranean diet with nuts (83 events), respectively, versus the control group (109 events). Among high risk subjects, a Mediterranean diet supplemented with extra-virgin olive oil or nuts reduced the incidence of major cardiovascular events [6].The Sydney Diet Heart Study included 458 men aged 30-59 years with a recent coronary event [7]. The intervention group received replacement of dietary saturated fats (from animal fats, common margarines, and shortenings) with omega 6 linoleic acid (from safflower oil and safflower oil polyunsaturated margarine). Controls received no specific dietary instruction or study foods and all non-dietary aspects were designed to be equivalent in both groups. An intention to treat, survival analysis approach to compare mortality outcomes by group showed that the intervention group (n=221) had higher rates of death than controls (n=237) (all cause 17.6% *v* 11.8%, hazard ratio 1.62 (95% confidence interval 1.00 to 2.64), P=0.05; cardiovascular disease 17.2% *v* 11.0%, 1.70 (1.03 to 2.80), P=0.04; CAD 16.3% *v* 10.1%, 1.74 (1.04 to 2.92), P=0.04). Inclusion of these recovered data in an updated meta-analysis of linoleic acid intervention trials showed non-significant trends toward increased risks of death from CAD (hazard ratio 1.33 (0.99 to 1.79); P=0.06) and CVDs (1.27 (0.98 to 1.65); P=0.07). It is remarkable that substituting dietary linoleic acid in place of saturated fats increased the rates of death from all causes, as well as CAD, and CVD. An updated meta-analysis of linoleic acid intervention trials showed no evidence of cardiovascular benefit. These are the complications of meta analysis and funded research. These findings could have important implications for worldwide dietary advice to substitute omega- 6 linoleic acid, or omega-3 fatty acids or polyunsaturated fats in general to alter the w-6/w-3 ratio to 1:1 (3).

A randomized, single blind, controlled trial was carried out on 406 patients with acute coronary syndromes (ACS) diagnosed following WHO criteria [8]. An experimental intervention group received Paleolithic style diet characterized by fruits, vegetables, whole grains, almonds and walnuts and the control group fat modified according to the National Cholesterol Education Program Step 1 (prudent) diet. Main outcome measures were compliance with experimental diets at one year and all cause mortality and its association with omega-6/omega-3 fatty acid ratio after a follow up of two years. The experimental group received significantly greater amount of fruits, vegetables and whole grains, almonds, walnuts, and mustard oil and lower amount of refined bread, biscuits and sugar and butter and clarified butter compared to control diet group at one year of follow up. Intervention group was also advised fish and forbidden red meat intake. Total adherence score to Paleolithic style diet and prudent diet were significant in both the groups. Omega-6/Omega-3 fatty acid ratio of the diet which was much higher before entry to the study (32.5±3.3), was brought down to significantly lower content in the Paleolithic style diet group A (n = 204, compared to control group diet B (n = 202) at entry to the study (3.5± 0.76 vs. 24.0± 2.4 KJ/day, p<0.001). The fatty acid ratio remained significantly much lower in the experimental group compared to control group after one year of follow up (4.4±0.56 vs. 22.3±2.1,KJ/day, p<0.001). Total mortality was 14.7% in the Paleolithic style diet group and 25.2% in the control group, after a follow up of two years. The association omega-6/omega-3 ratio of fatty acids with mortality showed a gradient in both the groups independently, as well as among total number of deaths.

A lower omega-6/omega-3 ratio of fatty acids from 1-10 was associated with a significantly lower mortality whereas increase in omega-6/omega-3 fatty acid ratio to more than 10 was associated with an increasing trend in mortality; 1.7% at ratio less than 5 and 19.9% at ratio 30. In a further study, randomly selected records of death of 2222 (1385 men and 837 women) decedents, aged 25-64 years, were examined [9]. The score for intake of prudent foods was significantly greater and the ratio of omega-6/omega-3 fatty acids of the diet significantly lower for deaths due to 'injury' and accidental causes compared to deaths due to non-communicable diseases (NCD). Multivariate logistic regression analysis revealed that after adjustment for age, total prudent foods (OR,CI: 1.11;1.06-1.18 men; 109;1.04-1.16 women) as well as fruits, vegetables, legumes and nuts (1.07; 1.02-1.12 men; 1.05; 1.99-1.11 women) were independently, inversely associated whereas Western type foods (1.02; 0.95-1.09 men; 1.00; 0.94-1.06 women); meat and eggs(1.00-0.94-1.06 men; 0.98; 0.93-1.04 women) and refined carbohydrates (0.98; 0.91-1.05 men, 0.95; 0.89-1.02 women) and high omega-6/omega-3 ratio of fatty acids were positively associated with deaths due to NCDs. Increased intake of high omega-6/omega-3 ratio Western type foods and decline in prudent foods intake may be a risk factor for deaths due to NCDs.

It is clear from above scientific evidence (1-9), that there is a robust case to develop functional foods by the food industry for prevention of NCDs. Such products should have similar effects on human nutrition and metabolism as natural Paleolithic style foods or Mediterranean style foods, without any adverse effects of foods additives and food precessing. There is need to educate the food industry to develop functional foods characterized with low glycemic index and optimal amount of soluble fiber, vitamins, minerals, antioxidants, essential and nonessential amino acids and a balanced ratio of omega-6/omega-3 fatty acids, similar to the notional Palaeolithic diet [15-17].In earlier studies, these foods and nutrients rich in the Paleolithic style and Mediterranean style diets have been found to be protective against NCDs [15-17].

Table 1, from a recent study showed that, there was a significant decrease in the prudent food consumption among victims dying due to NCDs compared to those dying due to other causes [4] (Table 1).

The nutrient consumption pattern during various stages of evolution is shown in Figure 1 which indicates that NCDs are the creation of *Homo economicus* societies. The food industry should explore the possibility of adding these nutrients which have been reported to have nitric oxide activating effects [15-22]. Dietary supplements such as resveratrol and epicatechin present in wines and fruit juices and flavanols rich in cocoa as well as ,w-3 fatty acids, can protect against adverse effects of food processing. Some of the spices such as turmeric, cumin, chillies, garlic and onion rich in antioxidants could be incorporated in the functional foods making them more functional and healthy.

They also provide additional beneficial effects on cardiovascular health, insulin resistance and memory dysfunction [14-17]. In a recent meta-analysis of randomized controlled trials, supplementation with probiotics in critically ill adult patients was found to be associated with a reduction in the incidence of ICU-acquired pneumonia and ICU length of stay [23].

Data from a total of 13 trials including 1,439 patients were included in the analysis. While probiotic administration was not associated with a significant reduction in ICU (OR=0.85) or hospital (OR=0.90) mortality, or shorter duration of mechanical ventilation (-0.18 days) or a shorter hospital length of stay (-0.45 days), probiotics were associated with a reduction in incidence of ICU-acquired pneumonia (OR=0.58) and shorter stay in the ICU (-1.49 days).

Table 1. Food Intakes and w-6/w-3 Fatty Acid Ratio of Diet in Relation to Causes of Death Based on Assessment by Dietary Diaries of the Spouse and Questionnaires filled by the Nutritionist

Causes of Death	Prudent Diet Western	Western Type Diet	Total Foods	Score	w-6/w-3 Ratio
n=1385	Men (Mean± Standard deviation) g/day				
Injury-accidents (n=215)	892±252	202±22	1094±302	7.93±2.8 31	31.3± 5.3
Communicable Diseases.(n=372)	806±237	256±28	1062±198	7.26±2.7	38.2± 6.6
NCDs					
Malignant (n=77)	715±241	412±53	1127±311	6.44±2.5	42.2± 6.8
Circulatory (n= 406)	757±245	437±47	1194±318	6.81±2.6	45.3±8.3
Chronic lung diseases (97)	705±202	405±41	1110±302	6.25±2.2	42.0± 7.4
Kidney diseases (n=163)	617±188	505±55	1122±325	5.72±1.8	41.8±6.1
Diabetes (n=23)	605±175	522±61	1127±334	5.65±1.7	41.6±5.7
Kendall's t	0.045*	0.048**	0.025	0.041*	0.042*
n=837	Women (Mean± Standard deviation) g/day				
Injury-accidents (n= 139)	822±234	186±23	1008±224	8.40±2.2	25.5± 5.7
Communicable diseases (n=194)	736±237	218±33	954±201	7.51±1.9	34.5± 5.6
NCDs(n=502)					
Malignant (n=54)	657±197	305±35	962±221	6.70±1.8	41.6± 6.8
Circulatory (n= 240)	655±205	332±41	987±218	6.68±1.7	44.5± 7.5
Chronic lung diseases (n=95)	660±180	382±48	1029±180	6.12±1.3	40.0± 6.5
Renal diseases (n=87)	565±155	380±130	995±165	5.57±1.1	39.7± 6.8
Diabetes (n=26)	553±146	387±135	940±153	5.53±1.1	40.0±6.5
Kendall's	0.041*	0.067**	0.024	0.043*	0.041*

Values are mean (Standard deviation), *=P<0.01,**=P<0.001. by comparison of food consumption among victims dying due to NCDs and with victims dying of injury and accidents among both sexes. Ref. [4].

This study indicates that this approach appears to be useful for the food industry to develop superfoods by further modifications of probiotics with omega-3 fatty acids, flavanols by adding cocoa, walnuts and black raisins.

In a systematic review of studies which included the results from 14 human studies (7 prospective, 3 cross-sectional, 1 controlled, 3 case-control) and 13 animal studies, found that dietary omega-6 to omega-3 fatty acid ratio can influence brain composition [24], Alzheimer's disease pathology, and behavior (as according to animal studies), and revealed an association between the omega-6 to omega-3 ratio, cognitive decline, and incidence of dementia. This review supports growing evidence of a positive association between the dietary omega-6/omega-3 ratio and the risk of Alzheimer's disease. A recent study involving 3,457 children aged 8 to 15 years, out of which 354 had reduced birth weight (25). Systolic blood pressure was found to be 1.1 mm Hg higher in those with reduced (less than 10th centile) compared with normal birth weight, and pulse pressure was significantly higher (3.4 mm Hg) in children with reduced birth weight. Those with reduced birth weight who were in the highest tertile of EPA and DHA intake were found to have significantly lower systolic blood pressure (-4.9 mm Hg) and pulse pressure (-7.7 mm Hg), than those with normal birth weight. These data are consistent with the hypothesis that long-chain omega-3 fatty acids reduce blood pressure in those with impaired fetal growth.

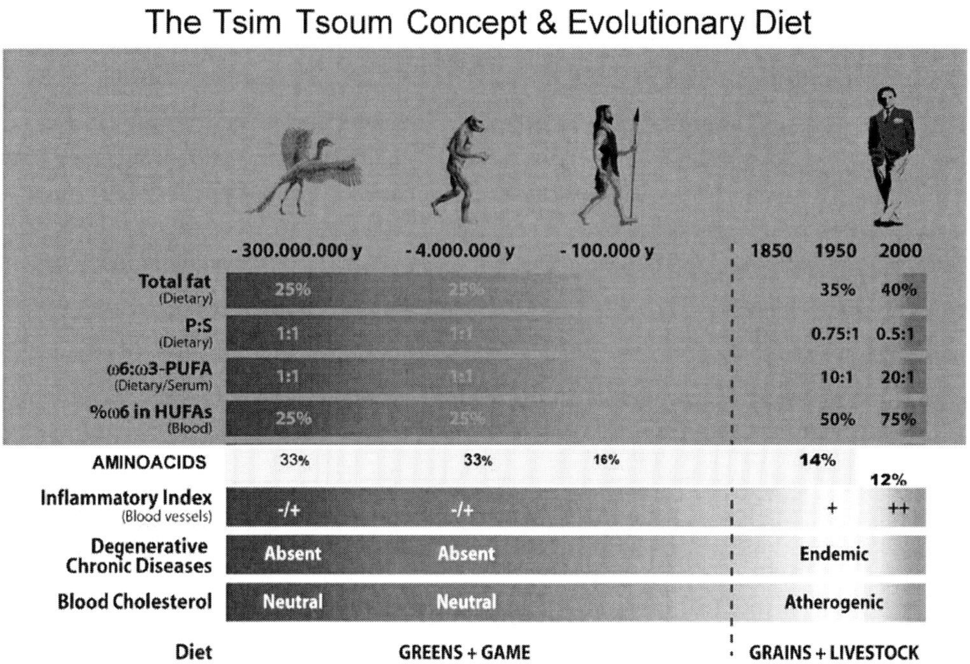

Figure 1. Nutrient intakes among *Homo sapiens, Homo erectus*,hunter gatherers and *Homo economicus* men, ref [3].

Further efforts should be made to educate the people and industry on protective effects of wines and fruit juices and to consider adding nitric oxide activating agents like epicatechin and flavones as well as omega-3 fatty acids in alcohol to provide additional beneficial effects. Some experts from European Union have proposed "Mind, Body Index= BMI" to address total health because physical, social, mental and spiritual health may depend on body composition. It is remarkable that animals including man in the wild do not suffer overweight but *Homo economicus* (Figure 1) do suffer [10]. Even modern husbandry animals do not do so. This is in contrast to companion pets. The human part – the clever mind – appears responsible for the disease due to poor social, mental and spiritual health in the western world. It is important to analyze facts as primary and secondary risk factors. Food is here secondary. It contributes, yet not causes the problem. Just as cholesterol contributes, but does not cause heart disease. (www.columbus-concept.com). Once understood and accepted, such a basic principle allows one to take the right decision about prevention of CVDs and other chronic diseases.

In brief, food industry needs drastic changes to produce and promote slowly absorbed functional foods containing low omega-6/omega-3 fatty acid ratio, vitamins, antioxidants, fiber, flavanols and essential and non-essential amino acids, preferable by using food stuffs rich in these ingredients. Because in earlier studies, supplementation with vitamins and antioxidants showed no beneficial effects on health and diseases. Public and industry should demand no tax subsidy to lower the cost of such foods.

Acknowledgments

To International College of Nutrition and The Tsim Tsoum Institute for providing logistic support to write this article.

References

[1] De Meester F, Zibadi S, Watson RR. *Modern Dietary Fat Intakes in Disease Promotion.* Humana Press, Springer,(www.springer.com) New York 2010.

[2] DeMeester F, Wild-type land based foods in health promotion and disease prevention: the LDL-CC:HDL-CC model. In: *Wild Type Foods in Health Promotion and Disease Prevention*, editors Fabien DeMeester and RR Watson, Humana Press, NJ 2008,p 3-20.

[3] De Meester F, Watson RR, Sherma Zibadi, Editors, Omega-6/3 Fatty acids; Functions, Sustainability Strategies and Perspectives. *Humana Press*, Spernger, (www.springer.com), New York 2012

[4] Rohrmann S, Overvad K, Bueno-de-Mesquita HB, et al., Meat consumption and mortality. Results from the European Prospective Investigation into Cancer and Nutrition. *BMC Med* 2013.

[5] Judd SE, Orlando Gutierrez O, Kissela BM, et al., Southern diet pattern increases risk of stroke while plant-based pattern decreases risk of stroke in the REGARDS study. International Stroke Conference 2013; February 7, 2013*; Stroke.* 2013; 44: A144

[6] PREDIMED Study Investigators. Primary prevention of cardiovascular disease with a Mediterranean diet. *N. Engl. J. Med.* 2013; February 25, 2013, DOI: 10.1056/NEJMoa1200303

[7] Ramsden CE, Zamora D, Leelarthaepin B, Majchrzak-Hong SF, Faurot KR, Suchindran CM, et al., Use of dietary linoleic acid for secondary prevention of coronary heart disease and death: evaluation of recovered data from the Sydney Diet Heart Study and updated meta-analysis. *BMJ* 2013;346:e8707. Cite this as: *BMJ* 2013;346:e8707

[8] Singh RB, Fedacko J, Vargova V, Pella D, Niaz MA, Ghosh S. Effect of low W-6/W-3 fatty acid ratio Paleolithic style diet in patients with acute coronary syndromes: A randomized, single blind, controlled trial, *World Heart J.* 2012;4:71-84.

[9] Fedacko J, Vargova V, Singh RB, Anjum B, Takahashi T, Tongnuka M, Dharwadkar S, Singh S, Singh V, Kulshresth SK, De Meester F, Wilson DW. Association of high w-6/w-3 fatty acid ratio diet with causes of death due to non-communicable diseases among urban decedents in North India. *The Open Nutra Jour.* 2012;5:113-123.

[10] WHO. Mortality and burden of disease estimates for WHO Member States in 2008. Geneva: World Health Organization, 2010.

[11] Singh RB, Takahashi T, Nakaoka T, Otsuka K, Toda E, Shin HH, Kyu Lee M, Beeharry V, Hristova K, Fedacko J, Pella D, De Meester F, Wilson DW, Juneja LR. Nutrition in transition from Homo sapiens to Homo economicus. *The Open Nutra J.* 2013;6: (in press).

[12] Moodie R, Stuckler D, Monteiro C, Sheron N, Thamarangsi , Lincoln P, Casswel S on Behalf of the Lancet NCD Group. Profits and pandemics: prevention of harmful effects

of tobacco, alcohol, and ultra-processed food and drink industries. The Lancet, Early Online Publication, 12 February 2013,doi:10.1016/S0140-6736(12)62089.

[13] The "Heart Disease and Stroke Statistics--2013 Update" and the Need for a National Cardiovascular Surveillance *System Circulation*. 2013;127:21-23.

[14] U. S. Health In International Perspective. Shorter Lives, Poorer Health. National Academy of Sciences, USA 2013. (For the report on US health disadvantage see http://www.nap. edu/catalog.php?record_id=13497;For the US prevention strategysee http://www.healthcare.gov/prevention/nphpphc/strategy/report.pdf.

[15] Editorial. Wealth but not health in USA. *Lancet* 2013;381:177. doi:10.1016/S0140-6736(13)60069-0

[16] Singh RB, Anjum B, Takahashi T, Martyrosyan DM, Pella D, De Meester F, Wilson DW, SSD Beehari, Keim M, Shastun S. Poverty is not the absolute cause of deaths due to non-communicable diseases NCDs. *World heart J.* 2012;4: (in press).

[17] Toda E, Toru T, Singh RB, Alam SE et al., Can low w-6/w-3 ratio Paleolithic style diet stop cardiovascular diseases? Tissue is the Issue. *Amer. Med. J.* 2012;3:183-193.

[18] De Lorgeril M, Renaud S, Mamelle N et al., Mediterranean alpha-linolenic acid-rich diet in secondary prevention of coronary heart disease. The *Lancet,* vol. 343, no. 8911, pp. 1454–1459, 1994.

[19] Singh RB, Dubnov G, Niaz MA, Ghosh S, Singh R, Rastogi SS, Manor O, Pella D, Berry EM. Effect of an Indo-Mediterranean diet on progression of coronary disease in high risk patients: a randomized single blind trial. *Lancet* 2002,360:1455-1461.

[20] Singh RB, Rastogi SS, Ghosh S, Niaz MA, Singh R, Verma R, Randomised controlled trial of cardio protective diet in patients with recent acute myocardial infarction results of one year follow up. *BMJ* 1992; 304: 1015-1019. PMID: 1586782.

[21] Singh RB, Kumar A, Neki NS, Pella D, Rastogi SS,Basu TK, Otsuka K, De Meester F, Wilson DW. Acharya SN, Juneja L, Takahashi T,Diet and Lifestyle Guidelines and Desirable Levels of Risk Factors for Prevention of Cardiovascular Disease and Diabetes among Elderly Subjects. A Revised Scientific Statement of the International College of Cardiology and International College of Nutrition-2011, *World Heart J.* 2011;3:305-320.

[22] Hristova K, Nakaoka T, Otsuka K, Fedacko J, Singh R, Singh RB, De Meester F, Wilczynska A, Wilson DW. Perspectives on Chocolate Consumption and Risk of Cardiovascular Diseases and Cognitive Function. *The Open Nutraceuticals Journal,*2012, 5, 207-212.

[23] Barraud D, Bollaert PE, Gibot S; Impact of the Administration of probiotics on mortality in critically ill Adult patients: A meta-analysis of randomized controlled trials. *Chest* 2013;143:646-55.

[24] Loef M, Walach H. The omega-6/omega-3 ratio and dementia or cognitive decline: a systematic review on human studies and biological evidence. *J. Nutr. Gerontol. Geriatr*, 2013; 32: 1-23.

[25] Skilton MR, Raitakari OT, Celermajer DS.High Intake of Dietary Long-Chain ω-3 Fatty Acids Is Associated With Lower Blood Pressure in Children Born With Low Birth Weight: NHANES 2003-2008. *Hypertension.* 2013 Mar 4. [Epub ahead of print]PMID:23460284.

In: Circadian Cardiology with Focus on Both Prevention ... ISBN: 978-1-63463-956-9
Editors: K. Hristova, A. Shehab, G. Cornelissen et al. © 2015 Nova Science Publishers, Inc.

Chapter 22

Singh's Verbal Autopsy Questionnaire for the Study of Prevalence of Behavioural Risk Factors among Decedents Dying Due to Cardiovascular and Other Causes of Death

Krasimira Hristova[1], J. P. Sharma[2], R. B. Singh*[2], Jan Fedacko[3], Miki Takanuga[4], Toru Takahashi[5] and D. W. Wilson[6]

[1]National Heart Hospital, Sofia, Bulgaria
[2]Halberg Hospital and Research Institute, Moradabad, India
[3]PJ Safaric University, Kosice, Slovakia
[4]Department of Food and Nutrition, Fukuoka Women's University, Japan
[5]Graduate School of Human Environment Science, Fukuoka Women's University, Japan
[6]School of Medicine, Pharmacy and Health, Durham University, UK

Abstract

Background: The pathways for development of non-communicable diseases (NCDs) are sedentary behavior, Western diet, tobacco use, psychosocial stress, and alcoholism among individuals and populations with underlying lack of general and health education. There is only scanty information about the prevalence of health behavior in the population. This study examines the prevalence of health behaviors among victims dying due to NCDs to find out the accuracy of the verbal autopsy questionnaire.

* Correspondence: Prof. Dr. R.B. Singh, MD, FICC Halberg Hospital and Research Institute, Civil Lines, Moradabad-10(UP) 244001, India. Email: rbs@tsimtsoum.net , drkk@dataone.in.

Subjects and methods: Randomly selected death records of 2,222 (1,385 men and 837 women) decedents, aged 25-64 years at time of death, were studied by verbal autopsy questionnaires. All the risk factors and protective factors; prudent diet, moderate physical activity, moderate alcohol intake, active prayer, no tobacco were assessed by questionnaires which was completed with the help of the victim's spouse and a local treating doctor practicing in the concerned field. Lack of knowledge on health education about the role of exercise, prudent diet, and adverse effects of tobacco use and alcoholism was studied by the validated questionnaires. Multivariate logistic regression analysis was conducted to demonstrate the association of protective factors with prevalence of deaths due to cardiovascular diseases.

Results. The prevalence of protective behavior; no tobacco were; 51.6% (n=715) among men and 83.9% (n=702) among women victims. Fruit, vegetable and legume (>250g/day) intake were observed among 51.4% (n=712) men and 50.4% (n=422) women. No subjects were consuming >400g/day of FVL. Western type food (>255g/day) intake were observed among 63.2% (n=875) men and 59.9% (n=502) women. Occupational and spare time physical activity were noted among half of the subjects. Regular practice of yoga/active prayer were observed among 40% of subjects.

Multivariate logistic regression analysis conducted after adjustment of age and body mass index, revealed odds ratio for physical activity (odds ratio, men 1.07(CI 1.02-1.12),P<0.001,women 1.05 (CI 0.99-1.11),P<0.001), prudent diet (men 1.12(1.07-1.19),P<0.001 women 1.08 (1.05- 1.17),P<0.001), active prayer(men1.01 (0.95-1.07),P<0.05,women 0.99 (0.94-1.05)P<0.05) were inversely associated with risk of deaths due to CVDs. No tobacco intake and moderate alcohol intake were significantly positively associated with decreased risk of death due to CVDs among men but not in women.

Conclusion. It is possible that protective health behavior can be accurately assessed by an extended verbal autopsy questionnaire administered to a spouse of the decedents among Indians.

Keywords: mortality, verbal autopsy, social autopsy, tobacco autopsy, dietary autopsy, nutrition, disease, sudden death, stroke, infections, cancer.

Responsibilities: Study design, questionnaire, writing of manuscript by RBS, collection of data by RBS. Others experts; JF, VM, DP, FDM helped in the analysis of data and writing of manuscript, calculations and review of references.

Introduction

The emergence of non-communicable diseases (NCDs) may have a sequence during transition from under-nutrition and poverty to over-nutrition and affluence [1,2]. Overweight and central obesity come first in conjunction with deficiency of certain lifestyle related biomarkers; hyperinsulinemia, angiotensin, oxidative stress and pro-inflammatory cytokines which are important in the pathogenesis and emergence of NCDs [1,2]. This sequence of development of NCDs may be altered by learning and alteration in health behavior which may be accomplished by alteration in public health policies for achieving total health including social, mental and spiritual health, apart from the physical health [3-5].

There has been a substantial increase in the burden of morbidity and mortality due to NCDs which are mainly, in the last few decades in the United States and other emerging countries [6-8]. Mortality and burden of disease estimates for WHO Member States in 2008 revealed that 36·1 million deaths per year occur as a result of NCDs which are mainly due to cardiovascular diseases (CVDs) [8]. These estimates represent almost two out-of-three deaths per year worldwide.

Approximately, two-thirds (63%) of premature deaths in adults (aged 15–69 years), and three-out-of-four of all adult deaths are attributable to NCDs [8]. A recent study from India shows that 57.0 % of deaths in adults (aged 25-64 years) were due to NCDs, (including 31.0% due to CVDs), 25.5% due to communicable diseases, and 15.9% due to injury and accidents, indicating that these findings on deaths due to NCDs are similar to WHO estimates [6,8]. It is clear that the populations having lowest income had lowest prevalence of deaths and those having highest income had highest prevalence of deaths due to NCDs Causes of mortality and behavioral risk factors are not commonly known in developing countries, because autopsy is not possible in many countries due to religious considerations. Therefore International College of Nutrition, International College of Cardiology and the Tsim Tsoum Institute have developed a verbal autopsy questionnaire based on WHO guidelines to find out the causes of deaths in these countries [6,9].

Risk factors of NCDs are significantly more common in the urban population of India [1-5] and among higher social classes 1-3, compared to lower social classes 4-5. Infections and under-nutrition are more common among lower social classes, which may be the commonest cause of death in developing countries [6]. In developed countries, lower social classes have greater cardiovascular and cancer mortality, as well as all-cause mortality than higher social classes [6-8]. Since the public health and medical care resources are insufficient in many countries, accurate information on overall and cause specific mortality are important to prioritize various activities related to prevention and management of diseases. Certification by an attending physician based on clinical manifestations and investigations is the standard method to find out the causes of mortality. Since many deaths occur at home without investigations and without autopsy, in many low and middle income countries, this approach may not be applicable in all countries. It seems that data on causes of deaths and behavior pattern may be potentially biased, based on hospital records [6-8]. Verbal autopsy is a simple and low cost alternative for determining the causes of deaths which is available in most of the developing countries [9].

The reliability of verbal autopsy questionnaires may be enhanced by examining the available medical records of the victims [10-15]. It is also an excellent alternative in other resourceful countries where autopsy is difficult due to various considerations and it could be better than routine death certificates [10-15]. Verbal autopsy could be extended to study the risk factors as well as unhealthy behavior causing CVDs, by interview of the spouse, family doctor and other relatives [16-18]. There is only scant knowledge worldwide, about adoption of healthy lifestyle behaviors among individuals with a coronary artery disease (CAD) or stroke event in communities across a range of countries. Majority of the earlier studies examining the role of diet and lifestyle modification of individuals after vascular events have included subjects within 2 years after the event [19-22]. Unfortunately, only small proportions of subjects could be motivated to follow recommended diet and lifestyle practices as reported in the recent studies [22,23].

In the present study, we report the estimates for the prevalence of protective health behavior, among decedents dying due to CVDs, based on modified verbal autopsy questionnaire which is completed with the help of spouse and available medical records of the victim, in an urban population of north India.

Subjects and Methods

Religious considerations in India, do not allow us to conduct autopsy to find out the exact causes of death among victims dying due to various causes in the hospital and elsewhere. Therefore, records available at the Municipal Board, based on death certificate were the only sources which might not be absolutely correct. In India, many people die at home without consultation and treatment by a qualified doctor and hence medical certificate and records of treatment, indicating causes of deaths remain unknown. These records are based on certificates issued by the doctors based on clinical diagnosis and laboratory examinations or by doctors not qualified in modern medicine and are open to bias.

We studied the records of randomly selected, 2842 adult victims, aged 25-64years, of which 2222 victim's relatives, could respond for this study, during the last two years, July1999 to July 2001 as reported earlier[6]. Of 2222 victims, 1385 were males and 837 were females. All the families of these 2222 victims were contacted individually to find out the causes of deaths by verbal autopsy questionnaire [6]. WHO experts have suggested that the causes of death may be obtained by detailed questionnaires administered to spouse and the doctors involved [9,15]. The head of every family was personally called after communication with the help of the local accessible doctor in the street. The calls were made via letter, telephone and personal contact via street doctor, before any subject or family was declared a non-contact or non-replier.

The verbal autopsy questionnaire was pre-tested by the concerned committee in roughly 30 to 60 families on various aspects concerned with causes of deaths. It was found that 10% of the families who came within reach of survey were declared non-contact or non-responder as observed in another survey [6]. Detailed interviews were possible among 90% of the families approached.

Clinical data on age, sex, height, weight, marital status, occupation, education, past and family history, history of hypertension, diabetes, stroke, heart attack, kidney disease, liver disease, alcohol intake, drug intake, tobacco intake, lung tuberculosis, bronchitis, asthma, cancer, mental diseases, diarrhea and dysentery, and brain, CVD, infectious diseases, malaria, dengue and accidents, etc., were recorded. Leading questions were asked to know the system involved as given in the appendix I. The questionnaire was completed based on medical record of the victim, death certificate issued by the doctor, interview of the doctor and the family doctor, interview of the spouse and other family members, with the help of a pretested verbal autopsy questionnaire, to accurately know the cause of death in each record of the victims (Appendix I). Verbal autopsy questionnaire allowed the interviewer to ascertain the clinical presentation of the victim during illness and up to death based on above documents. This information on probable cause of death related to each body system was reviewed by the internist physician to assess the final cause of death among all the victims. Verbal autopsy questionnaire was extended to study risk factors of the diseases causing death, as well as protective factors; prudent diet, physical activity, regular active prayer no tobacco and moderate alcohol intake, which may be important in the prevention of these diseases in other populations.

Criteria of Diagnosis of Healthy Lifestyle

Subjects who admitted to drinking of alcohol at least 5-10 drinks per week on alternate days were categorized as moderate alcohol-consuming. Physical activity was assessed by questionnaire based on occupational, household and spare time activity. Practice of yoga/active prayer was considered if it was practiced for at least 20 minutes per day, at least 5 days in a week. Prudent dietary habits were considered in presence of fruit, vegetable and legume intake of minimum 400g/day (WHO), No tobacco consumption was considered when there was no past history of tobacco intake and no tobacco consumption during the last 30 days.

Physical Activity Autopsy

The criteria for moderate physical activity were; walking minimum 14.5 km a week, or climbing more than 20 flights of stairs a week, during household or occupational activities or in presence of moderately vigorous spare time physical activity on five days a week. Physical activity autopsy was conducted by verbal autopsy questionnaire as given in the Appendix II.

Tobacco Consumption Autopsy

Tobacco intakes of the victims were studied by completing a questionnaire by asking probing questions to spouse. Smoking of cigarettes, beedies, hukka and pipe as well chewing of tobacco in the form of flavoured tobacco, khaini and gutka were assessed as given in the Appendix III.

Social Class Autopsy

The socio- economic status (SES) of the family was classified based on attributes of housing condition, education, occupation, per capita income and ownership of consumer durables like car, television, automobiles etc in the household [9, 10]. Per capita income was calculated by dividing the total income of the family, by the number of family members. Social classes were graded in to SES 1-5 depending upon scores as given in the Appendix IV.

Dietary Autopsy

Dietary intakes of the victim were obtained by trained interviewers by finding out the food intake of the spouse by using 3-day dietary diaries and filling of questionnaires, by asking probing questions about differences in food intake by the decedents as given in the Appendix V.

Food models, food measures and food portions were used by the dietitian to find out the exact food intakes of the victims. Salt intakes were assessed by finding out the amount of salt mixed in the food divided by the number of family members and then adding salt consumed by each member during eating.

Family doctor or spouse of the victim was asked to suggest a person with identical age, sex, height and weight as that of the victim. In those medical records, in which data on height and weight were not available, we collected these data from these matching subjects.

Subjects were classified according to BMI into; overweight (23-24.9kg/m3), obese (25-29.9Kg/m2) and obesity (>30Kg/m2). The diagnosis of risk factors was based on available records, inquiry from the spouse and doctors involved.

Statistical Analysis

We used the chi-square test for the comparison of frequencies of men and women in the two groups. Only P values <0.05 and the two tailed t-test were considered significant to find out the level of significance and difference. Multivariate logistic regression analysis was conducted, after adjustment of age and body mass index, to demonstrate the association of protective factors with risk of death due to CVDs.

Results

The results for demographic data as well as protective factors based on the records and by verbal autopsy questionnaire, among 2222 victims, aged 25-64 years, dying due to cardiovascular causes, are given in the Table 1. The mean age, and body weight were significantly higher among men compared to women victims. Spare time physical activity and moderate alcohol intake were significantly more common among male decedents compared to females. Regular active prayer and no tobacco were significantly more common among female decedents compared to males. However, the protective factors; spare time physical activity, prudent diet, no tobacco, active prayer were not very common among decedents dying due to CVDs. (Table 1). There was a high prevalence of obesity, tobacco intake among men compared to women (Table 2). Social classes 3-5 were comparable in the two sexes. The prevalence of overweight decedents was 29.4% (n=654) and obese 20.8% (n=461) and majority of the overweight and obese victims were from social class 1-3. Table 3 reveals the causes of deaths according to modified verbal autopsy questionnaire [6].

This modified questionnaire was able to diagnose 23.4% deaths due to heart diseases and 9.8% due to brain diseases, including stroke (7.8%) and inflammatory brain diseases (1.9%) and 5.8% deaths due to malignant causes. Thus, circulatory diseases were the cause of death among 31.2 %(n=695) of the victims. Accidental deaths according to modified questionnaires were much higher (14.0%). Renal diseases including acute renal failure and chronic renal failure were the cause of death among 11.2% of victims, pulmonary diseases in 22.3% and liver diseases in 4.8% of the victims. Miscellaneous or other causes of deaths were reported among 8.5 %. Results on social classes, tobacco intakes and food intakes have been reported in other publications [9, 10, 22]. Multivariate logistic regression analysis conducted after adjustment of age and body mass index, revealed that physical activity, prudent diet, active prayer were inversely associated with risk of deaths due to CVDs. No tobacco intake and moderate alcohol intake were significantly positively associated with decreased risk of death due to CVDs among men but not in women.

Discussion

Health behavior or protective factors are important in providing health and the prevention of CVDs whereas risk factors increase the risk of morbidity and mortality due to CVDs and other chronic diseases. This study shows that protective factors; prudent diet, moderate physical activity, no tobacco intake and active prayer were least common among male decedents dying due to circulatory diseases (Table 1,4). However, regular active prayer and no tobacco were quite common among female decedents compared to males. The lower rates of tobacco and alcohol consumption in women, most pronounced in the lower-middle income and low-income countries, are likely due to cultural factors and social stigma associated with women drinking and smoking in these societies. However, some reports suggest that as incomes, independence and affluence in these countries increase, smoking and drinking as well as sedentary behavior among women could also increase [24-27].

Multivariate logistic regression analysis in our study revealed, that in both sexes, physical activity and prudent diet were strongly and inversely associated with decreased risk of deaths due to CVDs (Table 4). Active prayer was also inversely but weakly associated with decreased risk of death due to CVDs in men and women. No tobacco and moderate alcohol intake were also inversely associated with decreased risk of deaths due to CVDs among men but not in women because these habits are not common among women. Our study shows that a large gap is present, between existing and desired status of key lifestyle behaviors among decedents dying due to CVDs and other chronic diseases.

Table 1. Clinical data and protective factors among decedents, based on records and answers given by spouse

Clinical Data	Male (n=1385)	Female (n=837)
	Mean (Standard deviation)	
Mean age	42.12** (13.02)	40.05 (11.60)
Body weight	60.12** (6.24)	53.10(6.95)
Body mass index (kg/m)	23.18* (2.18)	23.65 (2.46)
Prevalence of protective factors, n (%)		
Prudent diet (FVL>400g/d)	278(20.1)*	122(14.6)
Regular active prayer	472(34.0)	384(45.8)*
Occupational physical activity	458(33.0)	252(30.0)
Spare time physical activity	210(15.1)*	92(10.9)
No tobacco intake	715(51.6)	702(83.9)**
Moderate alcohol intake (4-10 drinks/ week)	277(20.0)**	29(3.5)
Social class 1,2.	611(44.1)	384 (45.8)

*=P <0.05,**=P <0.001, P value were obtained by Analysis of variance by comparison of mean(SD) and chi square test by comparison of frequencies in the two groups.

Table 2. Prevalence of risk factors among decedents dying due to various causes

Risk Factors, n (%)	Male (n=1385)	Female (n=837)
Obesity (BMI>25kg/m2)	277(20.0)	184(22.0)
Overweight (23-24.9kg/m2)	470(33.9)*	184(22.0)
Hypertension (>140/90mmHg)	457(33.0)	251(30.0)
Diabetes mellitus (known)	110(8.0)	58(7.0)
Social class 1-3	899(64.9)	484(57.8)
Social classes 4-5	486(35.1)	353(42.2)
Poverty (< US$ 300.00 per month)	508(36.7)	362(43.2)
Sedentary behavior	747(53.9)*	368(44.0)
Tobacco intake (>Once per week)	623(45.0)**	125(15.0)
Salt intake (>10g/day)	782(56.4)	491(58.6)
Western type diet	775(56.0)	455(54.4)
Alcoholism (>20 drinks/week)	65(4.7)**	1(0.12)
Lack of knowledge on health education	1204(86.9)	732(87.4)

*=P <0.05,**=P <0.001, P value were obtained by chi square test by comparison of frequencies in the two groups.

Table 3. Causes of deaths based on available records and verbal autopsy

Cause of death	Men (n=1385)	Women (n=837)	Total (n=2222)
1. Cardiovascular diseases (n=520, 23.4%))			
Coronary artery disease	137(9.9)	85(10.1)	222(10.0)
Sudden cardiac death	25(1.7)	20(2.4)	45(2.0)
Valvular heart disease	105(7.5)	55(6.5)	160(7.2)
Inflammatory cardiac disease	24(1.7)	20(2.4)	44(2.0)
Diabetic vascular disease	23(1.6)	26(3.1)	49(2.2)
2. Cerebrovascular disease (CVD)	115(8.3)	60(7.1)	175(7.8)
3. Malignant neoplasm			**(n=131, 5.8%)**
4. Injury- accidents			**(n=313, 14.0%)**
5. Renal diseases (n=250, 11.2%)	163(11.7)	87(10.3)	250(11.3)
6. Pulmonary diseases			**(n=495, 22.3%)**
7. Liver diseases			**(n=107, 4.8%)**
8. Miscellaneous			**(n=189, 8.5%)**
Total	1385(62.33)*	837(37.66)	2222(100)

*=P<0.05, by comparison of death rate in men and women by chi square test.

Recent studies indicate that following an acute coronary syndrome, those who adhere to a healthier diet and lifestyle have a lower risk of recurrent CVDs (28-30).Smoking cessation is associated with a lower risk of death and myocardial infarction, high-quality diets and regular exercise are associated with lower risk of death or recurrent CVDs events after a myocardial infarction [28-30]. Thus, avoidance of smoking or its cessation, improving diet quality, and increasing physical activity can cause significant reduction in morbidity and mortality due to CVDS. In a further study, among 153 996 adults, aged 35 to 70 years, the prevalence of the 3 healthy lifestyle behaviors; avoidance or cessation of smoking, eating a healthy diet, and undertaking regular physical activities, were examined at enrollment in participants who reported having had a CAD or stroke event in the Prospective Urban Rural Epidemiology

(PURE) study [21]. Among 7519 individuals with self-reported CAD, 18.5% (95% CI, 17.6%-19.4%) continued to smoke; only 35.1% (95% CI, 29.6%-41.0%) undertook high levels of work- or leisure-related physical activity, and 39.0% (95% CI, 30.0%-48.7%) had healthy diets; 14.3% (95% CI, 11.7%-17.3%) did not undertake any of the 3 healthy lifestyle behaviors and 4.3% (95% CI, 3.1%-5.8%) had all 3 (21). Overall, 52.5% (95% CI, 50.7%-54.3%) quit smoking (by income country classification: 74.9% [95% CI, 71.1%-78.6%] in HIC; 56.5% [95% CI, 53.4%-58.6%] in UMIC; 42.6% [95% CI, 39.6%-45.6%] in LMIC; and 38.1% [95% CI, 33.1%-43.2%] in LIC). The lowest prevalence of eating healthy diets was in LIC (25.8%; 95% CI, 13.0%-44.8%) compared with LMIC (43.2%; 95% CI, 30.0%-57.4%), UMIC (45.1%, 95% CI, 30.9%-60.1%), and HIC (43.4%, 95% CI, 21.0%-68.7%). It is clear, that the prevalence of healthy lifestyle behaviors was low, among patients with a CAD or stroke event from countries with varying income levels; being lowest levels in poorer countries. However, analysis within countries, may not confirm these results.

Table 4. Multivariate logistic regression analysis for association of protective factors with risk of death from cardiovascular diseases, after adjustment of age among men and women

Risk Factor Men	Women
Odds ratio,(95% confidence interval)	Odds ratio,(95% confidence interval)
Prudent diet 1.12(1.07-1.19)**	1.08 (1.05- 1.17)**
Physical activity 1.07(1.02-1.12)**	1.05 (0.99-1.11)**
No tobacco intake 1.01(0.94- 1.08)*	1.01 (0.95-1.07)
Active prayer 1.01 (0.95-1.07)*	0.99 (0.94-1.05)*
Moderate alcohol intake 0.98 (0.92- 1.01)*	0.95(0.89-1.12)
Social class 1,2 1.22 (1.11-1.32)*	1.16 (1.10-1.25)*

P value <0.01, ** <0.001. OR= Odds ratio.,FVL=Fruit,vegetable, legume.

In our study, newly developed extended verbal autopsy questionnaire allowed us to find nearly half of the decedents to be sedentary and were consuming unhealthy, deficient diets along with tobacco intake, more in men than in women which may be important in the pathogenesis of deaths due to NCDs (Table 1,2). In the PURE study [21], one fifth of individuals continued to smoke, only about one-third undertook high levels of physical activity, and only two fifths were eating a healthy diet. Substantial proportions of individuals did not have any of these 3 healthy lifestyle behaviors.

In the Indian Experiment of Infarct Survival, a randomized, controlled trial, tobacco consumption (74(36%) vs 70 (35%) and alcohol intake (45(22.0%) vs 50(24.7%) g/ 1 day) were similar in the intervention group A compared to control group B (28). Smoking was completely stopped in both the groups during admission to hospital. However, after discharge from the hospital 24 out of 72 smokers in group A and 18 out of 70 in group B started smoking again (4-10 cigarettes/day) during the one year follow up. The Indo-Mediterranean Diet Heart Study showed that approximately, half of the patients were consuming tobacco at entry to the study [29]. Tobacco intake was stopped at entry to study by more than 90% of the subjects at entry to the study, but after the follow up of two years, almost 20% again restarted the tobacco use. In both trials, there was a significant reduction in morbidity and mortality after the follow up two years.

This study shows that using the modified and extended verbal autopsy questionnaire, allowed us to diagnose most of the health behaviors; prudent dietary patterns, physical activity status, no tobacco consumption and active prayer status via interview of the spouse and family doctor. The questionnaires also allowed us to diagnose 31.2% deaths due to CVDs (23.4% deaths due to heart diseases and 7.8% due to stroke). No other study has used verbal autopsy questionnaire to examine the presence of health behavior among decedents dying due to various causes. Therefore, we cannot compare our results with other studies.

Verbal autopsy questionnaire allowed the interviewer to ascertain the clinical presentation of the victim during illness and up to death as well as the presence of health behavior and risk factors patterns of CVDs. These characteristics of behavior may be available in the medical records or may be obtained from the family members and doctors of the decedents. This information on probable presence of health behavior or risky behavior, related to each body system was reviewed to assess the final health behavior pattern with a view on cause of death among all the victims by the internist physician (RBS). Verbal autopsy extended to examine health behavior pattern, appears to be a valid and reliable supplemental method to assess the role of health behavior or risky behavior which may be important in the pathogenesis of the causes of death in a population where records of death are maintained in a municipal corporation in any country of the world. The most important reason for the success of our extended verbal autopsy may be the appropriate modifications made to adept it to the local setting, in the adult questionnaire suggested by WHO.

In many developing countries, verbal autopsy has been found to provide more valid causes of death compared to routine death certificates issued by the doctors [6, 10-15]. In low resource countries like India, verbal autopsy methodology was developed for national mortality surveillance systems [6,12,14] as well as for body weight autopsy [6], social autopsy [6], tobacco consumption autopsy [16] and dietary autopsy [17]. It is possible that extended verbal autopsy to find out unhealthy behavior associated with causes of death can be obtained with accuracy, which may be useful in the prevention of diseases.

All studies using verbal autopsy for finding out the causes of death among adults, did not find similar results (10-15). In Chinese, Yang et al. reported that verbal autopsy is not a very precise tool for assessing the leading causes of death among adults and provide only marginal support to this method [10]. It is likely that verbal autopsy is a developing method which needs further refinement to improve its accuracy. Therefore lack of accuracy in the questionnaires may be an important cause of inconsistency in results in some of the studies [10]. Verbal autopsy methodology and questionnaire appear to have many variations depending upon the responder's education, investigators qualification and knowledge of the experts [6, 10-15]. In some of the studies, ICD coding system for the causes of death may not have been used and hence WHO has recently published instructions to improve the quality and standards for application of this method [15]. The advantage of medical records and death certificates during the completion of verbal autopsy questionnaire is that in majority of these decedents, a prior hospital based diagnosis of the primary cause of death was available. Therefore in many cases, leading questions may be asked to know the health behavior patterns in association with the cause of death. Therefore, this extended verbal autopsy method appears to be a promising approach and of gold standard in identifying most precisely the exact health behavior pattern and cause of death which may be as good as history of present illness in a living patient and postmortem after death. Our verbal autopsy questionnaire extended to study health behavior pattern provide information about social class, tobacco intake, active prayer, physical activity and dietary patterns of the decedents, which may be of great help in planning the prevention programs in the populations.

Limitations

One important weakness in our study is that we could not compare our results with health behavior pattern and causes of death obtained by death certificate alone verses verbal autopsy combined with medical records by using kappa statistics. Other weaknesses are that only one internist physician reviewed the records of all the victims and only when diagnosis of cause was doubtful, a second opinion was taken from another internist physician. It seems that at least 3 internist physicians may be available to review the records to finalize the cause of death, if adequate resources are available. In future, use of MP3 players to compare audio (observed on computer) with paper record on a 5% population to determine the degree of concordance of the autopsy questionnaire can further improve the verbal autopsy. Our results may also underestimate the potential sensitivity and specificity of the verbal autopsy methods because questionnaires were completed by only one interviewer. Recent studies also confirm the role of health behavior pattern in the pathogenesis of deaths due to various diseases [16].

In brief, our results indicate that extended verbal autopsy for assessment of health behavior; prudent dietary pattern, physical activity, no tobacco, active prayer, social class and causes of death, of the victims, appears to be reliable and valid. These methods of extended verbal autopsy can be used in a large scale cohort study in any country of the world.

Acknowledgments

Due to International College of Nutrition, International College of Cardiology and The Tsim Tsoum Institute, Krakow, Poland for logistic support in preparing this article.

Conflict of interest: none declared.

Appendix I.

HALBERG HOSPITAL & RESEARCH INSTITUTE/ICN/ICC/TTI

VERBAL AUTOPSY QUESTIONNAIRE BASED ON DEATH CERTIFICATE, MEDICAL RECORDS AND INTERVIEW.

Name and post of expert filling the Questionnaire ....Tel....

History Given By: Spouse/Father/Mother/Sister/Son/Daughter/In-laws (tick) Education of person.........

Date....................Place......................

Time...........of Interview................ Unique ID..................................

GENERAL DATA:
Name of Victim…………..…………Surname…… Sex……Age…
Address……………………………………………………………………………
Tel No……………………………………………………………………………..
1. Religion: (a) Hindu (b) Moslem (c) Christian (d) Sikh
2. Final education of victim………Occupation of victim……………Family Type……
3. The date…………… Place ………………. and Time of Death………………..
4. Physically Matching subject with victim/Name, address………

CLINICAL DATA OF VICTIM. (By Medical Records)

5. Family history of diabetes, hypertension, cholesterol, heart attack, stroke, cancer, Bronchitis, asthma, obesity, suicide, HIV, depression, psychosis, alcohol intake, tobacco.
6. Any past history of above, illnesses (Name)
7. Height………cm ….Weight ….Kg, BMI… (Kg/M 2)
8. Waist circumference…… cm, maximum hip circumference………WHR……
9. Blood pressure………….. mmHg, Grey hair/Cataract/Wrinkles. (Tick)
10. History of surgery………
11. Chest Pain. …………..……………………………………………..
12. PROTECTIVE BEHAVIORS; Prudent diet, physical activity, active prayer, no tobacco, no alcoholism,

Table: Causes of deaths reported based on available records and verbal autopsy (modified from WHO) (Ask leading questions to know the system involved).

Cause of death Death Certificate Spouse Doctor Final

1. Circulatory diseases; Coronary artery disease, Sudden cardiac death, Diabetes, Vascular disease, Valvular heart disease, Inflammatory cardiac disease.
2. Nervous system diseases; Cerebrovascular disease, Meningitis, encephalitis.
3. Malignant neoplasm; Lung, oral, liver, stomach, breast, uterus, cervix, ovary.
4. Injury- accidents; Unintentional, road accidents, poisonings, fires, falls, drowning.
5. Renal diseases. Acute renal failure, chronic renal failure.
6. Pulmonary diseases; Chronic bronchitis, asthma, tuberculosis, acute pulmonary infection.
7. Liver diseases; Hepatitis, cirrhosis
8. Miscellaneous; Pregnancy and perinatal, suicide, congenital anomalies. Burns, Diarrhea/dysentery.

Appendix II. Physical Activity Autopsy

PHYSICAL ACTIVITY: OCCUPATIONAL (score for weekly activities)

a. If sedentary workers such as office and desk work, professional, shopkeeper (who climb fewer than 26 flights of stairs a week.)
b. If mild activity workers such as peons, postman, tailor, barbers, housewife looking after grownup children, part time house work.
c. If moderate activity workers such as drivers, conductors, artisans, housewife looking after small children, part time house work.
d. If heavy activity workers such as manual laborers, carpenters, house maids, all house work.

SPARE TIME ACTIVITY:

(a) If walking 14.0 km./week and without any vigorous at least 5 days in a week.
(b) If gardening in small plots, walking 14-35 km./week.
(c) If walking <35 km./week, jogging, cycling, sports <30 minutes including swimming and dancing.
(d) If walking <35 km./week, jogging, sports<30 minutes at a time including swimming and dancing.

ACTIVITY STATUS	
Sedentary	1-2
Mild activity	2-4
Moderate	4-6
Heavy	6-8

Appendix III. Tobacco Autopsy Questionnaire

1. Which type of tobacco consumption you usually do?
   (a) Cigarettes (b) Beeri
   (c) Hukka (d) Chillum

2. How many cigarettes, beedies or how many times you smoke Hukka pipe or another per day ……………..………………..
3. If you do not smoke, have you ever smoked before? (Yes/No)
   If **yes** then, when, for how much time ……………………..
4. Do you have tobacco chewing habit? Y/N
5. If yes, which form of tobacco chewing?
6. Pan with zarda/Pan without zarda/Pan parag/Khaini/ any other….
7. How much tobacco or how many times you chew per day. …..times. .. mg
8. NO TOBACCO INTAKE.

Appendix IV. Social Autopsy Questionnaire

SOCIOECONOMIC STATUS: (Scores constructed on each of these attributes)

a) Educational status

4 If all members of family had education beyond X.

3 If the subject had education beyond X and all other members' up to X

2 If at least any one member had high school education but no one above X

1 If no member had education beyond V.

0 If all members illiterate.

b) Occupational status

4 If all members of family were professionals (doctor, engineer, professor etc.) and executives.

3 If the subject was a professional or executive and all other member of household below this level.

2 If at least one member professional or business executive and all other member below this level.

1 If no member was beyond clerk.

0 If all laborers or unskilled workers.

c) Housing status

4 If housing as a bungalow or flat in a good locality.

3 If housing was in a medium class locality but built with excellent floors, walls and roof.

2 If house was in a slum but good floors, walls, roof.

1 If 1-2 room, accommodation in slum.

0 If one room in slum, public toilet, no electricity or fan.

Scores for rented house were same.

d) Ownership of consumer durables

4 If the subject or family had a motorcar.

3 If family had scooter/video/colour television.

2 If family had bicycle sewing machine, radio etc.

1 If family had bicycle, no other consumer durable

0 If no consumer durable of any kind.

e) Per capita monthly income

4 If very rich, (>Rs. 1000)

3 If > Rs. 600

2 If Rs. 300-600

1 If < Rs. 300

0 If irregular earning, too poor to buy enough food

f) Knowledge about health education

4 If health behavior in good practice.

3.If health behavior in partial practice.

2.If health behavior irregular practice

1. If only knowledge of health behavior.

0.No practice and no knowledge of health behavior.

Socioeconomic Status

SES groups	Total Score
SES-1	17-20
SES-2	13-16
SES-3	9-12
SES-4	5-8
SES-5	<5
Total score	

Appendix V. Dietary Autopsy Questionnaire

The following questions are about the foods that spouse usually eat (Difference from victim's food intake to be noted)

Please give the number of days per week on which you usually eat various foods.
Ring the appropriate answer like this:

If you eat a food 7 days a week, then ring the 7, thus:	(7) 6 5 4 3 2 1 M R
If you eat a food 3 days a week, then ring the 3, thus:	7 6 5 4 (3) 2 1 M R
If you eat a food less than one day a week but more than once a month, then ring M, thus:	7 6 5 4 3 2 1 (M) R
If you eat a food less than once a month or never, then ring R, thus:	7 6 5 4 3 2 1 M (R)

CEREALS: Please Answer Every Question (g/week) Office use (g/day)

Bread	Number/day	Thickness
12. White	7654321 MR	THICK: MEDIUM: THIN

CHAPATI:	Number/Day	
13. Paratha	7654321 MR	THICK: MEDIUM: THIN
14. Puree	7654321 MR	THICK: MEDIUM: THIN
15. Chapati	7654321 MR	THICK: MEDIUM: THIN
16. Dalia	7654321 MR	

ROOT TUBERS:

17. Potatoes boiled, baked, mashed	7 6 5 4 3 2 1 M R
18. Potatoes: Chips or fried from shop	7 6 5 4 3 2 1 M R
19. Potatoes: Chips or fried curry from home	7 6 5 4 3 2 1 M R
20. Sweet Potato	7 6 5 4 3 2 1 M R

LEGUMES:

21. Pulse-Red gram, green gram, Bengal gram	7 6 5 4 3 2 1 M R
22. Baked beans, lentils of Butter beans	7 6 5 4 3 2 1 M R
23. Peas, Kidney beans	7 6 5 4 3 2 1 M R

VEGETABLES/SPICES:

24. Radish	7 6 5 4 3 2 1 M R
25. Cabbage	7 6 5 4 3 2 1 M R
26. Cauliflower	7 6 5 4 3 2 1 M R
27. Green leafy Vegetables (grilled/fried/salad)	7 6 5 4 3 2 1 M R
28. Onion	7 6 5 4 3 2 1 M R
29. Garlic	7 6 5 4 3 2 1 M R
30. Chilli	7 6 5 4 3 2 1 M R
31. Black pepper	7 6 5 4 3 2 1 M R
32. Spices (total, g/day)	7 6 5 4 3 2 1 M R
33. Brinjal	7 6 5 4 3 2 1 M R
34. Carrots	7 6 5 4 3 2 1 M R
(g/week) Office use (g/day)	
35. Bitter gourd	7 6 5 4 3 2 1 M R
36. Jack fruit	7 6 5 4 3 2 1 M R

FRUITS: (g/week)	Number
37. How many bananas eaten per week?	…………
38. How many guavas eaten per week?	…………
39. How many apples eaten per week?	…………
40. How many oranges eaten per week?	…………
41. How many grape fruit eaten per week?	…………
42. How many musk melons eaten per week?	…………
43. How many slices of melons eaten per week?	…………
44. How many tomatoes eaten per week?	…………
45. How many mangoes eaten per week?	…………
46. How many other fruits eaten per week? (name)	…………
47. How many slices of papaya eaten per week? (name)	…………

MILK:

48. Do you take milk Y/N

49. If yes, what type of milk do you usually have (St-No-go)

50. How much milk (not skimmed or dried) do you drink a day in tea or coffee in milky drinks and in cereals

(a) Less than 100 ml
(b) 100-200 ml
(c) 200-500 ml
(d) More than 500 ml

CURD: How much curd do you take g…………/week,

FOOD HABIT- Vegetarian/non-veg./ eggitarian

EGGS and FISH

51. How many eggs do you eat per week?

52. Fish (Singhara/oily, fish fingers) 7 6 5 4 3 2 1 M R

53. Which fish do you take (River fish/Sea fish /both)?

MEATS:

54. Lamb, curry, fried 7 6 5 4 3 2 1 M R
55. Chicken or other poultry, curry fried 7 6 5 4 3 2 1 M R
56. Beef (including mined beef) curry, fried 7 6 5 4 3 2 1 M R
57. Pork curry fried 7 6 5 4 3 2 1 M R

FATS: Visible fat, (g/week)

58. How much of the following foods do the whole family and person concerned use on average per week? (g)

(Whole Family)/Week (Person Concerned/Week)

Clarified butter (ghee) g………………….. ……………………………
Butter g………………….. ……………………………
Lard and vegetable ghee/trans fat
Fat (specify) g………………….. ……………………………
Liquid vegetables oils;
Mustard oil
Ground nut/soyabean/corn/
sun flower oil g………………….. ……………………………
Cream g………………….. ……………………………
Sugar g………………….. ……………………………
Salt g………………….. ……………………………

CONFECTIONARY:

59. Ice cream, sweet yoghurt or chocolate 7 6 5 4 3 2 1 M R
60. Digestive biscuits or plain biscuits 7 6 5 4 3 2 1 M R
61. Milk sweet (Rasgula, Barfi) 7 6 5 4 3 2 1 M R
62. Non milk sweets (Jalebi, Laddu) 7 6 5 4 3 2 1 M R
69. How many persons normally eat in your household?

Number of adults (including yourself) ………………………………
Number of children 1-4 years old ………………………………

70. Are you on a special diet? No, If yes, then
 (a) Slimming diet suggested by your doctor
 (b) Diabetic diet
 (c) Vegetarian diet
 (d) Fruit and vegetable enriched diet
71. What type of milk do you usually have?
 (a) Whole Cow's, Buffalo, Goat's, Mixes (packet)
 (b) Skimmed
 (c) Other (Specify please)
72. How many times per week you eat and drink outside
73. Do you eat heavy breakfast (>1000 Cal.)……………..

Dinner (>1500 Cal)…………………(Yes/No)

74. Do you eat prudent foods in the breakfast?

References

[1] Singh RB, Beegom R, Mehta AS, Niaz MA, De AK, Mitra RK, Haque M, Verma SP, Dube GP, Siddiqui HM, Wander GS, Janus ED, Postiglion A, Haque MS. Social class, coronary risk factors and undernutrition, a double burden of diseases, in women during transition, in five Indian cities. *Int. J. Cardiol.*, 1999:69;139-147.

[2] Singh RB, Takahashi T, Nakaoka T, Otsuka K,Toda E, Shin HH, Kyu Lee M, Beeharry V, Hristova K, Fedacko J, Pella D, De Meester F, Wilson DW, Juneja LR. Nutrition in transition from Homo sapiens to *Homo economicus. The Open Nutra J.* 2013;6:6-17.

[3] Moodie R, Stuckler D, Monteiro C, Sheron N, Thamarangsi , Lincoln P, Casswel S on Behalf of the Lancet NCD Group. Profits and pandemics: prevention of harmful effects of tobacco, alcohol, and ultra-processed food and drink industries. *The Lancet*, Early Online Publication, 12 February 2013,doi:10.1016/S0140-6736(12)62089.

[4] Pednekar MS, Gupta R, Gupta PC. Illiteracy, low educational status and cardiovascular mortality in India. *BMC Public Health.* 2011:**11**;567.

[5] Tokunaga M, Takahashi T, Singh RB, Rupini D, Toda E, Nakamura T, Mori H, Wilson DW. Diet, Nutrients and Non-communicable Diseases. *The Open Nutra Jour.*, 2012, 5, 146-159.

[6] Singh RB, Fedacko J, Vargova V, Kumar A, Mohan A, Pella D, De Meester F, Wilson DW. Singh's verbal autopsy questionnaire for assessment of causes of death, social autopsy, tobacco autopsy, and dietary autopsy based on medical records and interview. *Acta Cardiol.* 2011:66;471-481.

[7] Editorial. Wealth but not health in USA. *Lancet* 2013;381:177. doi:10.1016/S0140-6736(13)60069-0.

[8] WHO. Mortality and burden of disease estimates for WHO Member States in 2008. Geneva: World Health Organization, 2010.

[9] Soleman N, Chandramohan D, Shibuya K. Verbal autopsy: current practices and challenges. Bull *World Health Organ* 2006; 84: 239–245.

[10] Yang G, Rao C, Ma J, Wang L, Wan X. Validation of verbal autopsy procedures for adult deaths in China. *Int. J. Epidemiol.* 2006; 35: 741–748.

[11] Khademi H, Etemadi A, Kamanger F, Nourats M. Verbal autopsy: reliability and validity estimates for causes of death in the Golestan cohort study in Iran. *PLoS ONE*, 2010; 5:1-6.

[12] Jha P, Gajalakshami V, Gupta PC, Kumar R,Mony P, Dhingra N, Peto R, RGI-CGHR Study collaborators. Prospective study of one million deaths in India: rationally, design and validation results. *PLoS Medicine* 2006; 3:191-201.

[13] Baiden F, Bawah A, Biai S, Binka F, Boerma T..Setting international standards for verbal autopsy. *Bull World Health Organ* 2007; 85:570–571.

[14] Gajalakshmi V, Peto R. Verbal autopsy of 80,000 adult deaths in Tamilnadu, South India. BMC Public Health 2004; 4: 47.

[15] Setal PW, Sankoh O, Rao G, Velkoff VA, Mathers C. Sample registration of vital events with verbal autopsy: a renewed commitment to measuring and monitoring vital statistics. *Bull World Health Org.* 2005; 83:811-817.

[16] Singh RB, Singh S, Chattopadhyay P, Singh K,Singh V, Kulshrestra SK, Tomar RS, Kumar R, Singh G, Mechorova V, Pella D. Tobacco consumption and causes of death in an urban population of north India. *Int. J. COPD* 2007; 2:177-185.

[17] Sanghamitra P, Singh RB, Fedacko J, Vargova V, Takahashi T, Tokunaga M, Juneja LR,De Meester F. Dietary patterns and causes of death due to cardiovascular diseases and other diseases among urban decedents in north India. *WHJ* 2012;4:123-134.

[18] Takahashi T, Singh RB, Meester FD, Wilson D. How the 'West' can overcome unhealthy behaviors to prevent chronic diseases. *J. Socialomics* 2013; 2: e114. doi:10.4172/2167-0358.1000e114

[19] Chow CK, Jolly S, Rao-Melacini P, Fox KA, Anand SS, Yusuf S. Association of diet, exercise, and smoking modification with risk of early cardiovascular events after acute coronary syndromes. *Circulation*. 2010; 121:750-758.

[20] Iqbal R, Anand S, Ounpuu S, et al; INTERHEART Study Investigators. Dietary patterns and the risk of acute myocardial infarction in 52 countries: results of the INTERHEART study. *Circulation*. 2008;118:1929-1937.

[21] Teo K, Lear S, Islam S, Mony P, Dehghan M, Li W, Rosengren A, Lopez-Jaramillo P, Diaz R, Oliveira G, Miskan M, Rangarajan S, Iqbal R, Ilow R, Puone T, Bahonar A, Gulec S, Darwish EA, Lanas F, Vijaykumar K, Rahman O, Chifamba J, Hou Y, Li N, Yusuf S; on behalf of the PURE Investigators. Prevalence of a healthy lifestyle among individuals with cardiovascular disease in high-, middle- and low-income countries: The Prospective Urban Rural Epidemiology (PURE) Study. *JAMA*. 2013;309:1613-1621.

[22] Taylor AW; Price K; Fullerton S. A survey to assist in targeting the adults who undertake risky behaviours, know their health behaviours are not optimal and who acknowledge being worried about their health. *BMC Public Health* 2013; 13, 120

[23] Cahill LE, Chiuve SE, Mekary RA. Prospective study of breakfast eating and incident coronary heart disease in a cohort of male US health professionals. *Circulation* 2013; DOI:10.1161/CIRCULATIONAHA.113.001474. http://circ.ahajournals.org/.

[24] Heart Disease and Stroke Statistics 2013 Update. A report from the American Heart Association. *Circulation* 2013;127:e6-e245.

[25] Teo KK, Ounpuu S, Hawken S, et al; INTERHEART Study Investigators. Tobacco use and risk of myocardial dial infarction in 52 countries in the INTERHEART study:a case-control study. *Lancet*. 2006;368:647- 658.

[26] Kumra V, Markoff BA. Who's smoking now? The epidemiology of tobacco use in the United States and abroad. *Clin. Chest Med*. 2000;21(1):1-9.

[27] Taylor RS, Brown A, Ebrahim S, et al. Exercise based rehabilitation for patients with coronary heart disease: systematic review and meta-analysis of randomized controlled trials. *Am. J. Med*. 2004;116(10):682-692.

[28] Singh RB, Rastogi SS, Verma R, Laxmi B, Singh R,Ghosh S, Niaz MA. Randomized, controlled trial of cardioprotective diet in patients with acute myocardial infarction : results of one year follow up. *BMJ* 1992;304:1015-9.

[29] Singh RB, Dubnov G, Niaz MA, Ghosh S, Singh R, Rastogi SS, Manor O, Pella D, Berry EM. Effect of an Indo-Mediterranean diet on progression of coronary disease in high risk patients:a randomized single blind trial. *Lancet* 2002, 360:1455-1461.

[30] Singh RB,Rastogi SS, Verma R, Bolaki L, and Singh R. An Indian experiment with nutritional modulation in acute myocardial infarction. *Am. J. Cardiol.* 1992, 69: 879-885.

In: Circadian Cardiology with Focus on Both Prevention ... ISBN: 978-1-63463-956-9
Editors: K. Hristova, A. Shehab, G. Cornelissen et al. © 2015 Nova Science Publishers, Inc.

Chapter 23

New Roles of Coenzyme Q10 in Cardiovascular Diseases, Discovered by a Single Group

Anna Gvozdjáková*[*1], Toru Takahashi[2], R. B. Singh[3], Fabien De Meester[4], Douglas W. Wilson[5] and Fred L. Crane[6]

[1]Pharmacobiochemical Laboratory of 3rd Department of Internal Medicine, Faculty of Medicine, Comenius University, Bratislava, Slovak Republic
[2]Department of Nutrition, Graduate School of Human Environment Science, Fukuoka Women's University, Japan
[3]Halberg Hospital and Research Institute, Moradabad, India
[4]The Tsim Tsoum Institute, Krakow, Poland
[5]School of Medicine, Pharmacy and Health, Durham University, UK
[6]Purdue University W. Lafayette, IN, US

Abstract

Ubiquinone (Coenzyme Q10) is naturally present in our diet and synthesized in all body cells especially muscles. The biosynthesis of ubiquinone from the amino acid phenylalanine, tyrosine and acetyl-CoA, is a multi-stage process requiring at least eight vitamins and several phytochemicals. Coenzyme Q10(Co Q10), is a fat soluble quinone with characteristics common to vitamins which is found in the organs of various animal species with highest concentrations in the heart, liver, kidney, muscles and pancreas. Festenstein et al. in 1955 named the substance ubiquinone while Crane et al. in 1957 chose the name coenzyme Q. Ubiquinone is a component of the mitochondrial respiratory chain, participating in electron transport in NADH-coenzyme Q reductase (complex I), succinate coenzyme Q reductase (complex II) and the cytochrome system. Folkers and his group determined the structure of the quinone moiety which was found identical to

* Correspondence: Anna Gvozdjáková, PhD, DSc. Professor, Pharmacobiochemical Laboratory of 3rd Department of Internal Medicine, Faculty of Medicine, Comenius University in Bratislava, Slovak Republic, Email: anna.gvozdjakova@fmed.uniba.sk.

that described by Morton and his team, and suggested the name "ubiquinone" referring to the ubiquitous occurrence of this compound in various tissues. In 1957, Crane et al. demonstrated that it has an important role as a redox carrier in the mammalian respiratory transport chain. In 1972, Littarru, of Italy and the late Prof. Folkers from Texas, documented a deficiency of CoQ10 in human heart disease, particularly among patients subjected to bypass surgery in Houston. CoQ10 deficiency has also been found in blood samples from patients with cardiovascular diseases (CVDs) and brain diseases compared with levels in healthy human subjects. Yamamura and his group were the first to use CoQ10 for the treatment of cardiovascular disease (CVD) in the 1960's. However, Folkers et al. presented the rationale for using CoQ10 in treating congestive heart failure (CHF) (19) and Mortensen et al. presented the largest trial of CoQ10 in heart failure. Gvozdjakova et al. correlated brain CoQ10 with other clinical conditions. Singh's group has demonstrated for the first time that CoQ10 can modulate plasma insulin, lipoprotein(a), serum creatinine and albuminurea, serum nitrite, serum IL-6,TNF-alpha, blood pressure variability, atherogenesis, myocardial dysfunction in myocardial infarction and neuronal degeneration, hence CoQ10 needs exploration for its supplementation in CVDs and other chronic diseases.

Keywords: Ubiquinone, ubiquinol, antioxidant, free radical, heart failure, heart, brain

Introduction

Cardiovascular Diseases (CVDs) have become a major cause of morbidity and mortality in the developed and developing countries [1,2]. The increased risk of CVDs appears due to rapid changes in diet and lifestyle [3-5]. The International Atherosclerosis Society (IAS) here updates its recommendations on treatment of high level of blood cholesterol and dyslipidemia for the purpose of reducing risk for atherosclerotic cardiovascular disease (ASCVD) [6]. The sum of accumulated evidence of multiple types supports the contention that elevated LDL-C is a major target of lipid-lowering therapy. But there is growing evidence that very low density lipoproteins (VLDL) likewise promote atherosclerosis. Thus VLDL cholesterol (VLDL-C) is another potential target of cholesterol-lowering therapy. VLDL-C is especially elevated in persons with hypertriglyceridemia. The sum of LDL-C and VLDL-C includes cholesterol in all atherogenic lipoproteins and is called non-high density lipoprotein cholesterol (non-HDL-C). Therefore non-HDL-C can be considered an alternative to LDL-C as target of therapy. Non-HDL-C is more reflective of atherogenicity in persons with elevated triglycerides. It also can be accurately measured in non-fasting serum whereas LDL-C cannot be.

Total cholesterol and low density lipoprotein cholesterol (LDL) as well as non-HDL-C, become atherogenic after oxidation due to endogenous antioxidant deficiency and deficiency of antioxidants in the diet. Oxidized form of non-HDL-C which includes LDL-C as well as VLDL-C, is highly attractive to macrophages in the development of atherosclerosis. Coenzyme Q10 (CoQ10) is a natural antioxidant and bioenergetic agent present in our body which is protective against CVDs and brain degeneration [7-10]. It is crucial in the prevention of oxidation of lipids along with antioxidant enzymes, catalase and superoxide dismutase and dietary antioxidants [7]. Oxidation of high density lipoprotein cholesterol (HDL-C) may decrease the protective potential of HDL-C which may be helpful in the mechanism of atherosclerosis. This review re-emphasizes on the role of CoQ10 in atherosclerosis and other chronic diseases.

Discovery of Coenzyme Q10

CoQ10, is a fat soluble quinone with characteristics common to vitamins which is found in the organs of various animal species with highest concentrations in the heart, liver, kidney, spleen, muscles and pancreas [11-13]. Festenstein et al. in 1955 named the substance ubiquinone [12] while Crane et al. in 1957 chose the name coenzyme Q [14].

Coenzyme Q_{10} (CoQ_{10}) exists in three redox states in organism: fully oxidized (ubiquinone), radical (semiquinone) and fully reduced (ubiquinol). (Figure 1.)

Mitochondrial CoQ10 function as a component of the mitochondrial respiratory chain, regulate electrons transport from NADH-coenzyme Q reductase (complex I), succinate coenzyme Q reductase (complex II) and passing them to complex III [14,15] , resulting in ATP synthesis. As an alternative coenzyme Q function is possible in regulation of permeability transition pore opening (in porin) and CoQ uptake through Voltage Dependent Anion Channel (VDAC) of outer mitochondrial membrane (Figure 2), [18, 19].

Folkers and his group determined the structure of the quinone moiety which was found identical to that described by Morton and his team, and suggested the name "ubiquinone" referring to the ubiquitous occurrence of this compound in various tissues including brain [12-16].

In 1957, Crane et al. demonstrated that it has an important role as a redox carrier in the mammalian respiratory transport chain [14]. A high concentration of CoQ10 observed in healthy human myocardium has led to the assumption that a myocardial deficiency of CoQ10 is detrimental to cardiac function [20]. In 1972, Littarru, of Italy and the late Prof. Folkers from Texas, documented a deficiency of CoQ10 in human heart disease, particularly among patients subjected to bypass surgery in Houston, USA [21- 23].

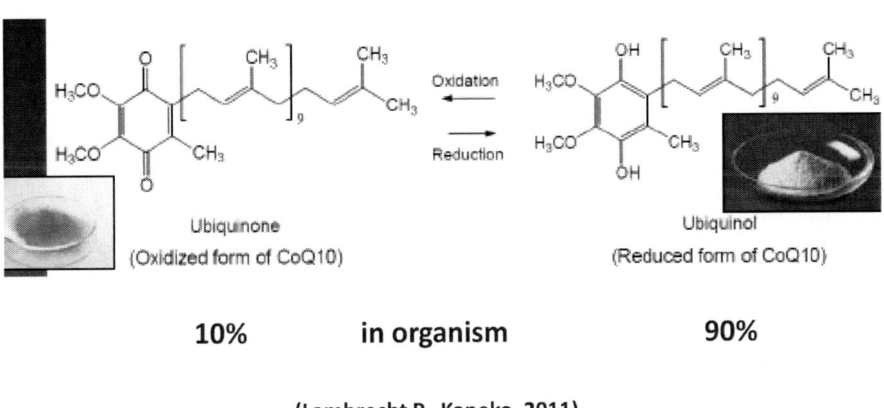

Figure 1. Chemical structures of CoQ10.

Brain coenzyme Q10 (CoQ10) concentration can influence the activity of several brain regions, including those which participate in the regulation of cardiovascular circadian rhythms, food intake, neuroendocrine stress response, activity and sleep regulation [16]. However, the effect of supplemented ubiquinol (reduced CoQ) into brain regions was not known. Previous our study determined baseline levels of ubiquinone (oxidized CoQ) in various rat brain regions and proved the bioavailability of the liposomal ubiquinol to selected brain regions after its administration into right brain ventricle. Our data indicate that administration of ubiquinol may create the basis for modulation of neuronal activities in specific brain regions (Figures 3-9), [17].

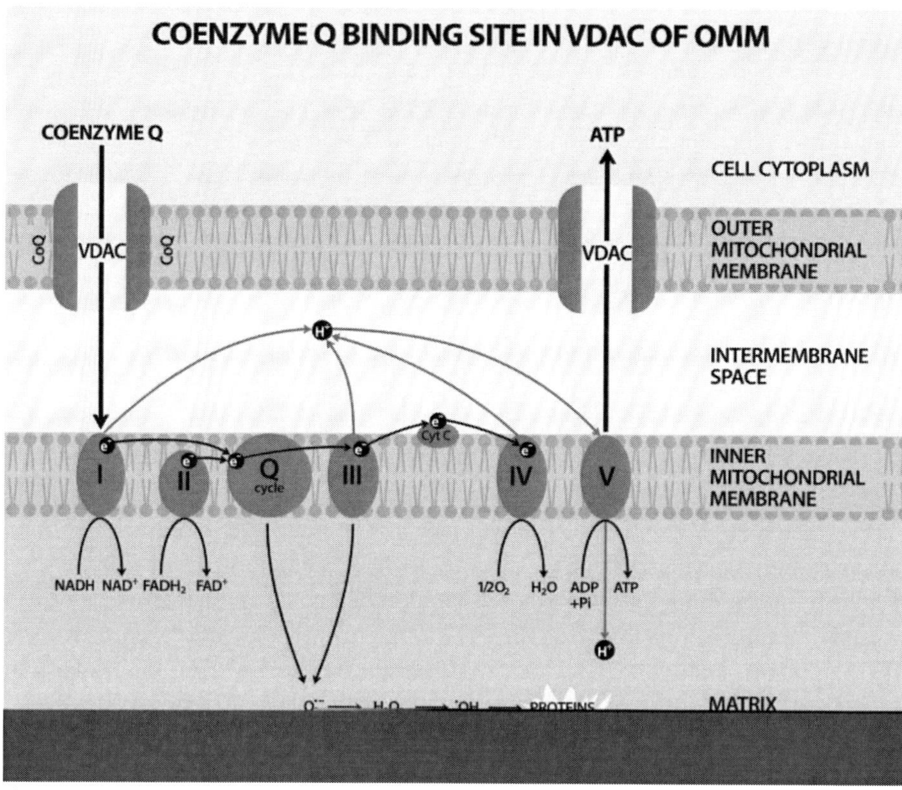

Legend:

CoQ – Coenzyme Q; VDAC – Voltage Dependent Anion Channel; ATP – adenosine triphosphate; ADP – adenosine diphosphate; Pi – inorganic phosphate; I, II, III, IV, V - respiratory chain complexes; H^+ - proton; e^- - electron; Q–cycle – Coenzyme Q cycle; cyt c – cytochrome c; NADH - reduced nicotinadenindinucleotid; NAD^+ - nicotinadenindinucleotid; $FADH_2$ – reduced flavinadenindinucleotid; FAD - flavinadenindinucleotid; $O_2^{\cdot-}$ - superoxide radical; H_2O_2 – hydrogen peroxide; $OH^{\cdot}$ – hydroxyl radical; H_2O – water; O_2 – oxygen;

Figure 2. Proposed novel Coenzyme Q binding site in Voltage Dependent Anion Channel of outer mitochondrial membrane.

Maximum CoQ10-OX concentration (compared with group without ubiquinol administration – time zero minutes) was found after 15 minutes in the following brain regions: nucleus supraopticus (NSO), nucleus suprachiasmaticus (NSCH), nucleus ventromedialis (NVM), nucleus dorsomedialis (NDM) and locus coeruleus (LC). After 60 minutes, the maximum CoQ10-OX concentration was found in basal ganglia (BG), nucleus paraventricularis (PVN) and cortex (CX), while after 120 minutes in organum vaculosum laminae terminalis (OVLT) and in substantia nigra (SN).

Administered ubiquinol increased CoQ9-OX concentration in OVLT, nucleus tractus solitarii (NTS) and A5 noradrenergic cell group (A5) after 120 minutes and had no effect on CoQ9-OX concentration in NVM, NDM and CX. These data show higher baseline concentrations of CoQ9-OX than of CoQ10-OX in rat brain regions BG, OVLT, SN, CX, NTS, A5, but lower in NSCH, NVM, NDM and LC. These different CoQ10 levels could be correlated with individual brain region functions. The benefit of therapeutic supplementation of ubiquinol in neurodegenerative diseases depends also on its bioavailability. It has been demonstrated for the first time that ubiquinol has beneficial effects on psychological manifestation in children with autism [16].

CoQ10 deficiency has also been found in blood samples from patients with cardiovascular diseases (CVDs) compared with levels in healthy human subjects [22-25]. Yamamura and his group were the first to use CoQ10 for the treatment of cardiovascular disease (CVD) in the 1960's [24]. However, Folkers et al. presented the rationale for using CoQ10 in treating congestive heart failure (CHF) [21] and Mortensen et al. presented the largest trial of CoQ10 in heart failure (25).

Ubiquinone is naturally present in our diet and synthesized in all body cells especially muscles. The biosynthesis of ubiquinone from the amino acid phenylalanine, tyrosine and acetyl- CoA, is a multi-stage process requiring at least eight vitamins and several phytochemicals [7]. It is reasonable that deficiency of any of these micronutrients may result into ubiquinone deficiency. Ramasarma studied the natural occurrence of ubiquinone and its distribution in the body [24]. Stocker and coworkers reported that ubiquionol-10 protects human low density lipoprotein more efficiently against lipid peroxidation than does alpha-tocopherol [26]. Since then, clinical reports worldwide have described favourable effects of both intravenous and oral CoQ10 in patients with CVDs of various etiologies [27-29]. CoQ10 is a potential antioxidant and bioenergetic agent. There is evidence that CVDs and other chronic diseases may be associated with oxidative stress and deficiency of antioxidant vitamins and minerals resulting in to increased risk of recurrent cardiovascular events and other non-communicable diseases [30-32].

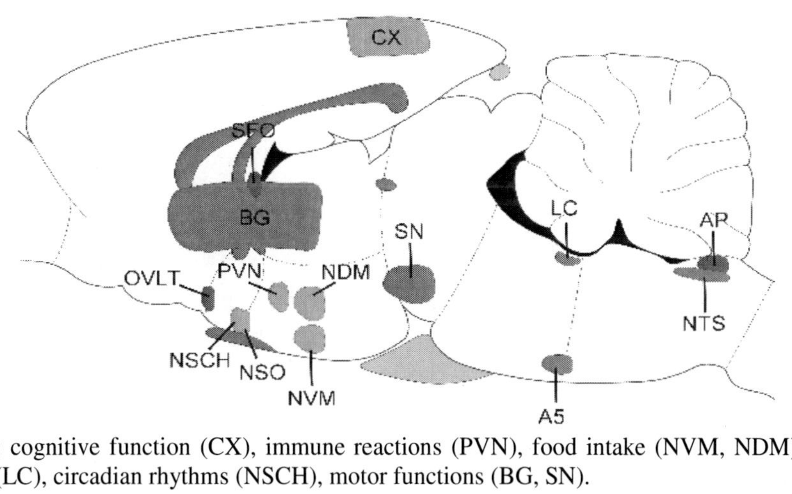

Regulation: cognitive function (CX), immune reactions (PVN), food intake (NVM, NDM), attention, stress (LC), circadian rhythms (NSCH), motor functions (BG, SN).

Figure 3. Bioavailability of Ubiquinol by selected rat brain regions [16].

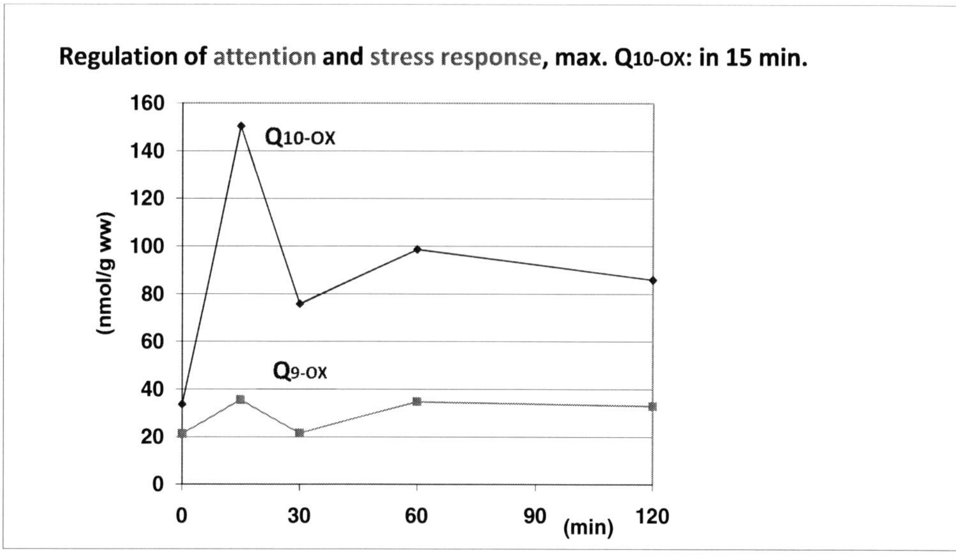

Figure 4. LOCUS COERULEUS (LC) [17] $Q_{10\text{-OX}}$, $Q_{9\text{-OX}}$.

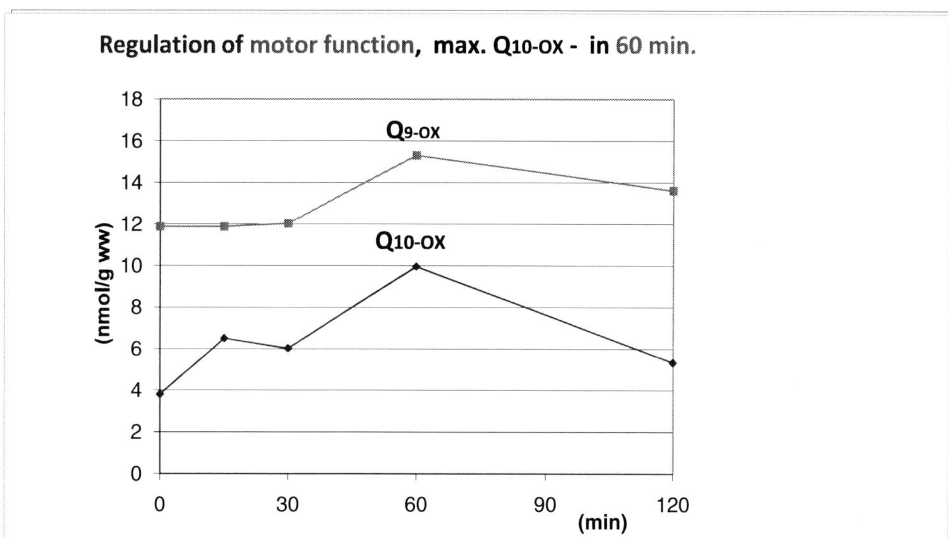

Figure 5. BASAL GANGLIA (BG) [17] $Q_{10\text{-OX}}$, $Q_{9\text{-OX}}$.

CoQ10 and antioxidant selenium were administered in patients with acute myocardial infarction in 1994 [33]. Finally, Singh et al. studied the effects of oral treatment with coenzyme Q10 (120 mg/d) compared with placebo for 28 days in 73 (intervention group A) and 71 (placebo group B) patients with acute myocardial infarction (AMI) [34]. After treatment, angina pectoris (9.5 vs. 28.1), total arrhythmias (9.5% vs. 25.3%), and poor left ventricular function (8.2% vs. 22.5%) were significantly ($P < 0.05$) reduced in the coenzyme Q group than placebo group. Total cardiac events, including cardiac deaths and nonfatal infarction, were also significantly declined in the Co Q10 group compared with the placebo group (15.0% vs. 30.9%, $P < 0.02$).

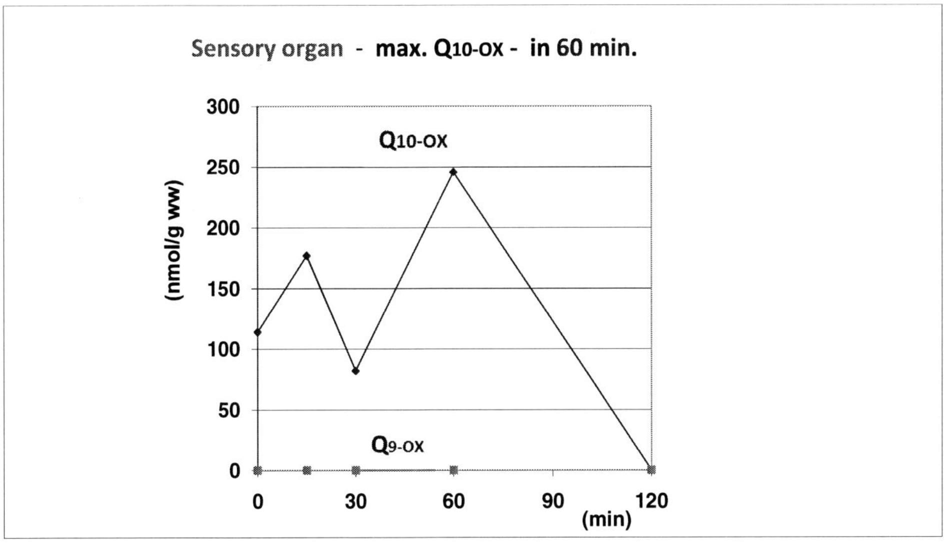

Figure 6. SUBFORNICAL ORGAN (SFO) [17] $Q_{10\text{-OX}}$, $Q_{9\text{-OX}}$.

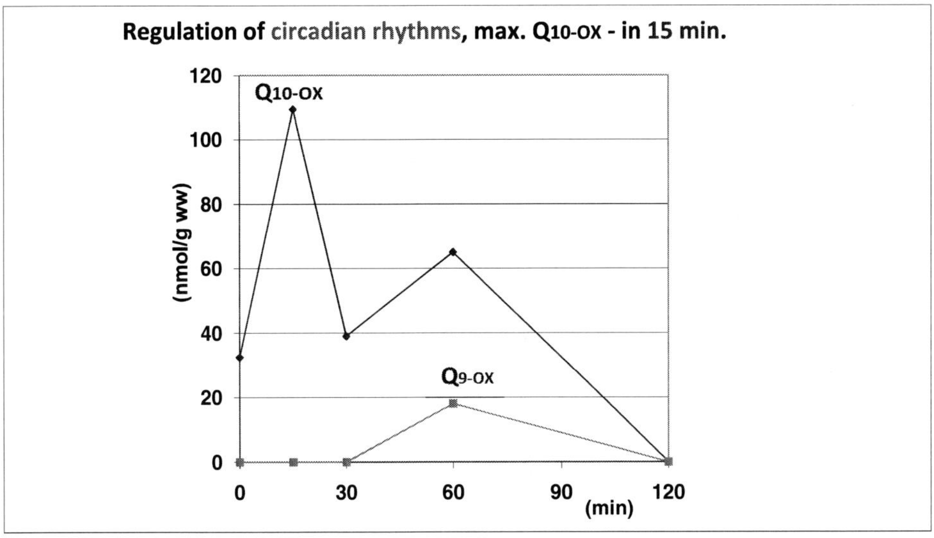

Figure 7. NUCLEUS SUPRACHIASMATICUS (NSCH) [17] $Q_{10\text{-OX}}$, $Q_{9\text{-OX}}$.

The extent of cardiac disease, elevation in cardiac enzymes, and oxidative stress at entry to the study were comparable between the two groups. Lipid peroxides, diene conjugates, and malondialdehyde, which are indicators of oxidative stress, showed a greater reduction in the treatment group than in the placebo group. The antioxidants vitamin A, E, and C and beta-carotene, which were lower initially after AMI, increased more in the Co Q10 group than in the placebo group. These findings suggest that Co Q10 can provide rapid protective effects in patients with AMI if administered within 3 days of the onset of symptoms. More studies in a larger number of patients and long-term follow-up are needed to confirm our results.

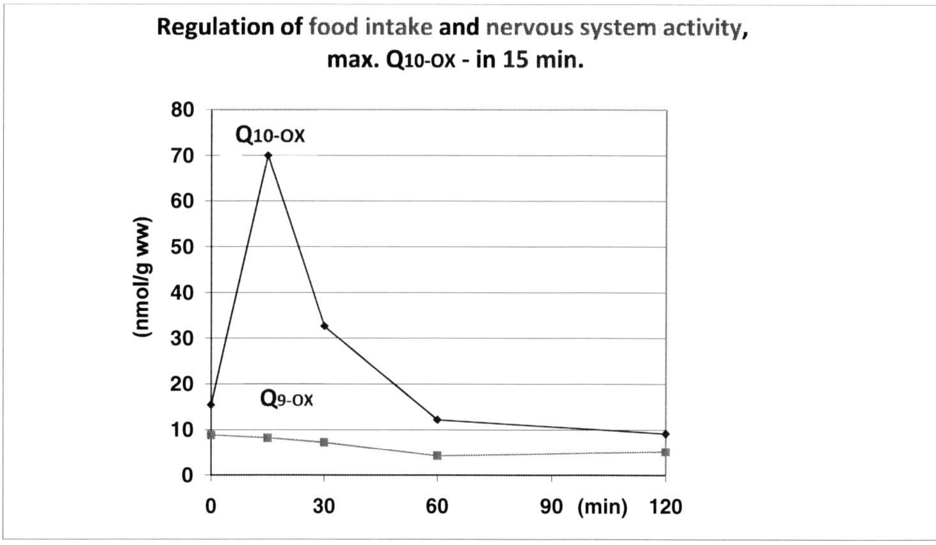

Figure 8. NUCLEUS DORSOMEDIALIS (NDM) [17] $Q_{10\text{-}OX}$, $Q_{9\text{-}OX}$.

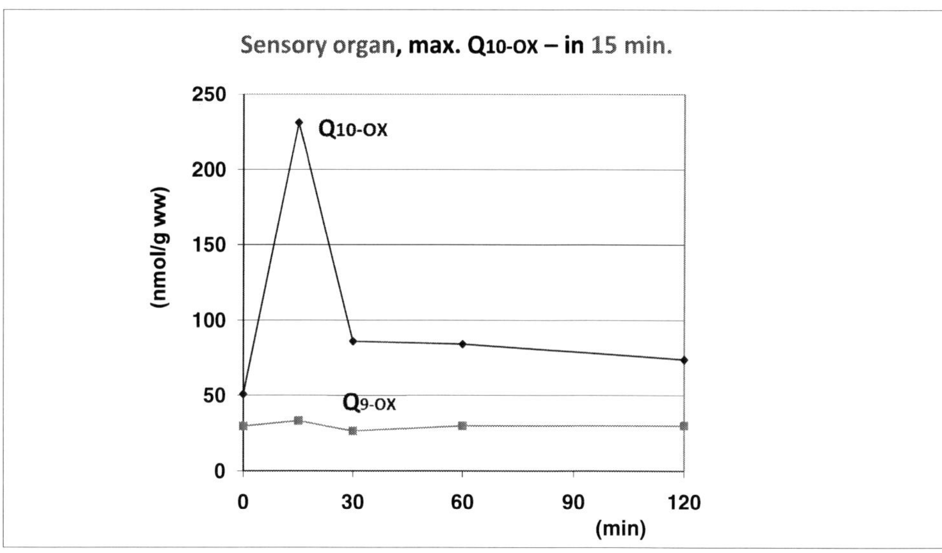

Figure 9. AREA POSTREMA (AP) [17] $Q_{10\text{-}OX}$, $Q_{9\text{-}OX}$.

Effect of hydrosoluble CoQ10 causing a decline in insulin levels in hypertensive patients with coronary artery disease have been reported by Singh's group for the first time in the literature [35]. Treatment with coenzyme Q10 (60 mg twice daily) were compared in a randomized trial, for 8 weeks in 30 (coenzyme Q10: group A) and 29 (B vitamin complex: group B) patients known to have essential hypertension and presenting with coronary artery disease (CAD). After the follow-up, the following indices were reduced in the coenzyme Q10 group: systolic and diastolic blood pressure, fasting and 2-h plasma insulin, glucose, triacylglycerol, lipid peroxides, malondialdehyde and diene conjugates. The following indices were increased: HDL-cholesterol, vitamins A, C, E and beta-carotene (all changes $P<0.05$).

The only changes in the group taking the B vitamin complex were increases in vitamin C and beta-carotene ($P<0.05$). These findings indicate that treatment with coenzyme Q10 decreases blood pressure possibly by improving insulin insensitivity by decreasing oxidative stress and insulin response in patients with known hypertension receiving conventional antihypertensive drugs.

A reduction in serum concentration of lipoprotein (a) among acute myocardial infarction (AMI) patients was also observed for the first time by Singh et al. in 1999 [35]. Subjects with clinical diagnosis of AMI, unstable angina, angina pectoris (based on WHO criteria) with moderately raised lipoprotein (a) were randomized to either coenzyme Q10 as Q-Gel (60 mg twice daily) (coenzyme Q10 group, n=25) or placebo (placebo group, n=22) for a period of 28 days. Results: Serum lipoprotein (a) showed significant reduction in the coenzyme Q10 group compared with the placebo group (31.0% vs 8.2% $P<0.001$) with a net reduction of 22.6% attributed to coenzyme Q10. HDL-C showed a significant increase in the intervention group without affecting total cholesterol, LDL cholesterol. Coenzyme Q10 supplementation was also associated with significant reductions in blood glucose, thiobarbituric acid reactive substances (TBARS), malondialdehyde (MDA) and diene conjugates, indicating an overall decrease in oxidative stress.

In view of the above beneficial effects of CoQ10 on vascular disease risk [37-40], Singh et al. demonstrated for the first time in rabbits, that CoQ10 can inhibit atherosclerosis and modulate the quality and chemical composition of atheroma [38]. The effects of the administration of coenzyme Q10 (3 mg/kg per day) (group A, n=10) and placebo (aluminum hydroxide, 3 mg/kg per day) (group B, n=10) were compared over 24 weeks in a randomized, single-blind, controlled trial. There were two groups of rabbits receiving a trans fatty acid (TFA)-rich diet (5-8 g/day) for 36 weeks. Oxidized rabbit chow with vitamin C plus ferric chloride was administered for 4 weeks in all rabbits. Intervention with coenzyme Q10 after feeding of TFA-rich diet was associated with a significant decline in TBARS, diene conjugates and MDA, and an increase in plasma levels of vitamin E in the coenzyme Q group compared to placebo group. These changes, which were indicators of a decrease in oxidative damage, were independent of lipid lowering. The aortic and coronary artery plaque sizes, coronary atherosclerosis index, aortic and coronary atherosclerosis scores were significantly lower in the coenzyme Q group than placebo group. Aortic and coronary plaque frequencies, as well as frequencies of ulceration, thrombosis or hemorrhage, and cracks and fissures, were also significantly lower in the coenzyme Q group, indicating a better quality of atheroma compared to those in the control group. Aortic cholesterol, triglycerides and sudanophilia were significantly lower and vitamin E significantly higher in the coenzyme Q group in comparison to the placebo group indicating that coenzyme Q10 can have beneficial effect on the chemical composition of atheroma. The findings suggest that antioxidant therapy with coenzyme Q10 may be used as an adjunct to lipid lowering for additional beneficial effects related to chemical composition and quality of atheroma independent of hypolipidemic agents.

This group also administered CoQ10 in other CVDs, as well as in conjunction with taurine [39, 40]. In a randomized, double-blind, controlled trial, the effects of oral treatment with coenzyme Q10 (CoQ10, 120 mg/day), a bioenergetic and antioxidant cytoprotective agent, were compared for 1 year, on the risk factors of atherosclerosis, in 73 (CoQ, group A) and 71 (B vitamin group B) patients after acute myocardial infarction (AMI). After 1 year, total cardiac events (24.6 vs. 45.0%, $p < 0.02$) including non-fatal infarction (13.7 vs. 25.3%, p < 0.05) and cardiac deaths were significantly lower in the intervention group compared to control group. The extent of cardiac disease, elevation in cardiac enzymes, left ventricular enlargement, previous coronary artery disease and elapsed time from symptom onset to infarction at entry to study showed no significant differences between the two groups. Plasma level of vitamin E (32.4 +/- 4.3 vs. 22.1 +/- 3.6 umol/L) and high density lipoprotein cholesterol (1.26 +/- 0.43 vs. 1.12 +/- 0.32 mmol/L) showed significant (p < 0.05) increase whereas thiobarbituric acid reactive substances, malondialdehyde (1.9 + 0.31 vs. 3.1 + 0.32 pmol/L) and diene conjugates showed significant reduction respectively in the CoQ group compared to control group. Approximately half of the patients in each group (n = 36 vs. 31) were receiving lovastatin (10 mg/day) and both groups had a significant reduction in total and low density lipoprotein cholesterol compared to baseline levels. It is possible that treatment with CoQ10 in patients with recent MI may be beneficial in patients with high risk of atherothrombosis, despite optimal lipid lowering therapy during a follow-up of 1 year. Adverse effect of treatments showed that fatigue (40.8 vs. 6.8%, p < 0.01) was more common in the control group than CoQ group.

Effects of coenzyme Q10 in new indications with antioxidant vitamins deficiency were also reported by Singh et al. for the first time, showing beneficial effects in patients with motor neuron disease and tuberous sclerosis as well as seizures and chronic renal failure (41). Congestive heart failure may be associated with CoQ10 deficiency as well as with carnitine deficiency [42, 43]. In a randomized, double blind controlled trial; Singh's group administered both the agents in patients with heart failure for the first time as metabolic treatment of heart failure [42, 43]. After a follow up period of 12 weeks, there was a significant improvement in ejection fraction and other parameters of heart failure [42]. Serum concentration of IL-6 and TNF-alpha, that are pro-inflammatory cytokines, showed a significant decline in the intervention group compared to control group which was again a novel discovery for CoQ10. Baseline serum CoQ10 (0.21±0.11 v/s 0.19+0.10ug/ml) was low, however, after 12 weeks, serum CoQ10 showed a significant increase in the carni Q-gel group compared to the control group (2.7±1.2 v/s 0.76±0.14 ug/ml) [42, 43]. Serum nitrite which is is a indicator of nitric oxide showed significant increase in the CoQ10 group compared to control group.

Long-term follow up, after treatment for 12 months, the quality of life visual analogous scale revealed that dyspnea, palpitation and fatigue and NYHA class II-III-IV, which were present at rest, in all the patients, at baseline, showed beneficial effects in the intervention group compared to the placebo group [43]. The deaths (3 vs. 8) and hospitalizations due to worsening of heart failure (2 vs. 11) among intervention and control group respectively, were significantly lower in the carni Q-gel group compared to the control group (5 vs. 19, P<0.02).Treatment with carni Q-gel was stopped after 12 months. Follow up after another 3 months (total 15 months) revealed that there was a worsening of NYHA class heart failure, as well as in the quality of life symptom scale and physical performance, assessed by 6-min walk test. There was a non-significant increase in hospitalizations in the intervention group after cessation of carni Q-gel softsules, compared to hospitalizations during the last 3 months. The findings indicated that treatment with ubiquinol + L-carnitine fumarate can cause a significant improvement in the quality of life, exercise capacity, as well as improvement in *The New York Heart Association (NYHA)* Functional Classification, which became worst after cessation of CoQ10.

Singh et al. also administered CoQ10 in chronic renal failure [44, 45]. Ninety-seven patients (mean age, 48 years) with chronic renal failure (serum creatinine > 5 mg/dl), with a history of declining renal function for at least 12 weeks, were randomly assigned to receive, in double-blind fashion, (CoQ10; 60 mg, 3 times per day orally) (Q-Gel) or placebo for 12 weeks [45]. The 45 patients who were receiving haemodialysis at the start of the study were encouraged to decrease the frequency or stop dialysis if there was an increase in urine output and a decrease in serum creatinine of more than 2 mg/dl. In the patients receiving haemodialysis and CoQ10, the mean serum creatinine concentration decreased from 9.5 to 6.7 mg/dl; mean *blood urea nitrogen* (BUN) decreased from 88.2 to 79.8 mg/dl; mean creatinine clearance increased from 40 to 54.9 ml/min; and 24-hour urine output increased from 1,300 to 1,920 ml. Renal function tended to worsen in haemodialysis patients receiving placebo, and the differences in the changes between groups were significant ($p < 0.01$ to $p < 0.001$). Significant improvements in each of these parameters relative to the placebo group were also seen in the non-dialysis patients treated with CoQ10. The number of patients receiving dialysis decreased from 21 to 12 in the CoQ10 group, and remained unchanged at 24 in the placebo group ($p < 0.02$). Eighty-one percent of the patients receiving CoQ10 had a positive response to treatment.

In a recent study, cerebrospinal fluid (CSF) concentrations of Co Q10 have been reported as breaking news finding in humans indicating that this biomarker can be used for diagnosis in the diseases of the brain [46]. Effects of Co Q10 administration in amyotrophic lateral sclerosis (ALS) have been reported again by Kawasaki et al. with beneficial effects [47]. In a further study, two patients presented with positive hepatitis B virus antigen reactivity [48]. Both the patients had medical records indicating clinical and biochemical manifestations of viral hepatitis. Treatment with coenzyme Q10 and w-3 fatty acids was associated with reversal of antigenicity causing negative hepatitis B antigen reactivity, an observation made for the first time in the literature. Endothelial dysfunction in type 2 diabetes and the possible impact on this condition of CoQ10, other antioxidants and nutritional supplements may be interesting [9, 49]. The effect of CoQ_{10} on endothelial dysfunction in ischaemic heart disease by Tiano et al, together with recent data highlighting that treatment with CoQ_{10} increases extracellular SOD activity indicate that this therapy can improve endothelial dysfunction [9]. In a randomized, double blind, placebo controlled trial, the effect of coenzyme Q10 (Co, Q10, 120 mg/day) (2cap BD, n=101) and placebo containing inert fibre (500mg, cap BD, n=99) were compared in patients with acute stroke, during a follow up period of 4 weeks [8]. The diagnosis of stroke was proven by computerized axial tomography (CAT) scan in all the patients. In 28 patients, plasma coenzyme Q was determined by HPLC showing low mean levels compared to healthy control subjects (CoQ 0.21 vs. 0.27ng/ml, $P<0.05$). The proportion of brain haemorrhage (27.7 vs. 25.2%, n=27 vs. 25) and cerebral infarction (73.2 vs. 74.7%, n=74 vs. 74) were comparable respectively. Approximately half of the patients presented with coma grade IV in association with hemiplegia in both the groups and rest half had hemiparesis. The proportion of deaths (13.8 vs. 18.2%,12 vs. 15) was slightly lower in the CoQ group and all accept 2 deaths in infarction,(control group), the deaths were in patients with brain haemorrhage, during the follow up of 4 weeks. These findings have also been confirmed in other studies (25, 47-50). Blood pressure variability appears to be a target for future cardiovascular protection which can be modulated by CoQ10. Coenzyme Q10 decreases all-cause mortality by half, according to the results of a multicentre randomized double blind trial presented at Heart Failure 2013 congress at Lisbon [23].

It is the first drug to improve heart failure mortality in over a decade and should be added to standard treatment, according to lead author Professor Svend Aage Mortensen (Copenhagen, Denmark). Singh's group recruited 110 patients from India to Q-SYMBIO Trial [23].

Table 1. Possible mechanisms of action of coenzyme Q10

1.	Repletion of CoQ10 deficiency.
2.	Antioxidant activity.
3.	Spares Vitamin E.
4.	Direct free radical scavenger via semiquinone species.
5.	Membrane stabilizing effect due to phospholipid protein interaction.
6.	Preservation of myocardial Na-K-ATPase activity and ATP manufacturing.
7.	Correction of mitochondrial "leak" of electrons during oxidative respiration.
8.	Induction of DT diaphorase which is an inhibitor of free radicals.
9.	Beneficial effect on prostaglandin metabolism.
10.	Inhibition of intracellular phospholipases.
11.	Stabilization of integrity of calcium ion dependent slow channels and possibly potassium channels.
12.	Spares vitamin C, A, and beta carotene by decreasing their cell consumption.
13.	Reduced insulin responses in myocardial infarction.
14.	Reduced lipoprotein (a) levels in myocardial infarction.
15.	Reduction in IL-6 and TNF-alpha in heart failure.
16.	Increased serum nitrite indicator of nitric oxide.
17.	Reduced serum creatinine in chronic renal failure.
18.	Reduced albuminuria in renal failure.
19.	Improved endothelial dysfunction
20.	Increase in coenzyme Q10 in cerebro spinal fluid in meningitis and encephalitis.
21.	Inhibition of atherosclerosis and stabilization of plaque.
22.	Proposed novel mechanism of Coenzyme Q binding site in VDAC of outer mitochondrial membrane

In brief, Singh´s group has demonstrated for the first time that CoQ10 can modulate plasma insulin, lipoprotein(a), serum creatinine and albuminurea, serum IL-6,TNF-alpha,serum nitrite, blood pressure variability, acute myocardial infarction and atherogenesis and neuronal degeneration, hence CoQ10 needs exploration for its supplementation in CVDs and other chronic diseases.

Conflict of interest has not been declared by the authors.

Acknowledgments

Due to International College of Nutrition and International College of Cardiology for logistic support to write this article. We are also thankful to Mr Raj Chopra, President, Tishcon Corporation, Westbury, NY for providing highly bioavailable CoQ10 for various studies, Grant Ministry of Education of Slovakia, VEGA 1/0614/12.

References

[1] WHO. Mortality and burden of disease estimates for WHO Member States in 2008. Geneva: *World Health Organization*, 2010.

[2] Singh RB, Fedacko J, Vargova V, Kumar A, Mohan A, Pella D, De Meester F, Wilson DW. Singh's verbal autopsy questionnaire for assessment of causes of death, social autopsy, tobacco autopsy, and dietary autopsy based on medical records and interview. *Acta Cardiol* .2011:66;471-481.

[3] Tokunaga M, Takahashi T, Singh RB, Rupini D, Toda E, Nakamura T, Mori H, Wilson DW. Diet, Nutrients and Noncommunicable Diseases. *The Open Nutraceuticals Journal,* 2012, *5,* 146-159

[4] Takahashi T, Singh RB, Meester FD, Wilson D. How the 'West' can overcome unhealthy behaviours to prevent chronic diseases. *J. Socialomics* 2013; 2: e114. doi:10.4172/2167-0358.1000e114 26.

[5] Singh RB, Kumar A, Neki NS, Pella D, Rastogi SS, Basu TK, Otsuka K, De Meester F, Wilson DW. Acharya SN, Juneja L, Takahashi T. Diet and Lifestyle Guidelines and Desirable Levels of Risk Factors for Prevention of Cardiovascular Disease and Diabetes among Elderly Subjects. A Revised Scientific Statement of the International College of Cardiology and International College of Nutrition-2011, *World Heart J.* 2011;3:305-320.

[6] IAS Position Paper: Global Recommendations for the Management of Dyslipidemia. http://www.athero.org/IASPositionPaper.asp , Accessed August 1, 2013.

[7] Fedacko J, Pella D, Fedackova E, *et al.* Coenzyme Q10 in Heart and Brain Diseases. *The Open Nutra J.* 2011; 4: 69-87.

[8] Singh RB., Sharma JP, Niaz MA, Mustafa F, Chandel JP, Littarru GP. Effect of coenzyme q10(qgel) in patients with stroke. Pr*esented in The 5th Conference of the International Coenzyme Q10 Association,* Kobe, Nov 9-12,2007.

[9] Tiano L, Belardinelli R, Carnevali P, Principi F, Seddaiu G, Littarru GP. Effect of coenzyme Q10 administration on endothelial function and extracellular superoxide dismutase in patients with ischaemic heart disease: a double-blind, randomized controlled study. *Eur. Heart J.* 2007; 18:2249-55.

[10] Alehagen U, Johansson P, Bjornstedt M, Rosén A, Dahlstrom U. Cardiovascular mortality and N-terminal-proBNP reduced after combined selenium and coenzyme Q10 supplementation: A 5-year prospective randomized double-blind placebo-controlled trial among elderly Swedish citizens. *Int. J. Cardiol.* 2012 May 22. PMID:22626835.

[11] Lehninger AI: Oxidation–reduction enzymes and electron transport. In *Biochemistry*, 2nd edition, p. 493–497. New York: Worth Publishers, 1975.

[12] Festenstein GN, Heaton FW, Lowe JS & Morton RA. A constituent of the unsaponificable portion of the animal tissue lipids (lambda max 272 mμ). *Biochem. J.* 1955;59:558-566.
[13] Linn BO, Page AC, Wong EI, Wong AL, Gale PH, Shunk GH, Folkers K: Isolation and distribution of coenzyme Q10 in animal tissues. *Am. Chem. Soc.* 1959;81:4007–4010.
[14] Crane FL., Hatefi Y, Lester RL & Widmer C. Isolation of a quinone from beef heart mitochondria. *Biochim. Biophys. Acta* 1957;25, 220-221.
[15] Wolf DE, Hoffman CH, Trenner NR, Arison BH, Shunk CH, Linn BO, McPherson JF & Folkers K. Coenzyme Q. Structure studies on the coenzyme Q group. *J. Am. Chem. Soc.* 1958;80: 4752-4758.
[16] Gvozdjáková A, Mravec B, Kucharská J, Lackovičová L, Ondičová K, Tkačov M & Singh RB. Increased ubiquinone concentration after intracerebroventricularly administered ubiquinol to selected rat brain regions. *Gen. Physiol. Biophys.* 2012; 31, 469–472.
[17] Gvozdjáková A, Mravec B, Kucharská J, Lackovičová L, Ondičová K, Mikulecký M, Tkačov M, Singh RB. Ubiquinol stimulates ubiquinone production in brain structures. Sixth Conference of the International Coenzyme Q_{10} Association, Brussels, Belgium, May 27 – 30, 2010. Abstract book: 85-88.
[18] Gvozdjakova A, Kucharska J, Babinska K, Ostrihonova S, Mego R, Nakladal D, Koplikova A, Ostatnikova D. *Effect of ubiquinol on oxidative stress, antioxidants and psychological manifestations in children with autism.* Seventh Conference of the International Coenzyme Q_{10} Association, Seville, Spain, November 8-11, 2012. Abstract book: 92-93.
[19] Gvozdjakova A, Kucharska J, Ostatnikova D, Babinska K, Nakladal D, Crane FL. Ubiquinol supportive therapy in children with autism. *Journal of Nutrition & Food Sciences,* 2013; 3/4: Abstract p. 115.
[20] GP Littarru, L Ho, K Folkers. Deficency of CoQ10 in human heart disease. *Int. J. Vit. Nutr. Res.* 1972; 42:291-295.
[21] Littarru GP, Ho L, Folkers Lehninger AI. Oxidation–reduction enzymes and electron transport. In *Biochemistry*, 2nd edition, p. 493–497. New York: Worth Publishers, 1975.
[22] Yamamura Y, Ishiyama T, Yamogami T, Morita Y, Ishio I, Kashiwamura S, Terader A, Tsukamoto N, Toyama S, Nakajima Y, Wada N: Clinical use of coenzyme Q for treatment of cardiovascular disease. *Jpn Circ. J.* 1967;31:168–169.
[23] Mortensen SA, Kumar A, Dolliner P, et al. The effect of coenzyme Q10 on morbidity and mortality in chronic heart failure. Results from the Q-SYMBIO study. Presented at Heart Failure Congress, May 25-28, 2013, *Final Programme* Number 440.
[24] Ramasarma T. Studies on ubiquinone. *J. Sci. Industr. Research* 27:147-64,1968.
[25] Yamamura Y. A survey of therapeutic uses of CoQ10. In *Biochemistry, bioenergetics, and clinical applications of ubiquinone.* Ed. G Lenaz, p. 479-505. Chickester: John Wiley & Sons, 1985.
[26] Stocker R, Bowry WE, Frei B. Ubiquionol - 10 protects human low density lipoprotein more efficiently against lipid peroxidation than does alpha-tocopherol. *Proc. Natl. Acad. Sci. USA* 1991; 88: 1646-1650.
[27] Takasawa K, Fuse K, Konishi T, Watanabe Y. Prevention of premature ventricular contract- ions with CoQ10 after coronary artery bypass grafting. In: Biomedical and

clinical aspects of coenzyme Q, Vol. 6. Ed. K Folkers, CP Littarru and T Yamagami, p. 357-359. Amsterdam: Elsevier. 1991.

[28] Dhalla AK, Hill MF, Singal PK. Role of oxidative stress in transition of hypertrophy to heart failure. *J. Am. Coll. Cardiol.,* 28: 506-514,1996.

[29] Serebruany VL, Ordonez JV, Herzog WR, Rohde M, Mortensen SA, Folkers K, Gurbel PA.. Dietary CoQ10 supplementation alters platelet size and inhibits human vitronectin (CD 51/CD61) receptro expression. *J. Cardiovasc. Pharmacol.* 29:16-22. 1997.

[30] Singh RB, Niaz MA, Ghosh S, *et al.* Dietary intake and plasma levels of antioxidant vitamins in health and disease; a hospital based case control study. *J. Nutr. Environ. Med.* 1995; 5: 235-42.

[31] Kuklinski B, Weissenbacker E, Fahnrich A. CoQ10 and antioxidants in acute myocardial infarction. *Mol. Aspects Med.* 199415 (Suppl): 143-147.

[32] Singh RB, Wander GS, Rastogi A, Shukla PK, Mittal A, Sharma JP. Randomized, double blind, placebo controlled trial of CoQ10 in patients with acute myocardial infarction. *Cardiovasc Drug Ther.* 1998 12:347-353.

[33] Singh RB, Niaz MA, Rastogi SS, Shukla PK, Thakur AS. Effect of hydrosoluble CoQ10 on blood pressures and insulin resistance in hypertensive patients with coronary artery disease. *J. Human Hyper* 1999 13: 203- 208.

[34] Singh RB, Chopra RK, Niaz MA, Kapoor R. Serum concentration of lipoprotein(a) decreases on treatment with hydrosoluble CoQ10 in patients with acute coronary artery disease. Discovery of a new role. *Int. J. Cardiol.* 1999 68: 23-29.

[35] Singh RB, Rastogi SS, Moshiri M. CoQ10 and its role in heart disease. *J. Clin. Biochem. Nutr.* 1999;26: 109-118.

[36] Singh RB, Shinde SN, Chopra RK, Niaz MA, Thakur AS, Onouchi Z. Effect of CoQ10 on experimental atherosclerosis and chemical composition and quality of atheroma. *Atherosclerosis* 2000;148: 275-282.

[37] Singh RB, Neki NS, Kartikey K, Pella D, Niaz MA, Kumar A, Thakur AS. Effect of coenzyme Q10 on risk of atherosclerosis in patients with recent myocardial infarction. *Mol. Cell Biochem.* 2003,246:75-82.

[38] Singh RB, Kartikey K, Charu AS, Niaz MA, Schaffer S. Effect of taurine and coenzyme Q10 in patients with acute myocardial infarction. In Taurine 5, Beginning the 21st century, Editors Lombardini et al, Kluwer Academic 2003, p41-48.

[39] Singh RB, Singh MM. Effects of coenzyme Q10 in new indications with antioxidant vitamins deficiency. *J. Nutr. Environ. Med.* 1999; 9: 223-8.

[40] Kumar A, Singh RB, Saxena M, Niaz MA, Joshi SR, Chattopadhyay P, Fedacko J, Machirova V, Pella D. Effects of carni Q-gel (carnitine plus ubiquinol) on cytokines in patients with heart failure in the Tishcon study. *Acta Cardiol.* 2007;62:349-354.

[41] Mechirova V, Kumar A, Singh RB, Sharma R, Chattopadhyay P, Niaz MA, Thakur AS, Fedacko J, Pella D and Juneja L. Randomized, double blind, placebo controlled trial of hydrosoluble ubiquinol and carnitine in patients with heart failure: longterm follow up results in the Tishcon study. *The Open Nutraceuticals Journal,* 2008, 1, 1-6.

[42] Singh RB, Khanna HK, Niaz MA. Randomized, double blind, placebo controlled, trial of coenzyme Q10 in chronic renal failure: Discovery of a new role. *J. Nutr. Environ. Med.* 2000;10:281-88.

[43] Singh RB, Kumar A, Niaz MA, Singh RB, Gujrati S, et al. Randomized, double blind, placebo controlled trial of coenzyme Q10 in patients with end stage renal failure. *J. Nutr. Environ. Med.* 2003,13:13-22.

[44] Singh RB, Naini A. Cerebrospinal Fluid (CSF) Concentration of coenzyme q10 in clinical practice, a breaking news finding in humans. *The Open Nutra J.* 2009; 2: 13-15.

[45] Kawasaki T, Singh RB, Cornelissen G, and Halberg F. Effects of coenzyme Q10 administration in amyotrophic lateral sclerosis (ALS). Report of a Case and Review. *The Open Nutra J.* 2012;5:187-192.

[46] Singh RB, Tokunaga M, Takahashi T, Wilson DW, Gerasimova E and De Meester F. Reversal of immunogenicity of hepatitis B antigen by treatment with nutraceuticals (OMEGA QGEL). *The Open Nutr. J.* 2013;6: (in press).

[47] Littarru GP, Tiano L, Belardinelli R, Watts GF. Coenzyme Q_{10}, endothelial function, and cardiovascular disease. *BioFactors* 2011;37:366-373.

[48] Halberg F, Cornelissen G, Singh RB. Timing nutriceuticals. *The Open Nutra J.* 2010;3:100-111.

[49] Cornélissen G, Halberg F, Schwartzkopff O, Gvozdjakova A, Siegelova J, Fiser B, Dusek J, Mifkova L, Chopra RK, Singh RB. Coenzyme-Q10 effect on blood pressure variability assessed with a chronobiological study design. *Abstract, Noninvasive Methods in Cardiology*, Brno, Czech Republic, September 14, 2005, p. 10.

[50] Singh RB, Fedacko J, Sharma JP, Vargova V, Sharma, Moshiri M, De Meester F, Otsuka K. Association of inflammation, heavy meals, magnesium, nitrite, and coenzyme Q10 deficiency and circadian rhythms with risk of acute coronary syndromes. *World Heart J.* 2010;2:219-228.

In: Circadian Cardiology with Focus on Both Prevention ... ISBN: 978-1-63463-956-9
Editors: K. Hristova, A. Shehab, G. Cornelissen et al.

Chapter 24

Arrhythmia in Elderly Subjects

***N. S. Neki*[1,*], *Miki Tokunaga*[2], *Toru Takahashi*[3], *Jan Fedacko*[4] and *Elena Gerasimova*[5]**

[1]Department of Medicine Government Medical College/Guru Nanak Dev Hospital, Amritsar, India
[2]Department of Food and Nutrition, Fukuoka Women's University, Japan
[3]Graduate School of Human Environment Science, Fukuoka Women's University, Japan
[4]PJ Safaric University, Kosice, Slovakia
[5]State Medical Academy of NN Burdenko in Voronezh, Russia

Abstract

The incidence of cardiac arrhythmia in the elderly is increasing due to enhanced longevity and risk factors. Coronary artery disease, diabetes mellitus, hypertensive heart disease, dilated cardiomyopathy, heart failure of both systolic and diastolic type and degenerative causes are main contributory factors. Systemic illnesses like infection, gastrointestinal upset, electrolyte disturbances, antidepressant drugs and alcoholism may trigger such arrhythmias in elderly. The investigation, management and clinical outcomes of atrial fibrillation (AF), symptomatic bradycardia, ventricular tachyarrhythmias and syncope that are pertinent to the elderly will be reviewed.

Keywords: Elderly; Arrhythymia; Intracardiac implantable defibrillator (ICD)

Introduction

The incidence of cardiac arrhythmia in the elderly is increasing due to enhanced longevity. Disorders of impulse formation and conduction are common in the elderly and may

* E-mail: drneki123@gmail.com.

result in symptomatic bradycardia or tachycardia. Aging is associated with decreases in myocardial cell density within the sinus node, loss of atrial myocardial fibres in approaches to the sinus node and internodal myocardium, and deposition of amyloid and interstitial fibrosis in the atria and specialized conduction system [1, 2]. Such abnormalities can lead to sinus node exit block, sinus arrest or transient or persistent AV block. Fibrosis in the atria may result in areas of slow conduction - one of the factors required for maintenance of atrial fibrillation (AF) [2].

There is evidence that elderly population may be associated with magnesium and potassium deficiency which are known to increase the risk of arrhythmias and sudden cardiac death [3, 4]. Elderly also have greater oxidative stress and deficiency antioxidant enzymes and vitamins which can predispose inflammation and arrhthmias [3, 4].In a cross-sectional survey [3], there were 595 subjects (314 males, 281 females) between 50-84 years of age inclusive. The overall prevalence of hypo-magnesemia was 11.8 per cent (n = 60) with a prevalence of 13.2 per cent (n = 33) in males and 10.6 per cent (n = 27) in females. The prevalence of hypomagnesemia showed significant declining trend in the concentration of serum magnesium, vitamin C, vitamin E and beta-carotene and a rising trend in lipid peroxides and diene conjugates with increase in age from 50-59 years to 70-84 years in both men and women. Multivariate logistic regression analysis showed that serum magnesium, vitamin C, vitamin E and beta-carotene were significant risk factors of ageing in both men and women. Antioxidant deficiency is also a risk factor for inflammation and arrhythmias. The findings suggest that some urban elderly populations of India can benefit by consuming higher dietary magnesium, potassium and antioxidant vitamins for prevention of ageing.

In another study, cross-sectional surveys were conducted to determine the association of magnesium with risk of coronary artery disease (CAD) [4]. There were 501 rural (270 men and 231 women) and 505 urban (250 men and 255 women) subjects between 50-54 years of age inclusive. The overall prevalence of CAD was three times higher in urban than in rural subjects (11.7 vs. 3.98) and the rates were comparable in both sexes. Dietary intake of magnesium was significantly ($P < 0.05$) higher in rural subjects in both men (520 +/- 58 vs. 415 +/- 47 mg/d) and women (432 +/- 40 vs. 316 +/- 38 mg/d). Dietary magnesium intake and serum magnesium were inversely correlated with CAD. The odds ratio for dietary magnesium intake indicates a higher prevalence of CAD at lower intakes of magnesium in both rural (0.67, 95 per cent confidence interval 0.51 to 0.86) and urban (0.72, 95 per cent CI 0.54 to 0.90) subjects.

Multivariate regression analysis showed that serum and dietary magnesium were significantly associated with CAD. Hypertension was not associated with CAD, and serum cholesterol, showed only weak association in both rural and urban subjects. The inverse association of dietary and serum magnesium with CAD shows that some urban populations of India may benefit from increased consumption of dietary magnesium and higher serum magnesium levels which are important in the pathogenesis of arrhythmias.

The high incidence of annular calcification, hypertension and ischemic heart disease in the elderly also contribute to the development of electrophysiological abnormalities that predispose to the development of conduction system disease or ventricular tachyarrhythmias. Elderly patient may present with palpitations, presyncope, unexplained falls, intermittent confusion, thromboembolic events, syncope or cardiac arrest. Some elderly patients are asymptomatic and the arrhythmia may be detected as an incidental finding on physical examination or during electrocardiographic monitoring.

Symptomatic Bradycardia

Symptomatic bradycardia is predominantly a disease of the elderly and is usually secondary to sinus node dysfunction (profound sinus bradycardia, sinus arrest and/or chronotropic incompetence) or to atrioventricular block (second degree AV block or complete AV block). The average age of a patient with symptomatic bradycardia requiring a permanent pacemaker is 73 years [5]. Patients may present with syncope, presyncope, intermittent confusion, an unexplained fall, fatigue or dizziness.

Investigations

The minimal investigations required in a patient with suspected bradycardia include a careful history and physical examination, an ECG and an echocardiogram to assess cardiac function. A rhythm strip should be performed during carotid sinus massage if no carotid bruits are heard on auscultation and there is no history of cerebrovascular disease. If symptoms are intermittent and an arrhythmia is suspected but not documented, additional investigations might include 24-48 hour ambulatory ECG monitoring, an event recorder, tilt table test, electrophysiologic study and/or an implantable loop recorder.

Sick sinus syndrome: This syndrome is manifestated through a variety of symptoms like dizziness, fatigue, syncopal or presyncopal episodes and arrhythmias such as sinus bradycardia, sinus pause(>3 sec), sinus arrest or exit block, sinus and AV node conduction abnormalities and episodes of tachycardia-bradycardia which are characterized by alteration of rapid atrial tachyarrythmias, atrial fibrillation and bradycardias. The mean age at which it occurs is 65 years. Symptomatic sinus node dysfunction is an absolute indication for pacing [6].

The indications for pacing in sick sinus syndrome according to the guidelines of the American College of Cardiology [7] are:

Class I. 1. Proven symptomatic bradycardia including frequent sinus pauses. 2. symptomatic chronotropic insufficiency

Class IIa 1. Bradycardia and presence of symptoms without a proven correlation between them. 2. Unexplained syncope in patients where severe sinus nide dysfunction is revealed later either spontaneously or during electrophysiological study.

Class IIb minimally symptomatic patients with slow rate in an alert state.

Class III Asymptomatic patients or when symptoms are not the result of bradycardia.

Atrial Arrhythmia

The prevalence of atrial premature beats increases with age particularly those older than 80 years. Brief salvos of supraventriular tachycardia occur in up to 50% of older subjects and the prevalence doubles between the seventh and ninth decades of life. These arrhythmias have not been shown to have adverse prognostic significance. Atrioventricular nodal re-entrant tachycardia is due to calcification of the atriventricular conduction system which impedes conduction of the stimulus thus facilitating orthodromic conduction via the bypass tract. Taking of medication often entails an increased risk of proarrythmia in the elderly. So the choice of ablation for the treatment of paroxysmal supraventricular tachycardia in the elderly has increased dramatically in recent years particularly so if there is underlying WPW syndrome.

Atrial Fibrillation

It is the most common sustained cardiac arrhythmia in the elderly patients. The prevalence of atrial fibrillation in elderly above 70 years is 2% and 7-8% above the age of 80 years [8, 9]. Long standing AF is associated, in the majority of patients, with organic heart disease. Independent risk factors for AF in the elderly include age, systemic hypertension, congestive cardiac failure, valvular heart disease, stroke, enlarged left atria size, mitral or aortic valve dysfunction, echocardiographic features of diastolic dysfunction, and raised serum levels of NT-proBNP [10, 11]. Hyperthyroidism is another cause of atrial fibrillation in the elderly and one study reports an AF incidence of 25% amongst hyperthyroid patients older than 60 years in comparison to 5% in those aged less than 60 years [12].

Atrial fibrillation results in the loss of atrial contraction and its contribution to the cardiac output,an irregular response, and frequently a poorly controlled rapid heart rate. The individual response to heart rate is quiet variable. Some patients remain asymptomatic, whereas others may be symptomatic. The strategy of AF management includes the prevention of thromboembolic episodes administration of antiplatelet or anticoagulant agent, control of arrhythmia, either by restoring and maintain sinus rhythm (rhythm control strategy) or simply by controlling the heart rate (rate control strategy).

Factors associated with rapid progression of AF (advanced age and structural heart disease) are also associated with a poor response to rhythm control strategy. Recently, the HATCH scoring system (hypertension, age older than 75 years, previous transient ischemic attack or stroke, chronic obstructive pulmonary disease, and heart failure) has been proposed to identify patients at risk of progression to persistent AF [13]; a HATCH score of ≥2 was associated with a high risk of disease progression despite antiarrhythmic therapy.

Control of rate and rhythm: The rhythm may be controlled by cardioversion and maintenance of sinus rhythm via antiarrythmic drugs Class Ia (quinidine,disopyramide) and Ic (propaphenone).and /or invasive method (pacing,ablation).

Rhythm control with antiarrythmic drugs, mostly amiadorone or sotalol, was compared with a rate control strategy with atrioventricular (AV) node-blocking agents. The AFFIRM trial 1 which included more than 4000 patients with Atrial fibrillation, mean 70±9 years, showed that the rate control strategy in combination with anticoagulation medication was at least as good as rhythm control, over a mean follow up period of 3.5 years [14]. On the other hand, patients in the rhythm control arm had higher rate of hospitalization and drug related adverse events. The superiority of the rate control strategy was most pronounced in patients > 65 years probably because of increased risk of proarrythmia in the elderly. The rate control strategy is therefore a clearly acceptable approach in elderly patients; the drugs of choice are beta blockers, calcium channel blockers (diltiazem, verapamil) and digitalis or a combination [15]. In those who are not active or who have heart failure, monotherapy with digitalis is an acceptable choice. The reduced glomerular perfusion in elderly people, even those with normal creatine values a factor that must be taken into account when digitalis is administered.

Amiadorone appears to be the safest and most effective antiarrhythmic drug for the restoration of sinus rhythm. Low dose amiadorone is particularly suited for the long term maintenance of sinus rhythm after initial cardioversion [16].

Dronedrone is a new antiarrythmic drug that is being developed for the treatment of atrial fibrillation. Dronedrone reduced the incidence of hospitalization due to cardioversion, events or death with atrial fibrillation. It should not be given in NYHA Class II or III heart failure with a recent decompensation requiring hospitalization or referral to a specialized heart failure clinic [17].

Unfavorable Conditions for Cardioversion

These include duration of atrial fibrillation of more one year, moderate to severe cardimegaly, echocardiographic left atrial dimension >45 mm, digitalis toxicity (contraindicaton), slow ventricular rate (contraindicaton), sick sinus syndrome (contraindicaton), mitral valve disease, congestive heart failure, chronic obstructive lung disease, recurrent atrial fibrillation despite antiarrythmic drugs.

Treatment in patients aged >80 years are equivocal, in spite of its efficacy. Coumarine therapy has many inherent problems, especially in elderly patients, and attempts are being made to find alternative medications. Ximelagartran is a new direct thrombin inhibitor. Data from clinical trials (SPORTIF) have confirmed that Ximelagartran is superior to warfarin in the prevention of embolic episodes. It is associated with fewer total episodes of major or minor haemorrhage but is feared to be haepatotoxic, Findings are awaited from studies with longer follow up in order to draw certain conclusions [18]. Patients at low risk or who have warfarin contraindications should take 325 mg of aspirin daily.

Atrioventricular nodal ablation and implantation of a permanent pacemaker: This is the therapeutic choice in the case of symptomatic patients with atrial fibrillation whose ventricular response cannot be controlled satisfactory by medications, or where the use of drugs causes unacceptable side effects. Various studies have shown that atrioventricular nodal ablation is particularly effective in controlling symptoms and improving exercise tolerance, while it also improves ejection fraction, especially in a patient of heart failure. Control of heart rate is achieved by implanting a permanent pacemaker, of a type that depends on the form of atrial fibrillation.

Atrial flutter: This is rather common arrhythmia in the elderly. It is due to re-entrant macrocercuit and can be treated by ablation of the slow conduction zone represented by the inferior cava-tricuspid annulus isthmus. The success rate of ablation for typical atrial flutter reaches 85% with a relapse rate of 10%.

Ventricular Arrhythmia in Elderly

Sustained ventricular tachycardia (VT) and ventricular fibrillation (VF) are the major causes of sudden cardiac death. These arrhythmias occur predominantly in the setting of left ventricular dysfunction. The average age of an ICD recipient is 65 years [19, 20]. Risk factors for sudden death secondary to VT/VF include left ventricular ejection fraction < 40%, frequent ventricular premature beats or non-sustained VT, low heart rate variability, increased

QT interval dispersion and inducible sustained VT in patients with coronary artery disease [21].

Investigations

The investigations required in a patient with sustained VT or VF include a careful history and physical examination, 12 lead ECG, review of laboratory data at the time of presentation including serum electrolytes and creatinine, magnesium level, troponin or CK, review of relevant rhythm strips at presentation and assessment of LV function. VT/VF due to a secondary cause including acute ischemia, acute hemodynamic deterioration, hypokalemia or hypomagnesemia, and Class I/III antiarrhythmic drug therapy should be excluded. Coronary angiography is generally recommended to determine whether significant coronary artery disease requiring revascularization is present.

Management

Therapy of coronary artery disease and/or congestive heart failure should be optimized. Several randomized clinical trials have demonstrated the superiority of the ICD over antiarrhythmic drug therapy for prevention of cardiac mortality in patients with symptomatic sustained VT/VF in the setting of left ventricular dysfunction [19, 20]. More recently, clinical trials have also confirmed that the ICD reduces mortality compared to medical therapy in patients at risk for VT/VF [22-24]. These trials were conducted in patients with significant left ventricular dysfunction (left ventricular ejection fractions < 30-40%). One subgroup analysis in Canadian implantable defibrillator study (CIDS) suggested that patients most likely to benefit are those > 70 years of age with symptomatic heart failure and LVEF < 35% [25]. However, the benefit of ICDs in patients > 70 years was not confirmed in an Antiarrhythemics versus Implantable Defibrilators (AVID) substudy [26]. Whether patients with severe left ventricular dysfunction (LVEF < 0.20) benefit from ICD therapy has been the subject of debate.

Clinical trials comparing ICD therapy to pharmacologic therapy have not been conducted in the very elderly population. Thus, it is unclear whether the ICD confers survival benefits to the very elderly. Other issues that may determine a recommendation for ICD therapy in the very elderly include their ability to travel for follow-up (every 6 months) and their ability to tolerate shock therapy from the device. Sinus tachycardia or atrial fibrillation with a rapid ventricular rate may result in inappropriate shock therapies in up to 20 % of ICD recipients within the first year of implantation.

Recommendations

- Elderly patients surviving cardiac arrest or hemodynamically significant VT (not within three days of acute MI and not associated with a reversible/correctable cause) should be considered for an ICD (Class 1, Level of Evidence A).

- Elderly patients with minimally symptomatic VT and LVEF < 35% should be considered for an ICD (Class 1, Level of Evidence A).
- Elderly patients with minimally symptomatic VT and LVEF greater than 35% should receive either pharmacological therapy or an ICD (Class 1, Level of Evidence C).
- Elderly patients with a left ventricular ejection fraction < 0.30 in the setting of coronary artery disease and a life expectancy of ≥2 years could be considered for an ICD for the prevention of sudden death (Class II, Level of Evidence B).

Syncope/Fall

Syncope and an unexplained falls occur frequently in the elderly [27-31]. In addition to brady or tachyarrythmias, these events may be due to postural hypotention, neurocardiogenic mechanism or carotid sinus hypersensitivity.

Investigation

The minimal investigation required in patients with syncope or falls suspected to be of cardiovascular etiology include careful history, physical examination, ECG and an echocardiogram. A careful history usually excludes seizures disorder and identify neurocardiogenic syncope. A rhythm strip performed during carotid sinus massage may identify patients with carotid sinus hypersensitivity and is contraindicated in patients with carotid bruits or CVA [32].

Management

Elderly patients with bifascicular block and recurrent syncope have been demonstrated to have intermittent high grade block based on rhythms documented at the time of recurrent symptoms following insertion of an implantable loop recorder [32]. This evidence supports the concept of empiric pacing to prevent recurrent syncope in patients presenting with syncope in the setting of bifascicular block. Neurocardiogenic syncope in the elderly may be associated with autonomic dysfunction. Although β-blockers have been advocated as an initial therapy, the efficacy of β-blockers awaits the results of randomized clinical trials [33]. Removal of agents likely to provoke hypotension is frequently problematic, as many of these patients have associated hypertension. Pacing may be beneficial in the prevention of recurrent syncope although the benefit of this therapy has not been specifically assessed in the elderly [34].

Orthostatic hypotension. It is a common problem in the elderly. Primary treatment should focus on non-pharmacological measures – adequate hydration, support stocking, and avoidance of standing stationary for prolonged period of time. When conservative management fails, available drugs include alpha1 adrenergic agonists (mainly midodrine) or plasma volume expanders (mainly fludrocortisone). These drugs may exaggerate existing hypertension and must be used cautiously.

Recommendation

1. Patients with recurrent syncope documented to be secondary to carotid sinus hypersensitivity (with a pause >3 sec. On carotid sinus massage) should receive a dual chamber pacemaker (Class I, Level of Evidence B).
2. Patients with syncope in the setting of bifascicular block should be considered for a pacemaker(Class II, Level B)
3. Elderly patients with a left ventricular ejection fraction <0.30 in the setting of coronary artery disease and life expectancy of 72 years could be considered for an ICD for the prevention of sudden death(Class II, Level of Evidence B).

Conclusion

As the aging population grows in size, the number of elderly patients with arrhythmogenic problems will also increase. The problems associated with use of antiarrhythmic drugs in patients with advanced age have led to a steady increase in the proportion of patients who are treated using invasive methods. Already elderly patients make up the vast majority of those who undergo treatment of AF, an arrhythmia that is particularly common in aged have been tested in recent years so far with equivocal results. Finally implantable defibrillator devices despite the lack of prospective studies that includes a large number of elderly patients appear to be the indicated method of treatment for malignant ventricular tachycardias especially in the presence of coronary artery disease. In dangerous VT/VF associated with severe heart failure, combined CRT and ICD therapy may be helpful although its cost effectiveness and long term survival benefits in the elderly are not established so far. The dosage and long term use of antiarrhythmic drugs in elderly should take into consideration the decreased absorption, increased clearance of proarrhythmic effects and also the presence of comorbid condition especially chronic kidney disease chronic obstructive airway disease and mentation.

References

[1] Gillis A. M. Sinus Node Disease In: Clinical Cardiac Pacing. 2nd Edition. KA Ellenbogen, GN Kay, BL Wilkoff, eds. Philadelphia: WB Saunders and Co. 2000; 405-425.

[2] Davies M. J., Anderson R. H., Becker A. E. Pathology of atrial arrhythmias in the conduction system of the heart. Butterworth & Co Ltd, London. 1983;203-215.

[3] Singh R. B., Rastogi V., Singh R., Niaz M. A., Srivastav S., Aslam M., Singh N. K. Magnesium and antioxidant vitamins status and risk of complications of ageing in an elderly urban population. *Mag. Res.,* 1996; 9: 299-306.

[4] Singh R. B., Niaz M. A., Ghosh S., Rastogi V., Raghuvanshi R. S., Moshiri M. M.. Epidemiologic study of magnesium status and risk of coronary artery disease in elderly rural and urban populations of north India. *Magnesium Res.,* 1996; 9:165-172.

[5] Sudlow M., Thomson, R., Thwaites B., et al. Prevalence of atrial fibrillation and eligibility for anticoagulants in the community. *Lancet,* 1998; 352: 1167-1171.

[6] Rodriguez R. D., Schocken D. D: Update on sick sinus syndrome, a disorder of aging. *Geriatrics,* 1990; 45: 26-36.

[7] Gregoratos G., Abrams G., Epstein A. E., et al: Guideline update for implantation of cardiac pacemakers and antiarrhythmia devices: The American College of Cardiology/American.

[8] Heart Association Task Force on Practice Guidelines. *Circulation,* 2002; 106: 2145-2161.

[9] Wolf P. A., Abbott R. D., Kannel W. B. Atrial fibrillation as an independent risk factor for stroke: the Framingham Study. *Stroke,* 1991; 22:983-8.

[10] Furberg C. D., Psaty B. M., Manolio T. A., et al. Prevalence of atrial fibrillation in elderly subjects (the Cardiovascular Health Study). *Am. J. Cardiol.*, 1994;74:236-41.

[11] Rosenberg M. A., Gottdiener J. S., Heckbert S. R., et al. Echocardiographic diastolic parameters and risk of atrial fibrillation: the Cardiovascular Health Study. *European Heart Journal,* 2012; 33:904-912.

[12] Patton K. K., Ellinor P. T., Heckbert S. R., et al. N-Terminal pro-b-type natriuretic peptide is a major predictor of the development of atrial fibrillation: the cardiovascular health study. *Circulation*, 2009; 120:1768-1774.

[13] Agner T., Almdal T., Thorsteinsson B., et al. A reevaluation of atrial fibrillation in thyrotoxicosis. *Danish medical bulletin*, 1984; 31:157-159.

[14] De Vos C. B., Pisters R., Nieuwlaat R., et al. Progression from paroxysmal to persistent atrial fibrillation. clinical correlates and prognosis. *Journal of the American College of Cardiology*, 2010; 55: 725-731.

[15] Wyse D. G., Waldo A. L., DiMarco J. P., et al. A comparison of rate control and rhythm control in patients with atrial fibrillation. *N. Engl. J. Med.*, 2002; 347:1825-33.

[16] Farshi R., Kistner D., Sarma J. S., et al. Ventricular rate control in chronic atrial fibrillation during daily activity and programmed exercise: a crossover open-label study of five drug regimens. *J. Am. Coll. Cardiol.*, 1999; 33:304-10.

[17] Lafuente-Lafuente C., Mouly S., Longas-Tejero M. A., et al. Antiarrhythmics for maintaining sinus rhythm after cardioversion of atrial fibrillation. *Cochrane Database of Systematic Reviews*, no. 4, Article ID CD005049, 2007.

[18] S. J. Connolly S. J., Crijns M., Torp-Pedersen C., et al. Analysis of stroke in ATHENA: a placebo-controlled, double-blind, parallel-arm trial to assess the efficacy of dronedarone 400 mg bid for the prevention of cardiovascular hospitalization or death from any cause in patients with atrial fibrillation/atrial flutter. *Circulation*, 2009; 120:1174-1180.

[19] Olsson S. B. et al. Executive Sterring committee on behalf of the SPORTIF III investigation. Stroke prevention with the oral direct thrombin inhibitor ximelagatran compared with warfarin in patients with non-valvular atrial fibrillation (SPORTIF III): randomised controlled trial. *Lancet,* 2003; 362:1691-98.

[20] The Antiarrhythmics versus Implantable Defibrillators (AVID) Investigators. A comparison of antiarrhythmic-drug therapy with implantable defibrillators in patients resuscitated from near-fatal ventricular arrhythmias. *N. Engl. J. Med.,* 1997;337: 1576-83.

[21] Connolly S. J., Gent M., Roberts R. S., et al. Canadian implantable defibrillator study (CIDS) : a randomized trial of the implantable cardioverter defibrillator against amiodarone. *Circulation,* 2000;101:1297-302.

[22] Windhagen-Mahnert B., Kadish A. H. Application of noninvasive and invasive tests for risk assessment in patients with ventricular arrhythmias. *Cardiol. Clin.,* 2000; 18: 243-63.

[23] Moss A. J., Hall W. J., Cannom D. S., et al. Improved survival with an implanted defibrillator in patients with coronary disease at high risk for ventricular arrhythmia. Multicenter Automatic Defibrillator Implantation Trial Investigators. *N. Engl. J. Med.,* 1996; 335:1933-40.

[24] Buxton A. E., Lee K. L., Fisher J. D., et al. A randomized study of the prevention of sudden death in patients with coronary artery disease. Multicenter Unsustained Tachycardia Trial Investigators. *N. Engl. J. Med.,* 1999; 341:1882-90.

[25] Moss A. J., Zareba W., Hall W. J., et al. Prophylactic implantation of a defibrillator in patients with myocardial infarction and reduced ejection fraction. *N. Engl. J. Med.,* 2002; 346:877-83.

[26] Sheldon R., O'Brien B. J., Blackhouse G., et al. Effect of clinical risk stratification on cost-effectiveness of the implantable cardioverter-defibrillator: the Canadian implantable defibrillator study. *Circulation,* 2001; 104:1622-6.

[27] Exner D. V., Sheldon R. S., Pinski S. L., et al. Do baseline characteristics accurately discriminate between patients likely versus unlikely to benefit from implantable defibrillator therapy? Evaluation of the Canadian implantable defibrillator study implantable cardioverter defibrillatory efficacy score in the antiarrhythmics versus implantable defibrillators trial. *Am. Heart J.,* 2001; 141:99-104.

[28] Moya A., Brignole M., Menozzi C., et al. Mechanism of syncope in patients with isolated syncope and in patients with tilt-positive syncope. *Circulation,* 2001; 104: 1261-7.

[29] Meredith P. A. Is postural hypotension a real problem with antihypertensive medication? *Cardiology,* 2001; 96(1):19-24.

[30] Brignole M., Menozzi C., Moya A., et al. Mechanism of syncope in patients with bundle branch block and negative electrophysiological test. *Circulation,* 2001; 104(17):2045-50.

[31] Faddis M. N., Rich M. W. Pacing interventions for falls and syncope in the elderly. *Clin. Geriatr. Med.,* 2002; 18:279-94.

[32] Kenny R. A. Neurally mediated syncope. *Clin. Geriatr. Med.,* 2002; 18:191-210.

[33] Kenny R. A., Richardson D. A., Steen N., et al. Carotid sinus syndrome: a modifiable risk factor for non accidental falls in older adults (SAFE PACE). *J. Am. Coll. Cardiol.,* 2001; 38:1491-6.

[34] Sheldon R. S., Raj S. R., Rose S., et al. Beta-blockers in syncope: the jury is still out. *J. Am. Coll. Cardiol.,* 2001; 38:2135-6.

[35] Connolly S. J., Sheldon R., Roberts R. S., et al. The North American Vasovagal Pacemaker Study (VPS). A randomized trial of permanent cardiac pacing for the prevention of vasovagal syncope. *J. Am. Coll. Cardiol.,* 1999; 33:16-20.

In: Circadian Cardiology with Focus on Both Prevention ... ISBN: 978-1-63463-956-9
Editors: K. Hristova, A. Shehab, G. Cornelissen et al.

Chapter 25

Can Elective Percutaneous Coronary Intervention Be Safely Performed without Surgical Backup?

Jan Fedacko[1,*], Viola Vargova[1], R. B. Singh[2], Eri Toda[3], Krasimira Hristova[4] and Daniel Pella[1]

[1]Faculty of Medicine, PJ Safaric University, Kosice, Slovakia
[2]Halberg Hospital and Research Institute, Moradabad, India
[3]Department of Cardiology,Tokai University Hachioji Hospital, Tokyo, Japan
[4]University National Heart Hospital,
Department of Noninvasive Functional Diagnostic and Imaging, Sofia, Bulgaria

Emergency coronary bypass surgical grafting (CABG) overpercutaneous coronary intervention (PCI) has become a rare event. This achievement of the angioplasters raises the question of whether availability of cardiac surgery on site is still necessary for the safe and effective performance of PCI? In view of the complications such as myocardial infarction (MI) after PCI, it is important to inform clinical practice as well as to place the findings in proper perspective for the purposes of diagnostic coding, insurance and epidemiological considerations, interpretation of clinical trials, and performance "scorecards." There may be emotionally charged arguments on scorecards, because of the potentially adverse consequences for operators, if PCI in their hands is perceived as being associated with high rates of MI. It is a excoriating situation for operators who are willing to take on high-risk PCI procedures. The utility and necessity for timely access to primary PCI has justified the expansion of PCI to hospitals without on-site cardiac surgery [1, 2]. However, controversy exists as to whether there should be a further expansion for non-emergent PCI to these hospitals? Since the risk/benefit ratio might be different in various hospitals and the need for timely access is less important to cardiovascular outcomes, such a possibility and availability can decrease the expenses of the set up of the hospitals and therefore cost of PCI.

* Correspondence: Jan Fedacko, Faculty of Medicine, PJ Safaric University, Kosice, Slovakia. Email, janfedacko@hotmail.com.

Recently, Massachusetts investigators demonstrated that there was no difference in major adverse cardiac events at 30 days and 12 months, when these patients were compared with those treated at hospitals with surgical capability [1, 2]. Most angioplasters agree that non-emergent PCI can be safely performed in community hospitals without on-site cardiac surgery. Each participating community hospital and operator met strict requirements, participating in this study regarding volume standards for hospitals and operators. Angioplasters were mandated to have performed a minimum of 75 PCI procedures annually to be included in the trial. The hospitals were also required to have access to a hospital with 24/7 surgical backup (patient arrival within 60 minutes). The centers had to perform a minimum of 300 diagnostic procedures in each of the previous two years, including 36 primary PCI procedures annually. In total, 2774 patients undergoing coronary angiography at a hospital without on-site cardiac surgery underwent PCI, and 917 were transferred for PCI to a hospital with surgical backup. The average patient age was 64 years old, approximately one-third had diabetes, and 30% of patients had undergone prior PCI. Baseline characteristics showed no significant differences. However, 24.1% individuals revascularized at the center without surgical backup had a prior MI compared with 20.2% of individuals treated at a center with on-site surgery ($p = 0.015$).In terms of the individual components of the safety and efficacy end points (death, MI, repeat revascularization, or stroke), there was no difference among those treated with and without surgery on site.

Table 1. Cardiac events in the two groups; no onsite surgery verses onsite surgery

Cardiac end points	No on-site surgery (n = 2774 (%)	On-site surgery (n = 917(%)
Death, MI, repeat revascularization, stroke at 30 day.	9.5	9.4*
Death, MI, repeat revascularization, stroke at 1 year.	17.3	17.8*

P for non-inferiority, $P < 0.001$.

Table 1 reveals the results of PCI done at centers; without onsite surgery with on-site surgery. It is clear that intervention performed at hospitals without cardiac surgery on site was noninferior to intervention performed with cardiac surgery on site with respect to the 30-day safety and 12-month effectiveness end points. These data suggest that PCI can be safely performed at hospitals without surgery on site. However, the criteria with formal programs and volume requirements should be followed honestly in these hospitals. This could be an acceptable option for patients presenting to such hospitals. However, the message is that the decision to perform elective PCI without surgical backup cannot be undertaken by community hospitals lightly.

Since pre-study preparation included a rigorous program with a long training period for operators and a long training period for hospitals with very clearly defined pathways to manage the patients, this message is important. It is also important to learn to select patients, train your operators, and establish a safe and reliable system for doing PCI, such as managing the patients and preparing for some of the potential disasters that can occur. Any complication, can suddenly change the mood and tenor of any enthusiastic program without on-site surgical backup. If this strategy sincerely followed, it may turn high-volume clinical centers and those hospitals that have on-site surgical backup, into low-volume centers, by displacing patients to community hospitals which may not be in the best interests of many patient. On-site surgery often serves as a surrogate for having other hospital services, including specialized care, as well as for having more mid-level practitioners seeing patients. These are programmatic issues that will have an impact on the national aspects of healthcare more than at the individual level.

The American College of Cardiology Foundation and American Heart Association have released new guidelines for bypass surgery and PCI featuring a common section on coronary artery disease (CAD) revascularization [3, 4] The new guidelines were partly a collaboration between surgeons and interventionalists, mirroring the "heart-team" approach to coronary revascularization decisions recommended in the guidelines. Mutual evaluation and mutual decision making will result in these guidelines being more often followed and lead to the best results for the patients. It seems, the most important thing about these new guidelines is the consensus that was established between cardiologists and surgeons over patient selection for those two main procedures. The recommendations are related to randomized clinical-trial data, and most of those data were achieved from patients where there was consensus between a cardiac surgeon and a cardiologist about decision on PCI or surgery. The guidelines strongly recommend (class I) that decisions about revascularization of an unprotected left main or complex coronary artery disease be made by a heart team, including a surgeon and an interventionalist. In case of skilled team, we need to provide as much as possible satelite PCI units out of the big centres to provide the access within time window for the emergent PCI.

The future guidelines may include decisions about doing PCI at community hospitals without surgical back up or at those hospitals with no onsite CABG. The results could be further improved if non emergent patients of CAD are prepared before undergoing the procedure. MI associated with PCI, has generated considerable controversy, about the mechanism and the prognostic implications [5].

MI after PCI may be; Type 1 (proximal type) related to the epicardial obstruction that occurs during the procedure. This infarction typically is either direct or indirect side-branch occlusion. Type 2 (distal type) is a result of downstream microvascular obstruction. It occurs as a consequence of embolization of atherothrombotic material that produces microvascular plugging, platelet activation, neutrophil activation, and release of active peptides that contribute to vasoconstriction in the setting of sympathetic stimulation. The risk factors for post-PCI MI include those that are patient related, lesion related, and procedure related, which may be interpreted differently by many investigators. Since coronary intervention is associated with marked increase in oxidative stress and inflammation and complications such as myocardial infarction (MI), efforts should be made to treat underlying abnormalities in these biomarkers for better outcome. This can be achieved by drug therapy as well as by foods and nutrients and lifestyle factors which have been found to induce cell signaling [6].

Acknowledgment

To International College of Cardiology for logistic support to write this article.

Conflict of Interest

Conflict of interest has not been declared by the authors.

References

[1] Jacobs A. K., Normand S. L. T., Massaro J. M. et al. Nonemergency PCI at hospitals with or without on-site cardiac surgery. *N Engl J Med* 2013: DOI: 10.1056/NEJ Moa1300610. Available at: http://www.nejm.org.

[2] Aversano T., Lemmon C. C., Liu L. et al. Outcomes of PCI with or without on-site cardiac surgery. *N Engl J Med* 2012; DOI: 10.1056/NEJMoa1114540. Available at: http://www.nejm.org.

[3] Hillis L. D., Smith P. K., Anderson J. L. et al. 2011 ACCF/AHA guideline for coronary artery bypass graft surgery: A report of the American College of Cardiology Foundation/American Heart Association Task Force on Practice Guidelines. *J Am Coll Cardiol* 2011; DOI: 10.1016/j.jacc.2011.08.009. Available at: http://content.onlinejacc.org.

[4] Levine G. N., Bates E. R., Blankenship J. C. et al. 2011 ACCF/AHA/SCAI guideline for percutaneous coronary intervention: A report of the American College of Cardiology Foundation/American Heart Association Task Force on Practice Guidelines and the Society for Cardiovascular Angiography and Interventions. *J Am Coll Cardiol* 2011; DOI:10.1016/j.jacc.2011.08.007. Available at: http://content.onlinejacc.org.

[5] Antman E. M., Morrow D. A. Biomarker Release after Percutaneous Coronary Intervention, A Message From the Heart. *Circulation: Cardiovas. Interven.* 2008; 1: 3-6 doi:10.1161/CIRCINTERVENTIONS.108.799858.

[6] Singh R. B., Choudhury J., de Meester F., Wilczynska A., Dharwadkar S. and Wilson D. W. Association of High W-6/W-3 Ratio Paleolithic Style Diets and Risk of Cardiovascular Diseases and Other Chronic Diseases: Is the Tissue the Main Issue? *World Heart J* 2012; 4: (in press).

In: Circadian Cardiology with Focus on Both Prevention ... ISBN: 978-1-63463-956-9
Editors: K. Hristova, A. Shehab, G. Cornelissen et al.

Chapter 26

Wealth Can Defeat Death and Achieve Heart Health in the United States

R. B. Singh[1,*], R. N. Mondal[2], Krasimira Hristova[3], Toru Takahashi[4], Daniel Pella[5], Fabien De Meester[6], Elena Gerasimova[7] and D. W. Wilson[8]

[1]Halberg Hospital and Research Institute, Moradabad, India
[2]Hypertension Research Center, Rangpur, Bangladesh
[3]Department of Cardiac Imaging, National Heart Hospital, Sofia, Bulgaria
[4]Department of Nutrition, Graduate School of Human Environment Science, Fukuoka Women's University, Japan
[5]PJ Safaric University, Kosice, Slovakia
[6]The Tsim Tsoum Institute, Krakow, Poland
[7] State Medical Academy of NN Burdenko in Voronezh, Russia
[8]School of Medicine Pharmacy and Health, Durham University, UK

Recently, former Vice President of the United States of America; Dick Cheney told an audience of surgeons, at the American Association for Thoracic Surgery 2013 Annual Meeting, (Minneapolis, MN, USA), if you have heart attack, accept it. He further emphasized that coronary artery disease (CAD) is just a part of "who I am" and I do not feel like a victim or feel sorry for myself. He also reiterated in the 'honored-speaker interview,' he never even regarded his heart ailments as impediments to his high-profile and often high-stress career. He never felt that he was in place where he could not do something because of his health issues and he led a very full and active life like some rich men.

He developed his first heart attack in 1978, when he was in the middle of his first campaign for Congress. His doctor's advice at the time set the tone for his approach to heart disease. He asked his doctor if this would keep him out of a political career, and what he took away from that conversation was his advice that 'stress comes from something you don't want to be doing, but hard work never killed anyone,' and that's the part he remembered. Although, no physician would give such advice but it could not have been better in terms of getting on

* Correspondence: Prof. Dr. R. B. Singh, MD, FICN Halberg Hospital and Research Institute, Civil Lines, Moradabad-10(UP) 244001, India. Email: rbs@tsimtsoum.net, drkk@dataone.in.

with his life, despite having CAD. What it meant to him was, when you have a problem, you deal with it and then you get to work. "You quit smoking and get back to work; you have a bypass and go about being Secretary of Defense."

He may be right for some of the physicians but all physicians would not agree with his views which are neither ethical nor very scientific. Below are given a few views about his attitude on his health in the Heart Wire (theheart.org, May 8, 2013)

Bravo to a Man Who Has Learned to Defeat Death

(Bahram Panahi, Heart Wire May 9, 2013) Regardless of his political ideas, some of us may salute him for not giving up the fight for survival. He may give more speeches to allay fears among heart patients. He is first a man and then a political man. Unfortunately, compassion and mercy has left our planet. I wish him well and a long happy life. History may, some day, judge him as a servant of his nation. Again, in my view, people should not mix issues. I wish quick recovery for every sick man, including Mr. Cheney. Just watch a short film of an open heart surgery, and then, try to rid yourself of hateful complexes. A patient is a patient, no matter who he is.

You Are no Bill Clinton, Mr. Cheney

(Neil Brandon, Heart Wire, May 8, 9, 2013) Bill Clinton made dramatic diet/lifestyle changes and got healthy. Cheney has cost the healthcare system hundreds of thousands of dollars with high-tech treatments. The former vice president and his physicians have acknowledged in the public record that he relied on medical technology, rather than treating cause of CAD by making diet and exercise changes, to address his cardiac disease. Despite his strong family history for CAD, he smoked heavily, was overweight, did not exercise, and led a high-stress life. The point, politics aside, is that this approach inarguably leads to a lot of wasted health care dollars. Whose approach to their heart disease was more cost effective, Cheney's or Clinton's? It is clear that Bill Clinton's approach is scientific and ethical whereas that of the former Vice President, appears to be a financial burden for USA as well as for the whole world.

The former Vice President looks back on his cardiovascular adventures dating back 35 years which included at least six myocardial infarctions, followed by bypass surgery, the implantation of a left ventricular assist device (LVAD), and finally a heart transplant. It's surprising, he is decidedly devoid of self-pity as revealed in his speech. He said, his LVAD was a low point and it was a most difficult surgery.

He survived three open-heart procedures. He was on ventilators for several weeks due to pneumonia in the intensive care unit. He remained in a dream state due to sedation in Tuscany and later on in a villa about 40 KM from Rome and his family was not sure about his survival. After he was discharged from the hospital, he lost 40 Kg of his body weight including muscle mass and in the mirror, he looked like his 80 year old father who was suffering from heart failure. The Vice President mentioned during his lecture that I could not even squeeze a tube of toothpaste to brush my teeth. The only bodily function he could perform without assistance was breathing. His wife Lynne has been very supportive to help him during his illness and recovery. He was placed on the transplant list after receiving the LVAD, and, after the LVAD ordeal, the transplant was almost a walk in the park. He had to regain a lot of strength, so that the transplant surgery becomes tolerable. He felt that in fact the easiest surgery he had was heart transplant surgery which is certainly easier than the LVAD. He said in response to an audience question, that he had compared notes on heart ailments with former President Bill Clinton, who has had major cardiovascular issues of his own.

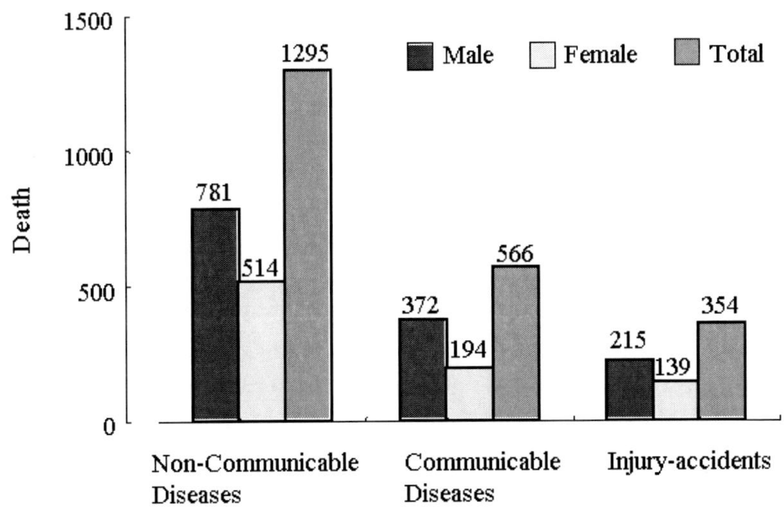

Figure 1. Causes of deaths due to non-communicable diseases (n = 1295), communicable diseases (n = 566) and injuries -accidents (n = 354) based on available records according to classification by World Bank (modified from World Bank 2008).

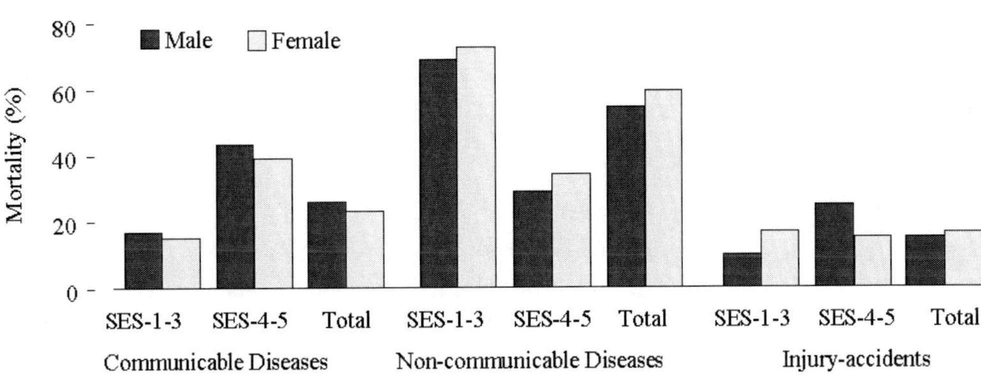

Figure 2. Causes of mortality due to broad causes of death according to income.

He is currently in the process of writing a book with the doctor who has been his cardiologist for years - Dr. Jonathan Reiner (George Washington University Medical Center, Washington, DC). Dr. Reiner writes as a doctor, and he as a patient, and both decided to take a crack at this in part because the history of his disease, dating back to 1975. It follows some of the key breakthroughs in technology that have made a big difference in reducing the incidence of death from heart disease in the US. We urge the former vice president to warn people not to follow him, but to follow Bill Clinton, the former president of USA so that the world gets ethical advice on prevention of CVDs.

The World Heart Federation and the International College of Cardiology advise the health educators that take the advantage of the time when a famous person gets heart attack. The public pays greater attention to the news when the problem occurs to a public figure. However, in this case the public has a wrong message regarding prevention of CAD. All physicians have some patients, who question the advice about prevention. They feel that money can provide everything including good health to them. In fact these views are responsible for increasing health budget in the world.

Cardiovascular Disease Mortality

The morbidity and mortality due to cardio-vascular diseases (CVDs) and other chronic diseases have become a global public health problem [1-4]. Mortality and burden of disease estimates for WHO Member States in 2008 revealed that 36·1 million deaths per year occur as a result of non-communicable disease (NCDs), which are mainly due to CVDs (2). These estimates represent almost two out of three deaths per year worldwide. Approximately, two-thirds (63%) of premature deaths in adults (aged 15-69 years), and three out of four of all adult deaths are attributable to NCDs in the world [2] (Figure 1). However, some experts misinterpret the WHO estimates and propose that, of total deaths, 22·4 million arise in the poorest countries, and 13.7 million in high-income and upper-middle-income countries and therefore poverty may be the major cause of NCDs (Figure 1) [2]. It seems that comparing mortality data among rich and 'poor' within countries, would not support these views about poverty and mortality (Figure 2) [1-7]. In India, high economic status is associated with greater mortality due to cardiovascular diseases. (Figure 2) [5].

Wealth and Deaths

The health budget in USA was US$2·7 trillion in 2011, which is $8700 for every person in the country, and represents 17.9% of the economy, that is far greater than any other economically advanced country [1, 3]. However, it made world news that despite spending more on health care per person than other high-income countries, many people in United States, die sooner, are least likely to reach the age of 50 years, and have higher rates of disease or injury [3]. Unfortunately, Americans are less healthy from birth to 75 years of age than people in 16 other economically wealthy countries, if judged by health alone. This health disadvantage has been getting worse for 30 years, especially among women, as disclosed in a report released on Jan 9 from the US National Research Council and Institute of Medicine [3]. This report compared comprehensive mortality and morbidity data from USA with affluent democratic countries; Australia, Canada, France, Italy, most of the Nordic countries, Spain, and the UK. Life expectancy is shorter at birth for American men than for men in any of the other 16 countries, and American women fare little better.

Surprisingly, Denmark is the only country that has a lower life expectancy for women at birth. In nine key areas of health; infant mortality and low birth weight; injuries and homicides; teen age pregnancies and sexually transmitted infections; HIV/AIDS prevalence; drug-related deaths; obesity and diabetes; heart disease; chronic lung disease; and disability, the United States fare least well, or is near the bottom of the tables. This health disadvantage applies to those with health insurance, a college education, higher incomes, and healthy behaviors as well as to those without [3].

Under these circumstances, it is difficult to believe how knowledge of health behavior and wealth alone can provide better physical health in general and total health in particular despite universal health coverage [6]. The universal massage by WHO is to follow; A - Avoid alcoholism, B - Be physically active, C - Cut down on salt and sugar, D - Don't use tobacco products, E - Eat plenty of fruits, nuts and vegetables. Moderate alcohol intake, particularly red wine may be a medicine as advocated by the International College of Nutrition.

Acknowledgments

Acknowledgements are due to International College of nutrition and International College of cardiology for logistic support to write this article.

Conflict of Interest has not been declared by the authors.

References

[1] Editorial. Wealth but not health in USA. *Lancet* 2013; 381:177.

[2] WHO. Mortality and burden of disease estimates for WHO Member States in 2008. Geneva: World Health Organization, 2010.

[3] *U. S. Health in International Perspective. Shorter Lives, Poorer Health.* National Academy of Sciences, USA 2013. (For the report on US health disadvantage see http: //www. nap. edu/ catalog.php? record_id=13497; For the US prevention strategy see http://www.healthcare.gov/prevention/nphpphc/strategy/report.pdf.

[4] Beaglehole R., Bonita R., Alleyne G. et al. for the Lancet NCD Group. UN High-Level Meeting on Non-Communicable Diseases: addressing four questions. *Lancet* 2011; 378: 449-55.

[5] Singh R. B., Anjum B., Takahashi T., Martyrosyan D. M., Pella D., de Meester F., Wilson D. W., SSD Beehari, Keim M., Shastun S. Poverty is not the absolute cause of deaths due to noncommuni- cable diaseses NCDs. *World heart J* 2012;4: 221-236.

[6] Vega J. Universal health coverage: the post-2015 development agenda. *Lancet* 2013; 381:179-180. doi:10.1016/S0140-6736(13)60062-8.

Index

B

C

D

E

F

G

H

I

J

K

L

M

N

O

P

Q

R

S

T

U

V

W

Y

Z